G·L·O·B·A·L S·T·U·D·I·E·S

EUROPE
SEVENTH EDITION

E. Gene Frankland

Ball State University

OTHER BOOKS IN THE GLOBAL STUDIES SERIES

- Africa
- China
- India and South Asia
- Japan and the Pacific Rim
- Latin America
- The Middle East
- Russia, the Eurasian Republics,
  and Central/Eastern Europe

McGraw-Hill/Dushkin
530 Old Whitfield Street, Guilford, Connecticut 06437
Visit us on the Internet—*http://www.dushkin.com*

STAFF

Ian A. Nielsen	Publisher
Brenda S. Filley	Director of Production
Lisa M. Clyde	Developmental Editor
Roberta Monaco	Editor
Charles Vitelli	Designer
Robin Zarnetske	Permissions Editor
Marie Lazauskas	Permissions Assistant
Lisa Holmes-Doebrick	Senior Program Coordinator
Laura Levine	Graphics
Michael Campbell	Graphics/Cover Design
Tom Goddard	Graphics
Eldis Lima	Graphics
Nancy Norton	Graphics
Juliana Arbo	Typesetting Supervisor
Larry Killian	Copier Coordinator

Cataloging in Publication Data
Main Entry under title: Global Studies: Europe. 7/E
    1. Europe—History—1945–. 2. Europe—Politics and government—1945–. 3. Europe—Civilization—1945–.
I. Title: Europe. II. Frankland, E. Gene, comp. ISBN 0–07–243357–4   ISSN 1059–2334

Seventh Edition

We would like to thank Digital Wisdom Inc. for allowing us to use their Mountain High Maps cartography software. This software was used to create the relief maps in this edition.

Printed in the United States of America   234567890BAHBAH5432   Printed on Recycled Paper

Europe

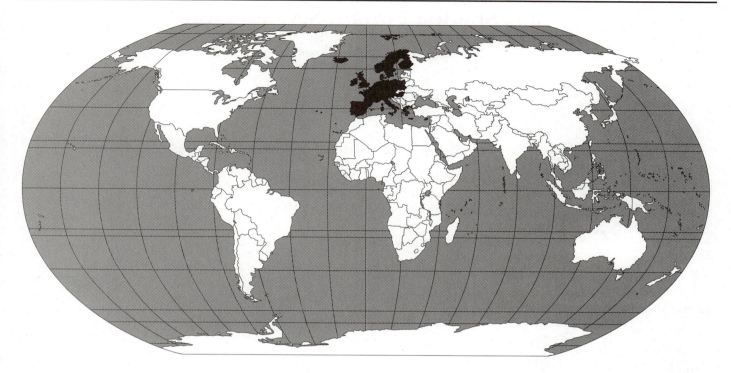

AUTHOR/EDITOR

Dr. E. Gene Frankland

Dr. E. Gene Frankland, the author/editor of *Global Studies: Europe,* is professor of political science and director of the European studies program at Ball State University, Muncie, Indiana. He received his B.A. degree in government from Southern Illinois University–Carbondale and his M.A. and Ph.D. degrees in political science from the University of Iowa. He teaches courses on comparative politics, Western European politics, the European Union, and environmental law and policy. During his career, Dr. Frankland has taught in overseas programs in Spain, Germany, England, and Austria. In spring 1996, he was a visiting fellow at Harris Manchester College, Oxford University. In spring 1997, he directed Ball State University's London Centre. His research has primarily dealt with environmental politics, especially the development and role of Green parties in Europe. He coauthored (with Don Schoonmaker) *Between Protest and Power: The Green Party in Germany* (1992) and is co-editing/co-authoring (with John Barry) *The International Encyclopedia of Environmental Politics* (forthcoming). Since 1989, he has regularly contributed chapters on the German Greens to edited volumes on German politics and on comparative Green politics. His articles have been published in *The Review of Politics, German Studies Review, Youth & Society,* and *Legislative Studies Quarterly.* Dr. Frankland has participated on numerous panels at regional, national, and international conferences.

PRIMARY CONTRIBUTORS

Dr. John K. Cox
Wheeling Jesuit University

Dr. Henri J. Warmenhoven
Virginia Commonwealth University

Contents

Global Studies: Europe, Seventh Edition

Austria Page 68

Denmark Page 81

France Page 92

Greece Page 112

Ireland Page 123

United Kingdom Page 171

Using Global Studies: Europe

THE GLOBAL STUDIES SERIES

The Global Studies series was created to help readers acquire a basic knowledge and understanding of the regions and countries in the world. Each volume provides a foundation of information—geographic, cultural, economic, political, historical, artistic, and religious—that will allow readers to better assess the current and future problems within these countries and regions and to comprehend how events there might affect their own well-being. In short, these volumes present the background information necessary to respond to the realities of our global age.

Each of the volumes in the Global Studies series is crafted under the careful direction of an author/editor—an expert in the area under study. The author/editors teach and conduct research and have traveled extensively through the regions about which they are writing.

In this *Global Studies: Europe* edition, the author/editor and contributers have written a broad regional essay, two smaller regional essays, and country reports for each of the countries included.

MAJOR FEATURES OF
THE GLOBAL STUDIES SERIES

The Global Studies volumes are organized to provide concise information on the regions and countries within those areas under study. The major sections and features of the books are described here.

Regional Essay

For *Global Studies: Europe,* the author/editor has written a regional essay focusing on the history and current characteristics of Europe. A regional map accompanies the essay. Two smaller essays, on the "expanding" Europe and the European mini-states, are also provided.

Country Reports

Concise reports are written for each of the countries within the region under study. These reports are the heart of each Global Studies volume. *Global Studies: Europe, Seventh Edition,* contains 20 country reports.

The country reports are composed of five standard elements. Each report contains a detailed map visually positioning the country among its neighboring states; a summary of statistical information; a current essay providing important historical, geographical, political, cultural, and economic information; a historical timeline, offering a convenient visual survey of a few key historical events; and four "graphic indicators," with summary statements about the country in terms of development, freedom, health/welfare, and achievements.

A Note on the Statistical Reports

The statistical information provided for each country has been drawn from a wide range of sources. (The most frequently referenced are listed on page 246.) Every effort has been made to provide the most current and accurate information available. However, sometimes the information cited by these sources differs to some extent; and, all too often, the most current information available for some countries is somewhat dated. Aside from these occasional difficulties, the statistical summary of each country is generally quite complete and up to date. Care should be taken, however, in using these statistics (or, for that matter, any published statistics) in making hard comparisons among countries. We have also provided comparable statistics for the United States and Canada, which can be found on pages viii and ix.

World Press Articles

Within each Global Studies volume is reprinted a number of articles carefully selected by our editorial staff and the author/editor from a broad range of international periodicals and newspapers. The articles have been chosen for currency, interest, and their differing perspectives on the subject countries. There are 17 articles in *Global Studies: Europe, Seventh Edition.*

The articles section is preceded by an annotated table of contents as well as a topic guide. The annotated table of contents offers a brief summary of each article, while the topic guide indicates the main theme(s) of each article. Thus, readers desiring to focus on articles dealing with a particular theme, say, the environment, may refer to the topic guide to find those articles.

WWW Sites

An extensive annotated list of selected World Wide Web sites can be found on the facing page (vii) in this edition of *Global Studies: Europe.* In addition, the URL addresses for country-specific Web sites are provided on the statistics page of most countries. All of the Web site addresses were correct and operational at press time. Instructors and students alike are urged to refer to those sites often to enhance their understanding of the region and to keep up with current events.

Glossary, Bibliography, Index

At the back of each Global Studies volume, readers will find a glossary of terms and abbreviations, which provides a quick reference to the specialized vocabulary of the area under study and to the standard abbreviations used throughout the volume.

Following the glossary is a bibliography, which lists general works, national histories, and current-events publications and periodicals that provide regular coverage on Europe.

The index at the end of the volume is an accurate reference to the contents of the volume. Readers seeking specific information and citations should consult this standard index.

Currency and Usefulness

Global Studies: Europe, like the other Global Studies volumes, is intended to provide the most current and useful information available necessary to understand the events that are shaping the cultures of the region today.

This volume is revised on a regular basis. The statistics are updated, regional essays and country reports revised, and world press articles replaced. In order to accomplish this task, we turn to our author/editor, our advisory boards, and—hopefully—to you, the users of this volume. Your comments are more than welcome. If you have an idea that you think will make the next edition more useful, an article or bit of information that will make it more current, or a general comment on its organization, content, or features that you would like to share with us, please send it in for serious consideration.

Selected World Wide Web Sites for Europe

Some Web sites are continually changing their structure and content, so the information listed may not always be available.

GENERAL SITES

BBC World Service—*http://www.bbc.co.uk/worldservice/europe/*—The BBC provides the latest news from around the world and Western Europe.

CNN Online Page—*http://www.cnn.com*—This U.S. 24-hour video news channel provides news, updated every few hours, text, pictures, and film.

Deutsche Welle Radio, English Service—*http://www.dwelle.de/english/*—Daily news reports, European press reviews, and features by Germany's international broadcasting network are provided on this Web site.

Environmental News Service—*http://ens.lycos.com*—This site presents environmental news from around the world, including frequent European reports.

Europa—*http://europa.eu.int*—Information on the European Union's goals and policies, history, institutions, publications, and statistics are available at this site.

International Network Information Center at University of Texas—*http://inic.utexas.edu*—Links to international sites, including United Kingdom and Western Europe are available here.

NATO—*http://www.nato.int*—This is the official site of the North Atlantic Treaty Organization. The site includes information on members, current programs, policy statements, and current relevant news.

New York Times—*http://www.nytimes.com*—This valuable site provides international news that is updated throughout the day and includes a separate page on Europe.

Penn Library: Resources by Subject—*http://www.library.upenn.edu/resources/subject/subject.html*—This vast site is rich in links of information, especially regarding Germanic, French, Italian, and Iberian studies.

Political Science RESOURCES—*http://www.psr.keele.ac.uk*—A dynamic gateway to sources available via European addresses.

Russian and East European Network Information—*http://reenic.utexas.edu/reenic.html*—This University of Texas site provides access to Balkans, Central and Eastern Europe, the Baltics, and former–Soviet Union republics Web sites.

Speech and Transcript Center—*http://gwis2.circ.gwu.edu/~gprice/ speech.htm*—This unusual site is the repository of transcripts of every kind, from radio and television, of speeches by world government leaders, and the proceedings of groups like the UN, NATO, and the World Bank.

United Nations System—*http://www.unsystem.org*—The UN's official Web site for the system of organizations. Everything is listed alphabetically.

U.S. Central Intelligence Agency—*http://www.cis.gov*—This site includes information about the CIA and its publications: *The World Factbook, Factbook on Intelligence, Handbook of International Economic Statistics,* and CIA maps.

U.S. Department of State—*http://www.state.gov/www/background_notes*—Country Reports, Human Rights, International Organizations, and other data can be found here.

Washington Post—*http://www.washingtonpost.com*—This site provides international news with separate regional pages for Western Europe, Central Europe, and Eastern Europe, as well as special European reports.

World Bank Group—*http://www.worldbank.org*—News [i.e., press releases, summary of new projects, speeches], publications, topics in development, countries, and regions.

World Trade Organization [WTO]—*http://www.wto.org*—Topics include world trade systems, data on textiles, intellectual property rights, legal frameworks, trade and environmental policies, and recent agreements.

SELECTED COUNTRIES

Austria: General—*http://www.austria.org*—The history, government, and statistics of Austria are listed on this Web site.

Belgium: Belgian Federal Information Service—*http://www.belgium.fgov.be*—This official site provides basic information "all about Belgium": its monarchy, government, history, economy, people, and land.

Cyprus: The "Cyprus" Home Page—*http://www.kypros.org/Cyprus/root.htm*—This non-governmental site provides an introduction to Cyprus, its republic, and its culture, as well as daily news reports and links.

Denmark: Royal Danish Embassy in Washington—*http://www.denmarkemb.org*—An interesting background on the Vikings, an extensive profile of Queen Margrethe II, and a guide for tracing Danish ancestry are provided at this Web site.

Finland: Virtual Finland—*http://virtual.finland.fi*—This site has background on government institutions, recent political events, and some information on Finnish culture.

France: French Embassy in Washington—*http://www.info-france-usa.org*—A profile of the French government and culture is available on this site.

Germany: German Embassy and German Information Center—*http://www.germany-info.org*—Many links to German topics such as facts, figures, statistics, news, history, and government can be found here. The site also includes current information about German life today.

Greece: The Greek Embassy in Washington—*http://www.greekembassy.org*—Information on Greek culture, history, and links to many Greek ministries and U.S.–Greek organizations are offered on this large site.

Iceland: Icelandic Foreign Service—*http://www.iceland.org*—A "wealth of information" about Iceland, its history, economy, culture, and policies is available on this site. A number of useful links are listed.

Ireland: Information on the Irish State—*http://www.irlgov.ie/frmain.htm*—Information about all aspects of the government of the Republic of Ireland can be found at this Web site.

Italy: Windows on Italy—*http://www.mi.cnr.it/WOI/*—This site examines Italian history and includes links to pages on some cities and towns.

Luxembourg: Luxembourg Tourist Office–London—*http://www.luxembourg.co.uk*—This site provides an overview of geography, history, economy, and population with links to other sources of information on Luxembourg.

Malta: Government of Malta—*http://www.magnet.mt*—Malta's official site introduces its government and its people with a comprehensive index of subjects on the Malta Government Network–Internet Service.

Netherlands: The Dutch Embassy in Washington—*http://www.netherlands-embassy.org*—Basic information on the Netherlands, its history, government, and foreign relations is available here.

Norway: Ministry of Foreign Affairs—*http://www.dep.no/odin/engelsk*—Norwegian news, facts and figures, press releases, and general information on a wide range of topics as well as links to other sources are available on this official site.

Portugal: The Presidency—*http://www.presidenciarepublica.pt*—This site, which is sponsored by the Portuguese presidency, provides basic information on Portugal, and links to complementary data, as well as information on the presidency, its development, and current incumbent.

Spain: Spanish Ministry of Foreign Affairs—*http://www.DocuWeb.ca/SiSpain*—Seeking to promote "free exchange of information on Spanish current affairs and its historical, linguistic and cultural development," this site presents basic information, other Web sites, and Internet courses.

Sweden: The Swedish Institute—*http://www.virtualsweden.net*—This "official gateway" provides basic information on Sweden's culture, economy, education, government, environment, and other policy areas. Links to Swedish governmental and nongovernmental sites are possible.

Switzerland: The Swiss Information and Communication Platform—*http://www.schweiz-in-sicht.ch*—This interactive gateway introduces Swiss geography, economy and science, people and society, and federalism and multilingualism. Access to documents, scholarly papers, maps, videos, and other links is possible.

United Kingdom: Britain in the U.S.—*http://www.britain-info.org*—Top stories of the British press are highlighted on this site, and it also links to information on politics, geography, and history.

ADDITIONAL COUNTRIES

Andorra—*http://www.washingtonpost.com/wp-srv/inatl/longterm/worldref/country/andorra.htm*

Liechtenstein—*http://www.news.li/navig/index.htm*

Monaco—*http://www.monaco.mc*

San Marino—*http://inthenet.sm/rsm/intro.htm*

Vatican City—*http://www.vatican.va/pcont2_en.htm*

The United States (United States of America)

GEOGRAPHY

Area in Square Miles (Kilometers):
3,717,792 (9,629,091) (about 1/2 the size of Russia)

Capital (Population): Washington, D.C. (568,000)

Environmental Concerns: air and water pollution; limited freshwater resources; desertification; loss of habitat; waste disposal

Geographical Features: vast central plain, mountains in the west; hills and low mountains in the east; rugged mountains and broad river valleys in Alaska; volcanic topography in Hawaii

Climate: mostly temperate

PEOPLE

Population

Total: 276,000,000

Annual Growth Rate: 0.91%

Rural/Urban Population Ratio: 24/76

Major Languages: predominantly English; a sizable Spanish-speaking minority; many others

Ethnic Makeup: 69.1% white; 12.5% Latino; 12.1% black or African American; 3.6% Asian; 0.7% Amerindian

Religions: 56% Protestant; 28% Roman Catholic; 2% Jewish; 4% others; 10% none or unaffiliated

Health

Life Expectancy at Birth: 74 years (male); 80 years (female)

Infant Mortality Rate (Ratio): 6.82/1,000

Physicians Available (Ratio): 1/365

Education

Adult Literacy Rate: 97% (official; estimates vary widely)

Compulsory (Ages): 7–16; free

COMMUNICATION

Telephones: 173,000,000 main lines

Daily Newspaper Circulation: 238 per 1,000 people

Televisions: 776 per 1,000 people

Internet Service Providers: 7,6000 (1999 est.)

TRANSPORTATION

Highways in Miles (Kilometers): 3,906,960 (6,261,154)

Railroads in Miles (Kilometers): 149,161 (240,000)

Usable Airfields: 13,387

Motor Vehicles in Use: 206,000,000

GOVERNMENT

Type: federal republic

Independence Date: July 4, 1776

Head of State: President George W. Bush

Political Parties: Democratic Party; Republican Party; others of minor political significance

Suffrage: universal at 18

MILITARY

Military Expenditures (% of GDP): 3.8%

Current Disputes: none

ECONOMY

Per Capita Income/GDP: $33,900/$9.25 trillion

GDP Growth Rate: 4.1%

Inflation Rate: 2.2%

Unemployment Rate: 4.2%

Labor Force: 139,430,000

Natural Resources: minerals; precious metals; petroleum; coal; copper; timber; arable land

Agriculture: food grains; feed crops; fruits and vegetables; oil-bearing crops; livestock; dairy products

Industry: diversified in both capital- and consumer-goods industries

Exports: $663 billion (primary partners Canada, Mexico, Japan)

Imports: $912 billion (primary partners Canada, Japan, Mexico)

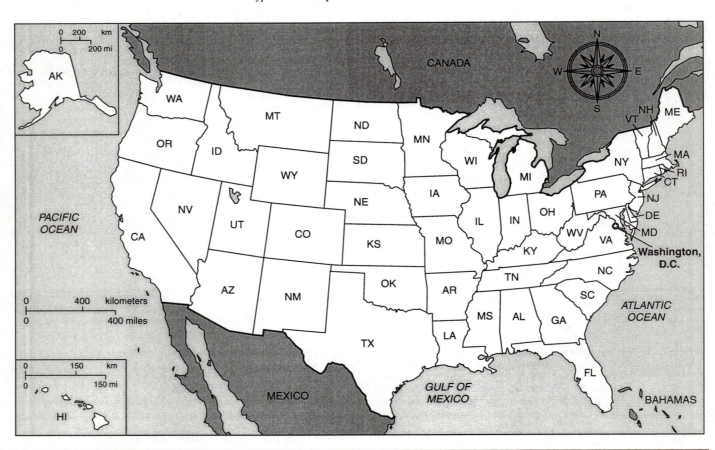

Canada

GEOGRAPHY

Area in Square Miles (Kilometers):
3,850,790 (9,976,140) (slightly larger than the United States)
Capital (Population): Ottawa (1,000,000)
Environmental Concerns: air pollution and resulting acid rain severely affecting lakes and damaging forests; water pollution; industrial damage to agriculture and forest productivity
Geographical Features: permafrost in the north, mountains in the west, central plains, and a maritime culture in the east
Climate: from temperate in south to subarctic and arctic in north

PEOPLE

Population
Total: 31,300,000
Annual Growth Rate: 1.02%
Rural/Urban Population Ratio: 23/77
Major Languages: both English and French are official
Ethnic Makeup: 28% British Isles origin; 23% French origin; 15% other European; 6% others; 2% indigenous; 26% mixed
Religions: 46% Roman Catholic; 16% United Church; 10% Anglican; 28% others

Health
Life Expectancy at Birth: 76 years (male); 83 years (female)
Infant Mortality Rate (Ratio): 5.08/1,000
Physicians Available (Ratio): 1/534

Education
Adult Literacy Rate: 97%
Compulsory (Ages): primary school

COMMUNICATION
Telephones: 18,500,000 main lines
Daily Newspaper Circulation: 215 per 1,000 people
Televisions: 647 per 1,000 people
Internet Service Providers: 750 (1999 est.)

TRANSPORTATION
Highways in Miles (Kilometers): 559,240 (902,000)
Railroads in Miles (Kilometers): 22,320 (36,000)
Usable Airfields: 1,411
Motor Vehicles in Use: 16,800,000

GOVERNMENT
Type: confederation with parliamentary democracy
Independence Date: July 1, 1867
Head of State/Government: Queen Elizabeth II; Prime Minister Jean Chrétien
Political Parties: Progressive Conservative Party; Liberal Party; New Democratic Party; Reform Party; Bloc Québécois
Suffrage: universal at 18

MILITARY
Military Expenditures (% of GDP): 1.2%
Current Disputes: none

ECONOMY
Currency (U.S.$ Equivalent): 1.53 Canadian dollars = $1
Per Capita Income/GDP: $23,300/$722.3 Billion
GDP Growth Rate: 3.6%
Inflation Rate: 1.7%
Labor Force: 15,900,000
Natural Resources: petroleum; natural gas; fish; minerals; cement; forestry products; wildlife; hydropower
Agriculture: grains; livestock; dairy products; potatoes; hogs; poultry and eggs; tobacco; fruits and vegetables
Industry: oil production and refining; natural-gas development; fish products; wood and paper products; chemicals; transportation equipment
Exports: $277 billion (primary partners United States, Japan, United Kingdom)
Imports: $259.3 billion (primary partners United States, Japan, United Kingdom)

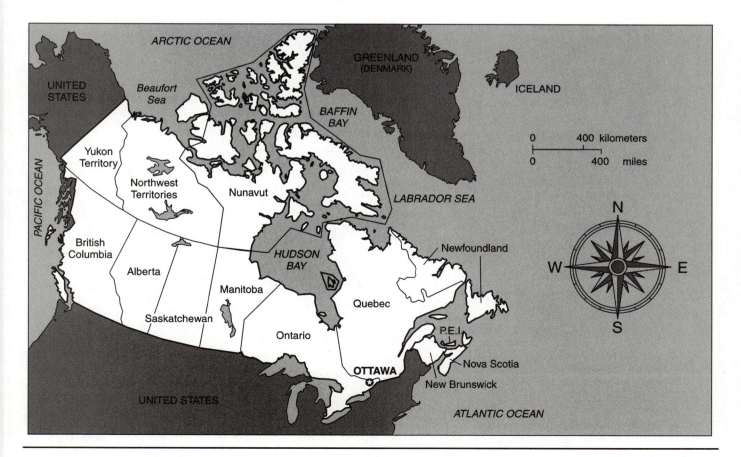

GLOBAL STUDIES

This map is provided to give you a graphic picture of where the countries of the world are located, the relationships they have with their region and neighbors, and their positions relative to economic and political power blocs. We have focused on certain areas to illustrate these crowded regions more clearly.

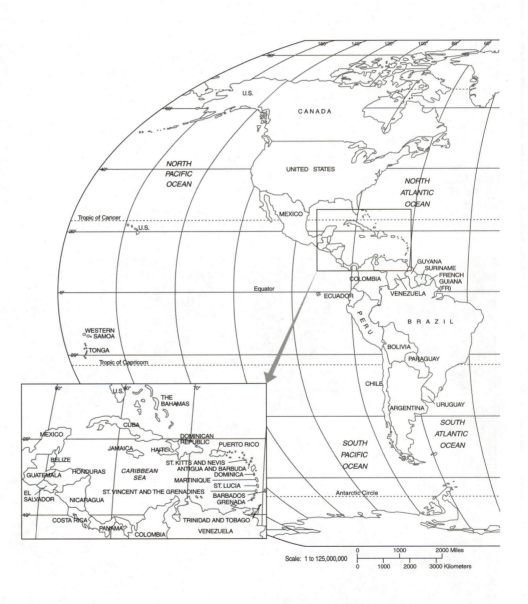

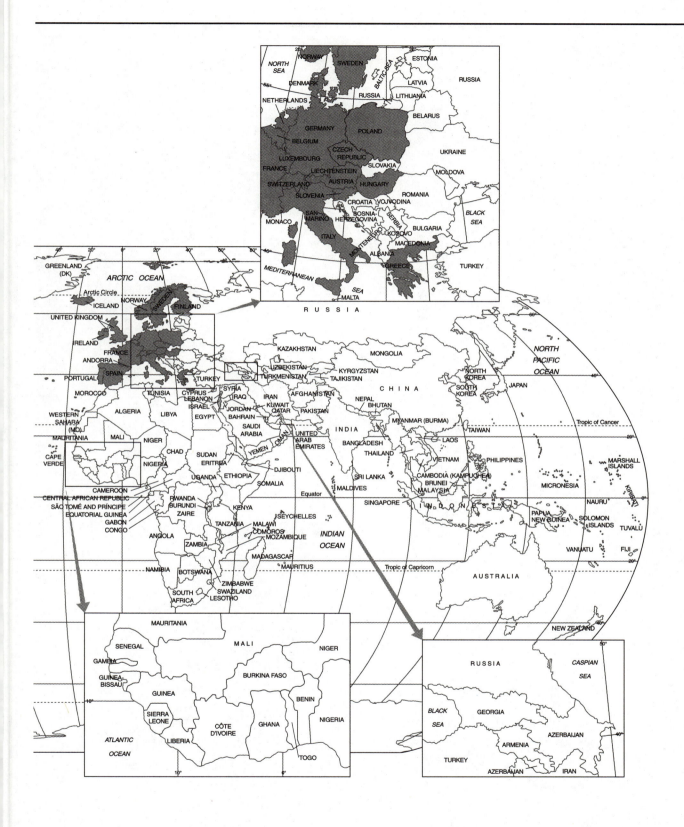

Europe

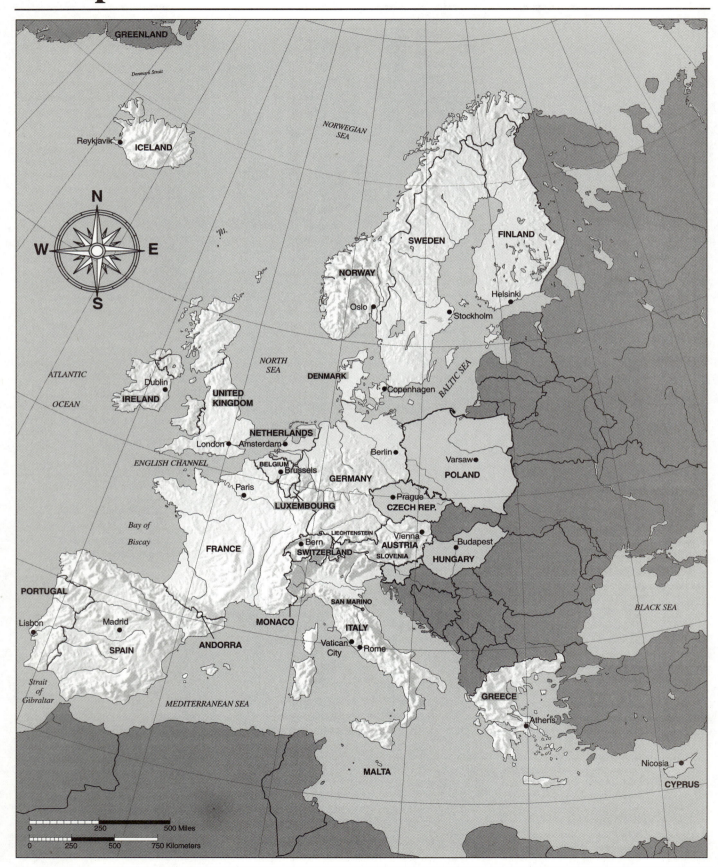

Europe: Centuries of Commonalities and Conflict

British historian Arnold Toynbee compared civilizations to living organisms. Civilizations, he said, are born, grow, flourish, and, after several centuries, die—that is, they fail to exert political or economic influence.

The European continent has spawned a variety of civilizations over the course of time. For many centuries, Europe was the equivalent of the known world, and its impact on global events has been truly formidable. World history was long defined from strictly European perspectives, a tendency that has been labeled *Eurocentrism*. However, two world wars (1914–1918 and 1939–1945) were to a large extent fought in the European theater, and the devastation that they wrought reduced the continent's prestige and influence. Europe's leading position in the world rapidly faded as a result.

In recent years, however, Europe has proven to be a reservoir, stronger and more durable than most analysts and observers had deemed possible.

EUROPE AND THE NEW WORLD ORDER

Although death and destruction during World War II were not confined to Europe, the continent took a thorough beating during those years. No wonder, then, that shortly afterward, disgust and revulsion with war emerged. These sentiments, in combination with observation of the stunning success of the United States' economic aid program for Europe known as the Marshall Plan, led a Frenchman named Jean Monnet to envision a European economic and political federation. His vision was translated by Robert Schuman, another Frenchman, into a declaration that aimed at Europe's eventual political and economic unification. The Schuman Declaration (1950) heralded the *European Communities;* and, although that term started to be used by the late 1980s in the singular (that is, the *European Community,* or EC), it gained spectacularly in substance. In the early 1990s, the EC entered one of its most critical phases, that of political union, a concept that would grant Europe a powerful impetus in world affairs. Indeed, it has since 1993 come to be called the *European Union,* or EU.

There is no certainty yet whether a European "superstate" is in the making as a result of centripetal forces. But a most remarkable coincidence is taking place: The "renaissance of Europe" (a phrase that in this context has political and economic overtones, in contrast to its fifteenth-century meaning) coincides with the disappearance of the rivalry between the states and the confrontation of the two Cold War superpowers, the Soviet Union and the United States. These superpowers emerged as a by-product of World War II. Nearly half a century later, in 1990–1991, one of them, the Soviet Union, crumbled and disintegrated. By the same token, the United States often appears to have lost its appetite for superpowership. The United States seems more willing to leave global issues and conflicts to international organizations, such as the United Nations. There can thus be little doubt that a new world order is emerging and that Europe is to play a large role in that construct.

The world as a whole still bristles with European concepts and ideas. This is well illustrated by the struggle for economic advancement and greater political autonomy in many countries in Asia, Africa, and Latin America—often called the *Third World* or *developing countries.* The belief in prosperity and the belief in self-government both derive from the European treasury of ideas (and their transplantation to the United States boosted them substantially). Another major change in world politics that illustrates the enduring influence of European ideas is the rise of Japan as a global economic power. Japan's current democratic system of government as well as its technology derive from European foundations, although both have also been influenced by indigenous traditions.

WHY STUDY EUROPE?

Three major sets of reasons explain why the study of Europe is critical. The first is cultural or philosophical and has to do with the significance of European ideas. The second is political and strategic and concerns the economic and political importance of Europe today as a group of countries that, even though they no longer control the known world, nevertheless exercise considerable economic and diplomatic power and influence. The third is scientific and analytical; it embraces the opportunity for comparative studies to learn from the modern democratic systems of Europe as they cope with similar challenges as evolving, postindustrial societies.

The cultural or philosophical reason for studying Europe focuses on the dominant role of European ideas around the world. The U.S. Constitution, to cite one example, is one of the finest expressions of European ideas about the limitation of power, freedom, and human rights; and the legal and political arguments in the United States concerning civil rights—what they are, and how they are best protected—are based on European ideas about society, citizenship, and freedom. In addition, there are the concepts of the "just war," the justification of self-defense, and peace, all issues that have been perennial questions in the American (and European) debate. Religious pluralism, arguments over the separation of church and state, and the relationship of religion to politics are also issues paramount in the United States that sprouted in Europe.

There are also important political and strategic reasons to study Europe: For half a century, Canada and the United States have been militarily allied with 14 Western European countries. The North Atlantic Treaty Organization (NATO, or the Atlantic Alliance) was founded in 1949, primarily as a collective-security alliance against possible expansion on the part of the Soviet Union. It has experienced a great many ups and downs. In addition to some intra–NATO quarrels, there has been recurring friction concerning the very superior position that the United States has taken in NATO as well as frequent squabbles about the relative contributions that members should give to the total effort. Until recently, this effort was, as NATO's Charter stated, in large part dedicated to the containment of

A harmonious blend of cultures and historical eras characterizes the Europe of today, as this view of Genoa, Italy, illustrates.

communism in the European theater. For decades, the alliance was far and away the most important foreign-policy commitment to which the United States was bound.

Interestingly, when the Soviet or East Bloc (a term referring to the Soviet Union and the Central/Eastern European countries) collapsed beginning in 1989—an eventuality that few analysts had foreseen or predicted—the Warsaw Treaty Organization (NATO's opposite number, also known as the Warsaw Pact) disintegrated as well. NATO, however, was left intact, although its future role has become uncertain, and the organization's mixed record in containing the various nationalist crises in the former Yugoslavia during the 1990s gave rise to considerable skepticism about its purpose and effectiveness. As of yet, however, no plans have been made to dismantle the Alliance; on the contrary, it was enlarged in 1999 by the admission of three countries from the former East Bloc: Hungary, the Czech Republic, and Poland. In 2000, EU countries agreed to form a rapid-reaction force of 60,000 troops, which some see as the seed of a European army. The multilateral force would be available after the year 2003 for deployment in crisis areas where NATO (and the United States) might choose not to intervene. Some U.S. officials are warning that a more autonomous European military force could undermine NATO's solidarity and capability. However, not only are the French and German governments supporting the plan, but so are the British and others.

For many North Americans, Europe conjures stereotypical biases. An example is the perception that Europe is fragmented and, in fact, irreversibly divided. Until recently, the main fault line was the so-called Iron Curtain, which divided Communist Central/Eastern Europe on the one hand and democratic Western Europe on the other. Other North American notions concerning Europe have reference to an abundance of internal continental differences, such as ethnicity, varieties of beliefs, and lifestyle and linguistic differences. One may also point to class distinctions, believed to be more endemic to Europe than to the supposedly egalitarian societies of Canada and the United States.

Thus, the notion persists that people in Europe generally favor democratic ideals but that their societies have on the whole remained stagnant. As a contrast, in the United States, democracy materialized, coming to expression in the 1787 Constitution. Even Thomas Jefferson, one of the most sophisticated Americans of his day—a man who had lived and traveled in Europe—at one point exclaimed, "We are ahead of Europe in political science." By this he meant that while the science of government might have originated in Europe, the United States had started to apply it, testing democratic tenets by putting them into practice. In Europe, on the other hand, much of the theory and philosophy of government had an aura of unreality, consisting of distant ideals.

Since North American values are closely related to those nursed and fostered in Europe, many Americans and Canadians assume that most Europeans are like them and certainly less different or interesting than Chinese, Arabs, Japanese, or any other powerful but non-European group whose activities *appear* to be more dramatic or *seem* to affect North America more immediately. During the first few decades after World

War II, Europe certainly diminished in political and economic importance. But it has clearly risen again, another major reason for studying the region.

Finally, study of Europe is valuable because of the similarities and differences that its varied political and socioeconomic systems have with those of North America. In one sense, this blend is useful in attempts to develop general explanations of political and socioeconomic changes. For example, theories about party and party-system changes, voting behavior, or generational value changes require cross-national data in order to be adequately tested. In other sense, this blend provides the opportunity for applied learning from the reform experiences of other "sister" countries. Americans could learn from the environmental intitiatives, campaign-finance laws, and drug policies found in various European countries. The classic example is the Scandinavian office of ombudsman (basically a citizen's advocate to deal with individual complaints about maltreatment by bureaucratic agencies), which has been emulated under various names in other parts of Europe as well as in North America. Some scholars argue that U.S. "exceptionalism" is on the wane. In any case, the possibility exists for increased theoretical and applied knowledge from paying closer attention to European countries, especially those at similar levels of socioeconomic development.

How Should "Europe" Be Defined?

During the Cold War, the continent was split into two artificial entities, "Western Europe" and "Eastern Europe" (or, in the terminological fashion that emerged in the 1990s, "Central/Eastern Europe"). Although historians may find echoes in the civilizational split between Catholic/Protestant Christianity and Orthodox Christianity, those earlier West–East borders were never as clear-cut as the "Iron Curtain" erected by Communist forces after World War II. (In this reconfigured Europe, Catholic Poland ended up in the East, and Orthodox Greece ended up in the West.) The post-1945 division was never strictly geographically descriptitve. To the far northeast, though neutral, Finland was generally considered by academic specialists as being within Western Europe. To the far southeast, despite a sizable Turkish community, Cyprus was more often than not viewed as an appendage of Western Europe. Vienna, the cosmopolitan capital of Austria, also perceived as a part of Western Europe though neutral, lay farther east than Prague, the capital of Czechoslovakia, and East Berlin, the capital of East Germany (officially, the German Democratic Republic), both loyal members until 1989 of the East Bloc. Thus, geographical terms were in actuality rooted in ideology. On one side of the postwar divide emerged a Europe generally characterized by democracy, market or mixed economies, and technological progress; and on the other side, a Europe generally characterized by communism, command and control economies, and technological backwardness.

During 1989–1990, the Iron Curtain and the Berlin Wall came down. However, "Central/Eastern Europe" did not disappear so quickly. As a Romanian explained to this writer in 1992, communism is gone, but the Soviet system remains. In those parts of Central/Eastern Europe at the forefront of the economic transition from communism to capitalism, participants in the democratic opposition movement often found themselves out of work while former Communist officials were "capitalizing" on their resources, connections, and knowledge to gain control of former state enterprises. Indeed the legacies of communism will be played out over generations. However, by the late 1990s, the evidence that "Europe" is being redefined, often painfully, had become compelling. This is illustrated most dramatically by the candidacies of the former Communist countries of Central/Eastern Europe and the Baltics for membership in the European Union. However, even those ex–East Bloc countries on the organization's "short list" (Hungary, Poland, the Czech Republic, Estonia, and Slovenia) are likely to face long and difficult transitions before they become EU members in the full sense like Spain and Portugal (which, as new democracies, joined in 1986).

With the circumstances still in flux, this book embraces for the most part a political, rather than a geographical, definition of "Europe." For example, it does *not* view Russia and the other (non-Baltic) successor states of the Soviet Union as politically a part of "Europe," though they may be members of the Organization for Security and Cooperation in Europe (OSCE) and the Council of Europe. The primary focus here is on the "Europe" of the EU member states and those countries that are closely aligned with the EU's economic regime through the European Economic Area (EEA) or, in the case of Switzerland, through bilateral treaty agreements. Obviously, aspiring members on the eastern and southeastern peripheries of "Europe" cannot be ignored, especially if they are major countries like Poland, Hungary, and the Czech Republic (which already are NATO members). It is most likely that over the next decade, the scope of "Europe" will expand significantly, and future editions of this book will do likewise: "Europe" is not yet finished.

HOW TO STUDY EUROPE

Given its variety of political, economic, and social arrangements, there is no best single way of looking at Europe. It is thus necessary to combine elements of four approaches. The first approach to studying Europe concerns the level of individual countries, which has been the traditional method of historical and area studies. The concept of the nation-state emerged when the Peace of Westphalia was concluded in 1648. That same era also gave birth to international law. As a result, Europe was rarely viewed as a consistent entity but, rather, as a conglomeration of countries, internally fragmented and competing for power among themselves. Two questions may thus be raised. First, why is Europe not more integrated? (One of the answers would undoubtedly be because ethnicity and culture have proven to be such strong variables.) Second, is it possible to speak of a European political system? (Here the answer would be, "Not yet.")

The second approach focuses on the supranational economic and security networks, such as the European Union and NATO, respectively. At this level, we learn of the broad limits of policy, how European leaders perceive their common interests, and how these countries relate diplomatically, economically, and militarily as a group to other actors on the world scene, the most important or dominant being the United States and Russia (the main successor state of the Soviet Union). Russia, of course, is no longer considered a global superpower. But in terms of geography, it is a large, neighboring power—with a huge nuclear arsenal. Several things should be kept in mind when examining Europe at this level. First, more is going on in Europe than merely the activities of organizations like NATO and the EU. Second, the memberships of these organizations do not include all the countries of Europe. (And, in the case of NATO, it includes non-European members.) Finally, ideologies and religions in Europe operate most distinctively at this level. Conspicuous examples are the environmental and peace movements of the 1970s and 1980s, respectively.

The third approach to the study of focuses on regions within countries or across borders. *Regionalism* is a term that has come up in the EU vocabulary, where it refers to some of the more outlying regions (such as Sicily and Scotland) among its members. These regions usually have been poor or less developed, and needing of special policies for economic activity. One has to bear in mind that regional issues often also involve ethnic issues—that is, on the fringes of nation-states, one tends to find ethnically divergent peoples. In recent decades, many ethnic groups have "rediscovered" their roots (such as the Welsh) and cross-border connections (such as German-speaking Italians). This has given rise to a range of activities, mostly cultural, but some political. In the future, these areas are bound to merit increasing attention.

The fourth approach to the study of Europe is truly comparative, concentrating on government institutions, policy formation, issues, and the all-important linkage mechanisms such as party systems and interest groups. This approach endeavors to find how governmental and nongovernmental institutions compare in several or all countries. Among the many issues that need to be studied and compared are the environment, social-welfare policy, unemployment, and immigration.

It is possible to construct a time frame that employs all four approaches. This time frame has six sections: 1) the pre-1945 background; 2) reconstruction, 1945–1951; 3) the consolidation of Western Europe and the Cold War, 1951–1963; 4) prosperity and détente, 1963–1973; 5) stagflation and insecurity, 1973–1989; and 6) since the end of the Cold War, 1989–present. While such a division is practicable and useful, one must bear in mind that the dates are somewhat tentative and that the periods overlap to some extent. There can, for example, be no doubt that both the consolidation of Western Europe and the Cold War started earlier than 1951, while détente was an intermittent phenomenon at best, punctuating an extended phase, much longer than the years specified.

PRE-1945 BACKGROUND

Europe owes part of its political organization to the Romans, whose empire started to dissolve between the fifth and ninth centuries A.D. This protracted dissolution was marked by a great deal of political fragmentation and instability, which in turn produced massive migrations by tribes across West Asia, North Africa, and Europe. As these movements gradually stabilized, the main centers of economic and cultural activity of Western civilization moved from the Mediterranean littoral to the river valleys of Western Europe, between the North Sea, the Alps, and the Western Mediterranean.

From the ninth to the thirteenth centuries, the settled, cultivated parts of Europe comprised roughly the same territory as Western Europe today. The major nationalities—including French, German, Italian, and English—emerged. Most Western European countries came into being as recognizable political entities during the Middle Ages. The most fundamental political conflict in Western Europe before 1945—that between French- and German-speaking peoples for control of Northwestern Europe and control of the political and cultural legacy of the Roman Empire—began in that period, with the breakup of the Carolingian Empire in the mid-ninth century. This empire had included the ancestors of both the French and the German nations and was the only successful attempt to unite these two ethnically different societies under one government. The eleventh century produced a landmark, in that the Norman invasion led by William the Conqueror changed England both demographically and linguistically.

In the fourteenth century, England, which was already a centralized monarchy under a single ruler, held large parts of France under its dominion. For a considerable period, England replaced a variety of Middle European principalities (most of whose people spoke Teutonic dialects) as France's main opponent. In that time, England reigned supreme in various parts of western France. In the fifteenth century, the French drove out the English, an effort that led to national unity and centralized political organization.

Early Modern Europe

Although feudalism and prefeudal conditions (e.g., the Roman Empire) greatly influenced the growth and development of Europe, the Peace of Westphalia, concluded in 1648, usually serves as an important historical watershed. It was then that the nation-state emerged as a form of political organization. Between the middle of the seventeenth century and the outbreak of the French Revolution in 1789, Western Europe developed its modern identity: a system of sovereign states. A number of countries, such as England, France, Spain, and

The Reformation had its seeds in the 1520 excommunication of Martin Luther by Pope Leo X. This illustration portrays some of the notable events of the Reformation era.

the Scandinavian countries (Norway, Sweden, and Denmark), assumed the size and adopted the character they have today. Germany, the Netherlands, the Alpine and Danubian lands, and Italy were divided politically into territories belonging to a loose political association that covered most of Central/Eastern Europe and, since the twelfth century, included the Holy Roman Empire on the one hand and a number of smaller nations on the other.

The Protestant Reformation, initiated in 1517, started to have a profound political impact after 1530. It divided Europe into hostile religious camps, each claiming the right to subject and convert the other. The old dichotomy of church and state, which had been strictly observed throughout the centuries of the Holy Roman Empire, now started to come apart. In addition, the old conflict between France and the diverse German states revived.

The combination of religious, ideological, territorial, and dynastic struggles led to the cataclysm of the Thirty Years' War (1618–1648). This was the greatest disaster in European history occurring between the Black Death of 1348–1349 and World War I. Nearly half the population of Germany, where almost all the action took place, perished as a result of the Thirty Years' War. It ended in the Peace of Westphalia, which more or less defined the nation-state and laid down rules for international relations. This drastically reduced the frequency and severity of wars in the period between the middle of the seventeenth century and the end of the eighteenth century.

The largest and most prosperous country for most of that period was France, which had 20 million inhabitants. The 12 million Germans had not found a home in a single state; rather, they were organized politically in a number of states, most of which in one way or another were affiliated with the Holy Roman Empire, an entity that dissolved in 1806. England, which, after annexing Scotland in 1707, had started to be called Great Britain, remained a much smaller and less important country until about 1800. However, as a result of the British victory over Napoleon Bonaparte, the enormous growth of its colonial empire, and its vanguard position in the

(The Bettmann Archive)

A depiction of the storming of the Bastille, July 14, 1789, the event that sparked the French Revolution.

Industrial Revolution, the nineteenth century was to become the "Century of Britain."

Political Development (1789–1914)

The term *sovereignty,* a product of international law, originally derived from the word *sovereign*—that is, the monarch. The French Revolution had no difficulty discarding the monarchy, but not the idea of sovereignty. Thus, the world witnessed the emergence of *popular sovereignty* as a joint product of the American Revolution and the French Revolution. Following John Locke, an English philosopher, the idea took hold in Europe that the people were to set the parameters of royal power. The republican and democratic concept of rule by the people through their chosen representatives turned out to be the first of four major and related factors that shaped European history in the nineteenth century.

A second strong force was *nationalism,* the belief that people who belonged to one ethnic strain, who spoke one language, and who fostered one culture were to cherish and maintain a common identity and that they ideally should live under one government. Nationalism, either in a democratic or an authoritarian format, held a particular appeal for politically divided peoples, such as could be found in what was to become Germany and Italy.

A third factor was *industrialism.* From about 1750, British inventors and entrepreneurs were discovering and organizing modern, industrial methods of production. The Industrial Revolution that resulted spread to the Continent in the second half of the nineteenth century and was accompanied by rapid population growth. The population of Western Europe expanded from 72 million in 1680, to 115 million in 1820, and to 200 million in 1900.

This was followed by the emergence of *socialism* as a powerful force. Initially, this new ideology represented little more than a response to the often unscrupulous methods increasingly applied in the latter stages of the Industrial Revolution. A number of thinkers found that government was not the only evil against which people had to be protected. One had also to watch against the economically powerful—the "capitalists," as they had started to be called in the late nineteenth century. They, too, could cause oppression, depriving people of a full life. During the late nineteenth century, first trade unions and subsequently political parties arose in countries where the Industrial Revolution made life miserable for the toiling masses. It was a matter of course that the platforms of these emerging political parties were geared toward a social transformation that would benefit the workers. Still, it proved hard to oppose and defeat the established order, which was based only nominally on democracy. Opportunities for input through voting were very limited (indeed, in nearly all of Europe, the average person was excluded from the vote). As a result, the new Socialist (and labor) political parties did not achieve a great deal. Until

World War I, progress was erratic at best, and none of the parties were able to gain decisive political power.

Last but not least, another wave of democratic revolutions, which involved both nationalist and socialist factions, made itself felt across much of Western and Central/Eastern Europe in 1848 and 1849. The revolutions failed, and the result was to strengthen the power of established governments. The unification under a single monarch of Italy in 1861 and of most German-speaking areas outside of Austria in 1871 marked the high point of the centuries-long process of political consolidation in Europe and of the more recent nationalist movement. Most educated people in the second half of the nineteenth century thought that the European system of states was stable and set for permanent, if gradual, progress in the struggle against poverty, sickness, illiteracy, and political repression. Indeed, that part of the nineteenth century has often been nicknamed the "Age of Optimism."

World War I (1914–1918)

In the period from 1848 to 1918, Europe reigned supreme in the world. Its major powers were Britain and France. However, Germany and Austria-Hungary were also being formed during that era, and Russia, which had hardly counted as a European power before the Napoleonic Wars, had gradually extended its diplomatic influence and strategic grasp after the Vienna Congress in 1815. By 1900, no European power could afford *not* to take Russian interests or policies into account in its own foreign policy. This Russian influence in the West turned out to be a permanent fact of twentieth-century European politics and was ultimately strengthened by the Communist Revolution, which, starting in 1917, transformed Russia and other nations nearby into the Soviet Union (U.S.S.R.).

There is no permanence in history. By 1914, the constellation of forces had changed to the extent that a major war, which would come to be known as World War I, could no longer be avoided. This "Great War" was to be a long and bloody struggle. (Some historians have adopted the view that there are so many connections between the two world wars that they should be jointly viewed as one protracted struggle, the "Second Thirty Years' War." In brutality and gruesomeness, that conflict easily compares to the Thirty Years' War of the seventeenth century.)

The leading cause of World War I was the growing power of Germany and the fear that this inspired in Germany's main rivals—namely, France, Russia, and Britain. After 1897, the rulers of united Germany moved from a policy of domestic consolidation and economic growth to a continental and ultimately global strategy aimed at giving Germany what they regarded as its rightful place as a world power, on a par with Britain. None of the other European powers could tolerate this ambition.

Furthermore, the old antagonisms, particularly the conflict between Germany and France, had not disappeared. Germany's defeat of France in 1870–1871 spurred the French governments of the Third Republic (1871–1940) to seek allies for a future war of revenge. Otto von Bismarck, the German chancellor from 1871 to 1890, knew that Germany would lose a general European war and sought to preserve

(The Bettmann Archive)

World War I: A German gun company is shown in battle at Darkehmen.

peace through a pact among Germany, the Austro-Hungarian Empire, and Russia. This fell apart due to Austro–Russian disagreements, however, and in 1894, Russia formed an anti-German pact with France. In turn, Great Britain, alarmed by Germany's growing military and economic power, broke with its long-standing policy of noninvolvement in Europe and signed agreements with France in 1904 and Russia in 1907, thus creating what became known as the "Triple Entente." On the other side, Germany remained allied with Austria-Hungary. In 1882, Italy joined this pact, which was called the "Triple Alliance."

On June 28, 1914, one of the most fateful days in history, an assassin killed the heir to the thrones of Austria and Hungary in Sarajevo, part of the Bosnian province of Austria-Hungary. During the next five weeks, threats and counter-threats escalated and triggered the mutual-support provisions of the Triple Entente and the Triple Alliance. By mid-August, Germany and Austria were at war with France, Great Britain, and Russia. At first Italy was neutral, but in 1915 it decided to enter the war as an ally of the Entente.

The rulers of Germany in 1914 knew that Germany's only chance was to defeat France quickly, before the United Kingdom could send strong forces to the Continent, and then to rush its forces east and defeat Russia in alliance with Austria-Hungary. The plan failed. The war in the west became a war of attrition between entrenched armies, wherein the side with the most men and the best supplies of food and ammunition ultimately prevailed. In the east, Germany finally defeated Russia in 1917–1918, thanks in large part to the Russian Revolutions of February and October 1917, which temporarily derailed Russia's military power. After a final German offensive in the west, in the early summer of 1918, the Entente (supported by the United States since April 1917) drove the German armies back almost to the German frontier. Revolution broke out in Germany and Austria-Hungary, and the emperors of the two countries abdicated. The government of the new German republic, composed of leaders of the Liberal and Socialist Parties, requested an armistice, which the Entente granted on November 11, 1918.

The fighting, which had cost more than 10 million lives, was over. The economic situation in Germany rapidly deteriorated after the armistice had been signed; and, when a peace treaty was signed in the following year, the conditions were by no means commensurate with German power at the time of the armistice.

The Interwar Years (1918–1939)

The impact of the war on politics, technology, religion, philosophy, social life, ideology, and the European and world economies culminated in the Great Depression of the 1930s. This was followed by World War II, the last act of the "Second European Thirty Years' War." In Germany and Italy, the seemingly continuous crises brought dictators to power; in Britain, Scandinavia, and the Low Countries (the Nether-

(The Bettmann Archive)

Otto von Bismarck, the "Iron Chancellor," founded the German Empire in 1871, thus introducing new equations into the European political and social arenas.

lands, Belgium, and Luxembourg), democrats remained in control and responded to the problems by instituting social-welfare programs and accepting that government had a responsibility for employment and living standards. In France, a strong right-wing movement threatened—but did not destroy—democracy.

The fundamental reason for the emergence of anti-democratic sentiment in Western Europe in the 1920s and 1930s was that a large and, in some countries, decisive proportion of the politically active population simply did not believe that democracy could deal with the economic, technological, and social problems of the time. They believed that the modern world demanded a new way of political life, a way involving national mobilization, paramilitary regimentation, and a leadership free to do what was necessary for national security and survival without being accountable to sectional interests.

Some of these antidemocrats went to the far right and joined nationalistic movements, of which the most important were the Nazi movement in Germany and the fascist movement in Italy. The leaders of these movements, insisting that terrible dangers threatened their peoples, promised glory and renewal for their nations. These dangers, the Nazis and fascists said, could be defeated only if the people shook off the shackles of democracy. Others, including many intellectuals, drifted to the far left and supported Communist parties, whose adherents promised equality and justice in a class-less society. Like the Nazis, they said that terrible dangers threatened this promise, which could be overcome only if citizens subjected themselves to the iron discipline of each country's Communist party, which was more or less controlled by the Soviet Union. Both Nazis and Communists viewed any thought or action (including murder) that served the movement as "good," any thought or action that might hinder it as "bad" and to be ruthlessly suppressed.

The Great Depression

The Wall Street stock-market crash of 1929 in the United States caused a worldwide depression. Unemployment soared in European countries, regardless of their political format and structure. World trade came to an end, in large part because of U.S. protectionism and fiscal policies. In the period 1930–1933, world manufacturing production outside the Soviet Union shrank by more than one third; the volume of international trade fell by two thirds; and more than one third of all laborers in Great Britain, Germany, Denmark, Norway, and France were unemployed.

In Germany, the situation was aggravated by the restrictions imposed by the Versailles Treaty. This allowed Adolf Hitler, who in the 1920s co-founded the Nazi Party after a failed coup, to make rapid gains. In 1933, his Nazi government was installed and the first contours of the "Third Reich" appeared. It could claim a measure of success by having built a superhighway system (the *Autobahn*) such as Germany had never had and by having launched a rapid rearmament program, which clearly violated the Versailles Treaty.

In fact, unemployment began falling on its own in 1934, because world trade and business investment were picking up throughout Europe. In the Scandinavian countries, alliances of Socialist and Liberal parties introduced social-security programs (which protected against unemployment), favored health insurance, and initiated programs of public works. The

British government did not introduce major social-policy reforms until 1945, but it succeeded in reducing unemployment somewhat in the 1930s by modest programs of public investment and other measures.

Unemployment and underinvestment were not the only problems facing Western European economies in this era. Most countries were also completing the transition from an economy based on agricultural employment and production to one based overwhelmingly on industry. Britain and Germany were furthest along in this process; the Scandinavian countries, Italy, and France did not become fully industrialized until after World War II. Spain finally shed its largely agricultural image after dictator General Francisco Franco's death in 1975.

Unlike Central/Eastern Europe, where the landed aristocracy and a small middle class dominated society, politics, and culture until World War II, Western and Northern European society had generated a middle class by 1920. The aristocracy was losing ground as cultural and political elites, but this retreat was gradual and continued into the post–World War II period. Most workers abandoned revolutionary ideology in 1917–1920 and followed reformist leaders of the Socialist parties, who wanted to work for social change and redistribution of income and wealth by democratic means rather than through the revolution that communism advocated.

France, Italy, and Spain had fewer industrial workers and proportionally more peasants. Italy and Spain were also restrained by a powerful Roman Catholic Church hierarchy that was arch-conservative, if not reactionary. Many Church leaders regarded liberal democracy as permissive and anti-Christian and thus lent their support to authoritarian movements. The main cause of their fears was the virulent anticlericalism of some liberal or left-wing movements in these countries.

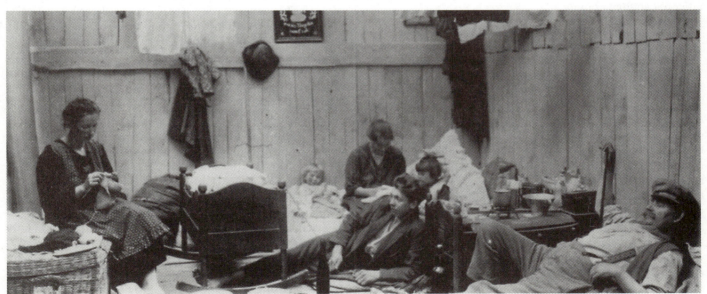

(The Bettmann Archive)

The Great Depression of the 1930s devastated the world economy, as illustrated by these unemployed refugees in a temporary workhouse in Germany.

The net effect of mutual distrust of liberal democrats and Catholics was to permit authoritarian groups to gain support. These groups, like Mussolini's Fascists, promised to protect the Church from leftist anticlericals, while in reality they exploited its social power and influence for their own ends.

The Spanish Prelude to World War II

In the early and mid-1930s, Spain experienced a high degree of instability, which was capped by the assumption of power on the part of a left-wing government based on the middle class and workers. The new government provoked a right-wing rebellion, which generated a civil war of extraordinary brutality, lasting from 1936 to 1939. Many governments in Europe took sides, though refraining from involving themselves officially. Even some North American volunteers joined the forces of the Spanish government vs. the rebels.

Some political analysts view the conflict as the dress rehearsal for World War II. Since the leaders of the rebels—Francisco Franco and his Falange movement—had a strong fascist orientation, they were able to count on active support from Germany and Italy. The republican (or left-wing) side gathered a measure of international sympathy, but it did not receive massive material support. It lost, and for nearly 40 years General Franco exercised unbridled power in Spain. One might have expected some gratitude on the part of the Spanish dictator for the German and Italian help, but Franco refused to join the Axis forces when World War II started, shortly after the end of the Spanish Civil War. Once France had fallen to Germany, Hitler and Franco met somewhere in the Pyrenees, on the French–Spanish border. The meeting generated an instant mutual dislike, which may have caused Franco to deny the German forces passage through Spain in order to capture Gibraltar.

Fascism and Nazism

The two basic antidemocratic movements, fascism and nazism, had a great deal in common. However, they are often erroneously treated as one and the same. Therefore, it is useful to compare and contrast them. Fascism was the first to arrive on the scene. Benito Mussolini, a journalist who had often published in *Avanti,* a socialist periodical, and who had been a leader of the Italian Socialist Party, suddenly switched from the left to the far right. Organizing bands of fighters and terrorists (these bands were called *fasci,* a Latin term for a bundle of wooden staves, the symbol of the office of Roman magistrates), Mussolini directed his speeches and addresses against communism in general and the Russian Communist Revolution in particular. In 1921, that broad movement, which by now was commonly identified as *fascism,* was reconstructed as a political party. The next year, Mussolini organized the historic March on Rome. In this huge protest rally, thousands of party members marched hundreds of miles in order to assemble in front of the main government institu-

tion in Rome. (Mussolini himself joined the marchers in the Italian capital.) The March on Rome was attended by a great deal of publicity. As a result of the media exposure, Mussolini was able to manipulate the king into naming him prime minister; but, Mussolini soon took to calling himself *Il Duce* (Italian for "The Leader"). In the mid-1920s, Mussolini seized full dictatorial powers. His laws and decrees banned all other parties. Dissidents were tried and sentenced. Those who were lucky were given internal exile; others, such as Antonio Gramsci, the leader of the Italian Communist Party, languished, and eventually died, in prison.

Hitler and Mussolini met several times before World War II began, concluding the Axis Pact; the two dictators expected their joint forces to steamroll through Europe. However, Mussolini did not immediately join Hitler once the war started in September 1939, preferring to sit on the fence for nearly a year. Only after the German forces had defeated Poland, Denmark, Norway, the Low Countries, and France did Il Duce declare war on the one remaining European country that stood up to Hitler: England.

Nazism and fascism differed considerably in ideological respect. Although Mussolini and the Fascist Party in general lauded and glorified Italy's past and forged an ancestral connection with the great Roman Empire, racial discrimination, highlighted by the Nazis' persecution of Jews, was conspicuously absent in fascist thinking. Only at a later stage of the war, after German forces had several times rescued Italy's military, did Mussolini oblige his Axis partner by rounding up Jews in Italy.

If the foundation of fascism in Italy had depended on the March on Rome and Mussolini's subsequent coup, the Nazis in Germany owed their success to circumstance, in particular the fragility of the Weimar Republic (1919–1933). This democratic system of government suffered from a number of intrinsic weaknesses. Electoral laws enabled many diverse political parties to be represented in Parliament. As a consequence, the government was always based on fragile coalitions of political parties. The parties that were not in power frequently voted down government proposals or programs; in many cases, the government was forced to resign as a result. Party leaders then came together to form a new government, which would soon be ousted in turn. Two parties in particular conspired to maintain this gridlock: the *Nazi Party* (short for National Socialist German Workers' Party) and the German Communist Party (KPD). Together they held a "negative majority": Although neither of them had the numbers to be an effective vehicle for government, jointly they could prevent all the other parties from gaining government power. Finally, on January 30, 1933, President Paul von Hindenburg was persuaded by advisers to resolve the deadlock by allowing Hitler, whose Nazis held a plurality but not a majority, to form a new government. Von Hindenburg was aged and tired and no longer seemed to care.

Hitler disliked his official title of chancellor (the German term for prime minister), and, like Mussolini, he had himself called "Leader" (in German, *Führer*). Just a few months after the Nazis gained office, the *Reichstag* (Parliament) building burned down. Arson was suspected, and a Communist (who had most probably been planted in the building) was tried, sentenced, and executed. Hitler then assumed emergency powers, an action that enabled him to shunt aside all the features of democracy—freedom of the press, freedom of assembly, freedom of speech, and so on. These emergency powers were never revoked and, in fact, lasted throughout the Third Reich (1933–1945). The KPD, the Socialist Party (SPD), and all other parties were banned, their leadership mercilessly persecuted. In less than three months, Germany became a totalitarian state. Hitler's rise to power, with the exception of his failed *putsch* (coup) in 1923, was completely constitutional. Article 48 of the Weimar Constitution gave the country's executive the power to declare an emergency, and the burning down of the Reichstag building could seem a legitimate opportunity to do so. Of course, Hitler did violate international agreements and treaties—first and foremost, the Versailles Peace Treaty that had been concluded in 1919.

The Road to the Holocaust

Hitler's regime was far more oppressive and violent than that of the Italian Fascists. One very important difference was the role of anti-Semitism in Nazi ideology. To the Nazis, Jews were irredeemably evil, like germs infecting an otherwise healthy body. Hitler and his government first took away the civil rights and liberties of Jews in Germany. In 1938, they started to confiscate what was left of Jewish property. By the outbreak of war, in September 1939, more than half of over 500,000 Jews who had lived in Germany in 1925 had left the country. Almost all those who remained perished in concentration and extermination camps during the period 1941–1945. Many who had fled Germany in the late 1930s were caught in countries that Hitler subsequently conquered.

World War II

Hitler's regime had not been in power for long when it began a program of armaments and expansion that destabilized European security and led to World War II. World War II began as a European war between Great Britain, France, and Germany. The war was a result not only of Nazi policies but also of the weakness of the French and British response, which allowed Hitler to think that there was nothing for which the French and British would fight.

Hitler had a secret as well as a public agenda in his foreign policy. His secret agenda was to make Germany all-powerful in Europe by destroying all rivals and to remove all Jews from Germany and occupied countries. His public agenda was to give Germany equal status with the other major powers. The first steps in this public agenda were to re-arm and to bring all ethnic Germans into a single state. Many Western politi-

(National Archives)

Adolf Hitler was named chancellor of Germany in January 1933. Within just a few months, Hitler removed all restraints on his power through the issuance of emergency decrees.

cians thought these wishes were reasonable, despite the aggressive ways in which Hitler presented them. Few objected, therefore, when, in 1938, Germany annexed Austria, which had existed as an independent German-speaking country since the collapse of the Austro-Hungarian Empire. Most Austrians, including Hitler himself, had wanted to join Germany in 1920, but the Allies had made that impossible. Later in 1938, Hitler annexed Sudetenland, the border areas of Czechoslovakia where the population happened to be mostly German. The British and the French accepted even this. Indeed, the British prime minister, Neville Chamberlain, went to Germany to conduct negotiations with Hitler. This "Munich Agreement" has come to stand for a policy of appeasement. In March 1939, Adolf Hitler's armies occupied most of what remained of Czechoslovakia, where there were no Germans to speak of. Finally, the Western powers realized that Hitler intended to dominate the whole of Europe.

When Germany invaded Poland, in September 1939, an immediate reaction followed: The Allies issued an ultimatum that if the German forces did not leave Poland within three days, a state of war would exist, between England and France on the one hand and Germany on the other. Hitler simply ignored the ultimatum. The war that followed lasted almost six years and cost 55 million lives. More than half of the

Thousands of Adolf Hitler's troops are shown here, massed in Germany at a Nazi rally in November 1934.

casualties were civilians, whereas in World War I, about 10 percent of casualties were civilians.

In order not to be fighting wars on two fronts, shortly before he invaded Poland, Hitler had concluded a "non-aggression" pact with the Soviet dictator Joseph Stalin. This treaty also made provisions for a division of Central/Eastern Europe. After the Germans defeated Poland, the Soviets occupied the eastern part of that country. Late in 1939, the Soviet Union attacked Finland, which had been within the sphere of the Russian Empire, as was agreed to in the pact. (In early 1940, the Finns were forced to cede the eastern part of their country to the Soviet Union.) Also in late 1939, the Soviets invaded and annexed the small Baltic countries of Estonia, Latvia, and Lithuania, which had gained independence from Russia in 1918.

Starting in 1940, Hitler conquered Denmark, Norway, the Low Countries, and France, and he threatened Britain with invasion. The Battle of Britain in August and September 1940 turned out to be an air war of unprecedented proportions. However, it was not to lead to an invasion of Great Britain. In April 1941, Germans suddenly turned toward Yugoslavia and Greece, conquering both countries after encountering fiercer resistance than expected. Shortly thereafter, on June 22, 1941, Hitler's Third Reich embarked upon "Operation Barbarossa," which was meant to liquidate the Soviet Union while England was still too weak to attack on the other side. Here Hitler committed the same fatal mistake that Napoleon Bonaparte had made about 130 years earlier. The Soviet territory turned out to be too vast, its winters too grueling, its people too determined. After some initial successes, the German war machine was driven back, slowly and steadily. The United States had up to this point remained neutral; but once Japan, Hitler's ally, attacked Pearl Harbor on December 7, 1941, the United States declared war on Japan, and Germany and Italy declared war on the United States three days later. Thus the European war became a world war.

Switzerland, Sweden, Spain, Ireland, and Portugal remained neutral throughout the war. The first two were long-established democracies and remained so, although both restricted civil liberties and free expression, in deference first to the Germans and later to the Western Allies and the Soviets.

Jewish women with crosses painted on their backs by Nazis. These women spent at least part of the war in forced labor at the German ammunition factory in Kaunitz.

The rest of the European continent endured various degrees of oppression and violence under German occupation. Although many British cities were bombed by German planes, the Third Reich never succeeded in invading Britain, where democratic government and civil liberties continued throughout the war.

Resistance Movements

Resistance movements arose in all occupied countries, which had come to include Italy, most of which the Germans occupied in September 1943, after the Italian government had surrendered to the Allies. In many cases, notably in France and Italy, the resistance had a Communist as well as a democratic wing, with radically opposed ideas for postwar politics. In Denmark, Norway, the Netherlands, and Belgium, the Communist movements had been much weaker before the war, and the democratic resistance was accordingly stronger. Its members regarded the old politicians as discredited. They aimed at a reconstruction of political life that would take power away from the old business and aristocratic elites, institute a welfare state, and equalize income. In Britain, similarly, a broad consensus of the political elite insisted that the people deserved a "New Britain" after the war, to include socialized medicine, vastly increased welfare provisions, subsidized housing, nationalization of heavy industry, and more equality of income and wealth. Few understood how impoverished the country actually was due to the war effort and how difficult it would be to pay for these reforms.

The Nazis' "Final Solution": Genocide

In 1941, after launching the attack on the Soviet Union, Hitler ordered his "Final Solution"—that is, the extermination, as opposed to deportation, of all Jews. Most European Jews lived in Poland, Czechoslovakia, Hungary, and the Soviet Union, and the Nazis killed more than 4 million Jews from these territories. In total, approximately 6 million Jews were murdered during the Holocaust (1933–1945). Nearly 5 million other people were also killed by the Nazis and their helpers.

By late 1942, the tide was turning against the German armies, which had suffered major defeats in Russia and in North Africa. The "Big Three"—U.S. president Franklin D. Roosevelt, British prime minister Winston Churchill, and the Soviet dictator Joseph Stalin—met on various occasions to plan the postwar shape of the world: for example, at Teheran, Iran, in November 1943; and at Yalta, in the Soviet Union, in February 1945.

Their agreement on establishing a "United Nations Organization" to secure world peace through mutual security concealed irreconcilable differences in interests and aspirations. Stalin wished to extend communism as far westward into Europe as possible; Roosevelt and Churchill believed that Stalin might forgo an expansionist policy if they invited him to join in peaceful world leadership. All three claimed to promote democracy and peace; but whereas the Western statesmen understood "democracy" to mean rule by the people and "peace" to mean rejection of policies of violence and expansionism, Stalin understood "democracy" to mean Communist rule and "peace" to mean Soviet global hegemony. The conflict that was later to grow into the Cold War loomed before the real war ended.

The most important bone of contention in Europe was how to deal with Germany. In 1944, the Soviets, the Americans, and the British decided to divide Germany. The Soviet Union, which had taken nearly 70,000 square miles from eastern Poland, wanted to compensate the latter by granting it 40,000 square miles of eastern Germany. Stalin also incorporated a much smaller part of eastern Germany into the Soviet Union. (These drastic territorial changes were officially recognized by the Helsinki Accords in 1975.) The remainder of Germany was divided into Soviet, U.S., British, and French zones of occupation. These zones closely corresponded with the military conquests at the end of the war.

In 1945, the Soviet, American, and British armies invaded Germany from the east and the west, crushing all resistance and forcing the unconditional surrender of the German armed forces on May 8. The Soviets, who had suffered far more than the other belligerents from Nazi atrocities, unleashed furious vengeance on the German civilians in their path. Almost the entire population of the part of Germany given to Poland—15 million people—fled westward in an attempt to escape the invaders. More than 2 million died as a result.

RECONSTRUCTION (1945–1951)
Western Europe recovered from the material destruction wrought by the war more quickly and completely than could

have been expected in 1945. Living standards had fallen drastically throughout the Continent during the war, the greatest drop being in Central/Eastern Europe. Shortly after the hostilities had ended, people all over Europe braced themselves for the task of removing the rubble of the war and rebuilding their countries. This impulse was reinforced by the European Recovery Program—popularly known as the Marshall Plan—which was initiated by the United States in 1947. Within a few years, the levels of production and living standards in all of Western Europe, except Germany, were at or slightly above prewar rates. The recovery marked the beginning of more than 20 years of economic growth, spreading affluence, and a resulting sense of optimism, which were the underlying reasons for an equally remarkable political revival.

Political Effects of World War II
World War II had two fundamental effects, one negative and one positive, that are still being felt today. The negative effect was the political division of Europe into West and Central/East. Western Europe was mostly democratic, whereas the countries of what people began calling Central or Eastern Europe were ruled by Communist dictatorships imposed by Soviet power. The dividing line between the two parts of Europe ran right through one country, Germany, and separated others, such as Austria and Hungary, that in the past had been united under a common government. The line was marked by a fortified border erected by the Communist regimes to prevent

(National Archives)

The cordiality apparent in this picture of Joseph Stalin, Franklin D. Roosevelt, and Winston Churchill (from left to right), meeting at Teheran in November 1943, soon dissolved into the mutual suspicion of the Cold War.

their citizens from escaping to the West. This demarcation emerged as soon as hostilities ended; by July 1945, it was fixed. In 1947, the wartime British prime minister Winston Churchill coined the phrase "Iron Curtain." The metaphor stuck, since what had been established divided Europeans from one another more ruthlessly and sharply than any earlier division in history, including the division into Catholic and Protestant camps following the Reformation of the sixteenth century.

One effect of the war was to discredit the belief of the interwar years that democracy was inefficient and inadequate, a belief that had led so many to turn to nazism, fascism, or communism for solutions to problems of politics and government in the modern world. After 1945, most Western Europeans, including Germans and Italians, came to regard the former contempt for democracy as the main indirect cause of World War II. Now they saw democracy as the only proper way of conducting political life and as the greatest guarantor of peace and prosperity.

This revitalization of democracy did not occur everywhere to the same degree. It was strongest in Germany, where defeat in war and the revelations of Nazi atrocities against Jews and other peoples turned the people against any totalitarianism; and in the countries of Northwestern Europe that had never surrendered voluntarily to antidemocratic ideologies—Britain, Scandinavia, and the Low Countries. In France and Italy, the antidemocratic right lost its credibility; but the antidemocratic left—that is, the Communist parties—remained influential. This occurred not only because they took the credit for defeating nazism and fascism, but also because many on the left continued to believe that democracy without socialism was politically and morally inadequate. In both France and Italy, Communist movements commanded the support of up to a third of the voters and presented a clear challenge to democracy until the 1970s.

The Cold War

The Big Three met for the last time, at Potsdam, in the middle of defeated Germany, in July–August 1945, to decide what to do about the future of the world. (By that time, Harry Truman had replaced Franklin Roosevelt, who had died in April.) Although the United States, Britain, and the Soviet Union agreed on a number of details, their differences in broader objectives and methods were now too clear to be papered over. However, many politicians in the West continued to hope for a period of peaceful collaboration with Stalin. By late 1946, few people in the United States and Britain shared that hope. East of the dividing line in Europe, Communist parties were seizing absolute power in ways that were not very different from Nazi tactics. Many feared that Stalinists were determined not only to crush democratic movements in Central/Eastern Europe but that they wanted to extend their control into Western Europe as well. The Soviet Army was much stronger than the skeleton U.S. and British forces re-

maining on the Continent, and the French and Italian Communist Parties, with their millions of supporters, were a formidable threat to the weak democratic institutions in those countries.

Countries in Western Europe faced two different, yet related, threats in the late 1940s: economic crisis, which might have led to civil war; and a Soviet attack to support a Communist bid for power. Economic recovery had begun in 1945; but the entire region badly needed raw materials and investment if this were to continue and not to end in inflation and unemployment, as had happened after World War I. Such an eventuality would present great opportunities to the Communist parties. Furthermore, economic growth was absolutely necessary if Western Europe were to re-arm sufficiently to deter Soviet military aggression. Finally, the Europeans would have to convince the United States to maintain and extend the military and economic ties to Europe that had developed during the war. U.S. isolationism after 1919 may well have contributed to the success of Hitler, who did not believe that the United States was interested in protecting democracy in Europe. Most Western Europeans understood in 1945 that this should not be allowed to happen again; for Western Europe by itself, in its impoverished condition, was no match for the Soviet Union.

Like the threat, the response had to be a double one, involving both the reorganization of the Western European economies to promote prosperity as well as a defense pact between the major countries of Western Europe and the United States. The Western European nations, especially France and Germany, would have to discard their old rivalries and understand that close economic, diplomatic, and military cooperation was the only way to preserve their newly regained liberty. Close collaboration between German chancellor Konrad Adenauer and French president Charles de Gaulle became the axis of a broader harmony in Europe. That cooperation took the form of a series of policies and institutions, established in 1947–1951, that have formed the framework of Western European politics ever since.

The Economic Framework

Immediately after the war, a number of European statesmen, ideologues, and writers expressed their hope that the European nations would join together in a "United States of Europe." Even Winston Churchill, Britain's wartime leader who as a rule held conservative views, mentioned this eventuality in his speeches in the Netherlands and Switzerland. Indeed, it could seem that such a political structure was less distant at that time than 50 years later. A European federation would have three advantages: a coordinated economic policy, permitting faster recovery; sufficient resources to defend such a union against any threat; and freedom from political control by either the United States, the Soviet Union, or a revanchist Germany, which would be an integral part of the federation and thus unable to become a threat to its neighbors.

By early 1947, it seemed clear that the Communist regimes in Central/Eastern Europe would never agree to joining such a political construction. That left the option of a federation of Western Europe alone. The leading spokesperson for a Western European federation involving a common economic market and united armed forces was Jean Monnet. From 1947 to 1954, Monnet and others worked toward this goal, but it continued to be beyond their reach. Western Europe remained divided into national states. Nevertheless, Monnet's ideas materialized in various supranational bodies for economic development as well as, to some extent, in the European contribution to Western defense.

All of Western Europe, including the western zones of occupied Germany, faced a serious shortage of dollar reserves in early 1947. They needed these funds to pay for imports from the United States that were essential to their continued recovery. Indeed, severe shortages of food in nearly all Western European countries made it necessary to retain rationing systems. In June of that year, U.S. secretary of state George C. Marshall (who as a general had participated in defeating Germany) announced the European Recovery Program. Responding to this offer of help, representatives of all the Western European nations except Spain and Finland, but including Sweden and Switzerland (which had kept out of the war), met in Paris to discuss the coordination of their economic policies.

As a first step, in October they signed the General Agreement on Tariffs and Trade (GATT), an undertaking to remove duties and other restrictions on free trade. The various European governments had concluded that one of the main causes of economic depressions had been "protectionism"— that is, a set of policies designed to protect domestic employment within individual countries. Protectionist measures provoke retaliation and, by reducing international trade, tend to dampen rather than protect domestic employment. GATT was a clear signal that the Western Europeans were inviting investment and promoting trade.

The Marshall Plan came into effect in the spring of 1948. It provided financial aid to all Western European governments (including West Germany, which did not possess its own government until late 1949) for a period of five years (1948–1953). The U.S. government had insisted that, in the execution of the Marshall Plan, it would not deal with individual European countries. Western European governments had accordingly met in Paris in 1947 in order to create the Organization for European Economic Cooperation (OEEC), which was to receive all aid coming from the United States and to manage and distribute it among themselves. When Marshall Plan aid ended, in 1953, the OEEC was not disbanded but was maintained as a monitoring agency for the economic policies and prospects of member countries. It later came to include as members the United States and Canada and, in 1960, was renamed the Organization for Economic Cooperation and Development (OECD).

France and Germany

The French feared Germany's potential economic power even after 1945. Before the war, Germany had produced almost half of the entire steel production in Europe and more than the other four major producers (France, Belgium, Italy, and Luxembourg) combined. Moreover, the German methods were more efficient, and their labor costs were lower. These conditions reappeared after the war. If Germany were allowed to produce freely again, the French worried, the Germans would soon regain their economic, if not political, dominance in Europe. To avoid that eventuality, the French wanted to exploit German coal and steel production for their own reconstruction while preventing German producers from becoming a future competitive threat.

In January 1947, the American and British occupation authorities granted some economic autonomy to the Germans under their control. The French interpreted this measure as a first step toward establishing an independent West German government. They also noted that German steel production, while still far below prewar levels, more than doubled in 1947–1948, and that there was a huge pent-up demand for steel both in West Germany and in Western Europe as a whole. They therefore insisted that the Ruhr—a German coal-mining and steel-producing area that was in the British zone of occupation—should either be removed altogether from Germany or at least be put under an international authority that would control production and distribution. The French were still thinking in terms of their ancient rivalry with Germany and believed that any German recovery could threaten them. Since Germany under Hitler had tried to conquer Europe, the French believed that they had a right to exploit German industrial power for themselves, if only to prevent it from becoming a danger in the future.

The Atlantic Alliance

The interest that the Americans and the British had in Germany in 1946–1948 was entirely different from that of other Western European nations. The United States and Great Britain wanted West Germany to recover economically so that the Germans could support themselves and help the rest of Western Europe in its economic recovery and resistance to the Soviet Union. Only a prosperous Western Europe would have the domestic stability and the resources to remain free; economic recovery and re-armament were two sides of the same coin. Discussions between the United States and Western Europe began at the same time as the Marshall Plan.

Since 1944, the British had been involved in a brutal civil war in Greece, where an army aided mainly by the Yugoslav Communists was trying to seize power. In the winter of 1947, the British told Truman that they could no longer afford the commitment and would have to withdraw. By then, the Cold War had started to dominate American foreign policy, and Truman not only took over from the British but also indicated explicitly that the United States would come to the rescue of

(Harry S. Truman Library)

U.S. president Harry S. Truman signed the anti-Communist Truman Doctrine on May 23, 1947.

foreign governments, resisting an overthrow by armed minorities if the latter were assisted from the outside. This became known as the "Truman Doctrine," specifically directed against communism's more subtle methods of aggression. Its promulgation heartened Western Europeans, who interpreted it as the first sign that the United States would help defend Western Europe against an aggressive Soviet Union.

In early 1948, Britain, the United States, France, and the Benelux countries agreed to permit the Germans to form a democratic, national government in the western zones of occupation. There was no chance of establishing a democratic government in all of Germany, because the Soviets would not accept a united Germany that was not communist. In March of that year, France, Britain, and the Benelux countries signed the Brussels Pact, which was an agreement for mutual cooperation and defense. It showed American political leaders that the Western Europeans had the will to defend themselves and needed help only with resources—which had to come primarily from the United States but secondarily from a reorganized and independent West Germany. In June 1948, the United States, Britain, and France—the occupying powers in West Germany—gave the green light to the Germans to start instituting a government.

When the Soviets realized that the occupying powers were moving toward the formation of a West German government, they retaliated by cutting off the land connections between West Germany and West Berlin. The former German capital city had been divided into four sectors, just as Germany as a whole had been; but, since Berlin was located deep inside the Soviet zone of occupation, the Western allies had to cross that zone to get to their sectors in Berlin. The Soviets hoped to drive the Western Allies out of Berlin. However, they had left the existing air corridors intact, not expecting that the Western Allies would ever be able to supply the millions of civilians as well as troops by air. The United States and Britain took up the challenge, and, for about a year, cargo planes provided West Berlin with all it needed. In May 1949, the siege was lifted.

The new government, the Federal Republic of Germany, came into being in 1949. It held its first free elections in August; and, on September 21, 1949, the American, British, and French military occupation authorities surrendered most of their powers to the new government, whose seat became Bonn, a somewhat sleepy German university city on the Rhine. Christian Democrat Konrad Adenauer, who had resigned from his post of mayor of Cologne when Hitler came to power, became West Germany's first chancellor. Until 1955, the Western powers retained ultimate authority through their occupation forces and their civilian High Commission, which was officially in charge of West Germany's defense and foreign relations. In fact, the Adenauer government had assumed control of West German foreign policy well before the occupation formally ended.

The Berlin crisis accelerated the dialogue between the Western Europeans and the Americans on defense. In early 1949, the United States, Canada, and the Brussels Pact countries (Britain, France, and the Benelux countries) invited Italy, Portugal, Denmark, Norway, and Iceland to join the proposed Western defense pact, which was signed on April 4, 1949. It was named the North Atlantic Treaty, and the organization to which it gave rise was the North Atlantic Treaty Organization. The original treaty ran for 20 years and was renewable by each country for periods of 10 years thereafter. No country has ever failed to renew, although France left the military organization in 1966 while remaining a treaty member. France also had the organization's headquarters moved from Paris to Brussels.

The Schuman Plan

In 1948–1949, the French government realized that its attempt to control German economic recovery was harming common Western European interests and that a prosperous Germany was a better partner than a poor and resentful one. France's foreign minister, Robert Schuman, who hailed from the Alsace, a French area bordering on Germany, established a close relationship with Chancellor Adenauer, who had come from the Rhineland, an area of Germany close to France. Adenauer himself was determined to make French–German reconciliation the cornerstone of his policy vis-à-vis Western Europe.

In 1950, Jean Monnet put to Schuman a plan to begin the unification of Western Europe by pooling French and German coal and steel resources in a "common market" open to other

On April 4, 1949, NATO was formed by nations pledged to the common defense of Western Europe.

countries in Europe. Schuman, in presenting the proposal publicly on May 9, 1950, expressed the hope that it would "create the first concrete foundation for a European federation which is so indispensable for the preservation of peace." Italy and the Benelux countries joined the talks, which resulted in April 1951 in the establishment of the European Coal and Steel Community (ECSC). Not surprisingly, the first chairperson of the High Authority (the ECSC governing body) was Monnet.

The founding of this Community proved to be of great importance. Subsequent Communities came to include the same six members. Although these nations were by no means integrated politically, people started to refer to "Little Europe" when discussing their joint ventures.

CONSOLIDATION (1951–1963)

NATO, the ECSC, and the new democratic government in West Germany were the three pillars of Western European economic growth and political stability in the 1950s. Spain and Portugal did not participate in the consolidation process, not only because they had not been involved in World War II, but also because their right-wing dictatorships were still holdovers from the war period.

Great Britain, Scandinavia, the Benelux countries, and, to a lesser extent, West Germany soon developed into modern welfare states, with high taxation and redistribution of income. France, Italy, and the rest of Southern Europe were poorer and tolerated more economic inequality among their populations.

Although all countries in Western Europe adhered to democratic tenets in varying degrees, there were differences in their

party systems and in the ideologies that were their moving force. The party systems of Britain and West Germany were somewhat similar, in that they both had two large parties (in Britain, the Conservative Party and the Labour Party; in West Germany, the Christian Democratic Union and the Social Democratic Party) plus a considerably smaller party that sometimes acted as balancer (in Britain, the Liberal Party; in Germany, the Free Democratic Party, also a liberal party).

The party systems of the Benelux resembled those of Scandinavia, in that the countries in both areas had multiparty systems that required coalitions. In both cases, center-left governments prevailed for many years; but Sweden, for decades in the vanguard of welfarism, adopted a center-right government in the early 1990s. In all Western European countries, with the exception of Italy, Communist parties had become quite small even before the collapse of the Soviet Union in 1991. In the 1986 elections in the Netherlands, for example, the Communists failed even to secure representation in Parliament. Another big change in this country was that most of the religious parties merged into the Christian Democratic Appeal, which was the mainstay of the government for many years but which in recent years has lost its plurality.

As far as political parties are concerned in Western Europe, the biggest change has not so much been that the leftist parties have receded, but that socialist doctrine itself has changed. It has moved to the right, and it has accordingly become difficult to tell the revised Socialist parties from the traditionally conservative parties. In the 1950s, Socialist parties had called for nationalization of heavy industry, increased employee

(UPI/Bettmann Newsphotos)

In a ceremony in Rome's Campidoglio Palace, statesmen of Belgium, France, West Germany, Italy, the Netherlands, and Luxembourg sign treaties for a European Common Market and European Atomic Energy Community, March 25, 1957.

rights, participation by trade unions in business decisions, and broad equality of income and wealth. Christian Democratic parties had supported private property and free enterprise but also had believed that Christian doctrine demanded that the owners of industry use their wealth for the common good. Thus they had supported the welfare state while ignoring the socialist redistributionist aims.

NATO and the prospect of a common economic market produced a favorable climate for investment and production, so most Western European countries enjoyed high economic growth rates, low unemployment, and low inflation during the 1950s. Living standards differed greatly between north and south, but, even in the more well-to-do north, they were substantially lower than for most Americans. Single-family homes, electric appliances, and automobiles were luxuries in most countries until the 1960s. Even indoor plumbing was by no means widely prevalent in the Mediterranean countries.

Western European culture during the consolidation years was anything but dull. Many writers and artists who had fled from the Nazis or kept silent enjoyed new popularity, while the younger generation of those who had actually fought in the war emerged with impulses and experiences of its own. The French and Italian Communist Parties exercised a great deal of control over culture—not by overt political force, as in Central/Eastern Europe, but through a sort of moral ascendancy that they held over many intellectuals who continued to regard the Soviet Union as a bastion of liberation and freedom from capitalism. Thus, people such as the philoso-

pher Jean-Paul Sartre, the author Simone de Beauvoir, and the painter Pablo Picasso became avowed Communists.

Defense
The North Atlantic Treaty was only the first step in securing the defense of Europe. Major problems included how much Europe could afford for its defense and how German resources might be used without allowing a German army—a reality that most people in areas occupied by Nazi Germany during the war (particularly in France) would find hard to swallow and that would in addition be unduly irritating to the Soviets. Even at the height of the Cold War, the vast majority of Western European politicians favored détente and therefore tried to avoid actions that the Soviets might regard as provocative.

In June 1950, Communist North Korea, supported by the Soviet Union and China, invaded South Korea. The United States decided to respond in kind, which meant that it could not devote as many of its scarce military resources to Europe, where the threat of a Soviet invasion seemed acute.

The problem of Europe's defense had been only partially solved by establishing NATO the year before. The Atlantic Alliance, after all, was there largely as a result of an American initiative and, to a great extent, at U.S. expense. Jean Monnet, chairman of the High Authority of the European Coal and Steel Community, wanted to kill two birds with one stone. On the one hand, he was very satisfied with the way in which the first Community (the ECSC) had developed, believing that the time had come to form other Communities. On the other hand, he felt that a "European Defense Community" could more conclusively solve the question of Europe's defense.

A New Community Emerges . . . and Fails

Thus, in February 1952, the six nations participating in the ECSC decided that, in view of its success and the apparent progress that was being made in the preparation of a new community, the European Defense Community (EDC) could now be formed. In May 1952, the design of the EDC was formally launched. Many French (and Italian) people were made uneasy by the prospect of German rearmament so soon after the war. In addition, there were concerns that Britain (Western Europe's strongest military power) was not willing to join the EDC, and there were doubts about the efficiency of an integrated military force. Following the cease-fire in Korea and the death of Stalin in 1953, the risks of the EDC seemed to outweigh the benefits for many French citizens. A clear majority of their representatives reflected these misgivings in August 1954, when the National Assembly (the lower house of the French Parliament) rejected the EDC.

The European Economic Community

When the EDC failed, the plans for the European Political Community were doomed. Western Europeans who viewed political integration as the way to peace and prosperity were dismayed. Monnet realized that the prospects were bleak for the ultimate supranational government that he originally had had in mind. Instead, he now proposed to work piecemeal toward integration, by means of a permanent dialogue between the leading ministers of the ECSC countries and possibly others in the future. Realizing that politics and economics are closely related, he trusted that a gradual harmonization of economic and social policies in widening spheres would eventually lead to political integration.

At their meeting in Messina, Italy, in June 1955, the ECSC foreign ministers accepted a six-point proposal for the further integration of member countries' economies in the areas of traffic, energy, social policy, and a common market for goods and services. Shortly thereafter, Monnet set up a committee of representatives from all the non-Communist political parties in Western Europe, except the French Gaullists, who opposed any surrender of national sovereignty. This Action Committee for the United States of Europe led the so-called *Rélance*—the "relaunching" of Europe.

This relaunching culminated in the Rome Treaties of March 1957, which established the European Economic Community (EEC) and the European Community for Atomic Energy (Euratom). The purpose of the EEC was to put the Messina proposals into practice in all possible areas. The purpose of Euratom was to promote nuclear energy rather than oil as a replacement for coal as the main source of electric power. Oil, after all, had to be imported, a fact that rendered Europe vulnerable to the vagaries of the Middle East.

West Germany

In May 1952, just before it joined in the discussion of the EDC, West Germany performed its first important foreign-policy act when it signed the Paris Accords with Great Britain, France, and the United States. These accords defined the relations between West Germany and the countries that had occupied it. They included provisions for West German participation in Western defense, but the defeat of the EDC prevented the provisions from coming into force as planned. In September 1954, at a conference seeking alternatives to the failed EDC, participating countries decided to admit West Germany to NATO and to permit West German armed forces under German officers. The Paris Accords were revised, and the formal end of the occupation of West Germany was determined in the following month. On May 8, 1955, 10 years to the day after the final surrender of Germany, the democratic government of Konrad Adenauer received virtually full sovereignty, including the right to have armed forces. West Germany was accepted into NATO.

The Second Berlin Crisis

In 1958, Berlin was still divided into a Soviet and three Western sectors. Access to West Berlin was either over land, through the territory of the former Soviet zone of occupation, now the Communist state of the German Democratic Republic (East Germany), or by air, along certain air corridors. Movement among the sectors within Berlin was free, which meant that East Germans who came to East Berlin could easily move to the Western sectors and thence fly out to West Germany. More than 3 million East Germans escaped to the West this way from 1945 to 1961. They represented the best and the brightest of the workforce. Both the Soviets and the East German regime wanted to stop the flow, but to do so would have violated standing agreements. The Soviets therefore had to produce a crisis in East–West relations so serious that the West would accept closing off access to West Berlin from East Berlin as a lesser evil.

The Soviet leader, Nikita Khrushchev, saw his chance to solve the Berlin problem while at the same time causing confusion in the camp and undermining European faith in the United States. In November 1958, he demanded that the Western Allies leave Berlin within six months; failure to do so would cause him to transfer the control of the access routes between West Germany and West Berlin to the government of East Germany. Had this ultimatum been successful, there is no doubt that the East German regime would have swallowed West Berlin. The Western Allies would have lost all of their support in West Germany, and the Germans would have been very tempted to leave NATO and try to make a deal with the Soviets.

Although the execution of Khrushchev's bold plans did not achieve their primary purpose of neutralizing Berlin, they did wreak confusion and disarray in the Western camp. Indeed, in 1959–1960, the Soviet leader almost succeeded in splitting the ranks of the NATO countries. British prime minister Harold Macmillan went to Moscow, without consulting the United States, to try to reach a settlement with Khrushchev. Like many Europeans, Macmillan had come to believe that the East–West confrontation was largely artificial and could be solved by negotiations and concessions. Adenauer regarded this as a betrayal of common interests, but by now he was very old and increasingly isolated, both in West Germany and in the Western alliance as a whole. And the new American president, John F. Kennedy, was not prepared to risk war to maintain the principle that the government of the German Democratic Republic had no jurisdiction in the Soviet sector of the city, as the Four-Power Agreement had stipulated. The East German government declared East Berlin its capital. Khrushchev gave the East German authorities the green light to erect a wall between the eastern and western sectors of Berlin. They began erecting the barrier on August 13, 1961. Until the Berlin Wall was finally breached, in November 1989, escape from East Germany across the wall or across the land frontier to West Germany was nearly impossible, and most of those trying it paid with their lives.

European Repercussions

The second Berlin crisis pulled the French and the Germans closer together. In 1958, Charles de Gaulle had returned to power in France with the promise of ending the civil war in Algeria, a French possession in North Africa. In Europe, he sought to keep Britain out of the European Economic Community while wooing the Germans away from their close alignment with the United States. This strategy culminated in the Franco–German Non-Aggression Pact and Friendship Treaty of January 16, 1963. The treaty prescribed semiannual consultations between the heads of government of the two countries and listed a range of economic and political issues for future collaboration. The treaty had great symbolic importance. To conclude and celebrate it, Adenauer went to Reims, a city in northeastern France where kings had been crowned when France was still a monarchy. In the medieval cathedral, Adenauer and de Gaulle knelt together to receive communion, a gesture that was reminiscent of the ancient unity of the French- and German-speaking peoples in the Carolingian Empire of the eighth and ninth centuries.

De Gaulle hoped that the treaty would not only be the beginning of a political and economic relationship but also forge a military relationship between France and Germany that would gradually supersede NATO and reduce U.S. influence in Western Europe. The Bundestag (the lower house of the German Parliament), however, unilaterally added a provision to the treaty saying that nothing in it would affect the Federal Republic's ties to the United States. Much later, in the late 1970s and 1980s, German and French leaders revitalized the treaty, and it began to take on the shape envisaged by de Gaulle: an alternative structure for Western European military cooperation.

(The Bettmann Archive)

French president Charles de Gaulle, December 1958.

Great Britain

Great Britain did not join the ECSC in 1951, and it took no part in the process of European integration during the 1950s. The British had a centuries-old tradition of noninvolvement on the European continent except as a response to a direct threat to Britain, and this tradition was still dominant in British foreign policy. The country preferred to rely on the "special relationship" with the United States and on the Commonwealth, its worldwide network of dominions and colonies that all had free access to the British market. If Britain joined the EEC, an organization that had done away with tariffs, it would have granted all those countries, most of them developing-world countries with cheap labor, a tremendous competitive advantage in Europe. Neither the ECSC nor Great Britain, therefore, was interested in British membership under the circumstances.

Toward the end of the decade, the British government decided to dissolve what remained of the British Empire. That entailed also the gradual termination of "imperial preference," as the trade relations with the Commonwealth were called. Britain then decided to move slowly toward involvement in European politics. Its first step was to organize the smaller European countries, which were not members of the EEC, into a free trade area. In 1959, Britain, Portugal, Austria, Switzerland, Sweden, Denmark, and Norway concluded a treaty in Stockholm that established the European Free Trade Association (EFTA).

The Failure of Political Integration

Monnet's hope that the "relaunching" of Europe in 1955–1957 would lead to political integration turned out to be illusory. A significant dilemma that faced the EEC's founders was summarized in the phrase *élargissement ou profondisse-*

ment ("widening or deepening"): Should the organization increase its membership, or should it enhance the quality of its integration? In 1961, the heads of government of the six EEC countries decided to restart the process of political integration. They began developing a European Political Statute, a kind of basic constitution, albeit a more modest one than the European Political Community had envisaged in 1952–1954. The statute would have strengthened the supranational institutions of the EEC, the European Commission, and the Council of Ministers, at the expense of national sovereignties. Although it could seem that the French were behind the plans for the statute, Charles de Gaulle was not interested in any proposal that would reduce national sovereignty; thus, its failure in August 1962 proved no great disappointment to him. There would be no further significant attempt to extend the integration process to the political area until the 1980s.

In August 1961, Britain officially applied for admission to the European Economic Community. In January 1963, de Gaulle vetoed the British application, arguing that the British industrial facilities were backward and obsolete. British membership would, in his opinion, drag down the performance level of the existing Community. Yet it is likely that de Gaulle's real concern was the U.S.–Britain special relationship. A renewed British application did not meet success until 1972, after de Gaulle had died.

PROSPERITY AND DÉTENTE (1963–1973)
The early 1960s marked a number of important turning points in postwar European development. First, the series of crises concerning Berlin ebbed after the Berlin Wall was built. Most people thought that Khrushchev and the East German Communists had achieved what they wanted in closing off access to West Berlin and that there was now reason to hope for some stability in East–West relations in Germany. Since that had hitherto been the main potential flashpoint of the Cold War, the achievement of stability was a momentous development, promising to turn the political division of Germany into a condition of, rather than a threat to, peace and stability in Europe.

Second, de Gaulle's veto of British membership in the EEC confirmed the existence of "two Europes" within Western Europe: a core consisting of the original six ECSC members; and a periphery of unconnected, less dynamic, and less integration-minded outsiders. Even after Britain and other countries joined in the 1970s and 1980s, the EEC remained unofficially a two-tier institution, consisting of a core of original members far more integrated psychologically as well as economically than the latecomers, and with a more positive attitude to the process of European integration.

Third, the European economy witnessed a dramatic rise. This was most notable in West Germany, where the phenomenon was labeled *Wirtschaftswunder* ("Economic Miracle"), but other countries both within and outside the EEC also enjoyed unprecedented prosperity.

Finally, a new generation had grown up in Europe, a generation that was not reflexively hostile to communism and consequently less Cold War–minded. This generation, moreover, had lost sight of the American achievements of winning the war and promoting European recovery. Indeed, the new generation that had started to people the universities appeared suspicious of American motives, particularly where Europe was concerned. Many European intellectuals criticized the often obsessive forms of American anticommunism (such as McCarthyism). Anti-American sentiment as a rule expressed itself in large demonstrations in front of American embassies. It was concerned with the U.S. involvement in Vietnam, with everything that revealed American imperialism in general, and with the U.S. nuclear arsenal and the world's armament race.

But the 1960s also changed Europe in general. A different set of mores developed, and it is possible to speak of a cultural revolution (although hardly in the sense of China's Cultural Revolution of 1966–1976).

Defense and Security
The Berlin Wall reminded the democratic majority of Western Europeans that there was little that they could do directly to change the existing East–West division of Europe. Accepting that Communist rule in Central/Eastern Europe might be a permanent fact of life, they began in the mid-1960s to view their relations with the Soviet bloc in a different way. Instead of supposing either that the Soviets were poised to attack Western Europe or that Western Europe and the United States should try to undermine Communist power, they asked whether it was possible to combine deterrence of the Soviets with an increase in economic and diplomatic ties that might give the Soviets a stake in peaceful relations and the Central/Eastern Europeans the means to regain some autonomy vis-à-vis Moscow.

By 1970, the major Western European governments had adopted the view that the way to preserve peace and their own independence was not primarily through a stronger defense but through closer relations with the Soviets and the Central/Eastern European regimes. In other words, Western European politicians came to believe that the Soviet leaders had relaxed their hostility to democracy and now shared the Europeans' own goals and values concerning peace. These assumptions continued to be the foundation of Western European policy toward the East until the breakup of the Soviet Union in 1991. U.S. administrations did not share them to the same degree, and these attitudes have been the source of numerous misunderstandings within the Atlantic Alliance.

The British, Germans, and Italians were the most committed to the new view of security. The French government under de Gaulle shared it but also believed that Western Europe must in any case develop its own policy toward the Soviet Union, because the United States could not be expected permanently to guarantee the defense of Western Europe.

(Photo No. ST–C230–37–63, in the John F. Kennedy Library)

U.S. president John F. Kennedy in Berlin, June 26, 1963.

President de Gaulle wanted Europe to be a "third force," headed by France. He therefore concentrated on the development of an independent French nuclear force and, in 1966, withdrew France from the integrated military structure of NATO. In doing so, de Gaulle weakened NATO in the short term but gave France greater autonomy and self-confidence, which strengthened Western Europe in the long term.

The last obstacle on the road to East–West stability in this period was the Cuban missile crisis of 1962. Ironically, that incident also made Western Europeans realize that the United States no longer had clear superiority over the Soviets in intercontinental nuclear forces. This was one reason why most people in Western Europe believed so strongly in East–West stability from the 1960s onward. If the United States could no longer threaten to destroy the Soviet Union in response to a Soviet threat or attack in Europe without the risk of being destroyed itself, then the NATO doctrine of massive retaliation was no longer credible. Instead, the NATO governments, in December 1967, declared a new doctrine, which contained two elements. The first was "flexible response," which meant that NATO would respond to a threat or an attack at any level, nuclear or conventional, without declaring beforehand to which level it would resort. The idea was that this uncertainty would itself be a sufficient deterrent. The second element was the principle that deterrence and détente—as long as they were based on negotiations with the Soviet Union for arms-control and true peaceful coexistence—were two sides of the same coin. Neither could be pursued without the other. These two elements were to be official policy for more than a quarter of a century.

The mutual dependence of deterrence and détente signaled the beginning of the era of arms-control negotiations in Europe. After 1968, talks took place almost continuously between the two chief protagonists, the Soviets and the Americans, concerning military forces in Europe. The Western Europeans were not directly involved in most of these talks. Nevertheless, arms control became a solid part of Western European political orientation, regardless of their actual effect on the numbers of Soviet troops and nuclear forces, which was negligible.

The first major arms-control agreement was the Nuclear Nonproliferation Treaty of 1968 (in force from 1970), whereby the United States, the Soviet Union, Great Britain, West Germany, and certain other countries undertook *not* to supply to third parties any nuclear technology that might be used for weapons. Some countries, such as France and the People's Republic of China, delayed their participation. India was indignant, arguing that the Nuclear Nonproliferation Treaty was highly discriminatory, or at least biased toward the status quo. In the course of 1992, rumors—occasionally corroborated by press reports—circulated to the effect that certain countries and China had violated the treaty as well as other international laws by selling nuclear materials to Iraq. And in 1994, a new rebel emerged: North Korea.

Germany's Ostpolitik (Eastern Policy)

The German term *Ostpolitik* originated in Otto von Bismarck's days (the early 1870s). The "Iron Chancellor" did not want to embark upon colonialism, believing that an overseas empire would be exceedingly vulnerable in times of war. Instead, he drew attention to the vast, almost vacant, spaces to the east of Germany. *Ostpolitik* in those days referred to policies that would enable Germany to acquire those lands as *Lebensraum* ("living space") for its population.

Shortly after the 1969 elections put the Social Democratic Party of Germany (SPD) into office, Willy Brandt, the new West German chancellor, revived the term in an entirely different context. Now *Ostpolitik* referred to attempts at reconciliation by West Germany with all its eastern neighbors, with the exception of East Germany. These countries had greatly suffered in World War II from the German onslaught. Brandt visited most of them, gave away large sums of money by way of compensation, and negotiated diplomatic relations with all of them. The so-called Eastern Treaties (which specifically referred to the Soviet Union, Poland, Czechoslovakia, and East Germany) became part of the permanent legal and diplomatic framework of West European politics as regards Central/Eastern Europe. In doing so, West Germany naturally had to recognize the status quo, which implied the acceptance of Soviet control of Central/Eastern Europe and, more important, Communist rule in East Germany. However, what Brandt and West Germany received in return was incalculable: A great deal of prestige accrued to the Federal Republic, while the condition of the German nation was henceforth described in the formula "one nation, two states."

EEC Problems

As has been mentioned, the EEC was faced with the dilemma: widening (having more members) or deepening (becoming better integrated). During the 1960s, the Community failed on both scores. Two major problems emerged that became worse over time.

The first was that even though the EEC was supposed to be a common market in goods and services with no internal tariffs, member countries were able to protect their national economic interests, as they perceived them, in ways other than by outright tariffs. There were two main methods. One was to introduce complicated national rules for the approval of industrial products of all kinds, which, in accordance with EEC rules, made it necessary for producers elsewhere to adjust their products to each country's specifications. Such technical-approval rules were not legally equivalent to tariffs, but they had the same effect of making it difficult for foreign producers to enter a domestic market. For example, West Germany had laws specifying the exact ingredients and regulating other requirements for brewing beer. Other European beers followed a variety of very different regulations or specifications. As a result, West Germany, referring to its beer-brewing laws, was able to prevent foreign beers made with different specifications and procedures from being sold in the country. The other method of protecting national economic interests was taxation. The EEC was a customs, not a tax, union. Each government levied taxes in whatever way it chose. High indirect taxes on business transactions or on imported consumer durables were an effective means of discouraging importation and consumption.

In the early 1960s, the governments of the ECSC introduced another kind of tax, one that had been in use in Great Britain and which has since been adopted by all Western European governments. This "value-added tax" (VAT) works like U.S. sales taxes, but the difference is that it is levied at each stage of the process of production and distribution. The producer as well as all intermediaries must collect the tax from the next person down the chain; they are then entitled to a refund of the tax that they themselves have paid. Only the final consumer, who actually uses the product, gets no refund. Rates of VAT varied from product to product and from country to country, as did the list of products or services liable to it. Gradually, all governments raised the tax rates and extended the scope of VAT. In the beginning, rates were typically around 5 percent; and many items, such as food, educational materials, and legal fees, were exempt. But by the 1970s, rates in some countries were as high as 33 percent for some products, and there were very few exemptions.

A second major stumbling block for the EEC in the 1960s was agriculture. By 1965, it had established a common market in agricultural products with common prices. The member governments discovered that the inevitable effect of this common market would be to drive farmers in most of France and

parts of West Germany out of business, since their costs of production were far higher than those of poorer farmers in Italy or more productive farmers in the Netherlands. Consequently, they instituted a system of price supports and payments from country to country to maintain agricultural employment and farmers' income at specified levels. This "Common Agricultural Policy" (CAP) soon became enormously expensive; by the mid-1970s, it was consuming more than 80 percent of the EEC budget as well as most of the administrators' time, thus effectively blocking any movement on substantive issues of common interest. The main benefits of the CAP were reaped by France, at the expense of West Germany.

In 1965, the ECSC, the EEC, and Euratom were united in a single administrative entity, governed in common by the Council of Ministers (one cabinet member from each country, depending on the subject that is being discussed), the European Commission (composed of top administrators who took oaths of national independence), and the European Parliament (delegates originally appointed by the parliaments of the various member states, and after 1979 directly elected). In 1972, shortly after de Gaulle retired as president of France, formal membership applications were accepted from Great Britain, Denmark, Norway, and the Irish Republic. The Norwegian voters rejected membership in a referendum, but the other three countries formally joined the EEC on January 1, 1973. Gradually, this more integrated entity became commonly known, simply, as the European Community, or EC. (In 1993, it became officially the European Union, or EU.)

A Cultural Revolution

The generation of statesmen who had lived as adults through Europe's "Second Thirty Years' War" of 1914–1945 and who survived to build democracy in Western Europe after World War II had to a great extent faded from the political scene by the mid-1960s. Some of the leaders of that generation were men remarkable for their stamina and longevity. Konrad Adenauer was finally forced out of office by his own party in 1963. Four years later, he died, at age 91. Charles de Gaulle, leader of the Free French in 1940–1944, of France in 1944–1946, the first president of the Fifth Republic in 1958–1969, and advocate of an *Europe des Patries* (an autonomous Europe composed of sovereign national states), died a year after leaving the presidency, at age 79. Alcide de Gasperi, Italian prime minister during the postwar years, was another giant on the European scene who actively participated in European integration. And finally, Harold Macmillan, who as British prime minister became known for his "Winds of Change" speech (which referred to the liquidation of the British Empire), having held the office from 1957 to 1963, died in 1986, at age 92. These statesmen regarded the outbreak of World War I in 1914 as the major disaster in modern European and world history from which all the other subsequent disorders stemmed: totalitarianism, genocides, the

THE EUROPEAN UNION: A CONFUSING NOMENCLATURE

To many people, the names as well as the abbreviations relative to the European Union (EU) have become more than a little confusing. It should be borne in mind that, in fact, three Communities were established in the 1950s. The first was the European Coal and Steel Community (ECSC), which was founded in 1951 by the Treaty of Paris. The original ECSC included six countries: France, the Federal Republic of Germany (West Germany), Italy, Belgium, the Netherlands, and Luxembourg. The intent was to pool the coal and steel resources of these countries—that is, to entrust these resources to some supranational agency.

The second and third Communities came into being through the Treaties of Rome, which were both signed at the Italian capital on March 25, 1957. These Communities were the European Economic Community (EEC), an economic-integrative entity and by far the most important Community, and the European Atomic Energy Community (EURATOM). The latter was meant to be a research body that would investigate the peaceful use of atomic energy. But it soon ran into serious difficulties and, as a result, it always remained very much behind the two other Communities. The Treaties of Rome covered the same countries as the Treaty of Paris.

The three Communities operated separately. However, since they had the same membership and were pursuing similar goals, the three Communities were increasingly viewed as a collectivity. In the late 1960s, moreover, Community institutions (the European Commission, the Council of Ministers, the European Parliament, and the Court of Justice) were, for the sake of economy, applied to all three of the Communities. The term *European Community* (EC) started to be used for all of them. With the Maastricht Treaty of 1991, that term was replaced by the *European Union*.

(Note: To avoid confusion, we generally use the term *European Union* in this volume.)

Another term that may require some explanation is *Common Market*. It was used for the first time in theTreaty of Paris, where it meant putting together all the coal and steel resources for the common use of the six member states. The term also came to be used as a synonym of the European Economic Community. In recent years, it has fallen into general disuse, although it is still sometimes heard in Britain and the United States.

European Integration

The European experience has demonstrated clearly that politics and economics are closely intertwined. Although the economic aspects of the old EEC remain important, many authorities believe that the political aspects of the EU may gain significance to the point where they may overshadow the economic ramifications—a contingency foreseen as early as 1957. The Single European Act, which came into force on July 1, 1987, to all intents and purposes constitutes a comprehensive amendment to the Treaties of Paris and Rome. It too underscores the politicization of European integration, and it therefore is no longer possible to view the Communities as vehicles solely of economic integration. An interesting side effect of the momentous changes that they have brought about is that, in the United States, the word *Europe* has almost become synonymous with what the EU has come to stand for, although Europe as such is much larger in area and is composed of a host of countries that are not—at least, not yet—member states of the Union.

INTEGRATIVE PROPOSALS AND PROCESSES

1947 The Marshall Plan is announced; Jean Monnet begins working for a common European economic market and defense

1950 The Schuman Declaration, aimed at supranationalism

1951 The Treaty of Paris establishes the ECSC

1957 The Treaties of Rome establish the EEC and EURATOM

1960s Adoption of the Common Agricultural Policy

1962 The Fouchet Plan coordinates foreign and defense policies

1969 The first summit in The Hague comes to be known as the "relaunching" of Europe

1970s The Lomé Conventions render many developing-world countries associate members; the Communities become known as the European Community

1979 The first direct election of the European Parliament

1980 The Mansholt Plan for Agriculture

1981 The Genscher-Colombo Plan is considered

1983 The Solemn Declaration of European Union

1984 An initiative concerning the Draft Treaty on European Union

1985 The Schengen Agreement, concerning the removal of border controls, initially not extending to all members

1986 The Single European Act is approved

1988 Bruges speech by British prime minister Margaret Thatcher favors a Europe of nation-states

1990 Britain enters the European Exchange Rate Mechanism

1991 The European Council adopts the Maastricht Treaty for a closer union; the EC becomes the European Union

1992 An Irish referendum endorses Maastricht; a Danish referendum rejects it; a French referendum approves it—barely; the Treaty on European Union further revises the founding treaties

1995 The EU pledges to start looking for membership candidates in Central/Eastern Europe

1997 The draft Treaty of Amsterdam proposes further revisions to the founding treaties; Agenda 2000 aims to strengthen the EU and prepare for enlargement

1999 Europe's currency union (EMU) is launched with issue of the Euro (coins and banknotes due in 2001)

2000 Humboldt University speech by German foreign minister Joschka Fischer favors a "Federal Europe"; the draft Treaty of Nice is approved, advancing institutional reforms in preparation for a wave of new members

attacks on democracy, economic disruption, and chaos. They did their best to prevent a repetition of these events.

Their departure from power symbolized a broader shift in Western European politics and culture—from concern with defense and peace in the most elementary sense, to concern with social issues, domestic politics, and new revolutionary ideologies. In many countries, left-wing parties came to power for the first time, with new agendas for extending the welfare state, cutting down military spending, giving more power to the common person, and redistributing income, wealth, and authority. Since these agendas required more, not less, government, their outcome usually was to expand bureaucracy and to centralize power. Also, government tended to be more technocratic. This produced disillusionment and cynicism, two tendencies that would strongly characterize European politics in coming years.

The most dramatic expression of the shift in European attitudes and culture in the 1960s was in the area of higher education. The number of university students had been growing explosively since the late 1950s. A university education was no longer an elite luxury in Western Europe but something that almost everyone who qualified might expect to have. A great many of the new students rejected orthodox communism but were oriented toward "new left" ideology, directed against the social and political establishment.

French education, at all levels, has always been extremely centralized and rigid. The university curricula provoked the indignation of university students all over France, but particularly in Paris. These curricula had become old-fashioned, if not obsolete. Emphasizing the classics and the humanities, they in no way met the requirements of the times—that is, they had not adapted to the emerging postindustrial world. Those shortcomings were clearly reflected in massive graduate unemployment. In May 1968, the situation came to a head. Paris then presented a revolutionary picture: Students established barricades on the Left Bank, close to the Sorbonne and Nanterre University. From these somewhat dubious fortifications they hurled cobblestones and Molotov cocktails (firebombs) at the police. Soon workers were also staging protest actions against de Gaulle's government. Although no one was killed, these scenes made a deep impression on French society, which has continued to refer to these radical activities as *les événements* ("the events"). Unfortunately, their outcome was disappointing. After temporary adaptations and adjustments, the education system suffered a retrenchment of obsolescence.

STAGFLATION AND INSECURITY (1973–1989)

Although economic growth had slowed down by the early 1970s and many Western European countries had significant budget or trade deficits, no one expected the sudden crisis of 1973–1974, which put a final end to what the French called "the thirty glorious years" (1944–1973) of increasing prosperity and even more rapidly rising expectations. The crisis began with oil-price increases imposed by the Organization of Petroleum Exporting Countries (OPEC). These increases triggered equally dramatic increases in the costs of many other commodities essential to the European economies. The result was an unprecedented combination of inflation and unemployment, which economists termed "stagflation." One of its most remarkable aspects was that unemployment and inflation no longer responded to traditional economic laws. In the past, they had related conversely—that is, if one went up, the other went down. But in the early 1970s, unemployment and inflation rose (and fell) simultaneously.

Barely had the Western Europeans dealt, with varying success, with the first wave of stagflation than a further crisis developed. In 1979, the Islamic Revolution in Iran produced a new round of oil-price increases, which hit Western Europe as hard as the first had done. Still, it seemed that Europeans had become more accustomed to these stresses. At the gas stations, there were no long lines of cars waiting to be filled; no country instituted "carless Sundays" (as the Netherlands had done during the first oil squeeze). Nevertheless, the period did generate some friction in the Allied camp. The Soviet leaders in the early 1980s offered to supply Western Europe with natural gas from Siberia. To that end, a pipeline had to be built across the Soviet Union, Poland, and East Germany. U.S. president Ronald Reagan feared that such a supply of energy would render Western Europe politically vulnerable, since the supplier, the Soviet Union, could always threaten to cut the supplies if Western Europe went against its wishes. However, Western Europe's internal energy sources were extremely limited, a fact that was bound to create a dependency on outsiders anyway. The acrimonious debates that ensued between the United States and its Western European allies with reference to the Soviet offer were known as "pipeline politics."

About the same time, the major NATO countries, believing that their security was threatened by new nuclear weapons, decided to update their own weaponry, at the instigation of the United States. The most controversial element of that modernization was a plan to deploy new U.S. intermediate-range nuclear missiles in Western Europe. This category of weapons posed the danger that Europe itself could be destroyed by nuclear weapons (in the past, the so-called strategic missiles were mainly intended to hit the two adversaries, the United States and the Soviet Union). The plan provoked a large wave of pacifist movements in Western Europe, which for centuries has been a battlefield for wars. Pacifism rocked the governments of various Western European countries. The peace movements in Central/Eastern Europe were controlled by the Communist parties.

Economic Slowdown

In Western Europe, oil provided the fuel for transportation as well as most of the energy for generating electricity. In 1973, Western Europe imported 90 percent of its oil. The import price of crude oil in September 1973 was $2.50 a barrel; by January 1974, it had jumped to $12.00.

The effect of the OPEC price increases was to transfer about 2 percent of the value of Western Europe's total annual production of goods and services to the oil-producing states. This meant that the Western Europeans had that much less to spend on their own goods and services, a factor that led to unemployment. At the same time, labor was politically strong enough in most countries to prevent employed workers from suffering any significant loss of real income. Instead, the net loss of wealth fell disproportionately on the young or unemployed. Unemployment, which had been virtually unknown in many countries since the early 1950s, rose in 1974–1975 to about 10 percent of the labor force in all the major countries, where it hovered for two decades or more.

The impact of the "oil shock" varied. The trade-surplus countries, above all West Germany, absorbed the increases far more easily than the trade-deficit countries like Italy. Norway and Britain had an advantage. In the 1960s, explorers had found petroleum underneath the North Sea, but, at the price level that prevailed at the time, it had not been economical to

extract this oil. After the oil shock, extracting the continental-shelf oil between Britain and Norway proved competitive, and, by the late 1970s, Britain had become an oil-exporting country.

In the second oil shock, in 1979–1980, prices rose to about $40.00 a barrel. This hit Western Europe especially hard, since the value of the U.S. dollar was rising against European currencies. Again, the surplus countries did better than the deficit countries. More important, all countries had been improving energy efficiency, so that much less oil was needed for the same effect. In addition, France and West Germany expanded their nuclear-energy programs. In the 1980s, nuclear fission produced about 70 percent of all the electricity used in France. Only in the early 1990s did France begin to face some difficulties in its reliance on nuclear energy. However, nuclear energy generated early and enduring opposition in Germany, which ultimately led to the government's decision in 2000 to begin phasing it out.

Terrorism and Military Security
Most Western Europeans were too young to remember the domestic violence that had been a factor in politics in the

(Mobil Oil Corporation)

North Sea oil was a boon to several Western European countries during the 1970s and early 1980s, but the bonanza was not so sweet after oil prices dropped dramatically in the mid-1980s.

1920s to the 1940s. They were all the more taken aback when terrorist groups began to kill civilians, to bomb public buildings, to hijack airplanes, and to disrupt the Olympic Games in Munich in 1972 with a massacre of Israeli athletes. Some of these terrorists were Europeans themselves, young radicals who had started out as student activists in the 1960s and who had joined gangs such as the Red Army Faction in West Germany and the Red Brigade in Italy. Others hailed from the Middle East. It took some time for the Western European governments and police forces to coordinate an effective response to them, but, in a series of dramatic actions in 1977–1979, most of the gang members were captured or dispersed. In West Germany, the government of Chancellor Helmut Schmidt had a number of antiterrorist laws passed that to many appeared too drastic, in that they went very much against the spirit of democracy.

Toward the end of the 1970s, some Western European leaders, notably Schmidt, concluded that the 1960s view that détente policies would encourage the Soviets to reduce their vast military arsenal was mistaken. The Soviets were building up both their conventional and their nuclear forces in Central/Eastern Europe, far beyond what they needed to suppress dissidence or maintain their own security. In 1977, Schmidt suggested that the United States either obtain a reliable arms-control agreement with the Soviets for disarmament in Europe or install new nuclear missiles in Western Europe to achieve a balance with the Soviets. In 1978, NATO decided to begin a long-term process of modernizing all its forces to maintain the credibility of the flexible-response strategy in the 1980s.

In December 1979, NATO decided to pursue arms-control talks with the Soviets regarding European-theater nuclear forces—but if those talks failed, the United States would

West German chancellor Helmut Schmidt (1974–1982), addressing a UN session on disarmament.

(UPI/Bettmann Newsphotos)

deploy 572 new missiles. This "double-track" measure encountered the opposition of hundreds of thousands of Western Europeans, mostly young persons who had no memory of the 1940s and 1950s, when many people feared a Soviet attack and were determined to defend democracy by force. The members of these new peace movements did not see the Soviet Union as a military or political threat. In the early 1980s, they demonstrated especially in the cities of West Germany, Britain, and the Netherlands against the projected deployments. There were no peace movements to speak of in France and Italy. Both countries had large Communist parties that had used peace slogans for their own purposes in the past. Many remembered this and regarded the new peace movements as witting or unwitting tools of Soviet expansionism.

The American negotiators failed to persuade the Soviets to reduce their nuclear superiority within Europe. Therefore, in late 1983, NATO started to deploy the U.S. missiles. The peace movements had failed in their immediate objective. In the longer term, though, they had changed the terms of debate on defense and security, above all by questioning the moral and political value of the NATO doctrine of flexible response, including the threat of a nuclear response to attack.

Strengthening the Core

One important result of the economic problems and the common Western European concerns over defense in the later 1970s was the revitalization of the French–German dialogue. The Friendship Treaty of 1963 prescribed such a dialogue, to take place chiefly in the semiannual meetings of the heads of government of the two countries; but, until the mid-1970s, the talks concerned mainly economic and trade issues. In 1974, the French elected as president the conservative Valéry Giscard d'Estaing. To the surprise of many observers, he formed a close friendship with Chancellor Schmidt, a social democrat. Giscard and Schmidt turned the semiannual talks into mini-summits, where they discussed issues of general European significance, including security. Both felt that Western Europe needed to establish its own position in international politics rather than remain dependent on the United States. They also believed that the United States no longer had the resources, the interest, or the ability to conduct a consistent worldwide foreign policy. Finally, they believed that if Western Europe was to have a voice of its own, that voice should come from the core countries—France and West Germany.

Peace Movements: A New Credo

The peace movements of the early 1980s convinced a sizable number of people in Western Europe that even the *threat* of using nuclear weapons in response to a Soviet attack was immoral. A factor that gave the peace movement added legitimacy was that it was in large part based on religion. Critics perceived the churches as trying to win back the young by

(Reuters/Bettmann Newsphotos)

A demonstrator attaches an antinuclear banner at a war memorial in Munich, Germany.

endorsing their various causes. In addition, most Western Europeans had over 20 years come to believe that there was probably no Soviet military threat and that any East–West problems could be solved by negotiations. A popular phrase of the 1980s had it that Western Europe and the Soviet Union (with Central/Eastern Europe) were bound together in a "security partnership," a notion that implied that each side's security depended on that of the other. A corollary was that Western Europe should do nothing that might provoke the Soviets or that might cause them to think that the West was planning to undermine Communist regimes in Central/Eastern Europe.

Most of the peace movements operated in Northern Europe, but throughout the democratic part of the continent, many politicians, mostly but not exclusively from Socialist parties, began calling for the United States to withdraw its troops. The United States then maintained naval, ground, and air forces in West Germany, Italy, Turkey, Greece, Spain, Portugal, and Great Britain. Greece and Spain had right-of-center governments, which were particularly insistent that the United States should either close its military bases or offer significantly more in return for the right to use them. In these countries, irritation over U.S. bases stemmed from a complex mixture of feelings of national inferiority and anti-Americanism.

Even among older West Germans, who had been overwhelmingly pro-American since the 1950s, there was more than a little apprehension about the U.S. troop presence. It had become customary to claim that peace in Europe would be better served if the Reagan administration abandoned its provocative attitude and reduced its forces. In Great Britain, the Labour Party took a strong stand against the U.S. presence and the country's own independent nuclear force, but the

party was too factionalized to stand a chance of gaining power in the early 1980s.

In 1986, the Americans and the Soviets began a new round of arms-control talks aimed at reducing nuclear missiles in Europe. Most Western Europeans considered these talks promising and worthwhile; others wondered whether the Soviets could be trusted to fulfill their part of the bargain or whether they were merely negotiating to remove weapons that they no longer wanted anyway.

The Beginning of Change

Many in Western Europe put faith in the Soviet leader who took power in 1985, Mikhail Gorbachev. They believed that he was sincere in wanting to reform the Soviet economic and political system; that, in fact, he must do so if the Soviet Union were to remain a global superpower; and that it was in Western Europe's interest to help him with loans and technological assistance. Few thought that a modernized Soviet Union might be a new danger rather than a friendly new partner of the West.

One of the early admirers of Gorbachev was British prime minister Margaret Thatcher, who, even before Gorbachev had made it to the top, observed that here was a man with whom the West could do business. In time, the West Germans became enthusiastic admirers of the Soviet leader, who, according to several public-opinion polls, was found to be "more likeable" than Ronald Reagan. In the 1980s, West Germany moved to regain Germany's historic role as the main trading partner of the Central/Eastern European countries. The German leaders thought that Germany could play a role in moving the Central/Eastern European regimes in the direction of democracy and human rights.

(UPI/Bettmann Newsphotos)

Soviet president Mikhail Gorbachev was in power from 1985 to 1991.

A Ballistic Missile Early Warning System site in Pylingdales Moore, England.

The French were more worried than the Germans about Gorbachev's intentions. Since de Gaulle's tenure in the 1960s, they had adopted their own stance in East–West relations and in Western European integration. By the 1980s, they had shed some of their anti-Americanism and had emerged as one of the more anti-Soviet of the Western European nations. There was the fear that Gorbachev simply wanted Western help to streamline the Soviet system and make it a more efficient threat to Western European liberties and prospects. They also feared that the Germans were likely to succumb to Gorbachev's diplomatic blandishments, chiefly because the Soviet Union held 17 million East Germans hostage. For several reasons, therefore, the French continued to push actively in the 1980s for a new European security system, based on the French–German alliance rather than on the weak institutions of the EU. First, falling back on de Gaulle's Eurocentric vision, they espoused the belief that Western unity would be the best guarantor of peace. In addition, they argued that unity must be based on Europe, asserting that the United States no longer had the will or the skill to maintain its superpower status vis-à-vis the Soviets. And finally, the only two relevant European powers were France and West Germany, with the latter likely to slide toward Gorbachev unless its tendencies in that direction were firmly countered by the French.

The Population Bust

All of Western Europe, but especially Germany, the Scandinavian countries, France, and Great Britain, were confronted by a new reality beginning in the early 1980s: declining population growth. This unprecedented development has serious implications. First, since Western Europe is above all a society of skilled, professional people, a population decline is bound to have a disproportionate impact on the global product of valuable goods and services. The whole world, not just Europe, benefits from Western European technology and production. Fewer Western European designers, entrepreneurs, and producers mean fewer high-quality goods for the whole world.

Second, societies progress as a result of the intellectual and economic investments and achievements of young people, whereas older people mostly repeat what they learned when young. Therefore, a decline in the numbers of young people is bound to have a disproportionate effect on inventiveness, productivity, and cultural continuity, with unpredictable but potentially catastrophic effects on the vitality of European society as a whole.

Third, all Western European countries had become advanced welfare states by the 1980s. The welfare systems operated on the assumption that there would be a sufficient number of young, productive people of working age to support the elderly and the disabled, the military forces, and people in education. In the 1980s, there were between three and four people of working age for every one welfare recipient, a ratio that could cause problems. It has been estimated that by the year 2020, that ratio will have deteriorated to perhaps about two to one in Germany. A declining number of working-age people will have to produce the revenue through their taxes to support the many pensioners. In addition, there are two other related factors that tend to jeopardize pension plans and social security systems: To combat unemployment, business and government have come up with early retirement

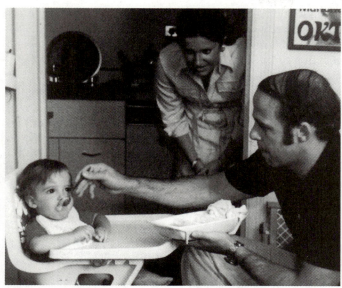

(United Nations/M. Faust)

European countries have made a strong commitment to both state-supported child care and paid parental leave.

offers. At the same time, life-expectancy rates have also risen in most industrialized countries. This means that pensioners will be dependent on public moneys for a much longer time. Raising taxes to pay for all this will have the effect of discouraging large families. Thus, the problem is bound to get progressively worse, escalating in an upward spiral. Experts predict that the whole welfare system on which political stability in Western Europe has rested since the 1940s is bound to disintegrate within a generation unless there is significant reform.

There is no single explanation for the problem, but several factors clearly contribute to it. One is the availability of birth control since the mid-1960s. This makes having children a matter of choice, and many people choose not to have them. Another is the welfare system itself, which means that people no longer need to have children to take care of them in old age or when they get sick. In addition, the economic cost of the welfare system is making it difficult for many families to have more than one or two children while simultaneously supporting the system through their taxes. Another reason is the psychological and cultural atmosphere that has prevailed in Western Europe since the mid-1960s, an atmosphere that puts immediate gratification and adult pleasures, such as foreign travel, over family life.

THE POST–COLD WAR ERA (1989–PRESENT)

Almost overnight, the Cold War, which had haunted world politics for more than four decades, came to an end. Gorbachev had unwittingly written the scenario of the Cold War's demise. In the mid-1980s, he introduced two new concepts: *glasnost* (Russian for "openness," which implied that the government was willing to lift its censorship) and *perestroika*

(which referred to the restructuring of the socialist economy). It was these two formulas that opened the Soviet Pandora's box, releasing forces that could no longer be contained. Gorbachev, who described himself as a "Leninist" sought to purge the Soviet Union of all the evil that had amassed over the course of seven decades. Also, he must have severely underrated the forces of nationalism that were stirred up by his reforms. Ethnicity in general and ethnonationalism in particular were dormant in the heyday of the Soviet Union; under Gorbachev's rule, they erupted. Not only the three Baltic nations (Estonia, Latvia, and Lithuania) but several other states, one by one, pulled away from the Union of Socialist Soviet Republics. Its collapse may best be described as an "implosion." In some republics, there were skirmishes; in others, serious fighting was attended by bloodshed and destruction. In late 1991, Gorbachev resigned as president of a Soviet Union that was no more.

In the late 1980s, Gorbachev visited the various heads of government in Central/Eastern Europe, leaving the message everywhere that the Brezhnev Doctrine was no longer in force. Soviet leader Leonid Brezhnev had in 1969 stated that there was a commonwealth of Socialist nations. If one of its members attempted to defect, the others had the obligation to prevent it. Since the Soviet Union was the most important member of the Warsaw Treaty Organization, the obvious implication was that the Soviet Union would intervene whenever an East Bloc nation tried to get rid of or to liberalize its Communist form of government. In fact, massive military interference had taken place in Hungary in 1956 and in Czechoslovakia in 1968. Gorbachev, considering his country's rock-bottom economy, told the East Bloc leaders that they would now be on their own, because the Soviet Union had far too many commitments.

Soon the Central/Eastern European nations became the scene of massive rallies, day after day. Thousands of peaceful demonstrators clamored for reforms, for better conditions, and for an improved economy. Ultimately, the leadership in these countries was faced with the dilemma to shoot or to step down. Since most leaders feared that the military could not be entrusted with the unsavory task of perpetrating a bloodbath, they chose the latter alternative. Surprisingly, there was no serious bloodbath in any of the countries except Romania. In Czechoslovakia, people prided themselves on the orderliness of their "velvet revolution." (A few years later, when the Czech Republic and Slovakia ceased to be a national entity, Czechoslovakia, this was referred to as the "velvet divorce.")

The German Reunification

The decline of the Soviet system and the unexpected events in Central/Eastern Europe had further repercussions. In East Germany, grassroots democratic movements surged against the Communist leadership. In addition, the season of flight had resumed, but, since the border between East and West Germany was heavily guarded, those wishing to leave East

Germany took recourse requesting asylum at the various embassies in East Berlin, the capital. Many left on Hungarian "vacation trips" in an attempt to enter West Germany via Austria.

As soon as neighboring countries allowed "transit tourists" to pass through without proper papers, a stampede commenced that could be stopped, first, only by promises on the part of the East German government that conditions would soon improve, and, subsequently, by unification overtures. On November 9, 1989, the Berlin Wall was breached, a day that Germans on both sides identified as the most significant day since the end of World War II. After the election of a non-Communist government in the East in March 1990, re-unification procedures were set in motion, and in October the two Germanys were officially united. The entire area is now called the Federal Republic of Germany.

However, the unification was not a source of universal joy. The government of the new Germany insisted that secret intelligence files be made accessible to the general public, particularly to those who were under surveillance during the years of repression on the part of the East German government. STASI (East Germany's Secret Service) files made it clear that a great many people had been overzealous in submitting reports on friends, family, and acquaintances to the generally hated organization. The fact that numerous reports appeared to be untrue naturally set bad blood among family members and friends.

The migration from East to West, which had been partly responsible for the drive toward unification, continued after the two Germanys had reunited. Many people were impatient, or at least unwilling to wait until the conditions in eastern Germany improved. The migration slowed to a trickle when Chancellor Helmut Kohl, who feared that the new Germany would be demographically unbalanced, indicated that, henceforth, the new arrivals would no longer receive the generous treatment that had traditionally been accorded to refugees from eastern Germany.

Changing the Focus of Confrontation

If superpowership has become a phenomenon of the past, the East–West confrontation that was, so to speak, its territorial expression, has also faded away since the fateful days of 1989. The Warsaw Pact was abolished almost immediately after the collapse of communism in Central/Eastern Europe. This was done at the explicit wish of its entire membership, including the former Soviet Union. However that may be, NATO has continued to exist. Indeed, it expanded in 1999 to include three new members—Hungary, Poland, and the Czech Republic—and further expansion is likely. When Yugoslavia fell apart in the early 1990s, the nearby and well-armed NATO forces remained inactive until 1995, when they became massively involved in Bosnia. In 1999, NATO launched air strikes against Serbia in retaliation for its refusal to sign a Kosovo peace plan.

While global attention was riveted on the new Germany that appeared to be emerging, Iraq occupied its oil-rich neighbor, Kuwait, on August 2, 1990. As the United States led coalition forces against Iraq in early 1991, the world's focus moved to the Middle East in general and to the Persian Gulf in particular. The European Union declared itself behind the United States; to some Europeans, Iraq's aggression against Kuwait may have evoked the specter of Hitler's *Anschluss* of Austria in 1938. Most Western European countries, however, did not go beyond an endorsement of American efforts. Except for Great Britain, France, and to a smaller extent Italy, they provided little in the way of a contribution, either by sending military personnel or materiel or by providing the financial wherewithal to sustain what was officially a UN attempt to restore Kuwait's sovereignty. Germany claimed that its Constitution, put together shortly after World War II, would not allow it to send troops. However, its constitutional qualms were overcome in 1999 as German aircraft and personnel participated in NATO's air war against Serbia.

The focus of the global confrontation is no longer East–West but, rather, North–South, although one should bear in mind that the latter dichotomy is based on economic premises. The phrase "North–South" has come to stand for the adversarial relationship between countries at opposite ends of the development concept: industrialized countries, with a high degree of economic advancement, versus the developing countries, which until fairly recently were parts of empires whose traditional societies had been left in a prolonged impasse. Once the latter had acquired political independence, they lacked the infrastructure and the know-how to make any headway. These new nation-states remain extremely poor. Most still do not have any industry to speak of. The South is very backward in economic respects; in addition, it is often riddled with corruption and, as a result, is politically highly volatile and unstable. The colonial experience has also rendered most of these new nations very sensitive with regard to Western interference.

A Peaceful Invasion

The South has started to confront the North in more everyday ways—that is, by entering North America and Europe in ever-larger numbers. International immigration, whether legal or not, has become one of the more critical issues since the Cold War ended because of its sheer numerical force. Among the huge numbers of illegal immigrants are North Africans crossing the Strait of Gibraltar in order to get into Spain, a country that may not be rich but that is at least up and coming as a result of its membership in the European Union; Spain also provides an avenue to other parts of Europe. Illegal immigrants have also streamed into Italy via sea from Albania and other countries, and many have headed north.

Germany as Core Country for Immigrants

The massive migrations from the South should not be taken lightly or as a phenomenon that is bound to blow over in due time. Originally a trickle, the numbers have soared and now literally run in the millions (official estimates of the total number of refugees all over the world approach 14 million).

In his book *The Call of the Toad,* Grass makes a Bengali rickshaw operator in Gdansk admonish his German audience in the following words:

> Please look upon me as a person with nine hundred and fifty million human beings, soon a round billion, behind him.

The modern *Völkerwänderung* (migration of nations) has come to Germany in three separate waves. The first group consisted of the so-called guest workers of the 1960s, who, like most guests, were supposed to spend a limited time (in this case, two or three years) with their hosts, in Western Europe. Germany, like most Western European nations, was short of labor. In their desperation, its corporations allowed many of the Turks and other guest workers from the Mediterranean basin to renew their contracts.

Then the mid-1970s produced, as an epilogue to the wars in Indochina, tragic groups of huddled masses, packed in unseaworthy vessels, who defied the elements in the Indian and Pacific Oceans in search of a new home. Whenever they landed on an island in one of the various archipelagoes in Southeast Asia, they soon found that they were not very welcome. Sometimes they were put into refugee camps, awaiting a further destination. Many Southeast Asians opted for resettlement in European countries, following the official channels. Some countries in Western Europe allowed a few hundred or so to settle.

Hardly had this relatively small flow ended than Southeast Asians adopted yet another modus operandi to enter Western Europe in general and Germany in particular. They boarded planes, with one-way tickets, to East Germany. Even during the Cold War, it was not difficult for non-Germans to get from East Berlin to West Berlin and from there to West Germany. Once on West German soil, they destroyed all their identification and traveling documents and asked for asylum. As a rule, it would take the German authorities some two years to find out to what extent their claims were genuine. Over time, the total of *Scheinasylanten* (literally, "fake asylum-seekers") came to exceed 1 million. In October 1992, the press reported that the flow into Germany of asylum-seekers appeared to have stabilized at 60,000 a month, more than to the rest of Europe combined. Many of them were not sent back or deported, even if their quest for asylum was officially rejected. The great majority appeared to consist of "economic refugees," persons thus who were not politically persecuted but who believed that they could make a better living elsewhere.

Critics have argued that the laws in Germany and elsewhere in Europe that govern immigration should be adapted to modern circumstances and thus be made considerably more rigorous. Germany has been very slow in getting to that point,

for two reasons. First, the German Basic Law mirrors the immediate postwar years. It was deliberately made very flexible with regard to immigration, a reaction no doubt to the harsh Nazi years. In addition, one should take into account that the only refugees who were expected to ask for asylum at that time (the late 1940s and the 1950s) were the sporadic refugees fleeing from Communist brutality in Central/Eastern Europe, including East Germany. It was certainly not necessary to make the Federal Republic impenetrable through the erection of legal barriers. If this argument harked back to the past, the second point referred to the future: The European Union would have to tackle the immigration issue, since persons and goods would be able to freely cross the borders within the EU. That being the case, there would be little point in having varieties of rules and regulations passed in one member state, since immigrants would be able to enter the Union through a member state with lax provisions, after which they could still move around and find their way to Germany.

The Re-migration of Expatriates

To complicate matters even further, Germany in the early 1990s became subject to yet another type of immigration. Hardly had the two Germanys been formally united when references were made to all the Germans who lived in Central/Eastern Europe and thus missed out on the joys of unification. The persons concerned were usually referred to as "ethnic Germans," a somewhat infelicitous phrase, since it could easily have been part of the Nazi lexicon. Historically, the *Drang nach Osten,* or drive toward the East, endemic in Germans concerned about the lack of *Lebensraum,* or living space, caused many of them to settle in eastern territories in the hope, perhaps, that some future German Manifest Destiny would cause them to be part of the Reich again. A large proportion of these ethnic Germans spoke heavily accented German or no German at all. Tens of thousands of them were repatriated (a term that was used officially) from Russia, Romania, Poland, and other Central/Eastern European countries at the expense of Germany's hard-pressed budget.

It is difficult to fathom why the chancellor of the Federal Republic made the repatriation scheme a priority while still maintaining that Germany was not a country of immigration. A great many people immediately responded positively to this call. With an economy that had turned sour, a large public debt for the first time in the existence of the Federal Republic, and diminishing resources, the German government was forced to reverse itself in repatriation matters. German consulates were given the task of making the Germans who qualified for repatriation stay where they were. Thus, the large-scale repatriation slowed to a trickle of emergency cases.

Public Reactions in France and Germany

Two countries—France and Germany—have been selected here to discuss their reactions to the massive invasion of foreigners, not so much because they happen to be the largest countries in Western Europe but because they are most affected. The National Front, led by Jean-Marie Le Pen, won as an ultra-nationalist party more than 15 percent of the vote in French national elections during 1995–1997 before splitting into contending factions. Germany has had three small, right-wing extremist parties that seek to make political capital out of foreigner issues. Many of the French and the Germans share the feeling of *Überfremdung* ("overalienization") that has caused the Swiss, who coined the term, to severely limit entry into their country. But thus far the governments of these two countries have not given in to the increasingly hostile sentiments of xenophobia. For example, Richard von Weizäcker, a former president of Germany, as well as various other top German authorities were greatly upset by the events in Rostock, in former East Germany, where Neo-Nazis and skinheads set fire to hostels that had been allocated to foreigners. As soon as the foreign women and children fled the fire (most of the men were at work), they were either killed or severely injured and in some cases were thrown back into the flames. The police arrived very late on the scene and thus left the women and children to fend for themselves, and the large crowd of onlookers did nothing to prevent the arson as well as the events that followed. The investigations into the behavior of the police in this instance have proven inconclusive. As far as the passive crowds were concerned, von Weizäcker, addressing the nation, solemnly stated: "We must never forget how the first German republic [the Weimar Republic] failed: not because there were too many Nazis too early but because there were too few democrats for too long. . . ."

Chancellor Kohl's government proceeded to push through restrictions on the constitutional right of asylum in 1993. A court challenge followed. In May 1996, *Die Zeit* reported: "First the legislators acted, now the Federal Constitutional Court has issued a ruling. What remains of the right to asylum is the memory of a heroic promise."

THE WELFARE STATE IN WESTERN EUROPE

Emerging as a human reaction to the excesses of the Industrial Revolution, the welfare state, particularly in Western Europe, has assumed a more comprehensive nature than in the United States. In 1883, German chancellor Otto von Bismarck, one of the most astute conservative politicians of the nineteenth century, introduced the first social-security measures, a comprehensive system that offered citizens insurance against accident, sickness, and old age. Establishing a social-security system in 1883 was by no means devoid of opportunism. Bismarck's opponents, the Social Democrats, had grown in number, and Bismarck felt that his political survival was at stake. To preempt their socialist state, he came up with the plan of comprehensive social security.

The Industrial Revolution and its aftermath gave, on the one hand, undeniable blessings to Europe in general, providing greater comfort and prosperity. On the other hand, however, they tended to divide European society. Laborers in those days were eking out an existence that was devoid of quality; frequently they were unemployed for prolonged periods without any compensation. If they were lucky, they could work long hours (10 to 16 hours a day) for pitiful wages. Economic and social justice were virtually nonexistent, and political democracy at that juncture was unable to remedy the situation. Bismarck's social-welfare measures, however slow and piecemeal their expansion over the continent may have been, became a beacon to the whole of Europe.

All modern welfare states provide unemployment and disability insurance in addition to national pension plans; on top of that, most provide sickness insurance and national medical care, which in some cases includes dental care and other benefits. While Europe has witnessed a wide variety in this progress toward the welfare state, the United States has been slow to follow suit, providing relatively minimal welfare programs. It is possible that, in the United States, too many people harbor suspicions that a welfare system is susceptible to manipulation and abuse. But it is certainly true that welfare states are extremely costly: The money to pay for all the benefits must come from some source and, as a rule, that source is the taxpayer. In countries that have a comprehensive welfare system, taxes are very high, not only because the benefits in question cost a great deal of money, but also because a larger bureaucracy is required to operate and oversee the numerous services provided by the government.

Denmark, the Netherlands, Norway, Sweden, and, to a somewhat lesser extent, Britain have all been in the vanguard of the development of programs characteristic of the modern welfare state in Western Europe. In many cases, the post-1945 growth of the welfare state progressed by leaps and bounds—the pace depending on the orientation of the government in power. However, gains were never reversed. Generally, the policies in the above countries may be identified as follows: 1) the provision of social services and transfer payments, which make up the welfare state proper; 2) the management of a capitalist market economy, aimed at maintaining optimal economic growth while minimizing unemployment; and 3) the regulation of the behavior of individuals, groups, and corporations, which reduces the costs of welfare programs by restricting the *need* for welfare. The ultimate aim of such measures is twofold: On the one hand, a decent standard of living is guaranteed to those worst off in society; on the other hand, the target is an increase of equality among socioeconomic groups without undercutting the dynamics of the market economy.

After economic growth slowed in the 1970s–1980s, the European social-welfare system has generally faced a difficult period of retrenchment. While the system has survived intact, it has become somewhat harder to obtain benefits than in the past. Also, there have been cuts in the cash benefits of the welfare system. However, most people still support the goals of the system. National statistics indicate a narrowing of income differentials; the rich are still wealthy relative to the poor, but they are less wealthy, and the poor are less poor, than they would be without the transfer payments and social services, and the taxes that pay for them.

As is the case in most Western industrialized countries in the past two decades, the United Kingdom has faced welfare costs that have been going up year after year. The population is aging, people are living longer, and less and less is being undertaken by private charity or through relatives. The country is, of course, a great deal richer than when William Henry Beveridge released his blueprint of the welfare state in 1945. The majority of families are doing fairly well for themselves: Home ownership has risen to approximately two thirds, for example. However, one could not overlook the lowest levels of the income range. A study by Peter Townsend of Bristol University calculated that the real incomes of the poorest 20 percent have dropped. Rendering uniform or universal benefits would thus be unwise and unjust, and the Department of Social Security, which designs and executes British welfare, will have to target the truly poor and concentrate on the area of pension reform.

It has not been easy to reform the welfare state, much less to cut it. In the first place, one has to take into account its size in Europe. Benefits levels have historically increased. But also, more benefits—that is, a larger variety of benefits—have been created, and some of them give one pause. In the Netherlands, for instance, the unemployed once received vacation money over and above their unemployment benefits, the reasoning being that unemployment does not really create leisure. In Germany, Chancellor Kohl's government sought in the mid-1990s to reduce labor costs and public spending with cuts in health care, dental care, sick pay, and retirement payments. Although hardly drastic cuts in Germany's generous welfare state, they turned out to be very unpopular and contributed to the defeat of his government in the federal election of 1998. One of the first things that his successor, Gerhard Schröder, did was to restore the cuts. However, his coalition of Social Democrats and Greens subsequently has had to seek ways of containing the costs of social programs while preserving their quality. For example, Schröder's government has pushed changes to reform the pension system, including incentives for private investments toward retirement. At the turn of the century, British prime minister Tony Blair's concept of the "Third Way" between the neo-liberal (laissez-faire) capitalism of Margaret Thatcher and the bureaucratic socialism of the old Left enjoyed popularity among Continental leaders of the center-left like Schröder and the center-right like Spain's José Maria Aznar.

The European Union Emerges

The original European Economic Community (the Six), did not expand from 1958 to 1973. It then admitted Britain (whose application for admission had been rebuffed by de Gaulle), Denmark, and Ireland. It admitted Greece in 1981, and, in January 1986, Spain and Portugal. During the early 1990s, Norway, Sweden, Finland, and Austria applied for membership. All but Norway joined the Union in 1995.

The original EEC countries remained the most eager for political integration. The last serious attempt before the 1980s to move forward toward political union was in 1962. In the late 1970s, the French–German dialogue provided an alternative forum for political coordination of the policies of the two most important continental countries, but this did not as yet engender the EU. At the same time, however, an institution known as the European Political Co-operation (EPC) was launched. The EPC was a weak reflection of the ambitious plans of the 1950s and 1960s for a "European Political Community." Still, the EPC may be viewed as the first intrusion of "high" politics into what had until now been largely economic concerns: It was intended to harmonize foreign-policy stands, or, at least, to allow the leaders of the member states to formulate common stands on political as opposed to purely economic issues.

It revealed three major handicaps. First, the EPC was intergovernmental and thus had no binding authority to commit member states. It was soon clear that those who had more recently become members were more reluctant to use it than were the original charter members. The former had joined largely for economic benefits to themselves; they did not like to be reminded that the original purpose of the European Economic Community, as expressed in the Treaty of Rome, included political integration.

Second, the EPC was able to act on its own only in areas where the major powers had no conflicts of interest with one another or with the United States or the Soviet Union. Thus, the only major policy initiative of the EPC was the Venice Declaration of 1980, accepting the Palestinian Liberation Organization (PLO) as the legitimate representative of the Palestinian (that is, Arab) population in Israeli-held territory in the Middle East. The purpose of the Venice Declaration, which the EPC countries made at a time when oil prices were at their peak, was to appease the Arabs in the hopes of securing favorable treatment from OPEC, which the Arabs dominated.

Third, distance played a role. When Yugoslavia was torn apart in 1992, it was obvious that Italy and Greece were much more affected than were Ireland and Denmark. It proved difficult to find a common position, not only vis-à-vis the enormous violence that had erupted but also regarding the recognition of the new states that were produced by the disintegration. Greece, for example, greatly objected to recognizing Macedonia—one of the offshoots—because there was also a Greek province by that name. Greece got its way, but the incident vividly illustrates that it would frequently be hard to arrive at compromises on foreign-policy positions.

In 1981, the German and Italian foreign ministers, Hans-Dietrich Genscher and Emilio Colombo, proposed a program for moving toward European political unity by 1990. The Genscher–Colombo plan harked back to the ambitions of the

1950s, reminding analysts that even the expanded EEC was supposed to have a purpose other than the making of deals and the transferring of money among member countries. Great Britain and Denmark objected to the plan's political implications, and these objections in fact foreshadowed similarly negative attitudes vis-à-vis the Maastricht Treaty in the early 1990s. The Council of Ministers did not accept the plan in full, but, in 1985–1986, all member countries did adopt various reforms designed to make the operations of the Council and the Commission smoother. The most important one was to remove the requirement that all important decisions of the Council be unanimous.

The Maastricht Treaty (officially the Treaty of European Union) of 1992 was a significant step forward in economic and political integration, though it was characterized by many compromises, including "opt-outs" for the British (and later the Danes). Like the Treaty of Rome, it embraced the goal of "an ever closer union among the peoples of Europe." At the insistence of British prime minister John Major (who headed the increasingly Euro-skeptic Conservative Party), the word "federal" had been banished from the text. Unlike its predecessors, the Maastricht Treaty was the focus of public controversy and failed in its first attempt at ratification in Denmark (and only narrowly survived the referendum in France). In 1993, after a favorable ruling by the German Constitutional Court, it could come into effect after ratification by Germany and all the other member states. Legally the European Communities became embedded within the "European Union," which became the new official name of the evolving supranational organization whose lineage goes back to the Schuman Plan of May 1950.

Although the Dutch and Belgian governments favored a step forward in political integration across the spectrum of policy areas, the Maastricht Treaty reflected a more cautious approach to the "ever closer union": the "three pillars" design. In the first pillar, the supranational aspects of the EU will continue to advance with more opportunities for qualified majority voting (QMV) instead of unanimity. The first pillar encompasses agricultural policy (which has consumed by far the largest portion of the budget through the years), economic and monetary policy, trade and competition policy, environmental and energy policy, regional development policy, scientific and technological policy, and assorted others. The second pillar is concerned with the common foreign and security policy. It thus supplants the EPC. However, decision making in this area remains intergovernmental; any member state can veto policies. The third pillar also is characterized by intergovernmentalism and deals with justice and home affairs, which involve sensitive questions of immigration, criminal justice, and policing cross-border criminal activities. The Treaty of Amsterdam (which came into effect in 1999) represented a refinement of Maastricht, rather than another big leap ahead. However, it did transfer several third-pillar issues into the first pillar (with a transitional period before majority voting would be routine). Above the three pillars would be the European Council, which has grown out of the informal summit meetings of the heads of government (and, in the case of France, the head of state); it began in 1961. The European Council would deal with thorny issues not resolved at the lower level and also seek consensus on the outlines of new departures in the development of the EU.

Maastricht left in place the four core institutions, whose origins can be traced back to the Treaty of Paris (1951), which launched the European Coal and Steel Community. The first core institution is the Council of Ministers, which historically has been the most powerful, in that it directly represents the interests of the member states. In actuality there are multiple specialized councils, which bring together the agricultural ministers, the environmental ministers, and so forth, as well as a general council of foreign ministers. Until Maastricht, the Council of Ministers had the final say on all EU legislation, making it in the effective sense the legislature. The Council is not accountable to the European Parliament or the national parliaments. Its meetings in Brussels are executive sessions rarely open to outsiders. The presidency of the Council of Ministers (and the European Council) rotates every six months among the member states. In early 2001, Sweden assumed the presidency, which it handed over to Belgium at midyear, which would give it up to Spain in early 2002, and so forth. The *troika* (Russian term for "three") of past president, current president, and future president cooperate in the cause of continuity; however, member states each take their term as the head of the EU seriously and work hard to advance their own agendas of issues. An important trend since the Single European Act of 1986 (actually a package of important amendments to the Treaties) has been for more Council decision making by qualified majority voting. The details of decision making are handled by subordinate bodies like the Committee of Permanent Representatives (senior civil servants and ambassadors from the member states), commonly known as COREPER.

The second core institution of the EU is the European Commission; historically it has been the "engine" of European integration. As of 2001, it consisted of 20 commissioners, with the big five countries in terms of population (Germany, France, Britain, Italy, and Spain) having two and each of the other 10 countries having one commissioner. The commissioners are appointed by their national governments for five-year renewal terms. They take oaths of independence from their countries, but they also must not alienate their governments if they wish to be reappointed. The European Parliament has the power to censure the entire Commission and compel its resignation, and since Maastricht it has voted on the installation of a new Commission. However, it cannot remove individual commissioners. In 1999, the European Parliament failed in an attempt to censure the Commission over charges of inefficiency and corruption. But in a surprise move, the Commission resigned as a result of the controversy.

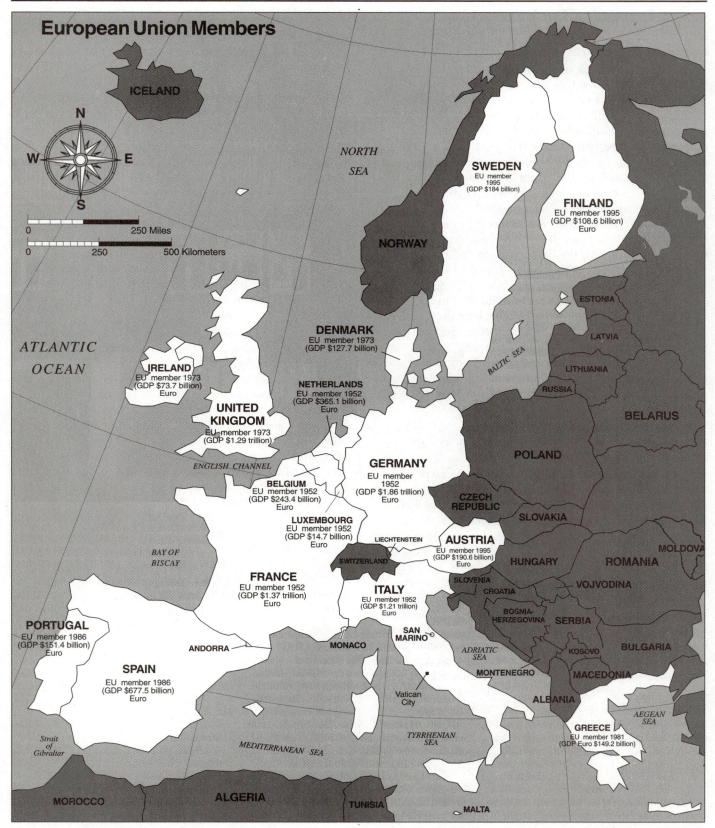

European Union Members

ICELAND

N
W E
S

0 ——— 250 Miles
0 — 250 — 500 Kilometers

ATLANTIC OCEAN

NORTH SEA

SWEDEN
EU member 1995
(GDP $184 billion)

FINLAND
EU member 1995
(GDP $108.6 billion)
Euro

NORWAY

ESTONIA
LATVIA
LITHUANIA
BALTIC SEA
RUSSIA

DENMARK
EU member 1973
(GDP $127.7 billion)

IRELAND
EU member 1973
(GDP $73.7 billion)
Euro

NETHERLANDS
EU member 1952
(GDP $365.1 billion)
Euro

UNITED KINGDOM
EU member 1973
(GDP $1.29 trillion)

ENGLISH CHANNEL

BELGIUM
EU member 1952
(GDP $243.4 billion)
Euro

GERMANY
EU member 1952
(GDP $1.86 trillion)
Euro

BELARUS

POLAND

CZECH REPUBLIC

LUXEMBOURG
EU member 1952
(GDP $14.7 billion)
Euro

LIECHTENSTEIN

AUSTRIA
EU member 1995
(GDP $190.6 billion)
Euro

SLOVAKIA

MOLDOVA

BAY OF BISCAY

SWITZERLAND

FRANCE
EU member 1952
(GDP $1.37 trillion)
Euro

ITALY
EU member 1952
(GDP $1.21 trillion)
Euro

HUNGARY

ROMANIA

SLOVENIA
CROATIA

VOJVODINA

PORTUGAL
EU member 1986
(GDP $151.4 billion)
Euro

ANDORRA

MONACO

SAN MARINO

BOSNIA-HERZEGOVINA

SERBIA

ADRIATIC SEA

BULGARIA

SPAIN
EU member 1986
(GDP $677.5 billion)
Euro

Vatican City

KOSOVO

MONTENEGRO

MACEDONIA

ALBANIA

Strait of Gibraltar

TYRRHENIAN SEA

MEDITERRANEAN SEA

GREECE
EU member 1981
(GDP Euro $149.2 billion)

AEGEAN SEA

MOROCCO

ALGERIA

TUNISIA

MALTA

The European Commission views enlargement of the EU as a long-term process. At the beginning of 1995, the European Union had 15 member states. Eleven of these participated in the implementation of the Euro on January 1, 1999; Greece joined the Euro Zone on January 1, 2001.

Romano Prodi, a highly regarded former prime minister of Italy, emerged as the new president of the Commission, and his new team obtained the support of the EP.

The Commission has both executive and legislative roles, thus departing from the U.S. scheme of separation of powers. It is the only body that can initiate legislative bills (in the form of directives as well as regulations). It also is the body that oversees the implementation of directives once passed by other bodies. More than 70 percent of the 28,000 EU full-time employees work for it; they are often referred to as "Eurocrats" and blamed by the national governments for most things that go wrong in administration, though policy implementation is mostly done by their national bureaucrats. Since Maastricht, the commission is empowered to seek fines for countries as well as firms that are in violation of EU directives or regulations. Of the Eurocrats, nearly 2,000 are translators. Although the working languages of the EU tend to be English, French, and German, altogether 11 languages have official status, thus necessitating translation of many unusual combinations. The Commission administers the EU budget and provides international representation, particularly in trade conferences. The recent trend has been for the Commission to lose influence compared to the European Council, Council of Ministers, and the European Parliament; however, outside of "high" politics, it remains a vital institution of integration.

The third core institution is the European Court of Justice (ECJ), which should not be confused with the European Court of Human Rights (in Strasbourg) or the International Court of Justice (in The Hague). The ECJ is based in Luxembourg. Its 15 judges are appointed by their respective national governments for renewable six-year terms. To date there has been no evidence of political appointees. The judges are assisted by nine advocates-general, who investigate the cases. The working language of the Court is French, and it has been heavily influenced by French legal tradition and practice. However, the judges act as the "guardians" of the Treaties because, in disputes, they have final say on the interpretation of provisions, as the U.S. Supreme Court does in reference to the Constitution. Some national governments have complained about the autonomy of the Court and would like to limit it in the future. The ECJ is not bound by precedent, and the opinions of dissenting judges are not published. The Court has not only asserted itself in regard to treaties (for example, in the early 1990s, it held up the European Economic Area until changes were made), but also in policy areas (for example, its decisions accelerated the harmonization of trade rules). The work load of the Court has grown steadily. National courts often have invited the ECJ to give a preliminary ruling on disputes involving EU law or the Treaties. Since 1989, a lower court, the Court of First Instance, has functioned to filter and ease its work load. Yet on points of law, all decisions by the Court of First Instance can be appealed. The ECJ has been portrayed by some authors as the most "federal"

of the EU institutions in its powers, as defined by the Treaties and practices.

The fourth core institution is the European Parliament, which though the most democratic institution historically has been the weakest. Indeed, until the Single European Act and the Maastricht Treaty, the EP was essentially a consultative body, the descendant of the Common Assembly of the European Coal and Steel Community. From an American viewpoint, it is still a bit peculiar as an aspiring legislature, in that it literally shuttles back and forth between Brussels and Strasbourg for its meetings, while its secretariat remains based in Luxembourg. Efforts to concentrate its plenary as well as committee activities in Brussels alongside the Commission and the Council of Ministers have run into a French veto. Prior to 1979, the members of the European Parliament (MEPs) were indirectly elected from their national parliaments. Observers generally discounted them as second-rate politicians, either not ready for prime time or well past their prime. Absenteeism tended to be higher than in national parliaments.

Since 1979, the MEPs have been directly elected in national elections in each member state, which is unique in the world for international bodies of politicians. National allocations of seats are roughly made according to population, but with small countries overrepresented. Thus, as of 1999, of the total 626 seats, the most populated country, Germany, had 99 seats, compared to six seats for tiny Luxembourg. Since 1999, after Britain reformed its electoral law, the MEPs have all been elected according to proportional representation, which allows small parties a chance to gain seats. The catch is that EP elections tend to be seen by the voters as "second-order elections," which has meant that turnout is significantly less than in most national parliamentary elections and that voters often use the EU election to send their governments a "message" on national issues rather than on European issues. MEPs sit as cross-national party groups rather than national delegations. These remain loose associations, far from being cohesive and disciplined transnational parties. In the June 1999 EP elections, the European People's Party (Christian Democrats and Conservatives) emerged as the largest party group, pushing the Party of European Socialists into second place. The Liberal Democrats (ELDR) had the third-largest party group. The Greens enjoyed gains that moved them into fourth place, ahead of the European United Left (Communists and Radical Leftists) and others. No single party group controls a majority of EP seats, thus necessitating cooperation among various groups.

Before Maastricht, the EP had the power on some issues of forcing the Council to override their opposition in a second vote by unanimity. After Maastricht, the EP gained the power of "co-decision" in select areas, which meant that if differences could not be negotiated with the Council, in the end it had a veto. The Dutch, Belgian, and German governments had favored even more power for the EP but had run up against opposition from the British and French. In the Amsterdam Treaty, co-decision was extended to additional policy areas. The EP operates with a series of specialized committees, which on some issues, such as the environment, have received publicity for their work.

Finally, the EU also includes several minor institutions. For example, the Court of Auditors had become more visible through the years, while the Economic and Social Committee has remained consultative and obscure. Maastricht added another consultative body, the Committee of Regions, to reflect the views of regional authorities and local authorities. However, the "Europe of the Regions" that some have envisaged as the nation-states progressively lose functions to Brussels and more national policy is devolved to subnational (and cross-national) regions still seems a long way off.

The Euro

Economically, however, the European Union has become a very powerful economic bloc, and protecting that status means pressing on with integration—first toward a single-currency, then to standardized tax policies and common economic policymaking, and ultimately to political union. The first of those steps, the single currency, has already been launched.

European economic integration began to take off toward the end of the 1980s. The fundamental long-term challenge facing Europe was to establish a strong and stable foundation for job-creating growth.

At the beginning of the 1990s, Europe was faced with deficits in three major areas that held it back from rapid and long-term growth; fiscal, jobs, and investment. Recently, there have been impressive progress made in the fiscal area, but there remain significant problems in both the job and investment sectors. In fact, in 1998 the average unemployment in the European Union countries was approximately twice what it was in 1979. (In 2001, the EU average unemployment rate had declined, but it was still twice that of the United States.)

It became evident that any improvement in employment levels and investment would require a major push toward the private business sector, and the key to both of these areas would be the establishment of basic structural changes that would provide an attractive environment for private investment. Europe, in the aggregate, is an economic superpower, and many Euro-enthusiasts believed that by developing a common monetary system, it would be able to flex this power and influence policy making in the broad global arena.

The Maastricht Treaty defined the features and set the timetable for the Economic and Monetary Union (EMU). It laid out the convergence criteria regarding budget deficits, national debts, and inflation rates that member states would have to meet in order to participate in the third and final stage of the EMU. However, protocols allowed the British and Danes to opt out.

The first action toward developing a common currency began in 1994 with the establishment of the European Monetary Institute (EMI), which would become the European Central Bank (ECB), based in Frankfurt, Germany. The next phase began in May 1995, when the European Commission adopted what was called the "Green Paper on the Single Currency," which outlined the parameters of how this currency would be implemented. In December 1995, the European Council of Madrid named this new Single Currency the *Euro* and established the date of January 1, 1999, for the single-currency policy to become a reality.

When the Euro came into being, 11 EU members participated in the launch: Belgium, Germany, Spain, France, Ireland, Italy, Luxembourg, Finland, the Netherlands, Austria, and Portugal. Britain under Tony Blair's first term government opted out, but indicated that it might join at a later point, assuming that the proposal gained majority support in a referendum. British polls have indicated that abandoning the pound sterling is going to be a hard sell for the government. Denmark also exercised its opt-out (and disregarding the advice of their leaders, Danes voted no in the 2000 referendum on joining the Euro Zone). The Swedish government chose politically to stay out for the time being, until public support would build for entry. Greece failed to meet the convergence criteria. (After doing so, it joined in 2001 as the Euro Zone's 12th member.)

World markets welcomed it and saw this consolidation of currency as a reflection of Europe's positive economic outlook. Initially, the Euro will be the currency of business and investment; actual Euros as hard cash, both bills and coins, will not be in everyday usage until January 1, 2002. With the launch of the Euro, the European Central Bank controls a market second only to that of the United States. Many see the role of the head of the ECB as comparable to that of Alan Greenspan, the head of the U.S. Federal Reserve. This tremendous centralized economic power is unique in history, and it will be the first time that most of Europe will have had one currency since the Roman Empire ruled nearly 2,000 years ago.

The first two years of the ECB were rocky ones. The Euro's value was pegged at slightly less than U.S. $1.20, but almost immediately the Euro suffered from weakness against the dollar. Within a year, its value had slipped below parity with the dollar. It reached the $0.83 level before starting to appreciate in late 2000, as the U.S. economy began to slow. The ECB's goal of price stability, defined as no more than a 2 percent annual inflation rate, encountered the challenge of soaring world oil prices in 2000. If one excludes direct energy prices, consumer prices have not (yet) risen in excess of the prescribed goal. Economists have emphasized the perils of a "one-size-fits-all" monetary policy, especially if European economies should slide into serious recession before the ECB has fully established itself. Its president, Wim Duisenberg (a Dutch former central banker), won't become the fixture that Alan Greenspan has in the United States, because, as a part of a deal that cleared away French objections to his appointment, he promised to step aside in 2002, long before his eight-year term expires, so that a French central banker could become president. Thus, at the outset, national interests cast a shadow on the ECB, whose independence on paper is greater than that of the U.S. Federal Reserve.

In a special report about the ECB's first two years, *The Economist* outlines three major problems that it must address. First is communications. The ECB needs to learn how to present and explain its monetary policy to the media and the financial markets much more effectively. Second is the relationship between the ECB and the national central banks (NCBs). Vital activities, such as bank supervision, are still under the control of the NCBs, which might not be sensitive to the wider impacts of bank failures on the Euro Zone. (The special report makes the case for some centralization before a crisis comes along.) Third, and most complicated, is the prospective entrance of Central/Eastern European countries into the Euro Zone. As is the case with the EU itself, the expansion of the membership of the EMU to include these relatively poor economies at various stages of transition poses many institutional and policy challenges. Adhering to the Maastricht convergence criteria would slow entry, allowing for incremental adjustments, but politics may drive the process, as it did in the late 1990s, when some of the original Euro Zone candidates were given credit for good-faith efforts despite statistical shortcomings. The "Eurosystem's" near future is likely to be more controversial and difficult than its recent past.

EUROPE'S FUTURE

In the immediate aftermath of World War II, the most destructive war in European history, federalists sought to seize the opportunity to transcend the nation-state. However, in the late 1940s, their initiatives to unite Europe faltered. The Council of Europe emerged as an intergovernmental organization with national vetoes intact. Jean Monnet, Robert Schuman, and others, while retaining their federal vision, shifted to a neo-functionalist strategy to bring about a pooling of national sovereignty in specific economic sectors that would be successful enough to cause spillover into progressively more areas, without provoking a nationalist backlash. The European Coal and Steel Community became the prototype of this strategy, while the abortive European Defense Community demonstrated the perils of moving too quickly into areas of "high politics." The ECSC led to the Treaties of Rome, and the European Economic Community and EURATOM began operations in 1958.

The six charter members committed themselves to a course of economic integration that led beyond creating a free-trade area. In the 1970s, the first enlargement brought in three countries seeking to benefit economically from expanding intracommunity trade and cooperation but with reservations about the supranational dimension of the undertaking. Even

before then the integrative process had not been smooth but, rather, had been characterized by periods of stagnation and spurts of activity. Political choices have been pivotal. "Widening" and "deepening" have hardly been disconnected; in the 1980s and the 1990s, both occurred. In fact, the prospect of enlargement has provided extra impetus to make reforms, which have expanded the policy scope of the European Union (as the European Communities became officially in 1993) and increased supranational aspects of its hybrid institutional framework. The "high politics" that the EDC had failed to integrate four decades earlier found its way into the EU as its second (intergovernmental) pillar for common foreign and security policy. The "common market" (later known as the "single market") became a full-fledged reality in the early 1990s; it was consolidated even further by the common currency and European Central Bank in 1999. Despite the pessimism in the mid-1990s, 11 out of 15 EU members qualified and joined the Euro Zone in 1999. Although still far from the "federal Europe" dreamed about in the immediate postwar years, European integration has produced a unique, still unfinished transnational political system that already rivals the United States in economic terms.

In the first decade of the twenty-first century, the EU, which has increasingly become synonymous with "Europe," faces a number of major challenges. Its responses to these will determine how it evolves. These challenges, which tend to be entangled, are the globalizing economy, the "fourth wave" of enlargement, and institutional (and ultimately constitutional) reform.

European Integration and Globalization

By all indicators, the single-market project of the European Union has been a tremendous success since 1958. Intracommunity trade has greatly expanded as barriers to the movement of goods, services, capital, and people have fallen, especially since the late 1980s. For example, rapid strides have been made in the harmonization of product standards; on the other hand, progress has been less impressive in the harmonization of tax policies. However, to compete globally in the new economy of informational technologies, obstacles to the start-up of new businesses and the creation of new jobs must be addressed at the EU level. Portuguese prime minister Antonio Guterres declared during his country's presidency of the EU in early 2000, "We want to transform Europe into the most dynamic and competitive knowledge-based economy in the world." However, Guterres also made it clear that the EU should strive to become more globally competitive without emulating the U.S. model of (less) social welfare for its citizens. While Tony Blair's government in Britain has not been reluctant to borrow ideas from the United States, many governments (most vocally, Lionel Jospin's government in France) have advocated a European drive toward full employment *and* greater solidarity.

The special EU summit in Lisbon, Portugal, in March 2000 set forth the strategic goal of the EU taking world economic leadership by 2010. Action was mandated in numerous areas: promoting a new information society open to all regardless of socioeconomic status, establishing an EU–wide area of research and innovation, providing a better environment for new and innovative businesses, liberalizing utilities and transport sectors, and promoting venture-capital markets. The outcome would be not just more jobs but, rather, "quality" jobs for young Europeans. EU leaders agreed to annual reviews of progress toward the Lisbon objectives; a monitoring system of "league tables" would name and shame countries according to their performance. The EU could draw encouragement from the world leadership in mobile-phone technologies by firms based in Finland and Sweden, countries that have encouraged innovation and yet have maintained social solidarity. There is also the case of the Dutch, whose economic and social reforms have brought the Netherlands' unemployment rate down to the U.S. level without abandoning its generous social-welfare system.

The demographic trends of the early twenty-first-century Europe, however, provide some sobering statistics. The populations of EU countries are aging; longevity has increased and birth rates have fallen in recent decades. According to the Agenda 2000 report, the number of EU citizens more than 60 years old will rise from 80 million to 91 million in 2010, with the fastest-growing category being those over 80. The surging number of senior citizens claiming benefits, coupled with the declining number of young workers to pay into the system, translates into an inevitable financial crisis for health-care and pension programs unless there are painful and politically unpopular reforms. Already one detects signs that questions of intergenerational equity could become politicized and threaten the social solidarity valued by current leaders. A number of governments are recognizing the need for immigrants to fill not only semiskilled jobs—unattractive to their own young people—but also to fill the high-tech jobs necessary to meet the demands of world competition. Many more immigrants paying into the social funds would help to secure the retirement of the old generations, lessening the burdens on the younger European generations.

However, to many Europeans, the prospect of the influx of millions of foreigners over the next decade threatens their national identities. Globalization entails far more than the lowering of barriers to trade and investment; it provides incentives and mechanisms for large numbers of people to migrate long distances from areas of economic deprivation to areas of economic promise. The international smuggling of people has been a major growth industry for organized crime. In February 2001, EU justice ministers were seeking an integrated policy approach to steming the tide of illegal immigrants (without shutting the door to genuine political refugees). Right-wing populist parties find fertile ground where citizens are anxious about the social and economic changes,

(Credit: UN/DPI Photo by Greg Kinch)

The momentum behind the conservative movement in Western Europe was demonstrated with the 1979 accession of Margaret Thatcher as Britain's prime minister and the election of her successor, John Major. However, in 1997, the swing toward a more moderate perspective was emphasized when the Labour Party, led by Tony Blair (pictured above), took power. Blair has said that, in contrast to Thatcher and Major, he wants Britain to be "at the heart of Europe."

and these politicians are hostile to the prospects of further European integration, either in the sense of deepening or widening.

Eastern Enlargement

In an interview with *Die Zeit,* Jacques Delors, former president of the European Commission and a driving force behind the integrative surge of the late 1980s and early 1990s, described the EU's expansion east as Europe's "most grand project" since the reconciliation between Germany and France in the years following World War II. The 1970s expansion had moved the EEC beyond the continental core of six to include three more peripheral but nevertheless well-established Western European democracies: Britain, Denmark, and Ireland. The 1980s expansion, which involved protracted negotiations, had given the European economic integration a southern dimension, incorporating three relatively poor countries. Political considerations were also evident: Greece, Portugal, and Spain had been brought aboard to help consolidate their new democratic regimes. The 1990s expansion came about more smoothly as three small countries with affluent economies and established democratic regimes joined. With the Cold War over, the neutrality of Sweden, Austria, and Finland was no longer an obstacle to their full membership, which extended the EU's northern and central flanks. The rapid implosion of Central/Eastern European Communist regimes during 1989–1990 and of the Soviet Union itself in 1991 had quite unexpectedly presented the historic opportunity to build a truly united Europe. Any euphoria, however, did not last long, as EU officials and Western European leaders came to fully appreciate the enormity of the challenges of eastern expansion. In the late 1990s, Poland, Hungary, and the Czech Republic became full-fledged members of NATO, but no former member of the Warsaw Pact had been allowed to join the EU, despite the moral imperative for enlargement expressed in the speeches of assorted Western leaders.

Though French president François Mitterrand proposed a confederation in 1991 to link the Central/Eastern European Countries (CEECs) together and to the EU countries (cool reactions in Prague and elsewhere aborted it), the course of the EU was to be more cautious and incremental. First, there was PHARE, an aid program to promote the economic restructuring of Poland and Hungary, which subsequently was extended to other countries. Next came the Europe Agreements with Poland, Hungary, and Czechoslovakia (later with the Czech Republic and Slovakia as well as with others), which were aimed at lowering trade barriers (but at the same time protecting sensitive industries and agricultural interests in the West), promoting economic cooperation, and encouraging political dialogue. There was no EU commitment to full membership within the Europe Agreements, to the disappointment of CEEC leaders.

At the 1993 Copenhagen summit, the EU countries for the first time articulated the general conditions that candidate countries would have to fulfill: stable democratic institutions, the rule of law and defense of human rights, special protections for minorities, and a modernizing economy with the capacity to cope with competitive pressures within the Single European Market (SEM). An additional condition was that the EU must be ready to take on new members without disrupting the European integration process. In 1994, Poland and Hungary became the first of the post-Communist countries to apply for membership. Subsequently, eight more joined the queue, along with Cyprus, Malta, and Turkey, which had applied previously. (Croatia is likely to be the next official candidate.) In the mid-1990s, German chancellor Helmut Kohl advocated the entry of Poland, Hungary, and the Czech Republic—all neighboring Germany—by 2000. However, the Commission favored a scheme that managed the growing list of aspiring candidates according to objective criteria.

Accordingly, in 1997, the Commission, viewing the countries' respective records on economic and political reform, designated Poland, Hungary, the Czech Republic, Slovenia, Estonia, and Cyprus as the "fast-track" candidates. The inclusion of Cyprus and Estonia gave expansion Southern and Northern components as well as the larger Central/Eastern

component. Poland's population of 38.6 million easily exceeds the combined populations of the other five. Poland also stands out as having by far the highest percentage of the national workforce (about 25 percent) employed in its agricultural sector, whose practices tend to be backward as compared to those in Western Europe. Thus, incorporation of Poland would represent a major challenge to the EU's Common Agricultural Policy, which has historically consumed more than half of the total EU budget to subsidize farmers. In 1997, the average per capita income in Poland came to only 14 percent of the EU's average. Especially the Germans have had concerns that Poles (and others) might flood into their country in search of higher wages and disrupt the labor market, if border restrictions were to come down abruptly. The Czech Republic and Hungary, both with about 10 million people, had higher average per capita incomes as figured against the EU average: 20 percent and 19 percent respectively. The small ex-Yugoslavian republic of Slovenia, benefiting from its connections with Austria, had better income statistics (42 percent of the average EU); while the small Baltic republic of Estonia, despite its Finnish connections, had worse income statistics (11 percent of the average EU). All five had experienced a series of free elections and their democratic institutions had weathered the post-Communist storms of politics. Cyprus, in comparison, was on a per capita income basis much closer (71 percent) to the EU average, but it is politically more problematic because of the partition of the small island country into Turkish and Greek sectors.

Bulgaria, Romania, Slovakia, Latvia, Lithuania, and Malta (which withdrew due to domestic politics and later reapplied) made up the second potential wave of candidates, although later the EU indicated that it did not mean to foreclose their chances for entry alongside the first six if their circumstances should improve rapidly. Also, the EU, responding to the displeasure of Turkish leaders at their treatment, later symbolically included Turkey (long a full NATO member and an associate member of the EC) among the candidate countries. Even setting aside the cultural/religious issue of its "European" identity, with its population of more than 60 million, low per capita income, and a problematic record on human rights, Turkey's full membership in the EU would seem to be decades in the future.

In March 1998, accession negotiations formally began with the six "fast-track" countries; they began with the other five CEECs plus Malta in 2000. Task forces were set up to screen each applicant's legislation and administrative capacity in all EU policy areas. The screening process has served to alert the national governments to the gaps between their current structures and practices and those of the EU. Joining the EU at this stage in its development is no simple matter; it involves a mind-boggling institutional transfer of the 80,000-page *acquis communautaire* (treaties, principles, directives, regulations, practices, obligations, and court decisions that have developed or have been agreed by the EU members). For example, political scientist Wade Jacoby estimates that the *acquis* currently includes 10,000 directives that the candidate countries have to respond to legislatively and administratively. Entry negotiations have involved many contentious issues, with the CEEC candidates seeking transitional exemptions from the full force of EU rules (such as regarding environmental protection, social policy, and state subsidies for industries) and with current member states seeking protection of their job markets and shares of various program funds. Although the major parties in the CEECs have remained committed to EU membership, there is survey evidence from Poland that public support for entry is not as enthusiastic as it was a few years ago. Similarly, the citizens in many EU countries have become more sensitive to the trade-offs that may come with enlargement, as indicated by the "Eurobarometer" poll in 2000 that found that only 20 to 25 percent of Germans, French, Finns, Austrians, Portuguese, and Britons viewed EU enlargement as a "priority."

In its November 2000 report, the European Commission assessed the progress of the individual candidate countries toward meeting the criteria for EU membership. Poland was seen as having made progress in economic modernization but lagging behind in agricultural reform and doing poorly in controlling its borders. (As an EU member, Poland's eastern border would become the EU's eastern border, and there are fears among its neighbors to the west about the consequences of its porous nature.) The Commission praised the Czech Republic for its progress in the treatment of the Romani (Gypsy) minority and its higher banking standards, but criticized its failure to move ahead on judicial and administrative reform. Hungary received generous praise in almost all areas, though the Commission recommended more independence for its National Bank. Estonia was singled out as the Baltic country making the most economic process, but it was urged to improve public administration, the judiciary, and the integration of noncitizens. The view of Slovenia was positive, but the Commission urged an acceleration of reforms in public administration. The Commission's report for the first time ranked the 12 "active" members (that is excluding Turkey) in terms of their economic prowess in face of the "competitive pressure" of EU membership. Cyprus and Malta were ranked the highest. Estonia, Hungary, and Poland followed, at the second level. The Czech Republic and Slovenia fell to the third level because of their slow pace of reforms. The fourth level was occupied by Latvia, Lithuania, and Slovakia; at the rear came Bulgaria and Romania.

The outcome of an independent study in late 2000 by PriceWaterhouseCoopers considering macroeconomic stability, productivity, infrastructure, and integration with Europe found (in rank order) Slovenia, Malta, Cyprus, the Czech Republic, Estonia, and Hungary more suitable for EU membership than current-member Greece! Poland followed behind, with Slovakia, Latvia, Lithuania, and Romania trailing it. (Bulgaria and Turkey received zero scores.)

The Nice summit in December 2000 postponed negotiations on the most challenging accession issues, such as free movement of labor, until 2002; however, the understanding was that new members could still be admitted prior to the European Parliament elections in 2004. When Sweden took over the EU presidency in early 2001, its view was that the negotiations should be speeded up and that each candidate should be admitted on its merits—not as first and second waves in which the laggards (Poland was singled out) hold up the others. The Germans, on the other hand, are determined that Poland be included in the first group, along with Hungary and the Czech Republic. Eastern enlargement promises to shift the EU's center of gravity much more than the previous enlargements. The EU will become more heterogeneous in cultural, political, and economic ways; as a result, "package deals" will become more complex. The implications of a larger and more diverse EU have made institutional reform inevitable.

Institutional Reform

The Maastricht Treaty included a provision for an intergovernmental conference in 1996 to consider additional treaty changes. During the mid-1990s, a growing concern was how to change the already unwieldy EU institutions to accommodate waves of new members in the early twenty-first century. Although it was a major topic in the deliberations that culminated with the Treaty of Amsterdam of 1997 (ratified in 1999), pre-accession reform was hardly the central feature of the follow-up treaty. In its text, qualified majority voting was extended to additional areas (thus, a veto by just one of a potential EU of 20 to 30 members could not bring things to a halt), and the size of the European Parliament was to be capped at 700 MEPs. A provision of the treaty makes it possible to suspend a member state's voting rights if its government departs from basic democratic values. (This provision was largely overlooked until the dispute with Austria during 2000 caused by the right-wing Freedom Party's sharing of federal power; Austria was not suspended, but for several months EU members signaled their disapproval via bilateral diplomatic sanctions.)

The most contentious issues relating to enlargement were relegated to a separate protocol. The European Commission was viewed by most observers as already too large, at 20 members, to be effective (the five big countries having two commissioners each and the smaller 10 countries one each). Also, the leaders of the big five were concerned about being outvoted after eastern expansion by coalitions of small countries in the Council of Ministers. In the protocol, the big five agreed to give up their second commissioners and be compensated by a future reallocation of Council votes. In addition, there would be a comprehensive review of the institutional framework before EU membership exceeds 20. Many observers were disappointed by the provisions of the Amsterdam Treaty, but the willpower to take another big step so soon after Maastricht was lacking among the major actors.

Soon yet another intergovernmental conference was under way to prepare another treaty in 2000 to deal with "leftovers." Although sufficient progress was declared by national leaders symbolically to open the door for the new members by 2004, there was a wave of critical reactions to the new draft treaty's provisions and how French president Jacques Chirac, who chaired the sessions, had managed the negotiations at the December 2000 Nice summit. Complaints streamed in from many editorialists, members of the European Parliament, and the president of the European Commission (which had been excluded by the French at key points). One negotiator declared to journalists, "This is the worst treaty the European Union has ever come up with." Basically, the Nice Treaty will not make the institutional processes of an enlarged EU less unwieldy. Not surprisingly, the stage was set for another intergovernmental conference in 2004 to deal with "leftovers." And, as *The Economist* observed, the Commission will continue to conduct its daily business "against a backdrop of institutional arrangements that are subject to constant negotiations and reinvention."

At Nice, the most contentious issue was that of re-weighting the votes in the Council to compensate for the willingness of the big countries to give up their second commissioners in 2005. The small countries objected to giving up their claim to a single commissioner each. The compromise was that the formula of "one country, one commissioner" would be maintained until the EU reached 27 members; after that, a (yet-to-be-defined) rotation scheme would be put in place. Germany and France clashed over the re-weighting of Council votes. Ultimately Germany accepted continued vote parity with France (and with Britain and Italy) though its population outnumbers each of the others by some 20 million, but it would be advantaged by the new requirement that a qualified majority vote represent 62 percent of the EU population. The new voting scheme would allow Germany plus two other big countries to block initiatives that they dislike, which upsets the Commission and the smaller countries. Spain ended up with 27 votes, two less than the big three; the trade-off for Madrid was retention of the national veto when it comes to distributing EU structural funds, which have facilitated its "economic miracle." Portugal and Belgium threatened vetoes over their share of the votes compared to neighboring Spain and the Netherlands, respectively. On the other hand, Poland was relieved to receive (when its entry comes) the same number of Council votes as Spain has; other candidates will also receive votes roughly proportional to their relative population sizes.

The ceiling on the membership of the European Parliament was raised to 732, in order to accommodate representatives from the new member states on their accession; this will require cuts in most existing national delegations (but not Germany's). Belgium, Portugal, and Greece each gained a couple of seats in exchange for their support for the overall deal. However, in contrast to the Maastricht and Amsterdam Treaties, the draft treaty did not increase the legislative powers of the European Parliament.

At the Nice summit, qualified majority voting in the Council was extended to about 30 of the 50 areas under consideration, ranging from trade to industrial policy. But in several sensitive areas, countries dug in and defended the continuation of national vetoes. For example, Britain, Luxembourg, and Ireland refused to bow to French pressure to depart from required unanimity in votes on taxation. However, the Nice Treaty, if ratified, will make it easier in the future for clusters of countries of eight or more to push ahead with "enhanced cooperation" (without the threat of national veto from laggards) in various policy areas, except defense. Other countries would be free to join up at a later time. A multispeed Europe is already reflected in the monetary union/common currency opt-outs granted to Britain and Denmark in the Maastricht Treaty. Under the code word "flexibility," the Amsterdam Treaty allowed a majority of members to move further in policy integration, if others did not feel so nationally threatened as to pull the emergency brake (veto). Nice thus goes further in this regard than Amsterdam, without fully embracing the model favored by various French and German politicians during the 1990s of a "core" of countries (presumably Germany, France, the Low Countries, and Italy) that moves faster and further to integrate than more "peripheral" members.

Like the Amsterdam Treaty, the Nice Treaty failed to reform the institutional framework in a way to make the EU's workings more intelligible to the citizens of Europe. Furthermore, the EU's "democracy deficit," long discussed by academics and practitioners, was hardly addressed by the treaty changes agreed to in Nice. Although national polls indicate that many people favor an increased role for the EU in tackling policy problems such as environmental pollution, the evidence for the overall emergence of "diffuse support" or popular "legitimacy" for EU institutions after 50 years of integration is still sparse. And the outcome of Nice is hardly likely to produce mass support for the elites-driven process of European integration, which, despite idealistic rhetoric, was characterized by messy deals. Belgian prime minister Guy Verhofstadt sharply criticized the big countries for putting their national interests far ahead of the common interest. Many critics of the Nice summit have been attracted to Finnish prime minister Paavo Lipponen's proposal, articulated in a November 2000 speech, for a convention that would include a wider range of representatives than just from national governments, to supplement the existing intergovernmental conference approach to making treaty changes. Lipponen's argument is that such a more inclusive approach would enhance the efficiency and the legitimacy of institutional reforms.

In May 2000, German foreign minister Joschka Fischer sought to stimulate a public debate over visions of Europe's future. In his Humboldt University speech, Fischer expressed his personal support for a European Federation based on a written constitution, a directly elected president, a competent and accountable government, and a bicameral legislature. He envisaged a multistage, decades-long process of development toward a federation in which the nation-states would play a stronger role at the European level than the 16 states of Germany play at the national level today. The French foreign minister, coming from a country without Germany's historical experience with federalism, was predictably less than enthusiastic about the EU's evolution into a federal parliamentary democracy in which Germany's larger population would give it even more leverage. Euro-skeptic British Conservatives raised the alarm about the emerging blueprint for a European "superstate," and Prime Minister Tony Blair countered that the EU could develop as a superpower without becoming a superstate. Nevertheless, many political actors, even those not enamored by federalism, have recognized that a comprehensive simplification and reorganization of the treaties is overdue to make EU institutions operate more effectively. Clearly the Nice Treaty did not do this, and thus contributed new impetus to the constitutional debate.

In the period leading up to the 2004 intergovernmental conference, one can anticipate a flurry of reports from the Commission and from academic and journalist observers urging a new constitutional settlement, one which, for example, would more clearly delineate the responsibilities and associated powers of the EU institutions and the national (and subnational) governments. A recent essay by Lional Barber, editor of *The Financial Times* of London, anticipates the adoption of a written European constitution with a bill of rights, superseding all of those impenetrable treaties, in the year 2020, but its nature turns out to be much more confederal than Fischer's vision.

CONCLUSION

The Maastricht Treaty's provisions established an EU citizenship to supplement existing national citizenship. For example, EU citizens, wherever they live, gained the right to vote and to be a candidate in European Parliament elections and local elections (assuming national enabling legislation has been passed), and they may petition the European Parliament and the EU ombudsman. Yet after nearly a decade, such new rights have done little to advance the supranational identity of citizens in the various EU countries. Journalist Peter Ford of *The Christian Science Monitor* recently observed that whenever one travels 20 or so miles outside of the capital city of any EU country, "the concept of 'Europe' evaporates." The EU will become much more tangible for ordinary people as Euro coins and banknotes begin replacing national currencies on January 1, 2002. A smooth transition could help to promote everyday European consciousness at least within the Euro Zone; a botched transition would cast new doubts on the integrative process itself.

While polls, even in Britain, indicate that the functionality of the EU is recognized in particular areas, affective ties to "Europe" remain eclipsed by those to one's nation and/or subnational region (Scotland, Flanders, Catalonia, Bavaria, and so forth). During 2000, a convention of 62 parliamentarians and governmental officials from the 15 member states labored to draw up the "Charter of Fundamental Rights" to show that the EU involves loftier goals than just the goods and services of a continental marketplace. Its 54 articles declared the basic political and social rights of the citizens of Europe. At the Nice summit, the Charter was solemnly proclaimed by the gathering of national leaders, but it was not included in the draft treaty, nor given any legal force. However, its proponents hope that over time it will evolve into the core of a European constitution, which would inspire public support.

Outside forces could also work to bring about "an even closer union," even as the EU doubles in size. To the east, the Russian Federation and other Soviet successor states pose uncertainties. Although former Russian president Boris Yeltsin floated the idea of Russian membership in the EU, his country's vast size, struggling economy, and uncertain commitment to European values like those proclaimed at Nice make it a much more remote prospect than Turkey. Authoritarian trends in Russia, Belarus, and Ukraine, and the re-emergence of a union among the Slavic successor states, could encourage greater common identity among the EU populations. Even if one sets aside the nightmare scenarios, it is still clear that the Russian Federation's military power (thousands of aging nuclear weapons) and its relative stability in coming years will have an influence on the EU's course of development.

Even more influential as an outside force will be the United States. European and U.S. economies have long been intertwined, and the merger mania of recent years has accelerated the process. At the sectoral level, the relationship has involved healthy competition, which at times has been marred by major disputes and even trade wars, notably regarding agricultural products. As an illustration of keen competition at the firm level, Airbus, a consortium of European firms, has emerged as a serious rival for Boeing in world aircraft production. Many European intellectuals consider U.S. neoliberal capitalism as a negative model because of its lack of concern for social solidarity. Human-rights activists are critical of the U.S. criminal-justice system, and especially its use of capital punishment. Furthermore, in (especially Western and Northern) Europe, leaders and citizens appear to be much more environmentally conscious than their American counterparts. The failure of The Hague climate-change conference in November 2000 dramatized the policy differences between the EU (where Greens share power in several national governments) and the United States even under the Clinton administration.

In 1999, EU leaders committed themselves to the development by 2003 of a 60,000-strong rapid-reaction force that could be deployed in circumstances where the United States did not want to get involved. The Clinton administration reacted coolly (and the Bush administration even more so) to the initiative, which American officials saw as an unnecessary duplication and a diversion of resources that would undermine NATO's defensive capabilities. The British government viewed the force as no threat to its "special relationship" with the United States and to NATO's preeminence. The French government saw the force as the embryo of a European army to back up an independent EU foreign policy, and the German government was caught in between. The Turkish government, as a NATO member, has balked at the idea of the EU (which has kept Turkey's membership application on hold since 1987) having access to NATO's collective assets. Relations with the new U.S. administration got off to an awkward start due to the Europeans' deep misgivings about President George W. Bush's plan to develop and deploy a (thin) nuclear missile defensive shield against attacks by so-called rogue nations. Europeans fear that the program would cause a new arms race, with the Russians and Chinese building and deploying new offensive missiles. The outlook is that the EU will continue to develop a common security policy, which will at times not mirror U.S. views.

Although a "United States of Europe" has not emerged over the past half-century, the European Union has continued to evolve imperfectly and unevenly as a transnational political system. Its complex set of institutions still defy the labeling efforts of political scientists. Czech president Vaclav Havel concludes, "What is arising in Europe is a new kind of Ensemble, significantly different [from a traditional confederation or federation]." Fifteen governments increasingly pursue national interests in the European arena where they have pooled sovereignty. But there is no European political culture, because the citizens of the EU have not transferred their support to its authorities and institutions. Over time, dual identities may emerge; but for the time being, the EU tends to be valued because of its functional relevance to problems, such as pollution, which cut across national boundaries. The EU's policy scope has not only expanded into more and more areas but also moved from "low politics" (e.g., commercial policy) to "high politics" (e.g., defense policy).

Since the 1950s, the EU has grown from six member countries to 15, and it is now on the verge of adding the first of the wave of 12 more. Thus, peacefully and democratically, the European Union is becoming a major political entity, however one labels it, with more than 500 million people and a total gross domestic product greater than that of the United States. The process is messy, tedious, and open-ended. Yet compared to the "old" Europe of warring nation-stations, the "new" Europe of integrating nation-states provides a positive model of unity in diversity for the countries and regions of the world.

An "Expanded" Europe

The breaching of the Berlin Wall in November 1989 marked the end of the Soviet empire in Europe. Subsequently, with the dismantling of the Warsaw Pact (the East Bloc collective-defense pact) and the collapse of the Soviet Union itself, Moscow's control over Central/Eastern Europe rapidly eroded. In the 1990s, all of the Central/Eastern European countries, as well as the western tier of former Soviet republics such as the three Baltic states (Latvia, Estonia, and Lithuania) and Ukraine, began a transition toward democracy and capitalism. These countries had different starting points on their path through the transition period, and their domestic conditions also varied widely. Some of the former East Bloc countries, such as Romania and Bulgaria, have not yet been able to make significant economic progress, and their foundations of political pluralism often seem shaky. The former Soviet republics of Moldova, Ukraine, and Belarus inspire even less confidence these days. The two former Communist countries of the region that were not under Soviet control—Albania and Yugoslavia—are in many ways the worst off: Albania because its central government barely controls the country's rebellious regions and organized crime, and ex-Yugoslavia because Serbian dictator Slobodan Miloševic drove most of it into war.

Bright spots in this picture, however, are the countries of Poland, the Czech Republic, Hungary, and Slovenia. The first three have already been accepted as members of the North Atlantic Treaty Organization, and Slovenia—held up by nearby conflicts in Bosnia and Croatia—is likely to join soon. Whether or not one agrees with the eastward expansion of NATO—many critics in the West and in the countries themselves argue that it unnecessarily provokes Russia, while stoking growth in local military establishments at the expense of social and infrastructure spending—the acceptance of former Communist countries into the Alliance does indicate general confidence in their progress toward stable democracy. All four of these countries also expect to join the European Union in the foreseeable future, as they have also made considerable progress in increasing efficiency and reducing government ownership in their economies.

POLAND

Poland is a large country with a long and distinguished history. In medieval times, Poland was a major European power, famous for its campaigns against the invading Ottoman Turks and for its relative religious tolerance. The Polish nobility was renowned for its fiery protection of its own

---

POLAND
(Republic of Poland)

GEOGRAPHY
Area in Square Miles (Kilometers): 120,700 (312,612)
  (smaller than New Mexico)
Capital (Population): Warsaw (2,219,000)

PEOPLE
Total: 38,650,000
Annual Growth Rate: −0.04%
Rural/Urban Population Ratio: 36/64
Major Language: Polish
Ethnic Makeup: 98% Polish; 2% Ukranian, Belarussian, and others
Religions: 95% Roman Catholic; 5% Eastern Orthodox, Protestant, and others
Adult Literacy Rate: 99%

GOVERNMENT
Type: republic
Independence Date: November 11, 1918
Head of State/Government: President Aleksander Kwasniewski; Prime Minister Jerzy Buzek
Political Parties: Freedom Union; Democratic Left Alliance; Peasant Alliance; Christian-National Union; Freedom Union; Conservative Party; Confederation for an Independent Poland; others
Suffrage: universal at 18

ECONOMY
Currency ($ U.S. Equivalent): 3.97 zlotys = $1
Per Capita Income/GDP: $7,200/$276,5 billion

GDP Growth Rate: 3.8%
Inflation Rate: 8.4%
Unemployment Rate: 11%

 http://lcweb2.loc.gov/frd/cs/pltoc.html

privileges. Having the right to elect and unseat kings, as well as to veto proposed laws that failed to gain unanimous support, the nobles kept Poland free of the kind of absolutist, divine-right monarchies that characterized most of the rest of Europe at that time. The famous Queen Jadwiga, who ruled from 1384 to 1399, linked the Polish throne to that of Lithuania, greatly expanding the country and founding the Jagiellonian dynasty. But Poland's leaders lacked the ability to form a common front against the country's enemies. By 1800, Poland had "disappeared" from the map of Europe, having been carved up and annexed by its strong neighbors Russia, the Habsburg Empire, and Prussia (the core province of the future imperial Germany).

Polish culture and nationalism, however, did not disappear. The nobility, spread among three empires, worked constantly for the re-creation of a Polish state. Some fought in the wars of independence of other countries. Under the rule of Napoleon Bonaparte and then under the Russian czars, Poles enjoyed some degree of political autonomy. These entities, called the Grand Duchy of Warsaw and Congress Poland, were important mostly because they kept the idea of an independent Poland alive. Poland's most famous poet, Adam Mickiewicz (1798–1855), agitated across Europe to raise support for the national cause. The city of Krakow, located in the province of Galicia in the Habsburg Empire, became the center of Polish intellectual activity due to the relative leniency of Austrian rule.

In 1830 and again in 1863, Poles in Russia launched major insurrections. The czar responded with great force, and autonomy was revoked in Congress Poland. As World War I approached, the two leading figures in Polish politics were Roman Dmowski of the National Democratic Party and Jozef Pilsudski of the Socialist Party. Dmowski wanted to use Russia as an ally to restore unity to the Polish lands; Pilsudski favored cooperation with Germany and Austria.

World War I resulted in the collapse of all three empires that controlled Poland. U.S. president Woodrow Wilson and the other framers of the peace settlement desired the re-creation of Poland, so the country "reappeared" on the map of Europe in 1919. Pilsudski became president of a country that was only 65 percent Polish but was understood to be the ethnic national state of Poles only. This led to many misunderstandings and conflicts with Poland's minorities, which included large numbers of Ukrainians, Jews, Belarussians, and Germans. The new democratic government quickly gave way to authoritarianism. Poland also had major problems with its neighbors. Germany was resentful that Poland had been given Upper Silesia and that the Baltic port of Danzig (Gdansk in Polish) was put under the control of the League of Nations. The new country of Lithuania was angry that it was not given the major city of Vilnius (Wilno), while the Soviet Union invaded Poland in 1919. A dispute with Czechoslovakia broke out over the region of Teschen (Cieszyn).

Adolf Hitler's Germany conquered Poland easily in September 1939, setting off World War II. Both Jews and Christians in Poland suffered tremendously under Nazi rule, but Polish Jews, who numbered close to 3 million, were targeted by the Nazis for total extermination. The Polish underground resistance was unwilling or unable to help many Jews. In addition to their concentration camps that they set up all over the Continent, the Nazis established a series of death camps in Poland. The most notorious of these was at Auschwitz-Birkenau. In the death camps, millions of Jews, from Poland or shipped in from other German-controlled countries, were murdered. The Polish capital of Warsaw, having been devastated by the Germans, nonetheless was the site of two major, bloody anti-Nazi uprisings: in the Jewish ghetto in 1943, and by the Polish Home Army in 1944.

After the war, the Soviet Union annexed eastern Poland and allowed Poland to annex part of eastern Germany. Several million people were expelled from the country in the 1940s—Germans to the west, and Poles hostile to communism to Siberia. The Yalta Agreement of 1945 gave the U.S. and British stamp of approval to Soviet leader Joseph Stalin's plans to dominate postwar Poland. U.S. president Franklin D. Roosevelt needed Stalin's cooperation in building the United Nations and in concluding the war against Japan, while Stalin was eager to build a buffer zone in Central/Eastern Europe to protect the Soviet Union from future German aggression.

The Polish Communist party, known officially as the Polish United Workers' Party (PUWP), ruled Poland from 1945 to 1989. Its first leader, Boleslaw Bierut, was a devout Stalinist. The government took control of the economy, except agriculture; political opposition was crushed; and the leading Catholic Church figure, the influential Cardinal Wyszynski, was arrested. In the late 1950s, however, Wladyslaw Gomulka, a reform Communist, came to power after a bold workers' protest. Stalin had died in 1953, and several of the East Bloc countries were liberalizing in the style of Joseph Broz Tito's maverick Communist government in Yugoslavia. Gomulka did not live up to his promise as a reformer, however; the economy stagnated, and restrictions on intellectuals and the Catholic Church increased. (Czeslaw Milosz, a Polish poet who won the Nobel Prize in 1980, captured the tenor of life under the PUWP in his book *The Captive Mind.*)

Demonstrations and strikes in 1970 brought a new leader to power, but economic problems and government crackdowns on strikers continued. In 1980, massive unrest in the Lenin Shipyards of Gdansk resulted in the creation of the Solidarity movement, which sought to legalize strikes and improve workers' standard of living. Solidarity also became famous for its struggle against governmental censorship and persecution of the Church. It was supported by Pope John Paul II, who was the former cardinal of Krakow, Karol Wojtyla. In 1981, with the Soviet Union threatening to invade, Polish general Wojciech Jaruzelski banned Solidarity and placed the country under martial law. There was an active underground press, and Solidarity leader Lech Walesa was awarded the Nobel Peace Prize in 1983. Finally, the rise of Mikhail Gorbachev in the Soviet Union took the pressure off Solidarity, and in 1989 the PUWP allowed Solidarity to run political candidates in the parliamentary election. The Communists were soon swept from power.

Solidarity leaders Walesa and Tadeusz Mazowiecki led the country through the early 1990s. Having positive relations with the newly reunified Germany was of massive importance. Thus, in 1990 the Stalin-era boundary along the Oder and Neisse Rivers was recognized as permanent, and German chancellor Helmut Kohl extended large amounts of financial assistance to Poland. In 1991, Walesa and the leaders of Czechoslovakia and Hungary formed the "Visegrad Group" to coordinate their countries' economic and foreign policies. The political scene inside the country quickly grew stormy, however, as long-time Solidarity intellectuals like Adam Michnik, the editor of the most influential newspaper, *Gazeta Wyborcza,* began to criticize Walesa, who was also being challenged by various populist and former Communist groups. Still, Walesa saw to it that the remaining Soviet troops left the country in 1993. He also set Poland on a course of rapid economic reform, called "shock therapy," aimed at privatizing government property and introducing free-market mechanisms. Inflation, prices, and unemployment rose alarmingly, as did the misery and anxiety of the Polish population.

But by the mid-1990s, unemployment was falling. Real income was up, and the economy grew at the heartening rate of 6 percent a year. Experts predicted that 80 percent of the economy would soon be in private hands. Western countries recognized Poland's economic progress and invested large sums in joint ventures, while the International Monetary Fund began granting loans. Existing international debts were restructured. The most innovative reform was a privatization program that distributed "vouchers," representing partial ownership of 7,000 different enterprises, to every adult in the country. Sixty percent of the vouchers in a given company went to the general public; 40 percent were distributed among the employees and the government. In this way, Poles sold or collected vouchers as Americans might manage their mutual funds; the government hoped that this method of decentralizing ownership would boost citizens' enthusiasm for the reforms, which also brought great hardship. The bleak spot in Poland's economic picture throughout the 1990s was the coal-mining sector, which is inefficient and a great polluter. Nevertheless, Poland is probably the greatest economic success story among the European countries "in transition" from communism. The economist Leszek Balcerowicz was instrumental in designing the reforms, and American experts were also brought in.

In November 1995, Walesa lost his bid for reelection to the presidency. His style had seemed increasingly imperial, and his close connections to the Catholic Church, especially on the touchy issues of abortion, sex education, and anti-Semitism, alarmed many voters. The winner was Aleksander Kwasniewski, a former Communist, now head of the Democratic Alliance of the Left. Many Poles at first wondered whether an ex–PUWP figure could behave in an honorable and democratic fashion, but Kwasniewski did so and remained very popular. Soon, however, Parliament (the *Sejm*) was controlled by a center-right coalition led by Premier Jerzy Buzek.

In 1997, NATO agreed to accept Poland as a member in 1999. The armed forces needed an expensive overhaul, and Russian president Boris Yeltsin grumbled at—but could not forestall—the NATO expansion. After steering Poland into NATO, Premier Buzek set his sights on a spot in the European Union. The EU, for its part, has significant concerns about Poland. The privatization process is not finished; its relatively inefficient agricultural system might require substantial subsidies; Polish workers could flood Western Europe's labor market; and Poland's eastern border must be secured against illegal immigration. Although 70 percent of Poland's foreign trade is with EU countries, its per capita gross domestic product in 1997 was only 31 percent of the EU average. At current rates of growth, it would take Poland until 2017 to reach an economic par with Greece, the poorest EU member. It remains to be seen if the EU will invite Poland to join in 2003.

Although stable and democratic, Poland faces two other ongoing political controversies: acrimonious debates between Jews and Catholics about the Holocaust; and "lustration" or "screening" laws that open police files to the public, resulting in accusations that some important government figures worked for the secret police in the 1980s. One prime minister has already had to resign because of such revelations; President Kwasniewski strongly opposed such a law but was himself cleared of charges of collaboration. The Holocaust issues capture the most attention in the international press, even though Pope John Paul II has done a great deal to smooth relations between the Jewish and Catholic communities outside of Poland. Throughout the 1990s, certain Polish Church leaders, including a famous anti-Communist cardinal, Józef Glemp, made remarks regarded by some as anti-Semitic; a priest named Tadeusz Rydzyk also ran a popular but highly inflammatory radio show that pandered to anti-Semitic sentiment. In early 2001, a new controversy broke out over commemorations of a 1941 massacre in the village of Jedwabne. The government decided to replace a plaque blaming Nazis for the massacre with one that points out the collaboration of Polish Catholics; conservative and nationalist groups in Poland responded with outrage.

The general mood in Poland by the year 2001 was rather optimistic, although the closing of the Gdansk Shipyard, where Solidarity got its start, and public controversy over the conservative politics of the Church and of most former Solidarity leaders divided the population. Two extensive visits by

Pope John Paul II in 1997 and 1999 sounded a more concili-atory note on social issues than his now-famous 1991 tour, during which he harshly criticized Poles for their permissive morals and lax faith. Also in 2001, the country's middle class was the biggest and strongest in Central/Eastern Europe, although this economic development sparked debate among Poles over retail credit and the power of advertising. After great debate, the country joined the Czech Republic and Hungary in abolishing the death penalty, in order to come into legal harmony with the states of the EU. That move, along with the new public debate about domestic violence, accom-panied by a graphic billboard campaign designed to stimulate discussion, shows that Poland is beginning to come to terms with human-rights problems more subtle than those caused by one-party rule.

HUNGARY

The Hungarians arrived in the Danubian Basin in the A.D. 800s. Their language, Magyar, has almost no relatives in Europe; it is part of the Uralic family and is distantly related to Finnish and Estonian and several Central Asian languages.

In 1000, their king, István, accepted Christianity. His name means "Stephen" in Hungarian, and the crown lands of Hun-gary's huge medieval kingdom are still called "the Lands of the Crown of St. Stephen." In 1222, the king issued a docu-ment called the Golden Bull, which ensured the power of the feudal nobility at the expense of the king. Indeed, the land-owning aristocracy dominated Hungary's culture and politics for centuries.

In 1102, the royal line in neighboring Croatia died out, and that country became subject to Hungary. After its defeat by the Turks at the Battle of Mohács in 1526, Hungary came under the rule of the Austrian Habsburg family. Austrian armies liberated the major city of Buda in 1686, but many Hungarians chafed under Austrian rule. Resistance was great-est in the mountainous, ethnically mixed eastern region of Transylvania, where insurgents known as *kurucz* mounted periodic rebellions against Vienna.

By the nineteenth century, Hungarians were the second-largest group (approximately 20 percent), after German-speaking Austrians, in the Habsburg Empire, but they were not even a majority in the Lands of the Crown of St. Stephen,

HUNGARY
(Republic of Hungary)

GEOGRAPHY
Area in Square Miles (Kilometers): 35,900 (92,980) (about the size of Indiana)
Capital (Population): Budapest (1,909,000)

PEOPLE
Total: 10,139,000
Annual Growth Rate: −0.3%
Rural/Urban Population Ratio: 35/65
Major Language: Hungarian
Ethnic Makeup: 90% Hungarian; 4% Romani; 3% German; 2% Serb; 1% others
Religions: 67% Roman Catholic; 20% Calvinist; 5% Lutheran; 8% Jewish, atheist, and others
Adult Literacy Rate: 99%

GOVERNMENT
Type: parliamentary democracy
Independence Date: 1001 (unification)
Head of State/Government: President Árpád Göncz; Prime Minister Viktor Orbán
Political Parties: Fidesz; Hungarian Socialist Party; Hungarian Democratic Forum; Alliance of Free Democrats; Christian Democratic People's Party; Hungarian Civic Party; others
Suffrage: universal at 18

ECONOMY
Currency ($ U.S. Equivalent): 294.33 forints = $1
Per Capita Income/GDP: $7,800/$79.4 billion

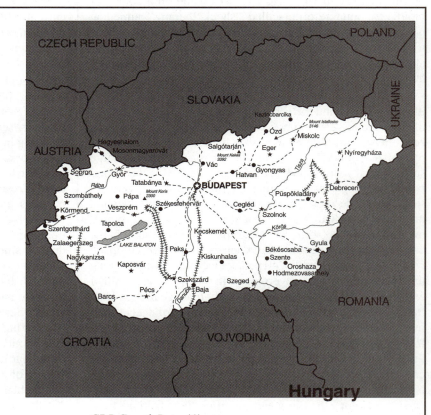

GDP Growth Rate: 4%
Inflation Rate: 10%
Unemployment Rate: 10%

 http://lcweb2.loc.gov/frd/cs/hutoc.html

which included, in addition to Hungary proper, the regions that are today southern Slovakia, eastern Austria (Burgenland), northern Serbia (Vojvodina), and, most important, western Romania (Transylvania). Still, the large Hungarian noble class dominated politics and culture in the eastern half of the Empire. Nationalism grew rapidly after 1800.

On March 15, 1848, in the spirit of the nationalist rebellions sweeping across Europe, Hungarians began an uprising against Austrian rule. Led by the nationalist editor and politician Lajos Kossuth (1802–1894) and the nobleman Lajos Batthyány, Hungary moved toward independence and the creation of a democratic government. But the government in Vienna was determined to fight to keep the Empire together, and the many national minorities inside Hungary were mistrustful, with good reason, of the Magyar desire to assimilate them. The new Habsburg emperor, Franz Josef, crushed the rebellion with great bloodshed. One of the most famous participants in the uprising was Hungary's fiery national poet, Sándor Petofi (1823–1849). He died fighting Austria's allies, the Russians, in Transylvania. His Romantic lyrics—whether about nature, patriotism, or love—are still quoted with great fondness.

After 1850, the vestiges of feudalism disappeared and Hungarian cities and industry grew rapidly. When Austria lost an important war to Prussia in 1866, Franz Josef decided to grant considerable autonomy to the Hungarians to keep them loyal. The new arrangement, called the *Ausgleich* (German for "settlement or compromise"), renamed the country Austria-Hungary and provided for the Austrian emperor to be crowned also as king of Hungary. Now the imperial government would limit itself to military, financial, and diplomatic affairs; everything else would be handled at the local level by the two capitals, Vienna and Budapest. This move greatly increased Hungarian self-confidence and power. A large urban working class (proletariat) took shape. Many Hungarian peasants emigrated to North America. The nobility still dominated politics, withholding suffrage from many Hungarians and threatening to assimilate large minority groups like the Croats and Romanians.

World War I destroyed the Habsburg Empire, leaving in its wake small separate nation-states like Austria, Hungary, and Czechoslovakia. By the Treaty of Trianon (1920), the traditional boundaries of Hungary were reduced by two thirds; the territories given to each of Hungary's neighbors contained millions of Hungarians in addition to other national groups. In fact, 28 percent of Hungarians now lived in other countries. Hungarians considered this treaty a great injustice, since Hungarians were in general not well treated in the Habsburg successor states; Trianon substituted one set of minority problems for another.

In 1919, a Communist government ruled Hungary for 133 days. Soon, like much of Europe at the time, the country fell into authoritarianism. A succession of prime ministers gravitated slowly toward fascism and cooperation with Nazi Ger-many, but the power behind the scenes was the conservative regent, Admiral Miklós Horthy (in power 1920–1944). Adolf Hitler arranged for a transfer of some of the Trianon territories back to Hungary in the two Vienna Awards (1938–1940), designed to reward Hungary for its loyalty to Germany. Hungary then participated in the massive assault on the Soviet Union in 1941. In the summer of 1944, the Nazis and Hungarian fascists overthrew Horthy, deported most of Budapest's large Jewish community to Auschwitz-Birkenau, and fought a ferocious, last-ditch campaign to prevent a Soviet takeover of the country.

By the late 1940s, the Hungarian Socialist Workers' Party, as the Communists were called, was in control. Ties with the Soviet Union were close until a major rebellion broke out in 1956. Reform Communists, led by Imre Nagy, hoped to end communism and join NATO. The Soviet government, led at the time by Nikita Khrushchev, responded with an outright invasion; thousands of Hungarians died fighting the Soviets. About 200,000 Hungarians fled the country, one of the biggest population movements in Europe since World War II. Hungary's new communist strongman, János Kádár, eventually steered the country in a surprising direction, however. He stimulated agriculture by allowing peasants to farm their own land, and he used the profit motive (or cost accountability) to improve industrial productivity. Hungary finished its evolution into an urban and well-educated society under Kádár, who also used the lure of "goulash communism"—easily available consumer goods and a relatively high standard of living—to maintain his popularity. Although not as liberal as Tito's communism in Yugoslavia, Hungary after 1956 had the least police oppression and censorship of any other Central/Eastern European country. The novels and essays of writer George Konrád (b. 1933) capture the moods and struggles of Hungarians in these decades.

Kádár retired in 1986, and the younger Communists allowed the creation of alternative political parties in 1989. Hungary was instrumental in cracking the control of Communists in East Germany, too, because it let thousands of Germans emigrate to the West by opening the Hungarian border to Austria. The new president (the respected writer Árpád Göncz) and prime ministers József Antall and Gyula Horn tried to stimulate and restructure the economy. But a deep recession and the lack of enthusiasm of many Hungarians for Western-styled "market mania" slowed the transition process. Hungary's large foreign debt scared off many foreign investors, and the political scene showed signs of instability with the rising popularity of an extreme rightist, István Csurka. Even the mainstream parties had significant differences among themselves, ranging from the conservative Democratic Forum to the Socialists. The Alliance of Young Democrats (FIDESZ) was the main party by the late 1990s, however. Its leader, Viktor Orbán, became prime minister and hewed to an increasingly populist and socially conservative line.

Privatization laws were passed only in 1995, and the pace of legal reform has been very slow, although thorough. Many Hungarians see the stock market as a kind of ignoble casino; others fear foreign control of their economy. Most important, Polish-style "shock therapy" goes against Hungarian conceptions of the supportive, symbiotic relationship that should exist between a government and its people. The Hungarian political scene is a place of vigorous competition between groups to the left and right of center. In an attempt to move toward closure on the Communist period, the names of a number of anti-Stalinist "enemies of the state" have been cleared; trials have also begun for border guards who killed dozens of fleeing Hungarians during the 1956 Revolution.

The political freedom of the 1990s has also had negative side-effects. There has been a revival of anti-Semitism and public prejudice against Romani (Gypsies) and Romanians. Far-right groups have protested the government's banning of translations of Hitler's book *Mein Kampf,* while a neo-Nazi rally in Budapest in February 1999 turned into a riot. Women, for whom the HSWP opened up many new career paths, have almost dropped off the political radar. Workers' benefits have been drastically cut, and prostitution and pornography are rampant. Budapest is also the scene of an increasing amount of criminal activity by the Russian mafia; a bomb blast in the city's main shopping district, thought to be gang-related, killed four and injured 25 in July 1998.

Hungary joined NATO in 1999, and even before that it allowed NATO to open a base at Taszár for use during the Bosnian conflict. Hungary is now seeking admission to the European Union, although, as in the case of the Czech Republic, many more economic reforms will probably be required first. Relations with Romania are officially smooth, but Romanian environmental problems, especially discharges of cyanide into rivers that flow through Hungary, are complicating the relationship. Hungarians remain concerned over the status of their 2 million co-nationals in Transylvania. A Hungarian-language university is in the planning stages in Kolozsvár (Cluj). In addition, the Hungarians of Slovakia have their own political parties but still feel unenfranchised, because the Slovak Constitution says that the country is the official home of Slovaks only. Until the fall of Yugoslav dictator Slobodan Milošević in 2000, the 350,000 Hungarians in northern Serbia were also very worried about assimilation and threats of "ethnic cleansing." Hungary's support of the 1999 NATO bombing campaign in the crisis in Kosovo worsened their situation, but the new government in Belgrade has promised to respect minority rights and to consider restoring Vojvodina's autonomy, abolished by Milošević in 1988. In 1999, Premier Orbán declared that the world should remember that all Hungarians belong to one indivisible nation, regardless of what country they inhabit. This statement might herald a more assertive stance toward Budapest's neighbors.

THE CZECH REPUBLIC

The Czechs are a Slavic people who are closely related to the Slovaks, with whom they shared the country of Czechoslovakia from 1918 to 1993. Since that date, the Czechs have had their own country, which is both larger and more prosperous than Slovakia. In two earlier periods of their history, around A.D. 900 and 1500, the Czechs had their own independent kingdom, but after 1526 they came under the rule of the Austrian Habsburgs. But they were almost constantly under pressure from the other members of the Holy Roman Empire, who were mostly large German-speaking feudal states. A century before the official start of the Protestant Reformation, the Czech lands of Bohemia and Moravia were the site of an important challenge to the hegemony of the Catholic Church, which was largely under German control. Native Czechs like the theologian Jan Hus (John Huss, 1370–1415) tried to reform the Church and also resisted the growing Germanization of the Czech culture. Hus was burned at the stake, but more than 200 years of religious and political unrest in the Czech lands followed. The notorious Thirty Years' War began in Bohemia in 1618; the Catholic Habsburg armies destroyed much of the region, and Czech, like Slovene in the southern part of the Empire, became a largely peasant language.

In the late 1700s, sparked by the rise of nationalism in France and by an absolutist government in Vienna that aimed to take away their remaining autonomy, a Czech national renaissance began. Its first great figure was Josef Dobrovsky (1753–1829), a linguist and historian. Although he, ironically, wrote in German and Latin, Dobrovsky's studies of language publications helped standardize the Czech language and started the field of comparative Slavic studies; his historical works provided a basis for the next generation of politicians and historians, who would be concerned with the restoration of self-rule to Czech lands. By 1848, another great scholar, František Palacky (1798–1876), had emerged as a leader of the Czech national-rights movement. He organized a Slavic congress in Prague that called for the Slavs of the Habsburg Empire to be given more power, though he did not seek independence. The Viennese government suppressed his movement by force, but the many rebellions of 1848 did result in the abolition of serfdom across the Empire. Palacky's ideas formed the basis of "Austroslavism," a long-lived federalist movement that sought to win autonomy and unity for Habsburg Slavs while remaining loyal to Vienna, which protected them from both Germany and Russia.

The next great figure in Czech history was Tomás Masaryk (1850–1937), a philosopher who advocated democratic reforms and the cooperation of related Slavic peoples. Like other Czech intellectuals, Masaryk emphasized that the Czech people were thoroughly Western in their outlook. He became the first president of the new country of Czechoslovakia in 1918, after World War I destroyed the Habsburg Empire. As in the case of the Serbs, Croats, and Slovenes of Yugoslavia, the Czechs and Slovaks had agreed—under some pressure from the victorious Allies—to live together after the war. Masaryk ruled until 1935, and Czechoslovakia was the only country in Central/Eastern Europe to remain democratic during the interwar period. The Slovaks were not entirely satisfied with the political and economic arrangements, however, and Adolf Hitler also had designs on a key region of the country known as the Sudetenland. Many Germans lived in this region, and Hitler used this fact as a pretext for dismembering and overrunning the country in 1938 and 1939.

During World War II, Nazi Germany ruled the Czech lands with an iron hand, while Slovakia became a satellite state run by a fascist puppet, Monsignor Jozef Tiso. The Czech industries, workers, and raw materials were important to the German war effort, and resistance activities were severely punished, as in the total destruction of the village of Lidice in 1942. Czech Jews were singled out by the Nazis for annihilation. After the war, the reunited Czechoslovakia briefly returned to democratic rule under Edvard Beneš, who carried out controversial measures, such as expelling 3 million Germans from Sudetenland. He also wanted close relations with the Soviet Union, while he blamed the West for selling out to Hitler in the infamous Munich Pact of 1938, which had given the Germans a green light to carve up the country.

In February 1948, the Czechoslovak Communist Party (KS), led by Klement Gottwald, took over the country by partially legal means. Once again the Slovaks were more or less subordinated to the Czechs, although their lands experienced significant growth in heavy industry. Even more important, both populations suffered greatly under the doctrinaire Stalinist rule of the KS. All political opposition was eliminated, and economic, cultural, and religious life was brought under the control of the party. As in Poland and Hungary, a vicious purge was carried out in the early 1950s against reform-minded communists, such as Rudolf Slánsky, whom Joseph Stalin feared would try to imitate the more liberal, maverick socialism of Josep Broz Tito in Yugoslavia. On the positive side, unemployment was eliminated, education and job prospects for women improved, and health care was made available to all citizens. Like in other East Bloc countries, discontent grew in the Czech underground. In 1968, it emerged in a reform movement called the Prague Spring, the goal of which was to create "socialism with a human face." Artists, students, and younger Communists demanded an end to censorship and the secret police, but not an exit from the Warsaw Pact, as the unfortunate Hungarians had demanded in 1956. Still, on August 20, 1968, a massive Soviet-led

THE CZECH REPUBLIC

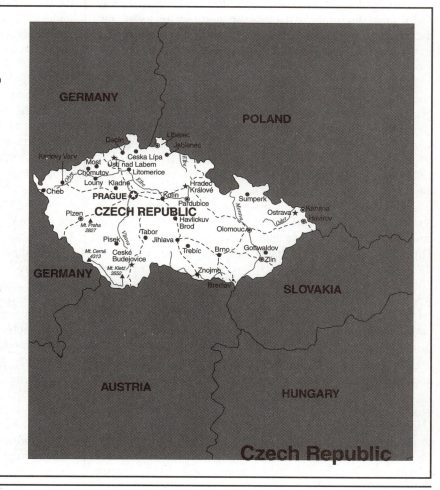

GEOGRAPHY
Area in Square Miles (Kilometers): 30,379 (78,703) (about the size of South Carolina)
Capital (Population): Prague (1,213,500)

PEOPLE
Total: 10,272,000
Annual Growth Rate: −0.08%
Major Languages: Czech; Slovak
Ethnic Makeup: 81% Czech; 13% Moravian; 3% Slovak; 3% others
Religions: 40% atheist; 40% Roman Catholic; 5% Protestant; 15% others
Adult Literacy Rate: 99%

GOVERNMENT
Type: parliamentary democracy
Head of State/Government: President Václav Havel; Prime Minister Miloš Zeman
Political Parties: Civic Democratic Party; Christian Democratic Union–Czech People's Party; Czech Social Democrats; Communist Party; others

ECONOMY
Currency ($ U.S. Equivalent): 40.00 korunas = $1
Per Capita Income/GDP: $11,700/$120.8 billion
Inflation Rate: 2.5%
Unemployment Rate: 9%

 http://www.cia.gov/cia/ publications/factbook/ez.html

military force invaded the country and toppled the reformist leader Alexander Dubcek.

Soviet leader Leonid Brezhnev justified the invasion by asserting the Soviet Union's right to intervene in any country where a socialist system was threatened. This "Brezhnev Doctrine" later provided the rationalization for the 1979 invasion of Afghanistan, and many people feared that it would lead to an occupation of Poland during the 1980s. In Czechoslovakia, a heavy-handed, pro-Soviet leader named Gustáv Husák was installed and the political and cultural scenes were "normalized," meaning that nascent resistance was once again crushed. Of the many excellent Czech writers who have treated the post–World War II decades in their works, Milan Kundera is arguably the greatest. His many novels, especially *The Joke,* have been widely translated and discussed. They provide a subtle and stimulating exploration of normal social questions in abnormal political conditions.

In 1977, 300 Czech intellectuals signed a famous public statement that became known as "Charter 77." It contained a critique of government human-rights abuses and demanded democratic reforms. Its signers were hounded mercilessly, but within a decade, the rise of Mikhail Gorbachev in the Soviet Union gave added impetus to reform. By 1988, younger and less ideologically hidebound KS leaders, such as Miloš Jakeš, had come to the fore. Economic changes were in the works, such as the loosening of central planning and the introduction of accountability for profits and losses in individual enterprises.

The events of the autumn of 1989, however, allowed the opposition to push the KS out of power. On November 24, a giant demonstration of 300,000 people took place in Prague; soon afterward, as Communist officials were resigning one after another, a highly regarded opposition writer named Václav Havel was named interim president. This has been called the "Velvet Revolution" because of how smoothly (and almost bloodlessly) the KS was sent packing. Havel had been a signatory to Charter 77 and was also well known for his absurdist dramatical works; he was also famous for his resistance essays such as those anthologized in *The Power of the Powerless.* Havel was elected president by a huge margin in the summer of 1990 and was still serving in that position in 2001. Although the position of premier has more day-to-day power than the presidency in the Czech Republic, Havel raised the country's profile considerably and came to exert considerable moral authority in European affairs, due to his principled positions and respect for human rights.

Havel's political party, the Civic Forum, soon split over economic issues. Finance Minister (and later Premier) Václav Klaus favored a more rapid transition than Havel, and one with less of a social safety net to ease the shocks for average Czechs. Political leaders also disagreed over whether or not to forbid former Communists from participating in politics. Havel here, too, favored a moderate approach, in the interests of national reconciliation and because he realized that former Communists were often the only politicians with administrative experience and connections with industry leaders. But former KS members were also criticized for being in a position to profit too much from the economic reforms.

In an effort to speed up the privatization of the economy, price controls were lifted in 1991 and the closing of unprofitable stores and factories began. Various methods of selling off state-owned enterprises were attempted, the most famous being the "coupon" method, whereby Czech citizens received shares in state companies and could keep, sell, or trade them as they saw fit. Agriculture was also privatized, with land being returned to the families from whom it was confiscated in the 1940s. By early 1992, Havel had cultivated a harmonious relationship with Chancellor Helmut Kohl of Germany and had negotiated the withdrawal of the last Soviet troops. The issue of reparations to the millions of Germans expelled after World War II still bedevils Czech–German relations, but it has not prevented Germany from becoming far and away the leading foreign investor in the Czech economy, providing an important source of technology and jobs. In a 1997 agreement, the Czechs apologized for the expulsions and the Germans apologized for the wartime occupation.

One of the most vexing problems facing the Czechs was the devastated environment, suffering from decades of abuse under the KS's fixation on heavy industry like steel foundries, chemical plants, and electricity production. Another grave and persistent problem was minority relations. There are more than 200,000 Romani in the country, and they have frequently been the victims of hate crimes and discrimination by local officials. In 1999, an enormous controversy broke out in the industrial city of Ústí nad Labem, where local officials tried to build a wall to separate the Romani and non-Romani parts of town. Non-European "guest workers," such as Vietnamese, who were "left over" from the era of cooperation with other socialist countries, were also attacked, sometimes by groups using Nazi slogans and symbols.

The final big problem was the secession of Slovakia. Czech premier Klaus, now heading his own Civic Democratic Party, and Slovak strongman Vladimir Meciar both favored splitting up the country; Havel was against it. Many ordinary Czechs and Slovaks were also against it, even though Slovakia was growing increasingly prone to a rather xenophobic form of ethnic nationalism and was not comfortable with a rapid pace of economic reform. But no referendum was ever held; the parliaments in the two halves of the country decided on separation themselves, effective January 1, 1993.

In 1998, Havel narrowly won reelection as president; Klaus was defeated as Czech voters swung away from his right-of-center politics and somewhat authoritarian personal style, but he became the speaker of Parliament. The new premier, Miloš Zeman, inherited a country in which the banking system had almost broken down, the stock market was stagnant, and the economic growth rate was much less than that in Poland or even Hungary (1.1 percent, as compared to 6 and 3.3

percent, respectively); moreover, the population was strongly divided over Havel's desire to win membership in the North Atlantic Treaty Organization. Many Czechs were far more concerned about economic growth and social stability than they were about future Russian aggression, and they viewed the massive upgrade of the Czech armed forces (which was required by NATO) as a waste of money. Although Havel supported NATO actions against the Serbian dictator Milošević, many Czechs did not support the 78-day air campaign against the fellow Slavic and European country of Serbia.

The Czech Republic did join NATO in 1999. Now many Czechs want to join the European Union. Their bid to join is being strongly supported by Germany, but it will drive an even bigger wedge between the Czechs and Slovaks. As part of the EU, the Czechs will have to maintain stricter control over their frontier with Slovakia, limiting trade and requiring visas for travel. This is sure to anger the government in Bratislava, just as Poland's accession would isolate its eastern neighbor, Ukraine. In the new millennium, the Czech political scene is growing increasingly turbulent, as Havel's health fails and as far-right, anti-Semitic, and racist movements gain momentum. Environmental concerns also continue to mount, since the considerable heavy-industry base was fostered by Communist leaders without regard for pollution control. Parts of the country are almost uninhabitable due to pollution from heavy metals and coal; furthermore, Austria strongly objects to the nuclear plant at Temelín, which it claims is unsafe and must be shut down before entry to the EU.

SLOVENIA

The Slovenes are a Slavic-speaking people who were part of the Habsburg Empire until the end of the Great War in 1918. They were one of the first non-German peoples to be incorporated into the Habsburg domains in the fourteenth century; they were also one of the smallest groups under Vienna's rule, numbering barely 2 million even in the twentieth century. Slovenes never had a famous medieval state that they could call their own, although they were part of a Slavic tribal alliance or kingdom known as "Samo's empire" in the A.D. 600s. The Latin name for the capital of this empire is the basis for the names of the two central regions of Slovenia, Kranj and Koroško (Carniola and Carinthia in English). Slovenes also predominated in the other Habsburg hereditary lands, such as Styria, Istria, and Gorizia. Their capital city, Ljubljana, was long known by its German name, Laibach, and most Slovenes were illiterate peasants for centuries.

Due to their important strategic location in the Habsburg Empire, between Vienna, the powerful cities of northern Italy,

SLOVENIA
(Republic of Slovenia)

GEOGRAPHY
Area in Square Miles (Kilometers): 7,834 (20,296) (about the size of New Jersey)
Capital (Population): Ljubljana (270,800)

PEOPLE
Total: 1,928,000
Annual Growth Rate: 0.12%
Ethnic Makeup: 88% Slovene; 3% Croat; 2% Serb; 7% others
Major Languages: Slovenian; Serbo-Croatian
Religions: 96% Roman Catholic; 1% Muslim; 3% others
Adult Literacy Rate: 99%

GOVERNMENT
Type: parliamentary democratic republic
Head of State/Government: President Milan Kucan; Prime Minister Janez Drnovšek
Political Parties: Slovene Christian Democrats; Liberal Democracy of Slovenia; Social Democratic Party of Slovenia; United List; Slovene National Party; Slovene People's Party; others

ECONOMY
Currency ($ U.S. Equivalent): 256.15 tolars = $1
Per Capita Income/GDP: $10,900/$21.4 billion
GDP Growth Rate: 3.5%
Inflation Rate: 8%
Unemployment Rate: 7.1%

and the Adriatic Sea, Slovenes were exposed to tremendous assimilation pressures. During the era of the Reformation, the modern Slovene language began to take shape as religious books were printed, mostly by Protestants, so that the local people could participate more in religion. The first Slovene dictionary also dates from the late 1500s. Catholic control over Slovene lands was quickly reestablished, however, and a national consciousness and a clear political agenda were very slow to develop among Slovenes. Napoleon briefly gave Slovenes self-rule when he conquered the region. But other subject peoples of the Habsburgs, such as Czechs, Italians, Hungarians, and Croats, moved much more forcefully toward the establishment of their own states.

By 1914, the Slovenes had produced several generations of scholars and poets who had cemented a national consciousness. One of them was France Prešeren (1800–1849), whose lyrics on folk and historical themes set the standard for written Slovene; one of his patriotic poems is now the Slovene national anthem. These scholars and poets had begun to formulate political demands in the name of all Slovenes. Their general political program centered on Austro-Slavism, which stressed the importance of cooperation among the Slavs of the Empire, which in turn served as a bulwark against German and Italian expansion. In more concrete terms, the Slovenes fought to preserve their language and to get all of the districts they inhabited put into one political unit or province. This kind of unity and autonomy under the scepter of the Viennese royal family, not complete independence or the establishment of a common Yugoslav state with related peoples like the Croats and Serbs, was the goal of all Slovene politicians. They struggled to achieve this goal by establishing a separate regional parliament, by gaining legal equality and cultural acceptance for their language, and by participating in ruling coalitions in Vienna whenever possible. The outstanding Slovene political leader of the time was Monsignor Anton Korošec, the leader of the Clerical Party from 1906 to 1938. The other main parties were the Liberals and the Social Democrats; all were equally, if moderately, nationalist. The nature of life in Habsburg Slovenia figures prominently in some of the works of the famous Austrian journalist and novelist Joseph Roth (1894–1939), especially *The Emperor's Tomb* and *The Radetzky March.* The irreverent, intensely personal, and politically explosive poetry and fiction of Ivan Cankar (1876–1918), along with the modernist architecture of Joze Plecnik (1872–1957), gave Slovenia a cultural status in Europe that far outweighed its small size.

After World War I and the collapse of Austria-Hungary, Slovenia became part of the new country of Yugoslavia. Slovenes believed that they would have unity and autonomy within Yugoslavia, and that living together with the Serbs and Croats would provide them with protection from their stronger Italian and German neighbors. Many Slovene-inhabited regions of these adjacent countries, however, were not included in Yugoslavia; and the central government in Belgrade proved to be authoritarian and incapable of stimulating economic growth. Still, the Belgrade government founded the first Slovene university and accepted Slovene parties into various ruling coalitions.

World War II brought great destruction to Slovenia, as it did to the rest of Yugoslavia. Most of the Slovene lands suffered under direct German annexation, and the rest were gobbled up by Italy. A resistance movement including Communists, Christian Socialists, and others played an important role in defeating the Axis forces and their local allies. After the war, the Slovenes again found themselves in a country called "Yugoslavia," but one that was truly federal in structure and led by Josep Broz Tito (1892–1980), a Communist who had directed the anti-Nazi resistance.

Tito fought against Serbian and Croatian nationalism, but he did not consider the Slovenes to be a threat. They were, furthermore, aided by the presence of one of their own—Edvard Kardelj—in the highest circles of the League of Communists. He was Yugoslavia's leading ideologist and Tito's right-hand man. Slovenia led the rest of Yugoslavia in literacy, per capita industrial production, and per capita income, and also enjoyed some economic benefits deriving from ready markets and sources of raw materials. Of course, Slovenes also paid much revenue into federal coffers, for the well-armed Yugoslav People's Army (JNA) and the development of poor regions like Kosovo and Montenegro. One of the heroes of unofficial postwar Slovene culture was the poet Edvard Kocbek, who had supported the partisans but was not a Communist; his graceful metaphysical lyrics are still revered by many Slovenes.

Slovenia's march to independence began in the late 1980s with calls for profound changes in the Yugoslav state structure, and then with the legal assertion of Slovene sovereignty. A weekly news magazine, *Mladina,* kept the public mobilized and informed by publishing frequent exposés about government corruption. Slovenia also had an active civil society, with many nongovernmental organization pushing for environmental protections, women's rights, and conscientious-objector status. There was also a lively "alternative" scene involving punk rock, modern art, and gay and lesbian movements.

In the 1990s, as Yugoslavia became increasingly dominated by Slobodan Milošević of Serbia, the League of Communists of Slovenia allowed the formation of non-Communist political parties. The new president was highly respected ex-Communist Milan Kucan, who was trained as a lawyer and now provided a steady hand at the helm. On June 25, 1991, Slovenia declared its independence from Yugoslavia. After a brief war, the JNA withdrew from Slovenia, which was thus spared the massive destruction that Serbian aggression visited upon Croatia, Bosnia, and Kosovo. Serbs have no historical claims to Slovenia, and the country has very few national minorities.

Since gaining independence, Slovenia has been characterized by relative stability, despite the presence of a large number of political parties, coalition governments, and a fair number of embarrassing scandals involving prominent politicians. Slovenia's first prime minister was Lojze Peterle of the Christian Democrats. With brief exceptions, the prime minister has been Janez Drnovšek of the party known as Liberal Democracy of Slovenia (LDS). From 1992 to 1997, Drnovšek's party—usually the biggest single vote-getter but lacking a majority—ruled in a center-left coalition. Since 1997, the LDS has ruled in a center-right coalition.

As in Poland, there has been considerable controversy over the political role of the Catholic Church. Slovenes are sharply divided over abortion and religious education in the schools. Much of the Catholic Church's property, like other property confiscated by the Tito government after World War II, has been returned or compensation paid. But the greatest internal challenge facing Slovenia remains the creation of stable political parties that reflect more than the personalities and ambitions of their prominent members. This development will be important for long-term economic development and foreign-policy consistency.

So far, Slovenia's economic progress has been heartening, even though privatization of government-owned businesses (called "socially owned property" in the former Yugoslavia) has been slow. Slovenes are determined not to allow great differences in wealth to erode their social cohesion, and most parties are committed to keeping a strong social "safety net," including pensions, protection against unemployment, national medical insurance, and subsidized education. What one might call Slovenia's "capitalism with a human face" has won the confidence of foreign investors; the European Union opened negotiations with Slovenia in 1997, and many expect that it will be the first of the Central/Eastern European countries to join the EU. Per capita, Slovenia's gross domestic product is three times that of Croatia and about a third greater than Hungary's. Still, it is only half that of Italy and one third as great as Austria's. But the economy grew in the 1990s at 3 percent annually, and the new currency (the *tolar,* a word derived from an old Habsburg currency, the thaler) has withstood inflationary pressures well. Slovenia is credited with having a disciplined and educated workforce, good roads, and many natural resources. Tourism, along the coast and at ski resorts, is a major source of foreign-currency earnings.

Slovenia has joined NATO's junior member program, called the "Partnership for Peace," but its full admission to NATO has been slowed by instability in the nearby Balkans. It is a member of many regional organizations, such as the Alps-Adriatic Working Community, founded in 1979. It was also the first former Yugoslav republic to occupy one of the rotating seats on the Security Council of the United Nations.

There are some problems, however, with neighboring Croatia, over the operation and finances of the Krško nuclear-power plant (in Slovenia, but near Zagreb) and over the status of the Slovene minority in Croatia. Most important, however, is the border dispute between the two countries on the Istrian Peninsula. Some actual territory is in dispute, but so are fishing rights and—very significantly from the point of view of the Slovenes, who have a very limited section of coastline—sea boundaries. Slovenia insists on direct access to Adriatic shipping lanes, a common-sense claim that any notion of "strategic rights" (such as the Croats themselves put forth in their dispute with Montenegro over the Prevlaka Peninsula in southern Dalmatia) would seem to support.

There are also disputes with Italy over former Italian property in the Trieste region. Although Yugoslavia had signed two treaties with Italy that supposedly settled this question, by the 1990s Italy was demanding that the issue of compensation from both Slovenia and Croatia be re-opened. The rightist Italian government of the early 1990s threatened to hold up Slovenia's admission to the EU and to abrogate the 1975 Treaty of Osimo, which guaranteed rights for the large Slovene minority in northeastern Italy. Although these issues have since faded, many Slovenes fear their revival if the political climate in Italy should change again.

The Austrian political scene is another source of concern for Slovenes. The well-known Slovene minorities in southern Austria, in the states of Carinthia and Styria, have been alarmed by the intolerant populist rhetoric of the Austrian Freedom Party (FPÖ). The FPÖ is now part of the ruling national coalition, and its leader, Jörg Haider, is the governor of Carinthia. Despite repeated controversies that cast him as an anti-Semite, Haider has a great deal of political power at the local and national levels and might soon be in a position to retract some of the Austrian Slovenes' minority rights.

On a more positive note, an important milestone for Slovenia was reached in June 1999, when U.S. president Bill Clinton visited Ljubljana. In addition to auguring well for decisive U.S. support for Slovenia on various international issues, Clinton met in Slovenia with Montenegrin president Milo Djukanovic, who was then a prominent opponent of Milošević; President Kucan proudly proclaimed Slovenia's role as good neighbor, intermediary, and role model. In September 1999, Pope John Paul II also visited Slovenia and announced the beatification of Bishop Anton Martin Slomšek, an important cultural figure from the nineteenth century.

The biggest challenges that Slovenia faces in the new millennium are domestic. Coalition-building seems to be the government's main preoccupation, and no party has enough votes to lead decisively. Economic restructuring must continue, and at a more rapid pace, while Slovenia must also prove that it is a reliable partner to other European countries by fighting organized crime (in drugs and prostitution) and carefully managing its nuclear-power industry. Meanwhile, the lack of scholarly consensus and public catharsis on the World War II–era conflicts between Slovenia's leftist and rightist forces, which resulted in large massacres by Communists in 1945, continues to bedevil national unity. Most worrisome of all, an upsurge in populism and right-wing politics, spearheaded by the mercurial former defense minister Janez Janša and Zmago Jelincic of the Slovenian National Party, is eroding the vitality of Slovenia's famed "civil society," whose members have advocated ethnic tolerance, nonviolence, and limitation of governmental powers.

CONCLUSION

The four countries discussed above represent fascinating case studies in "transition" out of European one-party systems with what were called "command economies." For many former Soviet allies and republics, it remains unclear just what the destination of the general post-Communist transition truly is. But Poland, the Czech Republic, Hungary, and Slovenia have strong historical connections to Western Europe; and their governments since 1989 have demonstrated a commitment to "rejoining Europe," by which they mean trying to catch up with countries like France, Germany, and Italy in economic and political terms. These countries do not seek to become satellites of the rich and powerful Western European states; nor do they necessarily envision the triumph in their countries of American-style economics, which is often referred to in Europe as "wild West" or "winner-take-all" capitalism. These four countries are following the general European model of constructing stable, politically pluralistic states in which the importance of free enterprise and private property is recognized but balanced with "social market" concerns for the general well-being of the populations.

Over the decade of the 1990s, these four states were widely viewed as the "not-if-but-when" countries as far as membership in NATO (now largely achieved) and the European Union were concerned. Generations of young people are now growing up for whom Communist rule is not even a memory; this can mean even greater political stability, but it seems to be bringing an increase in political complacency and passivity as well. The former bonds that held the peoples of Central/Eastern Europe together even seem to be weaking to some degree: Slovenes tend to know much more about Italian and Austrian politics and literature (not to mention American pop culture) than about trends in Serbia or even next-door Croatia, while young Czechs view Slovaks not as closely related if poorer cousins, but as total strangers. There is not much nostalgia for the days of Communist rule, but the recent staging of unique educational exhibitions in Poland and the Czech Republic is a reminder that the past cannot—and should not—be completely forgotten. In the fall of 2000, an exhibition entitled "The Art and Culture of the Communist Era" was opened in Warsaw. Even more poignant was the 1999 "Open-air Museum of Totalitarianism" on Wenceslas Square in downtown Prague. This display came at the same time that large demonstrations were taking place; the protesters were criticizing the government and Parliament for their lack of vision and were trying to rekindle Czech citizens' zeal for grassroots reform. The intriguing transformations in these four unique countries are far from over.

The European Mini-States

Europe has many countries that are small by North American standards. Belgium, the Netherlands, Luxembourg, Malta, and Slovenia, for example, are all similar in size to small U.S. states such as Rhode Island or New Jersey, though they tend to have much higher population densities.

There is a whole tier of European countries even smaller than these, however; they are popularly called *mini-states.* This term is distinct from *micro-state,* which is usually used to refer to new countries formed during the breakup of European empires in the Pacific in the mid-1990s. Micro-states are often just small islands. Mini-states, by contrast, tend to have deep historical roots (as internationally recognized countries, not just as societies), are usually landlocked and closely tied to a neighboring "host" or "guardian" country, have one or more major cities, and have considerably more banking and industrial concerns than micro-states. The European mini-states are Andorra, Liechtenstein, Monaco, San Marino, and Vatican City.

HISTORICAL ORIGINS

Andorra

The largest of Europe's mini-states in terms of both territory and population, Andorra is situated high in the Pyrenees Mountains, between France and Spain. Traditionally, Andorran society and culture have been strongly influenced by the neighboring province of Catalonia in Spain, which has its own culture and language (Catalán). Spanish and French political and economic influences have also been great.

Because of its remote mountainous location, Andorra was in a kind of political limbo until 1993, when it was fully recognized as a sovereign state. Since 1278, however, it has had its own local government, overseen by both a French count (nowadays the French president) and a Spanish bishop; these two officeholders have technically been considered the "co-princes" of Andorra. Executive, legislative, and judicial power was in the hands of their representatives for centuries.

In 1933, when democracies were failing all across Europe, an adventurer named de Skossyreff tried unsuccessfully to take over Andorra. The strange episode was ended by the intervention of France and the short-lived Spanish Republic. Andorra was subsequently neutral during World War II.

In 1966, a series of reforms was begun in which a kind of parliament was created and qualifications for citizenship were defined. Initially, individual political parties were outlawed, and women were denied the right to vote and stand for office. Just who counts as an Andorran citizen has remained a controversial issue, but the original, restrictive definitions have been eased. Women were granted suffrage in 1970. Andorra is the least developed, least democratic, and least well defined of the European mini-states.

ANDORRA
(Principality of Andorra)

GEOGRAPHY
Area in Square Miles (Kilometers): 170 (468) (2.5 times the size of Washington, D.C.)
Capital (Population): Andorra la Vella (22,000)

PEOPLE
Total: 66,850
Annual Growth Rate: 1.22%
Rural/Urban Population Ratio: 5/95
Major Languages: Catalán; French; Castilian
Ethnic Makeup: 43% Spanish; 33% Andorran; 11% Portuguese; 7% French; 6% others
Religion: Roman Catholic
Adult Literacy Rate: 100%

GOVERNMENT
Type: parliamentary democracy
Independence Date: 1278
Head of State/Government: representatives of France and Spain
Political Parties: Liberal Party of Andorra; Liberal Union; New Democracy; others
Suffrage: universal at 18

ECONOMY
Currency ($ U.S. Equivalent): the French franc and Spanish peseta are used
Per Capita Income/GDP: $18,000/$1.2 billion
Inflation Rate: 1.6%
Unemployment Rate: 0%

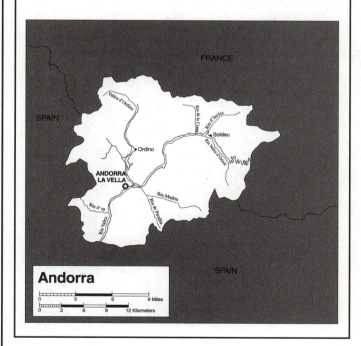

Liechtenstein

Liechtenstein is a small alpine country sandwiched between Austria and Switzerland. It is unique in that it is the only surviving monarchy among the German-speaking countries. Its history dates back to 1342, when Count Hartmann III became the ruler of the area around the city of Vaduz. Liechtenstein grew to its current size over the next century,

LIECHTENSTEIN (Principality of Liechtenstein)

GEOGRAPHY

Area in Square Miles (Kilometers): 62 (160) (nearly the size of Washington, D.C.)
Capital (Population): Vaduz (5,100)

PEOPLE

Total: 32,200
Annual Growth Rate: 1.02%
Rural/Urban Population Ratio: 79/21
Major Languages: German; Alemannic dialect
Ethnic Makeup: 88% Alemannic; 12% Italian, Turkish, and others
Religions: 80% Roman Catholic; 7% Protestant; 13% others and unknown

Adult Literacy Rate: 100%

GOVERNMENT

Type: hereditary constitutional monarchy
Independence Date: January 23, 1719
Head of State/Government: Prince Hans-Adam II; Prime Minister Mario Frick
Political Parties: Fatherland Union; Progressive Citizen's Party; The Free List
Suffrage: universal at 20

ECONOMY

Currency ($ U.S. Equivalent): 1.79 Swiss francs = $1
Per Capita Income/GDP: $23,000/$730 million
Inflation Rate: 0.5%
Unemployment Rate: 1.8%

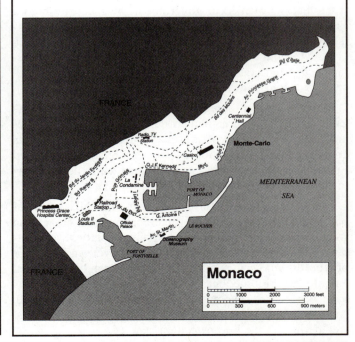

Liechtenstein

SWITZERLAND · Ruggell · Schaan · VADUZ · AUSTRIA · Samina · SWITZERLAND · Balzers

and the current royal family assumed control in 1712. After Napoleon's defeat in 1815, Liechtenstein joined the short-lived Germanic Confederation. In 1852, it entered a customs union with Austria, which lasted until the collapse of the Habsburg Empire in 1918. By 1921, Liechtenstein had found a new partner and patron in neighboring Switzerland. During World War II, both of these states were neutral, and they

continue to coordinate their economic and foreign policies closely to this day.

Monaco

Located on the Mediterranean, surrounded on three sides by southeastern France, the Principality of Monaco dates from the Middle Ages. It has been ruled by the Grimaldi family

MONACO (Principality of Monaco)

GEOGRAPHY

Area in Square Miles (Kilometers): 0.7 (1.9) (about 3 times the size of the Mall in Washington, D.C.)
Capital District: Monaco-Ville

PEOPLE

Total: 31,700
Annual Growth Rate: 0.4%
Rural/Urban Population Ratio: 0/100
Major Languages: French; English; Italian; Monégasque
Ethnic Makeup: 47% French; 16% Monégasque; 16% Italian; 21% others

Religions: 95% Roman Catholic; 5% other or nonaffiliated
Adult Literacy Rate: 99%

GOVERNMENT

Type: constitutional monarchy
Independence Date: November 19, 1419
Head of State/Government: Prince Rainier III; Minister of State Michel Leveque
Political Parties: National and Democratic Union
Suffrage: universal at 21

ECONOMY

Currency ($ U.S. Equivalent): 7.75 French francs = $1
Per Capita Income/GDP: $27,000/$870 million
Unemployment Rate: 3.1%

Monaco

FRANCE · Bd d'Italie · Av. Princesse Grace · Centennial Hall · Radio, TV Station · Monte-Carlo · Casino · Bd des Moulins · O.J.F. Kennedy Blvd · La Condamine · PORT OF MONACO · MEDITERRANEAN SEA · Bd du Jardin Exotique · Rue Grimaldi · Bd Rainier III · Princess Grace Hospital Center · Railroad Station · Av. du Port · Louis II Stadium · Official Palace · G. Antoine I · LE ROCHER · Av. St. Martin · Oceanography Museum · PORT OF FONTVIEILLE · FRANCE

0 1000 2000 3000 feet
0 300 600 900 meters

since the 1200s. Its independence, although more or less tolerated by the French for centuries, was formally guaranteed in 1861. After World War I, France renewed its guarantee but imposed several major restrictions on Monaco's sovereignty. The Grimaldis also agree that, should their dynastic line die out, France would have the right to annex the principality.

In 1911, Prince Albert approved Monaco's first Constitution. One of his successors, Prince Rainier, has ruled since 1949. Famous to outsiders for his marriage to the glamorous American actress Grace Kelly, Rainier also granted a new Constitution in 1962, which enshrined universal suffrage and the right for workers to organize in unions; it also abolished the death penalty and increased the power of the popularly elected National Council.

San Marino

San Marino claims to be the world's oldest republic. While the American and French Revolutions of the late 1700s established republics (indirect democracies), the powerful noble classes of Poland and England in the Middle Ages can be viewed as proto-democratic, because they resisted absolutist kings and fought to maintain a more decentralized political system. But San Marino, as well as the Netherlands and Switzerland, represent earlier examples of consolidated European democracies. Tradition has it that San Marino was founded in the fourth century A.D. by Christians (perhaps from Dalmatia) escaping the persecution of Roman emperors; in 1600, at any rate, the mountainous community drew up its first Constitution. Until the creation of the modern Italian state in the 1860s, the Italian peninsula was home to a confusing welter of city-states, principalities, and kingdoms; San Marino is the only one of these to have survived.

The Sammarinese, as residents are called, signed a treaty of friendship and cooperation with Italy in 1862. Despite revisions and a new Constitution, this treaty is the cornerstone of relations between the two countries. In 1939, San Marino joined a customs union with Italy and adopted the Italian lira as its own currency, in exchange for an annual subsidy from Rome. During World War II, Italian prime minister Benito Mussolini left the country unmolested, despite the fact that its mountains were used as hideouts by antifascist partisans.

Since World War II, San Marino has typically had coalition governments. The Grand Council, or Parliament, has 60 seats. For two substantial periods the Communist Party has been the leading power in coalition governments, making San Marino one of the few countries in the world to have had elected Communist participation in government at the *national* level. (Democratically elected Communists have played bigger roles in regional and local governments around the globe.) The other parties in the Council range from various types of Socialists on the left to Christian Democrats on the right.

SAN MARINO
(Republic of San Marino)

GEOGRAPHY
Area in Square Miles (Kilometers): 24 (60) (about 1/3 the size of Washington, D.C.)
Capital (Population): San Marino (4,600)

PEOPLE
Total: 27,000
Annual Growth Rate: 1.4%
Rural/Urban Population Ratio: 5/95
Major Language: Italian
Ethnic Makeup: 88% Sammarinese; 12% Italian (however, ethnic differences are virtually nonexistent)
Religion: nearly 100% Roman Catholic

Adult Literacy Rate: 96%

GOVERNMENT
Type: independent republic
Independence Date: 1600 (A.D. 301 by tradition)
Head of State/Government: co-regents are appointed every 6 months
Political Parties: Christian Democratic Party; Democratic Progressive Party; San Marino Socialist Party; Democratic Movement; Popular Alliance; Communist Refoundation
Suffrage: universal at 18

ECONOMY
Currency ($ U.S. Equivalent): 2,289 Italian lire = $1
Per Capita Income/GDP: $20,000/$500 million
Inflation Rate: 2%
Unemployment Rate: 3.6%

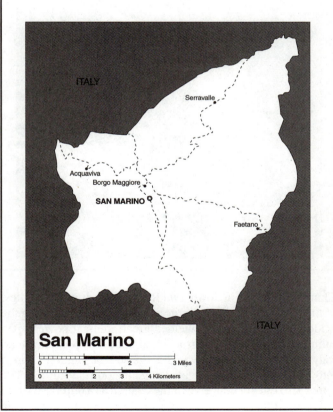

San Marino

Vatican City

Vatican City, also known as the Holy See because of its religious function as the headquarters of the Roman Catholic

THE HOLY SEE
(State of the Vatican City)

GEOGRAPHY
Area in Square Miles (Kilometers): 0.17 (0.44) (about 0.7 time the size of the Mall in Washington, D.C.)
Capital (Population): Vatican City (880)

PEOPLE
Total: 880
Annual Growth Rate: 1.15%
Rural/Urban Population Ratio: 0/100
Major Languages: Italian; Latin; others
Ethnic Makeup: 85% Italian; 15% Swiss and others

Religion: 100% Roman Catholic
Adult Literacy Rate: 100%

GOVERNMENT
Type: monarchical-sacerdotal state
Head of State/Government: Pope John Paul II; Secretary of State Cardinal Angelo Sodano
Political Parties: none
Suffrage: limited to cardinals under age 80

ECONOMY
Currency ($ U.S. Equivalent): 2,289 Vatican lire = $1 (at par with the Italian lira, which circulates freely)

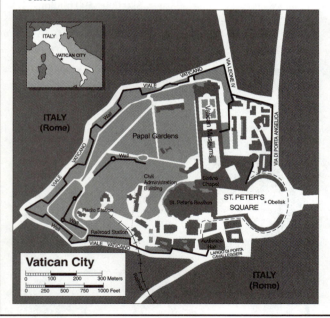

Vatican City

Church, is located in downtown Rome, Italy. It is the smallest of the mini-states in both territory and population. During the Middle Ages, the popes wielded great political power, even ruling large parts of Italy (the Papal States) as secular rulers. Italy's political situation was chaotic, and the Church stepped into the vacuum and provided organization and order. The popes also did not conceive of the Church as a purely spiritual force and they made great use of their connections to the Roman emperors and, later, the Holy Roman Empire.

In the eighteenth century, the Papal States began to shrink, as the power of secular Italian kingdoms grew and Italy moved toward unification. The Church resented the national-ism of the Risorgimento, as the movement to create a common (and more or less democratic) Italian government was called. By 1929, when Mussolini offered the pope the Lateran Treaties, less than one square mile of the Papal States remained. That area, around St. Peter's Basilica, is the modern-day Vatican City. These treaties, and a new concordat with Italy in 1985, made the Holy See a sovereign country, capable of receiving diplomats and sending out its own emissaries, called *nuncios,* who often play important roles in diplomatic negotiations and human-rights issues.

POLITICS AND GOVERNMENT

All of these states, except Vatican City, have some sort of parliament. In Monaco, Andorra, and Liechtenstein, the parliaments share power with a regent of some sort—usually a prince—while San Marino is a full-blown republic. These parliaments have a variety of political parties ranging from Social Democrats (Communists in San Marino) on the left to Christian Democrats and populists on the right. In Vatican City, the pope is the secular ruler.

The mini-states all have very close relationships with one or more of their large neighbors (the host or guardian states). Vatican City and San Marino are the most independent, although they have very close economic ties to Italy. Liechtenstein's international interests are represented by Switzerland, and Monaco agreed in 1918 to coordinate its foreign policy with that of France. Andorra's defense is provided for by both France and Spain, while its foreign policy is steered from Paris.

Of the five states, only Vatican City and Andorra maintain armies. The tiny Andorran military is mostly on show for tourists; while equally picturesque, the Swiss Guards in the Vatican do carry out real security functions. Since the 1981 assassination attempt on Pope John Paul II, Vatican officials have been even more security conscious. The Swiss Guards were very much in the public eye in 1998: For the first time, an officer of nonaristocratic lineage, Alois Estermann, was in line to become commander of the Guard, which was suffering from morale and recruitment problems. Then, this colonel and his wife were murdered in their Vatican apartment by a disgruntled guardsman, who then committed suicide. These were the first murders in recent memory in the Vatican.

The year 2000 was a big one for tourism and politics in the Vatican, since the Roman Catholic Church celebrated the year as "Jubilee 2000." Hundreds of thousands more tourists than usual jammed Rome for a variety of celebrations; the pope also toured widely and made many public pronouncements, including some startling admissions about the Church's mistaken policies and intolerance earlier in its history, such as during the Nazi era. These were designed to improve relations between Catholics and other religious groups around the world, notably Jews. Some of the preparations for the Jubilee involved the construction of a large underground parking deck on a site of significant archeological importance; over protests of Green Party activists and many others, the governments of Vatican City and Italy built the lot anyway.

Politics in Monaco is seldom big news, since the country is so closely aligned with France. But the very closeness of that relationship is beginning to worry Prince Rainier, who is 73 and whose only son, Albert, has no children. Women cannot occupy the throne in Monaco, so Rainier's two daughters cannot succeed him. As noted earlier, if the royal line should die out, the country will become part of France; the principality's residents, with their strong sense of Monégasque identity, are dismayed by this prospect.

The biggest news in all of the mini-states came from Liechtenstein, which has long been known as a tax haven for companies and individuals. Journalists and intelligence officials around Europe began claiming in 1999 that Liechtenstein's banks were laundering money for the Italian mafia, Russian gangs, and Latin American drug lords. Investigations soon showed that the country's justice system was very ineffective at detecting and preventing such crime, and a bank owned by Prince Hans-Adam II was implicated. Important international groups like the G-7 and the Organization for Economic Cooperation and Development put Liechtenstein (along with Panama, Lebanon, and the Bahamas) on their lists of trouble spots for financial crime. This move embarrassed the government into banning anonymous bank accounts in 2000.

Other developments in Liechtenstein included the adoption in 1998 of the *Rechtschreibreform,* or "Orthographic Reform," for German-speaking countries. This controversial package of new grammatical and spelling rules was also put into use in all schools and government offices in Germany, Switzerland, and Austria, although some newspapers, publishers, and writers have continued to protest against it.

THE ECONOMIES

The economies of all of the mini-states are fairly strong. Tourism plays an important role in each of them, especially in Andorra (shopping and skiing), Monaco (gambling and museums), and Vatican City (religious sites and museums). All of them also issue postage stamps, highly sought after by philatelists around the world because of their relative scarcity. Monaco's residents live tax-free, which means that many wealthy people choose to live there; but only Andorra and Liechtenstein are tax havens, so thousands of companies have resettled there to avoid corporate income taxes. As the European Union grows more comprehensive, one can expect the mini-states to come under pressure to rein in "offshore bank-ing" and to limit available tax breaks, since they all have special arrangements with the European Union or with individual member states.

Prince Rainier has deliberately diversified Monaco's economy, so that light manufacturing is now an important source of revenue; San Marino and Liechtenstein also have some industry. Andorra and Liechtenstein have small agricultural sectors as well. None of the mini-states has a fully independent currency. Monaco uses the French franc; Andorra uses the French franc and the Spanish peseta; and Liechtenstein uses the Swiss franc. Vatican City uses both the Italian lira and its own, equivalent lira.

Recently, the populations of Liechtenstein and Andorra have grown considerably, due to an influx of wealthy businesses from all over the world and of "guest workers" from poorer countries. Both kinds of immigration can create social tensions, usually centering on questions of identity and assimilation. Sometimes long-time residents also worry, as in the rest of Europe, that immigration will cause a rise in the crime rate, but so far in the mini-states, such fears have been unfounded. San Marino, Monaco, and Vatican City are experiencing little population growth.

CONCLUSION

Europe's mini-states have come a long way since the British writer Somerset Maugham said disparagingly of Monaco that it was a few hundred sunny acres populated by a lot of very shady characters. Gone also is the sense of uncomfortable vagueness about the mini-states that in 1961 prompted Max Frisch, a Swiss author, to entitle one of his controversial, soul-searching plays *Andorra.* One also hears much less today about the lack of "economic viability" of small countries, at least in Europe, where banking, high-technology industries, and tourism can all flourish when political borders are open.

Indeed, the success of the European Union might encourage some regions in existing countries to break away and form new mini-states—although most world leaders recognize that that would set a dangerous precedent for the rest of the world, where secessionist movements are often accompanied by great turmoil and violence. Far from being just pleasant stopping points for tourists and enhancements to the continent's business climate, Europe's mini-states are an important reminder that there was history before modern mass society and the nation-state; they bear intriguing witness to earlier forms of social organization.

Austria (Republic of Austria)

GEOGRAPHY

Area in Square Miles (Kilometers):
32,369 (83,835) (about the size
of Maine)
Capital (Population): Vienna
(1,540,000)
Environmental Concerns: soil and
air pollution
Geographical Features: mostly
mountains (Alps) in the west
and south; mostly flat or gently
sloping along the northern and
eastern margins; landlocked
Climate: temperate

PEOPLE

Population
Total: 8,131,200
Annual Growth Rate: 0.25%
Rural/Urban Population Ratio: 36/64
Major Language: German
Ethnic Makeup: predominantly
German
Religions: 78% Roman Catholic;
5% Protestant; 17% unaffiliated
or other

Health
Life Expectancy at Birth: 75 years
(male); 81 years (female)
Infant Mortality Rate (Ratio):
4.5/1,000
Physicians Available (Ratio): 1/289

Education
Adult Literacy Rate: 98%
Compulsory (Ages): 6–15; free

COMMUNICATION

Telephones: 4,000,000 main lines
Daily Newspaper Circulation: 465
per 1,000 people
Televisions: 336 per 1,000 people
Internet Service Providers: 35
(1999)

TRANSPORTATION

Highways in Miles (Kilometers): 124,000
(200,000)
Railroads in Miles (Kilometers): 3,796
(6,123)
Usable Airfields: 55
Motor Vehicles in Use: 4,105,000

GOVERNMENT

Type: federal republic
Independence Date: 1156 from Bavaria;
on May 15, 1955, Austria recovered
its sovereignty and independence
from occupation forces
Head of State/Government: President
Thomas Klestil; Chancellor Wolfgang
Schüssel

Political Parties: Social Democratic Party
(previously the Socialist Party);
Austrian People's Party; Freedom Party
of Austria; Communist Party; the
Greens; Liberal Forum
Suffrage: universal at 19; compulsory for
presidential elections

MILITARY

Military Expenditures (% of GDP): 1.2%
Current Disputes: none

ECONOMY

Currency ($ U.S. Equivalent): 16.26
schillings = $1
Per Capita Income/GDP: $23,400/$190.6
billion
GDP Growth Rate: 2%

Inflation Rate: 0.5%
Unemployment Rate: 4.4%
Labor Force: 3,700,000
Natural Resources: iron ore; petroleum;
timber; magnesite; coal; lignite; lead;
copper; hydropower
Agriculture: livestock; forest products;
cereals; potatoes; sugar beets
Industry: foods; machinery; vehicles and
parts; chemicals; lumber; paper and
pulp; tourism; communications
equipment
Exports: $62.9 billion (primary partners
EU, Switzerland, Hungary)
Imports: $69.9 billion (primary partners
EU, United States, Hungary)

 http://www.cia.gov/cia/publications/
factbook/geos/au.html

THE REPUBLIC OF AUSTRIA

Once the nucleus of a vast multinational governmental entity that ruled most of Eastern and Southeastern Europe, this small Alpine country assumed its current format when the Austro–Hungarian Empire collapsed at the end of World War I. In 1918, the Republic of Austria was established; two years later, a democratic Constitution, introducing a federal form of government, came into effect. However, as in neighboring Germany, the transition from an absolutist monarchy to a democratic republic proved extremely difficult in Austria. Loss of territory and the attendant economic disorientation caused by the war and its inflationary aftermath thoroughly destabilized the political and economic systems of the new Austria.

During the Great Depression of the 1930s, socioeconomic and political divisions within Austrian society continued, and the fragile political consensus sustaining the country's democratic order fell apart. In 1933, faced with an impending civil war at home and a growing threat to its national sovereignty from Nazi Germany, the Austrian government abandoned its democratic Constitution and reverted to authoritarian rule, in a futile attempt to stabilize the country and protect its independence.

In the following year, civil war did indeed erupt between the country's Lager subcultures. The term *Lager,* meaning "camp," is often used by students of Austrian politics to highlight the confrontational characteristics that the two major political subcultures—that is, Socialists and conservative Catholics—display. In the midst of worsening political violence, an extreme right-wing group of Austrian National Socialists took the maverick Chancellor Engelbert Dollfuss prisoner and murdered him in an aborted coup.

In March 1938, Austria succumbed to the threats and pressures on the part of Germany. Adolf Hitler, who had been an Austrian himself until six weeks before he was sworn in as chancellor of Germany in 1933, sent the Third Reich's armed forces into the country. The assorted Nazis (the Austrian ones as well as those from across the border in Germany) had actively prepared for this occasion, and the German troops and Hitler were hailed by much of the citizenry as they entered Vienna and other large cities. Records have revealed that Hitler initially intended to terminate the turbulence and to mold Austria into a National Socialist (Nazi) satellite. Soon, however, the appeal of *Anschluss* ("annexation") won the day, and for seven years Austria was submerged in the Third Reich as its *Ostmark* ("eastern province").

Before and during the war, there appeared to be little or no resistance among most Austrians to being part of Germany. On the contrary, Austria eagerly subscribed to all of Hitler's designs, including the persecution and murder of Jews in what would become known as the Holocaust.

THE POSTWAR YEARS

It was only natural that once World War II finally came to an end, the all-important question loomed: Was Austria an early *victim* of Nazi expansionism or a willing *belligerent,* and thus an ally of Germany? There was no consensus on this point among the Allied powers, which divided the country into occupation zones just like they had done in Germany. Nevertheless, greater leniency was applied to Austria than to Germany, in that the former was allowed its first postwar election as early as November 1945. This general election revolved around the two major parties of right and left—the Austrian People's Party (ÖVP) and the Socialist Party of Austria (SPÖ). These two blocs, which hailed from the pre-Anschluss days, fostered dogmas and philosophies that were highly antithetical to each other. Nevertheless, a grand-coalition government was forged.

It is hard to overstate the role that the Marshall Plan played in Western Europe, particularly in vanquished countries such as Germany and Austria. This large-scale injection of financial aid into Austria's economy greatly promoted the country's reconstruction and political stability.

The political situation was somewhat anomalous, in that Austria was encouraged to engage in the politics of independence while remaining occupied by the Allied powers. (Thus, "independence" implied only separation from Germany.) A breakthrough finally came in early 1955, when the Soviet Union announced that it was prepared to terminate the occupation of its zone, on the condition that the United States, Great Britain, and France would do likewise with their zones. Another condition was that the new Austria would sign a treaty that would commit it to "neutrality in perpetuity."

It is not easy to fathom the Soviet considerations behind this plan. Linkage has been suggested: West Germany was on the verge of joining the North Atlantic Treaty Organization, and the prospect of reunification of the two German zones might persuade it to follow the Austrian example of neutrality (which it did not).

In 1955, the Allied powers withdrew their armies from Austria; signed the Austrian State Treaty, which formally acknowledged Austrian independence; and made Austria commit itself to neutrality in per-

petuity. That commitment would prove not a little demanding to Austria, which had a capitalist economic system and close economic relations with its Western neighbors. When at one point the question arose whether or not Austria might become a member of the European Union (at that time known as the European Economic Community), the Soviet Union immediately vetoed any such plan, on the grounds that Austrian membership would constitute a breach of the neutrality clause in the Austrian State Treaty.

Consensus Politics

During the more than 20 consecutive years that the Socialist Party/People's Party coalition constituted the Austrian government, the gap was bridged that had long separated the country's Socialist and Catholic subcultures. A new, durable political and economic consensus came to be institutionalized. By the time the coalition fell apart, in 1966, the legitimacy of the country's democratic institutions was no longer in question. The single-party rule, initiated in that year through the election of a People's Party government, headed by Chancellor Joseph Klaus, provoked a great deal of confrontation during the campaign. As soon as the election fever subsided, rhetoric and hyperbole were replaced by consultation, if not cooperation, on the part of the ÖVP and the SPÖ.

In 1970, the Socialist Party scored gains in Parliament, which enabled it to defeat the Klaus government and form a minority government of its own, under the leadership of Bruno Kreisky. Minority governments are as a rule precarious and ephemeral, but Kreisky's government fared well; indeed, it gained an absolute majority in Parliament that the party would enjoy for the next 13 years. During the years of his chancellorship, Kreisky became a major spokesperson of the democratic left in Europe.

Political Conflicts

Kreisky's international stature notwithstanding, Austria's domestic problems accumulated. By the early 1980s, the government was no longer able to control the country's trade and budget deficits. In addition, an increasingly powerful environmental movement had started to sap the government party's strength. In 1978, environmentalists and leftist university students had mobilized sufficient popular support to bring about a national referendum in which a majority voted against the operation of the new Zwentendorf nuclear-power plant. (Austria has remained a nuclear-free zone ever since.) Two small Green parties, a moderate one and a left–alternative one, emerged to contest the 1983 elections. (In 1986, they allied and won eight seats in the lower house of Parliament.)

In the 1983 general elections, the Socialists were not defeated, but they lost their absolute majority in Parliament. Party leaders were reluctant to form another minority government and thus entered into a coalition government with the small Freedom Party (FPÖ). The FPÖ was formed in 1955 essentially out of the "Association of Independents," which had fought for the rehabilitation of former Nazis. The new party thus became the representative of the historical (smaller) German-Nationalist camp in Austria. However, it included both nationalist and liberal (in U.S. terms, economic conservatives) wings. Kreisky refused to lead the new coalition and stepped down. He was succeeded as chancellor by his Socialist vice-chancellor, Fred Sinowatz. Major disagreements on economic policy surfaced between the two coalition partners, and a series of political scandals made matters worse.

The situation was compounded by the controversy over presidential candidate Kurt Waldheim's Nazi ties during World War II. The allegations against Waldheim made headlines all over the world, as he had recently served as secretary-general of the United Nations. The presidential election left the country rife with unusually bitter conflict, which cut along party divisions. Waldheim won, but Sinowatz refused to serve under him. The SPÖ–FPÖ coalition disintegrated in 1986. In the ensuing general elections, Franz Vranitzky, a former minister of finance, led the SPÖ to victory. But since the party held a slim majority in Parliament, the Socialist government realized that it lacked sufficient votes to rule by itself. Accordingly, in 1987, Vranitzky formed a new coalition government with the Austrian People's Party.

In 1990, new elections were held. Against all expectations, the Socialist Party consolidated itself, gaining 43 percent of the vote. However, this victory reintroduced polarization, since the right-wing Freedom Party also made gains. In 1986, Jörg Haider had taken over leadership of the FPÖ from (liberal) Norbert Steger. Haider steered the small party on a course of right-wing populism, which brought the party 16.6 percent of the parliamentary votes in 1990 (up from 9.7 percent in 1986). In the 1994 and 1995 elections, the FPÖ became electorally the most successful far-right party in Europe, with around 22 percent of the national vote. Its gains came despite the departure of the party's popular 1992 presidential candidate Heide

(Photo courtesy of Joan Martin)

Austria has kept up with the evolving world economy, but many of its customs reflect its long history.

Schmidt and liberal leaders, who reconstituted themselves as the Liberal Forum (which gained, respectively, 6.0 percent and 5.5 percent of the votes in the 1994 and 1995 elections). In 1995, Vranitzky retained the chancellorship. (The Socialists were by then called the Social Democrats.) He remained at the head of the grand coalition until January 1997, when he resigned and handed over power to his finance minister, Viktor Klima. Klima endeavored to reinvent the SPÖ somewhat in the mold of Tony Blair's "New Labour" in Britain. In 1998, he benefited from a prospering Austrian economy, a successful six months' stint as the president of the European Council, and a series of scandals affecting Haider's party.

Nevertheless, the 1999 parliamentary elections saw the FPÖ support climb to 26.9 percent of the votes, allowing it (narrowly) to supplant the ÖVP as the number-two party. Not only the ÖVP but also the SPÖ (which won 33.2 percent) suffered their worst ever electoral results. The FPÖ ended up with more votes from blue-collar workers than did the left-of-center SPÖ. The Liberal Forum failed to clear the 4 percent parliamentary threshold, and Heide Schmidt resigned as its leader. The Greens won 7.4 percent of the votes, their best ever in national parliamentary elections, under new party leader Alexander Van der Bellen, a professor of economics. However, even if the inclination were there, the mathematics of parliamentary seats precluded the option of "red–green" government like Germany's.

Why did Haider's FPÖ do so well? International journalists have often dwelt on Haider's past pro-Nazi statements (for which he has since apologized). Even Haider's leading critics have emphasized that the majority of FPÖ voters are *not* right-wing extremists. Haider's charisma and skills in staging media events have set him apart from other Austrian politicians. He has had the knack of articulating the fears of the "common people" in terms that they understand. In contrast to the SPÖ and the ÖVP, Haider's party has been a critic of the European Union—at a time when many Austrians were having second thoughts about the costs and benefits of their country's EU membership. The FPÖ has long campaigned against the entrenched system of party patronage and resulting corruption associated with the governing parties. Indeed, polls indicated that 60 percent of the FPÖ voters cited the "anti-political class" rationale for their support, while the party's anti-immigrant stand was mentioned by 47 percent. During the campaign, FPÖ slogans had warned against *Überfremdung* ("too many foreigners"). (With 9 percent of the population, Austria ranks second highest among EU countries in terms of residents with foreign passports.) Finally, Haider's aggressive style and blunt words, regardless of what is in his party's electoral program, have continued to attract various malcontents on the right-wing of Austrian politics.

Coalition negotiations were complex and lengthy, with President Thomas Klestil playing an active role. The SPÖ refused to consider a coalition with the FPÖ. Efforts to revive the grand coalition failed. Efforts by the SPÖ to form a minority government also failed, which led to Klima's resignation as party leader. (In April 2000, 40-year-old Alfred Gusenbauer became the SPÖ's youngest leader ever.) In the face of international warnings and massive street demonstrations, the ÖVP and the FPÖ in February 2000 formed a government with former foreign minister Wolfgang Schüssel (ÖVP) as chancellor and equal numbers of cabinet ministers from the two parties. President Klestil prevailed upon Schüssel and

Haider to sign a joint declaration in support of European integration, human rights, democracy, and Austria's coming to terms with its Nazi era. For tactical reasons, Haider chose to stay outside of the government as state governor in Carinthia, and later resigned as national FPÖ leader (his successor was a loyal lieutenant). These moves did not stave off international sanctions.

FOUNDATIONS OF AUSTRIA'S POSTWAR SUCCESSES

In 1945, Austria resurrected the constitutional framework that it had forged immediately following World War I, which provided for a federal democratic republic made up of nine states (*Länder*), including the capital area of Vienna. Although the political constellation is formally structured as a federation, the division of powers has been such that the national government is by far the stronger. That government follows the traditional parliamentary model, consisting of a cabinet headed by a chancellor, a bicameral Legislature, and a president who functions as the head of state.

The Austrian president is popularly elected (in contrast to most presidents in parliamentary systems) for a six-year term. And the president has the constitutional basis to act as more than a ceremonial leader: He or she can intervene in the coalition-formation process and can, on his or her own authority, dissolve the National Council (the popularly elected lower house of Parliament) and dismiss the chancellor and cabinet.

The chancellor is appointed by the president from the dominant party within the powerful house, the National Council. However, the chancellor must maintain the support of a majority within the National Council in order to govern. The upper house, the Federal Council, representing the Länder, has somewhat limited powers, although it may review and delay legislation passed by the National Council. However, the National Council can overrule the Federal Council by a simple majority vote. The National Council elects three "People's Lawyers" (equivalents of the Nordic ombudsman), to whom any person can turn in case of administrative maltreatment. The Austrian Constitution includes instruments of direct democracy as well: popular initiatives to petition the Parliament for action and also for advisory and binding referendums. The power of the Federal Constitutional Court to declare legislation unconstitutional provides a check on the national and state parliaments.

During the 1920s and 1930s, the parliamentary framework came to be supplemented by corporatist arrangements that brought government officials together with representatives of major economic interest groups in an effort to enlist the latter's cooperation in meeting the political and economic crises of the period. Yet neither the parliamentary system nor its corporatist appendage was capable of containing the extreme social and political tensions that polarized Austrian society and wrecked the fledgling republic.

During the interwar period, the forces of the left (the Socialist Party and the trade unions) were effectively disenfranchised by the more powerful groups of the center and far right (the Christian Socialist Party, the Nationalists, and agricultural and industrial interests). The Socialist Party was outlawed in the 1930s. Austrian trade unions were poorly organized and incapable of exercising real power. The adoption of national socialism, as well as the Anschluss and its dreadful aftermath, discredited the right, which had generally supported the incorporation of Austria into the Third Reich. Conversely, the moral legitimacy and political power of the Socialist Party and of organized labor came out of the war greatly strengthened. Thus, the two major political forces—the Socialist and People's Parties and their affiliated interest groups—had become more evenly balanced.

After World War II, Austria's political and economic institutions were remarkably successful in cultivating political stability and economic prosperity. The reasons for this are complex. To be sure, Hitler's invasion and the Anschluss, the war, and the subsequent Allied occupation had been object lessons that clearly revealed to a majority of Austrians the high costs of domestic divisiveness. Furthermore, the immediate goal of restoring their own national sovereignty and independence, not to mention the formidable task of economic reconstruction, demanded a truce between the Läger subcultures. Over the long run, however, two changes wrought by the war turned out to be pillars of domestic tranquillity: the redistribution of power between the country's two major political camps, and the emergence of a new consensus regarding the role of the state in society and the economy.

Interestingly, this parity in strength produced a new consensus concerning the nature of political competition and the relationship between the state, society, and the economy, a consensus clearly reflected in the two-party, grand-coalition government and in its institutional counterpart, proportional democracy. The term *proportional democracy* refers to a system that allows public jobs or appointments from the cabinet level on down to be distributed among representatives of the various political parties or in proportion to their respective strengths in the legislature. The practice, which has permeated all aspects of Austrian public life as well as much of Austrian business and finance, survived the transition to one-party rule in 1966 and characterized subsequent grand coalitions.

More recently, political analysts have discussed Austrian politics in terms of consociationalism. Political scientist Arend Lijphart argued that Austria (like the Netherlands from 1917 to 1967) has become a *consociational democracy,* a system specifically designed to reduce conflict and promote compromise in highly fragmented societies. In a consociational democracy, the various segments of society—in the case of Austria, the Läger—are segregated through vertical organization. The actual wheeling and dealing of politics is conducted at the elite level. Thus, consociational democracy confines conflict to a less conspicuous place. Out of the glare of the media, the segments of society are bound together. For this "cartel of elites" to operate effectively, these segments must be able to cooperate and compromise without losing the support of their subcultures.

Another point of agreement was the need to create a mixed economy with a large public sector and a generous social-welfare state. Not surprisingly, the new political–economic trends were most strongly recommended by the Socialist Party and organized labor. However, the People's Party and its affiliated interest groups also accepted the greatly increased role of the state in the economy, in return for certain concessions. These included state protection and state subsidies for Austrian industry and agriculture as well as noninterference on the part of the state in the general management of state-owned enterprises.

The major organized interest groups also have some input into the policy-making process, and their role is fairly structured. First, there is Chamber government, a system of quasi-governmental institutions that represent the interests of various groups at the national and state levels. The Chambers are consulted by the government as well as the major political parties on policy questions affecting Chamber interests. A second institution is the Joint Commission, an ongoing incomes-policy committee in which representatives from the government, organized labor, and business meet to establish annual wage and price guidelines in accordance with projections of economic growth. Thus the Austrian model reveals a merger of consociationalism and functional representation.

THE SOCIAL PARTNERSHIP

Within the context of the rapidly expanding world economy of the 1950s and 1960s, Austria's social partnership yielded impressive results. By finding new outlets for Austrian exports, the postwar liberalization of world trade accelerated the modernization of Austria's economy. The country's manufacturing sector expanded vigorously as human and capital resources migrated from agriculture to industry.

The Austrian government played an important role in finding new outlets for Austrian exporters. Austria was a charter member of the European Free Trade Association, founded in 1960. It also started to cultivate a special commercial relationship with the European Union.

By the 1960s, in addition to its more traditional exports of timber and minerals, Austria had started to export a wide range of capital and consumer goods, including iron and steel, electrical and transportation equipment, chemicals, and textiles. By 1970, Austria exported nearly 20 percent of its gross national product and employed 30 percent of its workforce in manufacturing.

As long as world demand for Austrian exports and the productivity of the country's industry continued to grow at a reasonable pace, the Austrian economy had little trouble generating the surpluses needed to sustain standard-of-living increases for its citizens, the comprehensive social-security net, and the subsidies for less competitive domestic industries. Low inflation; a large, modern industrial base; a skilled indigenous workforce as well as an abundant supply of less skilled "guest workers"—all seemed to conspire to an indefinite continuation of the twin phenomena of economic prosperity and political stability.

Negative forces began to intrude, however. There were scandals connected with the Roman Catholic Church; it appeared that seminarians had been sexually abused by some of the nation's highest Church dignitaries. Another important issue was unemployment, which haunted Europe in the early and mid-1970s. It became clear that the subsidized and troubled industries could not stem its rise. The late 1970s brought a brief reprieve, but the fundamental problems facing the Austrian economy were suppressed only temporarily.

It would take the second oil shock of 1979 and the subsequent world recession to reveal clearly the full significance of the 1970s to the world economy and Austria's political–economic equilibrium. The events of the 1970s marked a watershed in the world economic order. Oil prices went down in the 1980s and were volatile in the 1990s, but world trade continued to grow slowly. A new generation of low-cost competitors emerged in the newly industrialized countries of Asia and Latin America. And most advanced industrialized countries, the General Agreement on Tariffs and Trade notwithstanding, sought to defend their domestic markets with new forms of protectionism.

Austria could do little to change its external economic environment. At home, however, the government began to explore various less expensive ways of recasting the old social partnership. After assuming power in 1986, the coalition government headed by the Socialist banker and former finance minister Franz Vranitzky cut state spending in an effort to reduce budget deficits. The government also embarked on an ambitious plan to restructure state-owned industries that have been a chronic drain on the public treasury during the past decade. Vranitzky's limited austerity program did not revolutionize or destroy the ongoing social partnership, but it marked the beginning of a gradual retrenchment, the trend to reduce the state's heavy obligations to the Austrian society and economy. The number and severity of commercial insolvencies reached unprecedented heights in 1995. Vranitzky resigned early in 1997.

Under Vranitzky's successor, Chancellor Viktor Klima, the Austrian economy started to grow faster than the EU average. Inflation was low, exports were booming, and personal incomes went up significantly in the late 1990s. Klima had to cope with distractions created by the FPÖ's anti-immigration agitation, and a banking scandal involving price fixing and sleaze. Allegations of political cronyism resulted in surprise visits by EU investigators to seven Austrian banks, in search of misbehavior.

The new Schüssel government announced in February 2000 a program of "renewal," which "safeguards prosperity, opens up chances for the future and guarantees social security." The government encouraged the ongoing privatization of state-owned banks and companies. It advanced reforms for the health-care system to contain costs but preserve benefits; similarly, pension reform was embraced. With Austria's budget deficit in 2000 the highest in the European Union in relative terms, Finance Minister Karl-Heinz Grasser (FPÖ), pursued major tax increases and spending cuts. Subsequently, his party saw its support slip in late 2000 in two state elections. At the end of 2000, Austria's unemployment rate was 3.2 percent (the third lowest level in the EU), and its GDP growth rate was 3.8 percent (above the average of the Euro Zone).

FOREIGN RELATIONS

Shortly after the ratification of the Austrian State Treaty of 1955, both houses of Parliament endorsed a constitutional amendment that committed the nation to permanent armed neutrality. Austria's defense forces are severely limited by the Four-Power State Treaty. Its army of approximately 55,000 ground troops is largely made up of draftees, who serve a compulsory six-month term and from then on may be called up for military exercises or emergencies. Austria is forbidden from possessing nuclear, chemical, and other weapons of mass destruction.

Austrian troops have participated in UN peacekeeping missions through the years. In 1995, Austria joined NATO's Partnership for Peace, and its relations with NATO have deepened in recent years. The SPÖ and the Greens have opposed Austrian membership in NATO. Although the ÖVP and the FPÖ have differed significantly in their views on the European Union (the latter party being Euro-skeptical), both parties of the new governing coalition tend to hold a positive view on joining NATO.

It has been neutral in the *military* respect, but Austria, unlike its neighbor Switzerland, has neither declared nor practiced *political* neutrality. The distinction may seem a fine one, but this has meant that Austria is not "non-aligned." In general, it sides with the West. It is not only a member of the United Nations but also of the Council of Europe and several Western-oriented international economic organizations. These include the Organization for European Cooperation and Development, the International Monetary Fund, the World Bank, and (until 1995) the European Free Trade Association. Numerous international agencies have their headquarters in Vienna, including, most notably, the International Atomic Energy Agency.

Because of its geographical position and an active commitment to promoting global peace and understanding, Austria has been able to play a role in international politics far out of proportion to its size. The country, particularly Vienna, has often been a meeting ground for East–West conferences.

The Conference for Security and Cooperation in Europe, formed by NATO and Warsaw Pact members in 1972 and re-created as the Organization for Security and Cooperation in Europe in 1995, has its

| The birth of Austria, resulting from the collapse of the Austro-Hungarian Empire **1918** | Austria in turmoil: civil war; Chancellor Engelbert Dollfuss is murdered **1934** | The *Anschluss:* Hitler's Germany annexes Austria, which becomes *Ostmark* **1938** | Austria is divided into four occupation zones; otherwise independent **1945** | Occupation forces withdraw; Austria signs the Austrian State Treaty **1955** | The question of Austria's role in the Nazi cause resurfaces during Kurt Waldheim's successful bid to become president **1980s** | The government cuts state spending to reduce budget deficits | Austria becomes an EU member; banking scandals; Austria joins the launch of the Euro **1990s** |

2000s

The EU imposes bilateral sanctions on Austria, but most Austrians still favor membership

Austria seeks to provide compensation for Nazi-era wrongs

secretariat and permanent council based in Vienna. And in 1981, Chancellor Bruno Kreisky, dedicated to the cause of international socialism, planned the North–South Conference at Cancún, Mexico. On the other hand, Kreisky also alienated the Jewish community and the State of Israel by selling arms to moderate Arab states in the Middle East and by supporting the Palestinian Liberation Organization's calls for a Palestinian homeland. It is also noteworthy that the Organization of Petroleum Exporting Countries (OPEC), which includes powerful Middle East oil producers, has chosen Vienna for the world oil cartel's headquarters.

In the last few years, Austria has also been forced to confront its role in the Holocaust and to respond to pleas for reparations and return of assets of Jews, including money placed in Austrian banks for safekeeping before and during the war, and artworks looted during that period. Austerity in budgeting did not hamper efforts to set up compensation funds for Nazi-era wrongs, such as slave labor and property seizures. In fact, all the parties represented in Parliament gave their approval in mid-2000. Although negotiations (as of early 2001) were not complete, the total sum offered by Austria was almost $1 billion. Leading members of the FPÖ have challenged neighboring countries to come to terms with the dark side of their post–World War II history, most notably the arbitrary expulsion of 2.5 million ethnic Germans by Czech authorities.

AUSTRIA AND THE EU

The collapse of communism in Central/Eastern Europe and the dissolution of the Soviet Union transformed the geopolitical landscape during 1989–1991. The Austrian government seized the opportunity to join the European Economic Area, which linked the economies of EFTA countries to the European Union. In June 1994, 66.6 percent of Austrian voters supported their SPÖ–ÖVP government's application for membership in the European Union; this was a higher level of support than shown by the electorates of Finland (57 percent), Sweden (52.2 percent), and Norway (47.8 percent). Austria, along with Finland and Sweden, came into the European Union in 1995. Ever since, the office of European commissioner for agriculture—an important one, given the fact that the EU's agricultural policy consumes the largest portion of its budget—has been ably occupied by former Austrian agricultural minister Franz Fischler. Environmentalists' concerns about the impact of expanding north–south road traffic in sensitive Alpine areas resulted in a special treaty allowing restrictions until 2003. Austrian governments have endeavored to raise EU environmental-pollution standards and have expressed special concerns about the safety of the nuclear-power plants of its eastern neighbors, now EU candidate members. In January 1999, Austria became one of the eleven EU members that joined the Euro Zone. Despite the anti–EU efforts of Haider, more Austrians than Swedes or Finns indicated in an October 1999 survey that they felt that their country has benefited from EU membership.

The year 2000 saw EU sanctions against the new government that included Haider's FPÖ. The 14 other EU member states agreed to curtail bilateral relations with the Austrian government, to decline to support Austrians for positions in international organizations, and to reduce contacts with Austrian embassies to the technical level. With far-right parties a major concern in their own backyards, French and Belgian leaders were the most vigorous in their criticisms of Austria. A message had been sent that the European Union stood for more than just economic values, but some of the smaller members worried about the precedent for being pushed around when the bigger members disagreed with the outcomes of democratic elections. The Austrian government threatened to hold a referendum to obtain popular backing for resistance to the European Union; the implication was that Austria might consider obstructing EU decision making on pressing issues. After seven months, EU (and U.S.) diplomatic sanctions were lifted. An independent report by three "wise men" (including the former Finnish president) had urged "particular vigilance" over the FPÖ's role, but had concluded that the sanctions "if continued would become counterproductive." In any case, there was no evidence of flaws in the new government's human-rights record, and, in fact, the Parliament had amended the Constitution to extend the special rights of ethnic minorities.

In late 2000, various anti–EU groups began a campaign to bring about a referendum on the withdrawal of Austria from the European Union. None of the parties represented in Parliament, including the FPÖ, endorsed it. The Schüssel government's program has favored "deepened integration." For example, it has supported the development of a European Security and Defense Union. The government has favored the timely enlargement of the European Union to include the candidate countries of Central/Eastern Europe, and has recognized the need for institutional reforms to contend with an enlarged Union. In little more than a decade, Austria has moved from being a Western outpost in a divided Europe to being at the center of a uniting Europe.

DEVELOPMENT

Approximately half of Austria's workers are involved in industrial jobs. Austrians are working hard to retain their footing in trade as the hub between Western and Central/Eastern Europe.

FREEDOM

Austria has a highly favorable human-rights rating. A special government appointee monitors women's rights. It extended the special rights of ethnic-minority communities in 2000.

HEALTH/WELFARE

Austria has a comprehensive welfare system at this time, but the current trend is to reduce the state's role in the Austrian economy and social-welfare sector. Still, social benefits remain generous.

ACHIEVEMENTS

Austrian theorists have reached great heights in economics (Austrian School of Economics). Vienna is internationally regarded as an important cultural capital. Several important international organizations are based in Vienna.

Belgium (Kingdom of Belgium)

GEOGRAPHY

Area in Square Miles (Kilometers): 11,781 (30,513) (about the size of Maryland)

Capital (Population): Brussels (948,000)

Environmental Concerns: water pollution; industrial air pollution

Geographical Features: flat coastal plains in the northwest; central rolling hills; rugged mountains of the Ardennes Forest in the southeast

Climate: temperate

PEOPLE

Population

Total: 10,242,000

Annual Growth Rate: 0.18%

Rural/Urban Population Ratio: 3/97

Major Languages: Flemish (Dutch); Walloon (French); German

Ethnic Makeup: 58% Flemish; 31% Walloon; 11% mixed or others

Religions: 75% Roman Catholic; 25% Protestant, others, or no affiliation

Health

Life Expectancy at Birth: 74 years (male); 81 years (female)

Infant Mortality Rate (Ratio): 4.7/1,000

Physicians Available (Ratio): 1/264

Education

Adult Literacy Rate: 98%

Compulsory (Ages): 6–18

COMMUNICATION

Telephones: 5,073,000 main lines

Daily Newspaper Circulation: 304 per 1,000 people

Televisions: 464 per 1,000 people

Internet Service Providers: 51 (1999)

TRANSPORTATION

Highways in Miles (Kilometers): 90,427 (145,850)

Railroads in Miles (Kilometers): 2,131 (3,437)

Usable Airfields: 42

Motor Vehicles in Use: 4,830,000

GOVERNMENT

Type: federal parliamentary democracy under a constitutional monarch

Independence Date: October 4, 1830 (from the Netherlands)

Head of State/Government: King Albert II; Prime Minister Guy Verhofstadt

Political Parties: Christian People's Party; Social Christian Party; Flemish Socialist Party; Francophone Socialist Party; Flemish Liberal Democrats; Francophone Democratic Front; Francophone Liberal Reformation Party; National Front; AGLAV (Flemish Greens) and ECOLO (Walloon Greens); others

Suffrage: universal and compulsory at 18

MILITARY

Military Expenditures (% of GDP): 1.2%

Current Disputes: none

ECONOMY

Currency ($ U.S. Equivalent): 47.69 Belgian francs = $1

Per Capita Income/GDP: $23,900/$243.4 billion

GDP Growth Rate: 1.8%

Inflation Rate: 1%

Unemployment Rate: 9%

Labor Force: 4,340,000

Natural Resources: coal; natural gas

Agriculture: livestock products; grains; flax; vegetables; fruits; tobacco

Industry: engineering and metal products; food and beverages; chemicals; basic metals; textiles; glass; petroleum; motor-vehicle assembly; coal

Exports: $187.3 billion (primary partner EU)

Imports: $172.8 billion (primary partner EU)

http://pespmc1.vub.ac.be/BELGCUL.html
http://www.odci.gov/cia/publications/factbook/geos/be.html

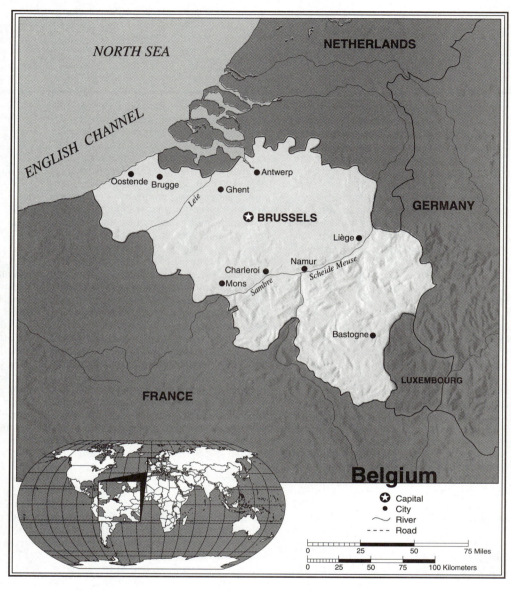

NORTH SEA

NETHERLANDS

ENGLISH CHANNEL

Oostende • Brugge • • Ghent • Antwerp

Leie

★ BRUSSELS

GERMANY

Liège •

Charleroi • Namur • Scheide Meuse

• Mons Sambre

Bastogne •

FRANCE

LUXEMBOURG

Belgium

★ Capital
• City
～ River
- - - Road

| 0 | 25 | 50 | 75 Miles |
| 0 | 25 | 50 | 75 | 100 Kilometers |

THE KINGDOM OF BELGIUM

In geographic terms, Belgium is one of the smaller countries in Europe. Apart from the Ardennes, a low-altitude mountain range that covers the southeastern part of the country, the land is fairly flat. However, Belgium, unlike its northern neighbor, the Netherlands, is not significantly below sea level. (Both are often jointly referred to as the "Low Countries," along with Luxembourg.) Belgium is located on the crossroads of more powerful European nations, and it has for that reason been nicknamed the "cockpit of Europe." Tragically, it has served as a battlefield for many European wars.

To the west, Belgium's 60-mile coastline adjoins the North Sea. The seacoast, however, does not afford a large port; the only important Belgian port is Antwerp, which is located not on the coast but, rather, at the mouth of the river Scheldt. There are few harbors for fishing boats. Contrary to the Netherlands, Belgium never developed into a seafaring nation.

Flemish, spoken in the western and northern parts of the country, is close to Dutch. Indeed, there is a tendency among Flemish people nowadays to call their language Dutch rather than Flemish. Walloon, spoken in eastern and southern Belgium, differs considerably less from French than Flemish does from Dutch. German, although spoken by only a tiny minority in southeastern Belgium, has been declared one of the country's official languages. Thus, Belgium has become officially trilingual.

Belgium has a temperate climate. Its summers are mild and its winters not too severe. There is a great deal of precipitation. The rivers Scheldt and Meuse are major commercial waterways. The Scheldt leads to Antwerp, which is the second-largest port in Europe.

The country is populous, certainly, considering its territorial size. In fact, its population density is one of the highest in the world.

Belgian industry traditionally was found largely in the central and eastern parts of the country; the western and northwestern parts were rural and agricultural. This contrast started to change after World War II, when industries were established west of Brussels. Currently the only distinction that can be made is that the central and eastern parts harbor older industries, some of which have become obsolete and consequently less competitive.

ETHNIC DIVERSITY

Belgium is a divided country—so much so that Julien Destrée, a nineteenth-century author, exclaimed, "There is no Belgian nation!" Of the 10.2 million citizens, 58 percent are Flemish. Walloons (the French-speaking, or Francophone, Belgian people) constitute 31 percent. Other groups make up the remaining 11 percent. Over the course of history, Belgium's main battles were fought between the Flemish and the Walloons. In numerical terms, the Walloons were traditionally a minority, but what they lacked in numbers was made up for by the prestige their language carried. Since French was a world language and Dutch was not, the Walloons believed themselves to be superior.

A rural exodus, however, had a devastating effect on the Flemish–Walloon ratio. Flemish people tired of life in the countryside left for the cities, learned French, and forgot their native language within a generation. This pattern changed after World War II. With the establishment of industries in Flanders, there was no need to seek fame and fortune in cities that were Francophone. The Flemish mentality also changed: It became less submissive, less willing to accept the presumed inferiority of Flanders.

In 1971, a new Constitution rendered the country quasi-federal. If Belgium had been officially bilingual, it now became *regionally* monolingual—that is, communities were created in which only French or Flemish was spoken. Communal problems continued to exist, however, and for many years Belgian authorities worked toward further constitutional reform, which was accomplished with the new Constitution in 1993. This time, a real federation was created. There are three autonomous regions: Flanders, Wallonia, and Brussels, where both cultures meet and mingle. The late King Baudouin was scrupulously bilingual in his monarchical duties.

HISTORY

Julius Caesar, a Roman general who subsequently became emperor, made mention of Belgians (whom he deemed very valorous) in his famous book *De Bello Gallico* ("On the War in Gaul"). But although Belgians may have existed for many centuries, they did not achieve nationhood until the 1830s, after they seceded from the Netherlands.

The area that is now Belgium (as well as Holland, historically often referred to as the Northern Netherlands) was once ruled by the Spanish King Charles V, who resided in Brussels. Abdicating in 1555, he left the throne to his son Philip, who preferred to rule from Madrid. The Spanish environment conditioned the new king, who was considerably less tolerant than his father had been. The Reformation was then rife in Western Europe. In the Northern Netherlands, this implied religious warfare (the fight for the new Protestant creed), which invigorated the struggle for political independence. The war was fought along political and religious lines. While it took the Northern Netherlands 80

(Courtesy Belgian Tourist Office)

The beauty of Belgium is exemplified by this photograph of a canal in Bruges.

| Belgium is under Spanish rule A.D. 1555–1713 | Austrian rule 1713–1795 | Annexation by France 1795 | The Treaty of Vienna adds Belgium to the Netherlands 1815 | The start of the Belgian Revolution 1830 | The Treaty of the 18 Articles regulates the separation of Belgium and the Netherlands 1831 | Belgium is invaded by German troops on their way to France 1940 |

years (from 1568 until 1648) to achieve independence, the southern part, remaining Catholic, did not rise against Spain. The Southern Netherlands (that is, Belgium) was subsequently transferred to Austrian rule; later, in Napoleonic times, it was incorporated in the French Empire. So was Holland, which by then had been a sovereign state for nearly two centuries.

Once the turmoil created by Napoleon Bonaparte had come to an end, rulers and statesmen from all over Europe convened in Vienna in order to reestablish Europe as much as possible in the way it had existed before Napoleon's conquests. Prince Clemens von Metternich, an Austrian statesman, presided. Viewing France as a monster that should be kept at bay, he believed that a strong buffer should be created in its immediate vicinity. England was on the other side of the English Channel and consequently was unable to prevent a French resurgence. Germany and Italy did not exist yet as sovereign nation-states. But if the Southern Netherlands (which thus far had never been independent) were to be added to the Northern Netherlands, a strong buffer would develop. The Northern and Southern Netherlands were thus merged by the Treaty of Vienna (1815).

The Congress of Vienna failed to take into account the reality of a centuries-long alienation: The Northern Netherlands (which had been independent) and its southern counterpart had grown apart, developing into two different cultures. The new Dutch monarch, King William I, did not make things easier: Since France had become unpopular as a result of the Napoleonic conquests, he saw fit to ban the use of French in public, despite the fact that a large number of his new subjects spoke only French. There were other grievances as well. Finally, in 1830, a revolution started. At that point, King William I had Belgium invaded by his Dutch army. The great powers, oblivious of the original rationale behind Dutch–Belgian unity, sided with the underdog—the new state of Belgium. After Belgium recruited a king as its chief of state, Belgium and the Netherlands started their separate destinies.

However, Belgium's self-determination was flawed from the beginning, since it was a bicultural state. The area of Flanders that had not been part of the Netherlands prior to the Congress of Vienna took up the Belgian cause and joined Belgian independence. The discord, which had never amounted to very much as long as Belgium had been a dependency, now began to be very pronounced. Subsequent interethnic relations in the Belgian state have long been troublesome. However, both Flemish and Walloon separatists have shared an aversion to violence.

THE MONARCHY

In spite of the fact that the revolution had been directed against the power of the Dutch king, Belgium opted for a monarchy, following the general trend in Europe. At the time, the country did have an aristocratic class, but the revolutionary leaders preferred to have the new king recruited from outside.

The Belgian monarchy is just as ceremonial and symbolic as its European counterparts; the Constitution does not grant any real powers to the king. Its royal house has not been spared personal mishaps. When the Nazi Germans were about to occupy the Low Countries in 1940, the Dutch queen fled. However, the Belgian king, Leopold III, preferred "in these dark times to remain with his people." This could have been heroic were it not for some of his wartime activities. He voluntarily paid a public visit to Adolf Hitler in Berlin. In addition, Leopold III, whose wife had died in a car accident before the war, remarried during the occupation, and, considering the austerity of the war years, the wedding was awkwardly sumptuous and costly. As if this were not enough, the father of the bride appeared to be a prominent Nazi sympathizer.

The immediate postwar years witnessed popular outrage against Leopold, who wanted to resume his royal activities after the war as if nothing had happened. The country was in an uproar. The increasing unrest finally forced the king to resort to a referendum that would decide whether or not he would abdicate. Its outcome favored the continued role of Leopold as king of the Belgians, by a small margin. However, the unrest persisted, and the king decided to abdicate in favor of his eldest son, who ruled Belgium for 42 years as King Baudouin I. Those decades were in many ways instrumental to the return of respect for the monarchy, although

Baudouin's reign, too, was not without mishaps.

Although monarchies elsewhere have also witnessed slumps in their prestige as a result of unseemly royal behavior, one cannot but conclude that the Belgian royal family has never been as popular as its counterparts in other European countries. Because Baudouin and his wife Fabiola did not have any children, the Constitution was changed to allow the throne to pass to Baudouin's younger brother Albert, or to Albert's eldest son. Since Baudoin's death in 1993, Albert has reigned, as King Albert II.

THE PARLIAMENTARY SYSTEM

In Belgium, effective executive power is in the hands of the government—that is, the prime minister and his or her cabinet ministers, who hold their positions because they are supported by a majority of the members of the lower house of Parliament, the Chamber of Representatives. Its 150 members are directly elected at least every four years. Voting is compulsory in Belgium. Elections follow the rule of proportional representation, whereby parties receive seats reflective of their share of the votes cast. As a result, several parties tend to win seats, which has made the formation of coalitions challenging and has led to fragile governments. All parties, from left to right, have their counterparts in the other linguistic community. Thus, for example, there are Flemish Socialists and Walloon Socialists. Belgium is unique in Western Europe in that it has no national parties.

The upper house of Parliament, the Senate, has 181 members, who also serve four years; some are indirectly elected, but most are directly elected. The Senate normally approves, without amendment, the laws passed by the Chamber of Representatives. Acts of Parliament cannot be declared unconstitutional by the high courts.

Despite social cleavages, Belgian elites have a tradition of compromise and cooperation. Furthermore, major interest groups are customarily consulted before the cabinet introduces legislation in Parliament.

ECONOMIC CONDITIONS

The Belgian economy has not always been sound. Socialist-led governments pro-

Belgium becomes a charter member of NATO **1949** ●	Leopold III formally abdicates **1951** ●	Belgium becomes a founding member of the European Union **1957** ●	Zaire, formerly a Belgian possession as the Belgian Congo, gains its independence **1960** ●	New Constitution **1971** ●	King Baudouin dies and is succeeded by his brother, Albert. Another new Constitution takes effect; Belgium is federalized; Belgium joins the launch of the Euro **1990s** ●

2000s

Belgium adjusts well to membership in "Europe"

Belgium holds the EU presidency for part of 2001 and pushes for reform.

moted the welfare state, which caused governmental expenditures to grow. The 1980s witnessed a degree of retrenchment, but the October 1987 U.S. stock-market crash had a devastating effect on Belgium. The Belgian economy appeared to bottom out at that point, however, and there has been some recovery since.

Today, about half of Belgium's entire production is sold abroad. The larger part of Belgian products are machine-made, mass-produced items, which can be made anywhere. Some people urge that Belgium concentrate on goods that can uniquely be identified as Belgian. Labor-intensive projects have shown themselves to be profitable in recent years, but businesses have become very apprehensive about hiring more people. Another frequent complaint is the very high level of taxation, particularly on personal income.

The Belgian economy certainly cannot be compared to that of the group of poorer (or "marginalized") countries within the European Union, but the recession of the 1980s–1990s clearly harmed it. In recent years, Belgium's unemployment rate has begun to decline. Nevertheless, in 2000, it was still about 9 percent, higher than in neighboring Luxembourg, the Netherlands, France, and Germany.

POLITICAL CONDITIONS

Ever since ethnic friction reemerged after World War II, Belgium has been preoccupied with establishing political, cultural, and linguistic units inside its body politic. Before Belgium became a completely federal state in 1993, numerous changes were made to the Constitution that caused Belgium to be divided into regions and communities—a division highly confusing to Belgians and non-Belgians alike. Furthermore, there are 10 provinces and 589 communes (local governments that are, in effect, municipalities). With the exception of Brussels, there are hardly any bilingual

communes. Nevertheless, some cities within the Dutch-speaking community have enclaves of French-speakers. Antwerp provides an example: It is overwhelmingly Dutch-speaking, but, in a certain area of the city, there are still some French-speakers who have refused to assimilate.

The government of Prime Minister Jean-Luc Dehaene, elected in 1992, was soundly defeated in the June 1999 parliamentary elections. Dehaene had been a popular figure, viewed as the "plumber" who fixes problems, both within Belgium and outside (he had been a strong contender to replace Jacques Delors as European Commission president, until vetoed by the British government). He oversaw the transition of Belgium to a fully federal state and trimmed the budget deficit so that Belgium could qualify for membership in the European Monetary Union (Euro zone).

However, Dehaene's coalition government was scarred by scandals, most notably the serial murders of children and charges of police incompetence in 1996, and bribes accepted by a cabinet minister from arms dealers in 1998. The fatal blow was the government's inept handling of food-safety issues prior to the 1999 elections: Dioxin had found its way into Belgian foodstuffs, which were subsequently banned as exports by the European Union. Christian Democrats found themselves out of the government for the first time in 41 years. Socialists also suffered losses. The Liberals were the big winners, along with the Greens. Flemish Liberal leader Guy Verhofstadt, an admirer of British prime minister Tony Blair's "Third Way," formed a coalition with his Walloon Liberal counterparts, the two Socialist parties, and the two Green parties (which entered Belgian national government for the first time).

One disturbing feature of the election results was the gains enjoyed by the Flem-

ish nationalist party Vlaams Blok, or VB (its Walloon counterparts did not do so well). The VB now holds seats at all levels of Belgium's complex federal system. Electorally, the VB is the strongest party in the major city of Antwerp; but it is not in power, because no other parties will form coalitions with it. The party's program is pro-independence for Flanders, anticrime, and anti-immigration. (A 1998 survey indicated that racist feelings tend to be higher in Belgium than in other EU countries.)

Belgium is a charter member of both the North Atlantic Treaty Organization and the European Union. In fact, Brussels houses the headquarters of both of these important regional organizations. Some of Belgium's prime ministers have been instrumental in the promotion of supranationalism in Europe. In 2000, Belgian foreign minister Louis Michel was out front in the EU campaign to sanction Austria due to the entry of the far-right, xenophobic Freedom Party into the governing coalition in Vienna. Polls indicate that clear majorities of Belgians see their small country as being advantaged by the process of European integration. National identity and pride have been dwindling, while individual attachments to locality, language, and Europe have grown. Nevertheless, despite its weak institutional structure, Belgium's demise as a (bicultural) state does not seem to be just around the corner.

DEVELOPMENT

Belgium is highly developed and has entered the postindustrial phase. In the 1980s and 1990s, however, its economy experienced difficulties as a result of recession. Belgium has had one of the highest national debt/GDP ratios in the EU.

FREEDOM

Belgium ranks high as a democracy and in regard to compliance with human rights, at least in its own borders. It granted independence to the Belgian Congo (Zaire) in 1960, where King Leopold II's horrific colonial rule had in earlier decades led to the deaths of millions of people.

HEALTH/WELFARE

Belgium has an advanced social-security system and favorable health statistics, including high life expectancy rates and low birth and death rates. In 2000, Belgium ranked 7th on the UN's Human Development Index.

ACHIEVEMENTS

Belgium has an excellent educational system. It hosts many students from developing-world countries. It also benefits economically and culturally by being the headquarters of NATO and the de facto capital of the European Union.

Cyprus (Republic of Cyprus)*

GEOGRAPHY

Area in Square Miles (Kilometers):
3,571 (9,250) (about half the size of Connecticut)

Capital (Population): Nicosia (186,400)

Environmental Concerns: water scarcity; water pollution; coastal degradation; loss of wildlife habitats

Geographical Features: central plain with mountains to the north and south; scattered but significant plains along the southern coast

Climate: temperate; mediterranean

PEOPLE

Population

Total: 758,500
Annual Growth Rate: 0.6%
Rural/Urban Population Ratio: 45/55
Major Languages: Greek; Turkish; English
Ethnic Makeup: 78% Greek; 18% Turkish; 4% British, Armenian, Maronite, and others
Religions: 78% Greek Orthodox; 18% Muslim; 4% others

Health

Life Expectancy at Birth: 75 years (male); 79 years (female)
Infant Mortality Rate (Ratio): 8.0/1,000
Physicians Available (Ratio): 1/667

Education

Adult Literacy Rate: 94%
Compulsory (Ages): 5½–15; free

COMMUNICATION

Telephones: 404,700 main lines
Daily Newspaper Circulation: 135 per 1,000 people
Televisions: 160 per 1,000
Internet Service Providers: 5 (1999)

TRANSPORTATION

Highways in Miles (Kilometers): 7,659 (12,765)
Railroads in Miles (Kilometers): none
Usable Airfields: 15
Motor Vehicles in Use: 343,000

GOVERNMENT

Type: republic
Independence Date: August 16, 1960 (from United Kingdom)
Head of State/Government: President Glafcos Clerides is both head of state and head of government
Political Parties: Greek area: Progressive Party of the Working People; Democratic Rally; Democratic Party; United Democratic Union of the Center; others; Turkish sector: National Unity Party; Communal Liberation Party; Republican Turkish Party; others
Suffrage: universal at 18

MILITARY

Military Expenditures (% of GDP): 5%
Current Disputes: intense opposition among the Greek Cypriots to the Turkish occupation of 37% of the island

ECONOMY

Currency ($ U.S. Equivalent): 0.68 Cypriot pound = $1; 1,160,000 Turkish lira = $1
Per Capita Income/GDP: $15,400/$9 billion (Greek); $5,000/$820 million (Turkish)
GDP Growth Rate: 3% (Greek); 5.3% (Turkish)
Inflation Rate: 2.3% (Greek); 66% (Turkish)
Unemployment Rate: 3.3% (Greek); 6.4% (Turkish)

Labor Force: 289,400 (Greek); 80,200 (Turkish)
Natural Resources: copper; pyrites; asbestos; gypsum; timber; salt; marble; clay earth pigment
Agriculture: potatoes; vegetables; grapes; citrus; barley; olives
Industry: food; beverages; textiles; chemicals; metal products; wood products; tourism
Exports: $1.1 billion (Greek); $64 million (Turkish)
Imports: $3.5 billion (Greek); $374 million (Turkish)

*Note: There are great disparities between the Greek Cypriot area and the Turkish Cypriot area in many of these statistics.

http://www.kypros.org/Cyprus/root.html
http://www.odci.gov/cia/publications/factbook/geos/cy.html

Map

MEDITERRANEAN SEA

Kyrenia
UN Buffer Zone
Morphou Bay
Turkish Cypriot-administered area
Khrysokhou Bay
NICOSIA
Famagusta Bay
Polis
Famagusta
UN Buffer Zone
Area controlled by Cyprus Goverment (Greek area)
Larnaca
Eastern Sovereign Base Area (U.K)
Paphos
Limassol
Vasilikos
Episkopi Bay
Akrotiri Bay
Western Sovereign Base Area (U.K.)
MEDITERRANEAN SEA

Cyprus

⊗ Capital
● City
〰 River

0 10 20 30 Miles
0 10 20 30 40 Kilometers

THE REPUBLIC OF CYPRUS

The Republic of Cyprus is situated on an island in the northeastern part of the Mediterranean Sea, south of Turkey. It is the largest independent island in the Mediterranean. Cyprus lies about 50 miles south of Turkey and 260 miles east of the Greek island of Rhodes. Its anomalous position—populated in large part by Greeks but located far from the Greek archipelago—has generated tremendous geopolitical strife.

Cyprus has a long recorded history, and the island has often been a bone of contention. Greeks and Turks have fought for its possession from times immemorial. During the late Bronze Age (1600–1050 B.C.), Greek traders and settlers roaming through the entire Mediterranean area expanded Hellenic (Greek) culture into Cyprus. From 700 B.C. onward, the island was successively dominated by Assyrians, Egyptians, and Persians, until it became part of the Roman Empire in 58 B.C. When, four centuries later, the split between the West and East Roman Empires took place, Cyprus fell to the latter. For some 800 years, it remained part of Byzantium (as the East Roman Empire came to be known), during which period it was attacked frequently. Richard the Lionhearted did so during the Crusades, and the English king briefly held the island before it came under Frankish control in the late twelfth century. Cyprus became an outpost of the Venetian Republic in 1489 and finally fell into the hands of the Ottoman Turks in 1571.

The Ottomans applied the *millet* system to Cyprus—a system that incorporated the Islamic belief that "like over like is mercy." This implied that non-Muslim minorities were governed by their own religious authorities. "Like over like" did not just promote the cohesion and internal solidarity of the ethnic Greek community; it also greatly strengthened the position of the Greek Orthodox Church. In fact, the Church became important in secular matters, at least as much as Turkish rule would allow.

Late in the nineteenth century, Great Britain assumed control of the island. Cyprus was formally annexed in 1914 and became a British Crown colony in 1925. The Mediterranean had by then in effect become a British sea—Gibraltar and Suez holding the entrances, with Cyprus and Malta the strongholds in the eastern and western parts.

INDEPENDENCE

Even before the island fell to Britain, strong ethnonationalism in the Greek community had started to generate unrest. The strife only intensified under the British colonial administration. Most inhabitants of Greek stock identified so strongly with the ancestral land that they came to view *enosis*—the notion that Cyprus should be part and parcel of Greece—as the ideal solution. After a brief lull during World War II, agitation resumed.

The situation in Cyprus was very complicated. It was a colony of Britain; but on the island itself, there was a measure of *internal* colonialism, in that the living conditions of the Turkish Cypriots were clearly inferior to those of the Greek Cypriots. Indeed, the former were as a rule regarded as second-class citizens.

Lengthy conferences on the subject took place. Finally, the idea of an independent Cyprus appeared to gain strength. On August 16, 1960, sovereignty was transferred from Britain to a Cypriot government in Nicosia, the new capital. But it did not take long for new communal violence to erupt, at which point both the United States and the United Nations became involved. These ethnic tensions gradually affected the island's demography, as the Greek and the Turkish peoples began to live in separate enclaves. This segregation augured ill for the island, foreshadowing long-lasting ethnic friction and hostility.

THE CRISIS OF 1974

The independent Cypriot government, headed by President (and Archbishop) Michael Makarios, kept the situation under control for some years. Indeed, it made some sincere efforts to lessen the ethnic tensions. However, the government of Greece, which since 1967 had consisted of a military dictatorship (a junta of right-wing colonels), was apprehensive about Makarios, whom it viewed as left-leaning, if not clearly Communist. In July 1974, the junta suddenly announced that it favored enosis as a solution to Cypriot troubles—that is, it wanted to render the island part of Greece. In making this statement, the Greek government in effect reneged on the treaty to which it had been party shortly before Cyprus became independent. The announcement, which amounted to an outright rejection of all that Makarios stood for, triggered an immediate response on the island, where the enosis ideal had simmered for a long time: Makarios and his government were overthrown.

They were replaced by the leader of the revolt, Nicos Sampson, a reputed enosis fanatic. It was only natural, then, that the extremely apprehensive Turkish Cypriot community reacted in turn. Turkey, no longer feeling bound by the treaties it had concluded when Cyprus was about to become independent, launched what it called the "Attila Peace Operation," landing troops on the island. Heavy fighting followed.

The foreign ministers of the guarantor powers (Greece, Turkey, and Great Britain) met in late July 1974. The following month, Turkey renewed its offensive, and by the time hostilities ended, it appeared in control of more than one third of Cyprus. The dissociation of the Greek and Turkish communities now became more pronounced. It seemed like an "iron curtain" descended on the island, in effect severing whatever communication the two groups had been able to maintain.

The northern part (where the Turkish troops had landed and operated) enjoyed a degree of self-rule, if only by default. Turkey also started to transmigrate relatively large numbers of Turks from the mainland into northern Cyprus. A large proportion of these "immigrants" hailed from Anatolia and similarly undeveloped areas of Turkey. This deliberate effort to change the island's demographic composition was viewed with the utmost suspicion by Greek Cypriots; even their Turkish counterparts did not favor this addition to their numbers.

On June 8, 1975, Turkish Cypriots voted overwhelmingly to allow a separate Turkish Cypriot state to emerge. Incumbents were elected for newly established executive and legislative offices. Some 200,000 Greek inhabitants of the Turkish-held area were forcibly removed, and a prolonged impasse followed. The confrontation escalated when, on November 15, 1983, northern Cyprus made an effort to secede. Turkish Cypriot leader Rauf Denktash issued a unilateral declaration of independence and proclaimed the new "Turkish Republic of Northern Cyprus." Thus far, Turkey has been the only country to recognize it as a distinct entity.

THE CHASM ENDURES

The two parts of Cyprus are separated by a buffer zone patrolled by the United Nations Force in Cyprus (UNFICYP). There is little movement of people and essentially no movement of goods or services between the two parts of the island. In 1988, the two leaders of the estranged areas on Cyprus, President George Vassiliou and Rauf Denktash, met without any preconditions. Turkish Cypriot leaders previously had refused to negotiate before a variety of conditions had been met, whereas the Greek Cypriots were willing to negotiate only once the Turkish troops had left the island.

In meetings in November 1988 with UN secretary-general Javier Pérez de Cuellar, the first round of intercommunal talks (begun in September) was critically reviewed. President Vassiliou indicated that a "wide gap" continued to separate the two sides. However, the two Cypriot leaders agreed to hold a second round of talks under the sponsorship of the UN secretary-general.

Hellenic culture comes to Cyprus 1600–1050 B.C.	Cyprus is incorporated into the Roman Empire 58 B.C.	Ottoman Turk influence begins A.D. 1571	Cyprus is conquered by Great Britain 1875	Cyprus becomes a British Crown colony 1925	Independence 1960	Turkey invades northern Cyprus 1974	Preliminary talks are held between the leaders of the two parts of Cyprus 1988	The European Union pressures Turkey and Cyprus to resolve their political and ethnic problems 1990s

2000s

Glafcos Clerides retains the presidency of the Republic of Cyprus

Rauf Denktash remains the leader of the Turkish Cypriots

This round was to focus on an integrated comprehensive approach to a solution. Vassiliou emphasized that agreement on three basic freedoms—movement, settlement, and property ownership—would be fundamental to any solution. He pointed out that Europe, to which Turkey aspired to belong, had based the whole of its post–World War II policy on those principles.

There is a strong possibility that the nation-state of Cyprus, however small in area as well as in populace, may ultimately receive a federal format, the two subnational entities being ruled by Greek Cypriot and Turkish Cypriot governments, respectively. But even a federation is not a guarantee of peaceful coexistence. And in any case, to date the deep conflict between the Greek Cypriots and Turkish Cypriots remains unresolved.

Generally, Greek Cypriots continue to resent that hundreds of their compatriots are still missing, or at least unaccounted for. (The Cypriot government has submitted a list that identifies 1,619 such persons.) The forced population movements in 1974 and 1975 appear to have been attended by terrorism and, on occasion, massacres.

In the mid-1980s, a new concern emerged among the Greek Cypriots. It came to light that the Turkish occupation authorities (civilian as well as military) had resorted to looting centuries-old monasteries and churches under their control. The Cyprus government and the Church repeatedly protested to UNESCO, the World Council of Churches, and other international bodies against the destruction and smuggling abroad of ancient and religious treasures by the Turks.

Apart from the pillage and plunder, the illegal occupation by Turkey has caused international organizations such as the European Union to be highly critical. In November 1993, the EU urged Turkey "to use all its influence to make a contribution." Several weeks later, the EU and Cyprus began talks meant to lead to negotiations for the island's accession to the Union. The European Union may thus ultimately hold the key to the solution of the Cyprus issue. There can be little doubt that the EU will never consider Cypriot membership before the island's political and ethnic divisions have been resolved.

THE PRESENT

There are democratic governments in both sections of Cyprus. However, discrimination and violence against women are serious concerns, and there are also occasional cases of police brutality. Glafcos Clerides of the conservative Democratic Rally Party was elected president of the Republic of Cyprus in 1993; he won a close reelection bid in 1998. The president in Cyprus is both the head of state and head of government; the office of vice-president, which is reserved for a Turkish Cypriot, is currently vacant. In 1995 and 2000, Turkish Cypriots reelected Rauf Denktash as their leader.

The "Green Line," as the border between the Greek and Turkish communities is called, is one of the most tense borders in the entire world. Nearly every year there are incidents in which military forces exchange fire or insults; sometimes civilians, often protesters, are killed there, as in 1996 when a Greek tried to pull down the Turkish flag from a border post and was shot to death. NATO can put pressure on Athens and Ankara—which have other persistent disputes—to try to pressure the local Cypriot Greeks and Turks. As mentioned above, the European Union also holds some leverage over the island as a whole. But a dispute over Greek Cypriot weapons purchases in 1997 proved too fiery even for veteran American special envoy Richard Holbrooke (the architect of the 1995 Dayton Peace Accords, which ended the war in Bosnia) to mediate. President Clerides threatened to purchase surface-to-surface missiles from Russia, and Turks in Cyprus and Ankara reacted with anger. (Turkey still has about 30,000 troops on the island.) Then Clerides said that he would acquire anti-aircraft weapons; after another uproar, he stated that he planned to deploy the weapons on the nearby Greek island of Crete instead of on Cyprus, a move that would have had repercussions for NATO. Finally, in late 1998, the weapons purchases were scrapped. But UN talks aimed at promoting more cooperation between the two sides continued into 2000, with little progress to show.

Two other international controversies continue to trouble the island's reputation as well. Greek Cypriot businesspeople and politicians have been accused of helping the former Milošević regime in Serbia launder money and avoid UN sanctions, while Turkey has been charged with promoting the immigration of large numbers of Anatolian farmers to Cyprus to change the ethnic composition of the island.

There remain profound disparities in the standards of living in the Greek and Turkish areas of Cyprus. Both are free-market economies (with significant administrative controls), but the Turkish zone is basically dependent on subsidies from Turkey; agriculture is important, and there is only a small tourism base. By contrast, the Greek zone has a robust, service-oriented economy, highly dependent upon tourism and trade. As long as the deep political chasm between Greek and Turkish Cypriots endures, it is unlikely that the differences in quality of life for people in the two areas will diminish significantly.

DEVELOPMENT

Cyprus is working with representatives of the World Bank and International Monetary Fund to address problems related to its foreign debt and other economic matters. In 1997, Turkey signed a $250 million economic cooperation accord with the Turkish Cypriot area to support education, tourism, and industry.

FREEDOM

Freedom has become restricted as a result of the internal hostilities. President Glafkos Clerides is working to reunite Cyprus within a federal system of government. There are some human-rights concerns, especially regarding treatment of women.

HEALTH/WELFARE

Infant mortality is low and adult literacy is very high in Cyprus. In 1988, smoking was banned in all public places. The Turkish Cypriot economy has about 1/3 the per capita GDP of the Greek Cypriot economy.

ACHIEVEMENTS

Cyprus was represented at the Seoul Olympics by a 9-member team that included participants in track and field, sailing, judo, and shooting events.

Denmark (Kingdom of Denmark)

GEOGRAPHY
Area in Square Miles (Kilometers):
16,638 (43,076) (about twice the
size of Massachusetts)
Capital (Population): Copenhagen
(633,000)
Environmental Concerns: air
pollution; North Sea pollution;
drinking- and surface-water
pollution
Geographical Features: low and
flat to gently rolling plains
Climate: temperate

PEOPLE

Population
Total: 5,336,000
Annual Growth Rate: 0.31%
Rural/Urban Population Ratio: 15/85
Major Language: Danish
Ethnic Makeup: Scandinavian;
Inuit (Eskimo); Faeroese; German
Religions: 97% Evangelical
Lutheran; 3% other

Health
Life Expectancy at Birth: 73 years
(male); 79 years (female)
Infant Mortality Rate (Ratio):
5.1/1,000
Physicians Available (Ratio): 1/358

Education
Adult Literacy Rate: 100%
Compulsory (Ages): 7–15

COMMUNICATION
Telephones: 3,465,000 main lines
Daily Newspaper Circulation: 308
per 1,000 people
Televisions: 516 per 1,000 people
Internet Service Providers: 12 (1999)

TRANSPORTATION
Highways in Miles (Kilometers):
42,960 (71,600)
Railroads in Miles (Kilometers):
1,773 (2,838)
Usable Airfields: 118
Motor Vehicles in Use: 2,097,000

GOVERNMENT
Type: constitutional monarchy
Independence Date: became a
constitutional monarchy in 1849
Head of State/Government: Queen
Margrethe II; Prime Minister Poul
Nyrup Rasmussen
Political Parties: Social Democratic Party;
Conservative Party; Liberal Party;
Socialist People's Party; Progress Party;
Center Democratic Party; Social Liberty
Party; Unity Party; Danish People's
Party
Suffrage: universal at 18

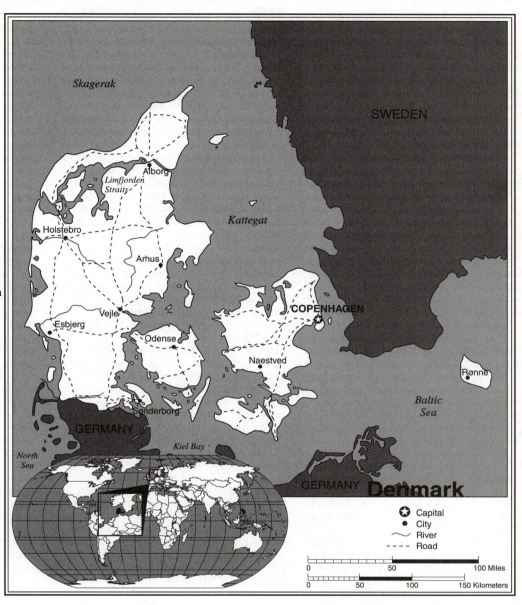

MILITARY
Military Expenditures (% of GDP): 1.7%
Current Disputes: Rockall continental-
shelf dispute with Iceland, Ireland,
and the United Kingdom

ECONOMY
Currency ($ U.S. Equivalent): 8.81 kroner = $1
Per Capita Income/GDP: $23,800/$127.7
billion
GDP Growth Rate: 2.3%
Inflation Rate: 2.5%
Unemployment Rate: 5.4%
Labor Force: 2,896,000
Natural Resources: petroleum; natural gas;
fish; salt; limestone; stone, gravel, and
sand

Agriculture: highly intensive; specializes
in dairying and animal husbandry;
grains; root crops; fish
Industry: food processing; machinery and
equipment; textiles and clothing; chemical products; electronics; construction;
furniture and other wood products;
shipbuilding
Exports: $49.5 billion (primary partners
EU, Norway, United States)
Imports: $43.9 billion (primary partners
EU, Norway, United States)

 http://www.cia.gov/cia/publications/
factbook/geos/da.html

A BLEND OF PAST AND PRESENT

Practically every person who visits Copenhagen will see the "Little Mermaid," a bronze figurine immortalized by Hans Christian Andersen's fairy tale. She sits on a rock in the northern part of Copenhagen's harbor. But on the coastline surrounding the harbor, the fairy tale yields to today's realities, exemplified by the massive construction hall and cranes of the Burmeister and Wain shipyard.

These contrasts of a past glittering with royal splendor and a more down-to-earth, democratic present, of a modern industrial society with agrarian roots, are pervasive in Denmark. Tivoli, Copenhagen's famed amusement park, was carved out of what were once the city defense works; the lake there used to be part of the city's moat. The *Folketing* (the Danish Parliament) meets in the Christiansborg Palace, the magnificent royal residence for four centuries. Thus, past and present offer a fascinating balance in the Kingdom of Denmark.

HISTORY

Denmark was once a great world power. In Viking times—that is, in the ninth and tenth centuries A.D.—a Danish kingdom extended to the east of England. This was known as *Danelaw*. The Danish king, Canute, and his son ruled as kings of England from 1017 to 1050.

Dynastic union with Norway and Sweden in the late fourteenth century, forged by the so-called Union of Kalmar, brought all of Scandinavia under Danish rule, including Finland as well as the North Atlantic islands of the Faeroes, Iceland, and Greenland. The Swedes ultimately revolted; there were adjustments in the balance of power that resulted in a military and diplomatic rivalry between Denmark–Norway and Sweden–Finland that lasted until the early nineteenth century. The Swedes wrested Skane (Norwegian Sweden) from the Danes and annexed Norway in 1814. Defeat at the hands of the Prussians and Austrians half a century later cost Denmark the duchies of Schleswig (in Danish, Slesvig) and Holstein. The Danish recovery of northern Schleswig (now called Southern Jutland) after World War I gave the country its current European borders. In the Caribbean, the Virgin Islands held by Denmark were sold to the United States during World War I. Americans feared that these islands might be used as a refuge for German submarines.

Iceland, another Danish colony, appeared to be moving toward full independence when World War II broke out. In April 1940, the Nazi German Third Reich conquered Denmark and Norway.

At this point, Great Britain became apprehensive that the Germans would consider colonies as nothing more than the spoils of war, in which case the strategically located island would also become part of the growing German empire. The British felt forced to make a preemptive move by occupying Iceland. As soon as the United States joined the British war effort (shortly after Pearl Harbor), it assumed a share of Iceland's protection.

The Faeroe Islands, despite a claim to independence (about a thousand years ago), are nominally Danish. The 17 *inhabited* islands have a total population of about 50,000; the land area of all the islands is 540 square miles. But although the Danish influence is considerable, the islands have enjoyed a great deal of autonomy for some time. They have, for instance, their own flag and their own currency, which carries Faeroe markings, which most Danes find hard to understand. Thus it could happen that, in the early 1970s, their popularly elected Assembly, the *Lagting,* decided to remain outside the European Union (then called the European Community), while Denmark acquired EU membership. The local government of the Faeroe Islands confined itself to establishing a trade agreement with the European Union some five years later. The islands, it turned out, were unhappy with that agreement. Thus the 1990s witnessed attempts to replace it with a broader free-trade treaty. The stumbling block has been EU resistance to a Faeroe demand for free export of fish to the Union. The rugged islands, which have little beyond fish, fear that their political influence in the Union will be marginal, since they would be represented by Denmark.

GREENLAND

Greenland, which has also been one of Denmark's colonies, is the world's largest island (840,000 square miles, of which more than 708,000 square miles are covered with ice. Located far northwest of Denmark, its population, now numbering approximately 60,000, has enjoyed home rule for some time, expressing its political will through an elected Assembly, the *Landsrad*. When Denmark was in the process of becoming a member of the European Union, the Greenlanders (mostly Inuits widely scattered over a vast area) had no objection to being included. Unlike the Faeroe people, they assumed that their fishing grounds would be inaccessible, or at least too remote to be vulnerable to interlopers and poachers from fellow EU members. That assumption soon proved to be erroneous, in that modern trawlers had

no difficulty in entering the rich fishing grounds close to Greenland. Soon the island's fishing industry was swamped by competitors. As soon as the Greenlanders realized their error, they started to make preparations to get out of the EU. In a referendum, a majority voted in favor of this unprecedented step, which was effected on February 1, 1985. It is the only area thus far to secede from the European Union. Ironically, Greenland remains an integral part of Denmark.

DEMOCRATIC INSTITUTIONS AND POLITICAL PARTIES

Democratic political institutions developed relatively late in Denmark. In fact, the Danish monarchy established its absolute supremacy in 1660, at a time when England was gradually embarking on the road toward a constitutional monarchy. Royal absolutism persisted in Denmark until the mid-nineteenth century, when King Frederik VII was forced to issue a liberal Constitution, which extended voting rights to all adult males. This Constitution was adopted on June 5, 1849. Although it has since been amended, its main principles still form the basis of the structure of government in Denmark. But it took another half-century before parliamentary supremacy was fully established, in 1901.

Constitutional reform in 1915 produced changes in the electoral system. Universal suffrage was granted, and from then on, nearly all adult men and women were able to participate in elections. This same reform introduced the proportional-representation system, which typically allows smaller political parties to survive. The system basically implies that each political party will be accorded the same proportion of seats in Parliament that it has secured of the total vote. To illustrate this principle, since the Folketing has 179 seats, a party that secures 10 percent of the total vote must be accorded 18 seats. However, a party must win at least 2 percent of the votes to be represented.

Constitutional reform in 1953 abolished the upper house, the *Landstinget,* which historically had been representative of the aristocracy. (Other Scandinavian countries have followed Denmark's lead, rendering their parliaments unicameral.) That same reform made female succession to the throne possible, an addition for which many Danes have been grateful since Margrethe II became queen, in 1972.

As a constitutional monarch, the queen's role is largely ceremonial. Nevertheless, due to the fragmentation of the Danish party system, she and her advisers play a nonpartisan role in the negotiations to

(UN/DPI Photo by Evan Schneider)

Poul Nyrup Rasmussen, leader of the Social Democratic Party, was appointed Denmark's prime minister in 1993 and has remained prime minister ever since. He is shown here addressing the UN General Assembly in 1997.

form a new government, especially when uncertainty exists about which combination of parties would be most acceptable to Parliament's majority. Polls indicate that the queen, as a person, and the monarchy, as an institution, enjoy the support of an overwhelming majority of Danes.

The election of 1973 has gone down in history as the "earthquake election." The number of parties represented in the Folketing was suddenly doubled, to 10. Two of the new parties were antitax parties; a third was a Christian party protesting what it perceived as a decline in society's moral standards. While these protest parties have since become less vehement and less influential, they continue to play a role in Danish party politics.

During 1982–1988, Danes referred to their government as a "four-leafed-clover government," as the coalitions comprised the Conservative Party, the Liberal Party, the Center Democratic Party, and Christian People's Party. These were not the largest parties, but they were clearly very accommodating toward one another. The Social Democratic Party (a working-class party linked to the trade-union movement) did not belong to this cluster, although it has invariably won the largest share of votes. On the other hand, its former ally, the Liberal Party, representing agrarian interests, has now become part of the nonsocialist establishment. The Liberal Party split in 1905, and the Social Democrats then allied themselves with the new Radical Liberals. That alliance lasted for some six decades. When it broke, the Social Democratic Party became odd man out again—that is, the largest party but not aligned except in ad hoc coalitions.

One political party that has lost a great deal of influence is the Danish Communist Party. It steadily lost its appeal in the period 1950 to 1990, possibly as a result of its links to the Soviet model. Its place on the extreme left of the spectrum has been filled by the more independent Socialist People's Party, founded in 1958.

Election campaigns in Denmark are short, generally lasting around six weeks. They largely revolve around television programs in which journalists put sharp questions to the candidates. The electorate also watches debates between party leaders. No television advertising is permitted. Voter turnout is high, often between 85 and 90 percent. Denmark, like most Western European countries, has an excellent registry system. No advance registration is required; all citizens are automatically registered as voters upon becoming 18. People who change residence are obliged to notify the civil registry of the municipality they leave as well as the civil registry of the municipality they enter.

Danish political parties are characterized by mass membership and considerable discipline. The Social Democratic and Liberal Parties have close ties to interest groups. Once elected, party representatives will nearly always vote their party line, on the national as well as local levels. If they do not, they may risk losing their renomination, because the party leadership controls the nominating process.

Danish parliamentary politics has tended to polarize into a two-bloc pattern since 1966. The parties on the left (the Social Democratic Party, the Socialist People's Party, and others) are squared off against the "bourgeois" or "middle-of-the-road" parties (the Conservative, Liberal, and Radical Liberal Parties), plus right-wing parties.

Denmark was governed from 1982 to early 1993 by Conservative prime minister Poul Schluter's coalition of the Conservatives, the Liberals, the Christian People's Party, and the Center Democrats, with support on domestic issues from the Radical Liberals. December 1992 witnessed the election of the center-left coalition, composed of the Social Democrats, the Center Democrats, the Radical Liberals, and the Christian People's Party, along with two of the four so-called North Atlantic members from the Faeroes and Greenland, which gained a total of 90 seats, giving them a majority of the 179 seats in the Folketing. This forced Prime Minister Schluter to resign after more than a decade in power. On January 25, 1993, Poul Nyrup Rasmussen, the leader of the Social Democratic Party, was appointed prime minister. Rasmussen has remained prime minister ever since, through a period of economic recession, a shrinking of the welfare state, and an electoral backlash against European integration. After the 1994 elections, his government consisted of Social Democrats, Radical Liberals, and Center Democrats (the latter until 1996).

With the economy doing well in 1998, Rasmussen caught the opposition off balance with his "snap" election call. The main opposition party, the Conservatives, handicapped by an internal leadership battle, suffered its biggest setback (losing 11 seats) since 1977. The Social Democrats did better than expected, but their partner, the Radical Liberals, lost ground. However, Rasmussen was able to reconstruct the governing coalition with the support of leftist and Faeroe parliamentarians.

The biggest gains in 1998 were made by the Danish People's Party (DF), which originated in a split within the antitax Progress Party in 1995. Campaigning in favor of old Danish values and against immigration, the DF won 7.4 percent of the votes and 13 seats (making it the equal of the Socialist People's Party). Its leader Pia Kjaersgaard's xenophobic and "euro-skeptic" appeals have translated into public support of more than 15 percent in recent polls. Fewer than 5 percent of the population are foreigners. However, though the Danish political culture has traditionally been characterized by tolerance, many immigrants, particularly Muslims, have not been integrated into society; more than half are unemployed and dependent on state support.

THE ECONOMY

Sixty-two percent of Denmark's land is arable—a very high percentage. The countryside is primarily flat and, unlike much of its Nordic neighbors, not rocky but fertile. Bearing in mind this factor, which predisposes the nation toward agriculture, as well as the absence of raw materials, one need not be surprised that industrialization has not made such vast strides in Denmark as in other Western European countries. Denmark is among the world's foremost agricultural countries. This characteristic still prevails, although industry has overtaken agriculture with respect to exports.

The structure of Danish agriculture has experienced a huge transformation over the last few decades. There were more than 200,000 farms only 20 to 30 years ago. Today, after a great many closures and mergers, that figure barely exceeds 100,000. Another indicator of the decline in agricultural pursuits is that only about 4 percent of the population are now employed in agriculture; a quarter-century ago, the figure was about 20 percent. But if these figures express quantities, Danish agriculture—or, more generally, farming—has *qualitatively* improved. Farmers are considerably better trained and better informed than they were in the past, thanks to the numerous agricultural colleges. Farmers in Denmark were among the first to experiment with cooperative societies in dairy produce, and other communal business ventures, as well as in the field of exports.

Industry is overwhelmingly made up of small- and medium-size companies, most of which have a penchant for strict quality control. Denmark has concentrated on light industry, another side effect of absence of raw materials. These enterprises are very labor-intensive. Examples are the production and export of furniture, handcrafts, medical goods, automatic cooling and heating devices, and precision instruments. Danish beer has been world-famous since the early 1950s. The export of machinery, textiles, and electronics has also risen. In a highly competitive global market, Danish industry has been able to find niches for particular specialties. For example, in the world of the child, Lego blocks have become well known.

Not so long ago, journalists were citing Shakespeare's line "There is something rotten in the state of Denmark" to describe the country's economy. Unemployment, which had begun rising in the 1960s, reached 12.5 percent in 1994. This high employment may be, in part, explained as a consequence of adjustments to changing conditions in the international market and high labor costs, which affected the competitiveness of some Danish exports. Trade deficits had become the order of the day. In mid-2001, the unemployment rate stood at 5.4 percent. Denmark has also enjoyed a trade surplus. Its economic growth rate was projected by *The Economist* to be 2.2 percent during 2000–2001, twice the rate of a decade ago. Furthermore, Rasmussen's government could point to a budget surplus, declining national indebtedness, and low inflation.

Fortunately, the Danish North Sea oil and natural-gas exploitation has reduced the country's dependence on imported fuels. The opening of Central/Eastern Europe also bodes well. Denmark looks across the Baltic Sea toward Estonia, Latvia, and Lithuania, with which it has had ties since the fifteenth century. Copenhagen was quick to open the first international air links with the Baltic nations and began ferry services with Riga, Latvia's capital, as soon as Central/Eastern Europe opened up in the late 1980s and early 1990s, and this initiative has paid off in economic terms.

Mid-2000 saw the opening of the 10-mile-long Oresund bridge-tunnel-artificial island link for cars, trucks, and trains traveling between Copenhagen and Malmö (Sweden's third-largest city). It is the hope of the two governments, which spent nearly $2 billion and five years on the world-class technological achievement, that it will strengthen cross-border economic ties so that the region will become a powerhouse rivaling Hamburg, Berlin, and Amsterdam.

THE WELFARE STATE

Denmark is one of the world's best-run and most generous social-welfare states. All working persons are entitled to an annual five-week paid vacation. Employees temporarily out of work as a result of sickness,

Gorm the Old initiates the Danish royal dynasty ca. A.D. 900	The Kalmar Union unites three Scandinavian countries 1397	The Reformation arrives in Denmark; people almost universally adopt the Lutheran faith 1530s	Absolute monarchy is introduced in Denmark 1660	Abolition of serfdom 1788	Denmark loses Norway to Sweden 1814	End of royal absolutism; Denmark's first Constitution 1849

disability, unemployment, or maternity/paternity leaves are accorded up to 80 percent of their wages. Danish workers are entitled to up to 2½ years of unemployment compensation—and the government has to come up with job offers. This system has of course proven to be very expensive during periods of high unemployment.

A variety of special programs benefit children—the so-called family allowance for each child under age 18, day-care facilities, and special housing allowances for low-income families. Public education is free through the university level. And the cornerstone of the welfare state is national health care: All Danish residents are covered by a plan that provides free hospitalization and medical care and covers most dental expenses.

The Danish welfare state has also been on the forefront of the struggle for equal rights for gays and lesbians. In 1989, Denmark became the first country in the world to allow same-sex couples to register their partnerships. A decade later, homosexual couples enjoyed all the legal and social rights of heterosexual couples, except for child adoption.

Welfare programs in Denmark enjoy virtually universal support, since ultimately everyone benefits from them. But they require a great deal of money. The government collects about half the total income in the country in taxes, the bulk of which goes to meet the costs of social services. That not all people approve was shown in the 1973 "earthquake election," when antitax parties jointly secured almost a quarter of the total vote. Income taxes peaked the next year; indirect taxes also witnessed a slow and continuous rise. Like the other Nordic countries, Denmark has been forced to trim its welfare state in order to put a brake on taxation as well as to protect the overall budgetary health of the economy.

Despite the cuts of the 1980s and 1990s, the Danish welfare state remains generous in its benefits. Its comprehensive social programs have wide public support despite high taxes (second in the world only to Sweden). In fact, one of the major concerns of voters who opposed their country's joining of the Euro Zone in the September 2000 referendum was fear that pressures for harmonization would undermine the social quality of life that the egalitarian Danes enjoy.

FOREIGN RELATIONS

Denmark was one of the Oslo States, an organization that furthered economic cooperation during the Great Depression. However, this was not a military alliance, and, being a small country, Denmark doubted that it would be able to protect itself against a large aggressor. The Danes decided, on the eve of World War II, to abolish their army and to rely on a small police force for domestic emergencies; Denmark was thus the only country to succumb to Nazi aggression without offering any military resistance at all. After five years, the German surrender forced the Nazi forces to withdraw. But the Danes had learned the lesson that proclaiming oneself "neutral" was not enough to prevent war and occupation.

When the East–West tension rose to a high pitch in the late 1940s, the United States took the initiative to establish a collective security system to deter the Soviet Union from invading and occupying Western Europe. It was not hard to find allies. At that time, Sweden, Norway, and Denmark were trying to establish "bloc neutrality," which implied that the countries concerned would be neutral unless one of them was attacked.

However, as soon as the North Atlantic Treaty Organization loomed, Norway abandoned the plan and went over to the Atlantic Alliance. Denmark followed suit, though it did not join NATO without reservations. The Danish government made it a condition of joining that no nuclear weapons would be stored on Danish territory in peacetime. This restriction caused difficulties when, 30 years later, NATO decided to install nuclear missiles in each of the European member-states. The Danish government was unable to secure a majority vote in favor of lifting its restrictions concerning the deployment of nuclear arms on Danish soil.

In 1945, Denmark was one of the 50 nations to sign the Charter of the United Nations, and generally the country has been considerably enthusiastic about this world body. Although its importance in the United Nations has been limited by its relatively small economy, Denmark has played a role in the organization out of proportion to its size. Contributions for development aid as a proportion of Denmark's gross national product are among the highest in the world, and the country has contributed to various UN peacekeeping forces. Denmark is also a member of all specialized UN agencies. The Danish delegation has emphasized the promotion of respect for human rights.

THE EUROPEAN UNION

Denmark has always been ambivalent with regard to its membership in the European Union. Initially, its economy was almost exclusively geared toward the Nordic Council. A shift took place in the early 1960s, when Denmark became a member of the European Free Trade Association. But when in 1972 Great Britain, the EFTA's foremost member, started negotiations for EU membership, Denmark felt that the time had come to consider Union membership as well. Both Denmark and Norway were involved in the negotiations that were held in Brussels and that would make the EU membership jump from six to 10. However, in September 1972, the Norwegian referendum revealed an opposition of 54 percent. It was believed that the Danish vote would be influenced by this rejection, but a few weeks later, the Danish Folketing voted favorably on accession by a surprising 141 versus 34 votes.

It is possible that the Danish public (including its politicians) were after all poorly informed about the advantages and disadvantages that came with EU membership. However that may be, the Danes often felt shortchanged. They repeatedly grumbled and occasionally threatened to withdraw. The real crucible, however, came in the early 1990s, during negotiations that led to the so-called Maastricht Treaty. At that time, the British were expressing reservations concerning political integration as well as financial union. They were allowed to "opt out" of specific Maastricht accords. The Danish suggestion that this privilege be accorded to other EU members as well was deemed irrelevant, since all other members had already expressed agreement.

Shortly after the Maastricht meeting, Denmark announced that its ratification of the treaty would be contingent on a referendum to be held in June 1992. From then

Parliamentary
supremacy is
established
1901

Universal suffrage
1915

The first Social
Democratic
government
1924

German
occupation
1940–1945

A constitutional
amendment
allowing female
succession to the
throne
1953

Danish
membership in
the EU (then the
EC) becomes
effective; the
earthquake
election
1973

The Maastricht Treaty
is first rejected, then
approved; Denmark
battles a few years of
recession; gay
couples are accorded
full legal rights
1990s

2000s

The bridge-tunnel
between Copenhagen
and Malmö opens

Danish voters reject
joining the Euro Zone
in September 2000

onward, Danish politics was at a fever pitch. Approval and disapproval of the treaty was not entirely consistent with party lines. There was, nevertheless, an expectation that there would be a majority of people favoring Danish accession to the economic and political union. However, it turned out that a slim majority (50.7 versus 49.3 percent) favored rejection. The outcome stunned the European Union. In view of the closeness of the vote, the Danish government proposed another referendum, to be held in June 1993, to focus on similar issues (to have another referendum on exactly the same issues was believed to be offensive to the Danish electorate).

Meanwhile, the European Union recovered from its narrow defeat by Danish voters. Subsequently, Ireland and Greece easily ratified the Maastricht Treaty, and French voters approved it, by the narrowest of margins. After their government had obtained "opt outs" similar to those of the British, the Danes, in a second referendum in June 1993, approved the treaty. In May 1998, the less controversial Amsterdam Treaty, which extends cautiously the Maastricht initiatives, received the support of 55.1 percent of Danish voters. While polls have indicated public misgivings about trends toward a federal Europe, there is also evidence that a large majority of Danes feel that their country has benefited from membership in the EU.

Although Denmark met all the conditions for membership in the Euro Zone, the government did not join the launch of the Euro (the single European currency)

on January 1, 1999. Rather, it chose to wait for opinion to shift after the monetary union was up and running and its advantages would be more evident. Not only the governing parties, but also major opposition parties, trade-union leaders, and corporate elites favored Denmark's membership in the Euro Zone. However, in the September 2000 referendum, Danes rejected the Euro 53 percent to 47 percent, dealing a personal setback to Rasmussen, who had spearheaded the pro-Euro campaign. However, the Liberals, who aspire to supplant the Social Democrats in the next election, also failed to deliver the "yes" votes of many of their supporters. The "no" camp was quite heterogeneous. It included those on the right concerned with the loss of national sovereignty, and those on the left concerned with the undermining of participatory democracy. There were also the nonideological concerns of those who saw threats to the benefits that they would receive from the Danish welfare state. Furthermore, the xenophobic and nationalistic Danish People's Party pointed to the European Union's campaign of sanctions against Austria in 2000 as a warning of what small countries would likely be facing within a more unified Europe.

While public concerns about immigration have grown in recent years, Danish leaders nevertheless favor the eastern expansion of the EU (and NATO), particularly to include the Baltic countries, from which the country is well positioned to gain in economic

terms. Of the current 15 EU member states, Denmark is one that has "punched above its weight." One policy area where the small country has been out front in the EU has been environmental protection. During the 1980s, no Green party emerged in Denmark largely because the established parties were already presenting themselves as "green" to the environmentally conscious electorate. In recent years, the Danish government has been advocating taxes for hazardous chemicals, bans on ocean dumping of radioactive wastes and on lead products, and recycling programs for electronic equipment.

DEVELOPMENT

The first few years of Denmark's membership in the EU were disappointing, since that period coincided with a severe recession. Economic recovery has been increasing but not yet complete. Denmark meets all criteria for membership in the Euro Zone, but it has not joined.

FREEDOM

Denmark has one of the highest human-rights ratings in the world. The country used to have an extremely liberal policy with respect to political asylum, but the tide of refugees has been stemmed by changes in the law. An anti-immigration right-wing party has emerged despite the tolerance of the Danish political culture.

HEALTH/WELFARE

Denmark has one of the most comprehensive welfare systems in the world. There have been times when Danes have tired of paying the high taxes required to maintain the system, but when the chips are down, most Danes feel that nobody should be without the assistance he or she needs. UN statistics indicate that Danes rank first in the world in average caloric intake.

ACHIEVEMENTS

Denmark was one of the first countries in the world to institute the office of ombudsman, in 1955. This office was created to protect the individual citizen against wrongdoing or even indifference on the part of the bureaucracy. A second ombudsman for consumer protection was instituted in 1975. Many countries have followed suit, instituting ombudsmen of their own.

Finland (Republic of Finland)

GEOGRAPHY
Area in Square Miles (Kilometers):
130,160 (337,113) (about the
size of Montana)
Capital (Population): Helsinki
(532,000)
Environmental Concerns: water
and air pollution; habitat loss
Geographical Features: mostly
low, flat to rolling plains
interspersed with lakes and
low hills
Climate: cold temperate

PEOPLE

Population
Total: 5,168,000
Annual Growth Rate: 0.17%
Rural/Urban Population Ratio:
36/64
Major Languages: Finnish; Swedish
Ethnic Makeup: 93% Finnish; 6%
Swedish; 1% Sami (Lapp) and
others
Religions: 89% Evangelical
Lutheran; 9% unaffiliated; 2%
Greek Orthodox or other

Health
Life Expectancy at Birth: 74 years
(male); 81 years (female)
Infant Mortality Rate (Ratio):
3.8/1,000
Physicians Available (Ratio): 1/371

Education
Adult Literacy Rate: 100%
Education Compulsory (Ages):
7–16; free

COMMUNICATION
Telephones: 2,841,000 main lines
Daily Newspaper Circulation: 464
per 1,000 people
Televisions: 372 per 1,000 people
Internet Service Providers:
36 (1999)

TRANSPORTATION
Highways in Miles (Kilometers): 46,669
(77,782)
Railroads in Miles (Kilometers): 3,515
(5,859)
Usable Airfields: 158
Motor Vehicles in Use: 2,231,000

GOVERNMENT
Type: republic
Independence Date: December 6, 1917
(from Russia)
Head of State/Government: President Tarja
Halonen; Prime Minister Paavo Lipponen
Political Parties: Social Democratic Party;
Center Party; National Coalition Party;
Left Alliance; Swedish People's Party;
Finnish Christian League; Green
League; Rural Party; others
Suffrage: universal at 18

MILITARY
Military Expenditures (% of GDP): 2%
Current Disputes: none

ECONOMY
Currency ($ U.S. Equivalent): 7.03
Finnish markka = $1
Per Capita Income/GDP: $21,000/$108.6
billion
GDP Growth Rate: 3.5%
Inflation Rate: 1%
Unemployment Rate: 8.9%
Labor Force: 2,533,000

Natural Resources: timber; copper; zinc;
iron ore; silver; forests
Agriculture: animal husbandry, especially
dairying; cereals; sugar beets; potatoes;
fish
Industry: metals; shipbuilding; timber;
pulp and paper; copper refining;
foodstuffs; chemicals; textiles; clothing
Exports: $43 billion (primary partners
EU, United States, Russia)
Imports: $30.7 billion (primary partners
EU, United States, Russia)

http://www.vtourist.com/Europe/
Finland/
http://virtual.finland.fi

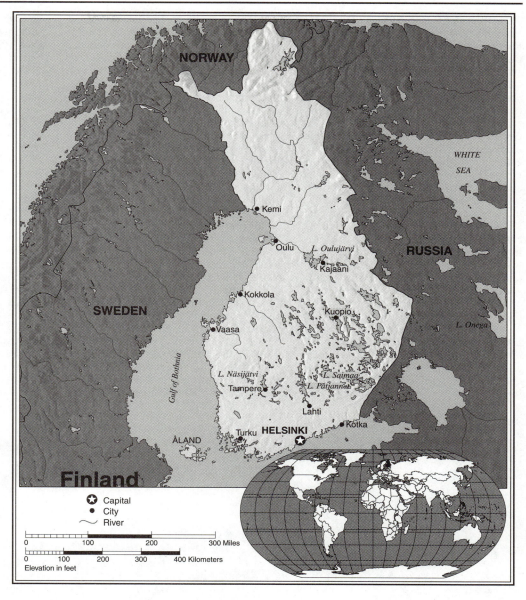

THE REPUBLIC OF FINLAND

In some of the frozen parts of Northeastern Europe, artifacts have been found that suggest that humans first made their appearance in what is now known as Finland some 6,000 to 8,000 years ago. However, not before about 100 B.C. did people arrive (probably from the central Russian area) who may properly be identified as Finnish. Except in its southern part, Finland remained sparsely populated, with clusters of communities that were nomadic (as were most of the Lapps) close to the polar circle and seminomadic in the center.

Finland's climate is not as arctic as its location would suggest; indeed, it is actually less cold than its neighbors Sweden, Norway, and northern Russia. But nature has left the separation between land and water somewhat indistinct. There are some tens of thousands of lakes, mostly small, which in total make up more than 10 percent of Finland's area. Approximately three quarters of the country is forested.

In view of these geographical conditions, it is not surprising that during most of its history, Finland was little more than an appendage to one of its politically better-developed neighbors. Indeed, its chief function may have been that of a *cordon sanitaire*—a buffer of sorts that prevented, or at least minimized, wars between Scandinavians on the one hand and Russians on the other. Surprisingly, neither of these neighbors overwhelmed Finland. And although Finland has always been tolerant of its minorities, there was no assimilation, which may be attributable to the uniqueness of Finnish language and culture. Finns descend from the Finno–Ugric group, found nowhere else in Europe. Even so, since they lived scattered over a relatively large area and did not have well-developed political institutions of their own when their country was occupied, nationalism was slow in coming. Sweden could thus rule the Finns for more than 500 years without engendering unrest of any significance.

Under Russian occupation, the area was converted into a Grand Duchy, headed by Czar Alexander II. During the last years of Czarist rule, however, some nationalist stirrings could be noted. The Finns seized the opportunity to declare themselves independent when Russia was in turmoil as a result of the 1917 Bolshevik Revolution. Nevertheless, that independence was hard to realize, since political divisions were also evident in the Finnish area. A civil war developed between the Whites (those who stood for an independent Finland) and the Reds (those who wanted Finland to be a part of the emerging Soviet Union).

The political turbulence was to last for several years, since other nations intervened. That the Whites prevailed was largely due to the single-minded heroism of General Mannerheim, whose exploits have become legendary in Finland. Finnish independence in 1917 was thus the result of skirmishes between political factions rather than of a revolution against foreign oppressors.

Although the Soviet Union had acknowledged Finland's sovereignty, in 1939 it started to make war on the fledgling nation. In this Russo–Finnish war, generally termed the Winter War, the Finns' valor and persistence proved to the world that their independence had not been solely a matter of rhetoric. While Soviet troops moved slowly through the snow in areas alien to them, Finnish ski brigades moved swiftly to their rear, fighting a guerrilla war that proved costly to the Soviet Union. In the end, however, the Finns lost the Winter War and were forced to cede part of Karelia, a province bordering on the Soviet Union.

(Sky Foto)

Finland boasts some 187,000 lakes, which make up more than 10 percent of its geographic area. Historically, the Finnish lakes have acted as a buffer between Scandinavia and Russia.

(Photo courtesy Lisa Clyde)

Turku is Finland's oldest city, dating back to the thirteenth century. It was the capital under Swedish and Russian rule until 1812. In 1827, Turku was almost entirely destroyed by fire; after it was rebuilt, it prospered until once again it burned during World War II. Shown above is a view of the courtyards of Turku Castle, dating from the medieval period, which today houses a historical museum, the Swedish Theater, and a Greek Orthodox church.

The settlement of the Winter War was followed closely by Operation Barbarossa, the invasion of the Soviet Union by the Nazi German war machine on June 22, 1941. At least initially, the carefully planned attack by Adolf Hitler posed a dilemma to Finnish nationalism. On the one hand, as a democratic nation, the Finns abhorred Nazi totalitarianism and violence; but on the other, the Soviet Union had clearly been their most recent enemy. Although under the circumstances it might have been possible to maintain a precarious neutrality, it did not take the Finns long to decide in favor of resuming the war against the Soviet Union. The retrieval of the part of Karelia that they had been forced to cede may also have been a motivation. The Finns insisted that there was no question of an alliance with Germany—that theirs was a separate war, unrelated to the general conflagration in the rest of Europe.

However that may be, it was soon obvious that their choice caused them to become de facto allies of Germany's Third Reich. German troops were allowed to move freely throughout Finland so as to attack the Soviet armies from the north. In part, this might have been achieved by crossing the Norwegian–Russian border, Norway having been occupied a year or so earlier; but using Finland as a base facilitated the German attacks from the north. German victories in the Soviet Union were hailed by the Finns. But reverses came after a time and, in 1944, Finland was forced to sue for a separate armistice. It was granted on the condition that Finland rid itself of all German troops, numbering more than 200,000, mostly in northern Finland. Thus Finland started its third war, fighting the Germans on its own soil. Some German troops escaped along the Norwegian route, but most were killed or taken prisoner and handed over to the Soviet armies.

If the Finns had expected that their belated cooperation with the Soviets would have ingratiated them to Moscow, they were mistaken: The Treaty of Paris, which settled the war in 1947, included harsh terms. Karelia was now entirely ceded to the Soviet Union, including the important city of Viborg. The loss of territory amounted to more than 11 percent of the country's total area and, very significantly, an even larger percentage of its arable land. Also, Finland was henceforth supposed to maintain only token military forces. (This stipulation was amended in 1963, when an agreement granted Finland more leeway in establishing an army commensurate with its size.)

On the positive side, Finland was not occupied by Soviet troops after World War II, as were numerous countries in Eastern Europe, and no force was applied to convert Finland into a Soviet satellite. Nevertheless, Finns were painfully aware that their small country existed in the shadow of its vast Soviet neighbor. Thus, the country signed a mutual-assistance treaty with the Soviet Union in 1948. The treaty was renewed for another 20 years in 1955, the same year that Finland joined the United Nations.

THE ECONOMY

World War II devastated Finland, not only because of the war against the Soviets but even more so by the subsequent fight against the Germans. It proved extraordinarily hard to recover from the destruction that had been wrought, especially since Finland did not receive any war reparations, having been classified as a belligerent on the side of the vanquished nations. Although the European Recovery Program (better known as the Marshall Plan) was extended to Finland, the Soviet Union did not allow it to accept the aid.

Since Finland had hardly embarked upon industrialization before the war, it is not surprising that it clung to sources of primary industry for a considerable period. It thus attempted to concentrate on agriculture, timber, and mining. Mining was insignificant in the 1950s; today, however, mining is an important part of Finland's economy.

Tourism has increased in importance, although the revenues from this industry are much lower than the amount of money Finns themselves spend as tourists abroad. This imbalance may decrease somewhat now that European travel agencies have started to focus on Finland as a tourist destination.

A major challenge for Finland emerged with the gradual collapse of the trade markets of the Soviet Union and other Communist neighbors in the late 1980s and early 1990s. The Finnish economy fell into a deep recession in the early 1990s. Unemployment soared, from 3.5 percent in 1991 to 18.5 percent in 1993. The currency was devalued, enterprises went bankrupt, and state indebtedness climbed.

After 1995, however, Prime Minister Paavo Lipponen's broad coalition government was able to stabilize prices, cut social spending, and raise taxes in time for Finland to meet the Maastricht Treaty's convergence criteria for joining the Euro Zone. Despite the Finns' high wages, taxes, and social-security contributions, their economy experienced rapid growth in the late 1990s; gross domestic product growth of 4.6 percent was forecast for the year 2001. Unemployment, while higher in other Nordic countries, declined officially to 8.9 percent in October 2000, as employment increased in business services and social services.

Finland is among the elite group of trend-setting countries of the new information economy. Pioneering uses of "smart phone" digital technology, the Finns rank number one in per capita ownership of cellular phones in 2000. Finland's Nokia surpassed the United States' Motorola to become the world's leading manufacturer of mobile phones; its sales accounted for nearly one fourth of Finland's exports. In 2000, the Finns ranked sixth in the world in the percentage of the population that "surfed" the Internet. Early on, the government intervened to push research and development in the new technologies. Observed the head of one of the country's booming scientific business parks (devoted to electronics, biotechnology, and environmentally friendly business): "These days Finns trust computers, telecoms and technology."

THE POLITICAL SYSTEM

Although Finland was part of Czarist Russia at the time that it declared itself independent, Swedish influence in Finland's society and culture has been considerably more pronounced and durable. Indeed, the country still is home to a sizable minority (6 percent) of Swedes who do not consider themselves expatriates. Swedish is also one of the two official languages. Finland's political culture reflects a Swedish background, and in many other ways the country can be clearly identified as Western European. In its anxiety to pass the acid test that Woodrow Wilson proclaimed in matters of national self-determination, the Finns adopted a democracy that rested on pluralism and consensual politics.

Finland has a dual executive, consisting of a president as head of state and a prime minister as head of government. Although not as powerful as the U.S. and French presidents, the Finnish president, especially during the tenure of Urho Kekkonen (1956–1981), has not been merely a ceremonial figure, as is the norm in most European parliamentary republics. As a result of constitutional reform in 1991, the president lost the power to dissolve the Parliament without the prime minister's request. Although the new Constitution of 2000 reduces the president's role in domestic politics, the president remains the commander-in-chief of the military and retains an effective role in foreign policy. President Martti Ahtisaari, a career diplomat, distinguished himself as a mediator in the Kosovo crisis of 1999 (and in 2000 as a private citizen played a role in the lifting of European Union sanctions against Austria). Since 1994, the Finnish president, who serves a renewable six-year term, has been directly elected by the voters. If no candidate receives 50 percent, as occurred in 2000, there is a run-off election between the top two vote-getters. In the 2000 run-off, Tarja Halonen, the Social Democratic foreign minister, edged out (by 2 percent) Esko Aho, a former prime minster and current leader of the Center Party. It is noteworthy that in the first round of voting, four of the six presidential candidates were women.

The prime minister—since 1995, Paavo Lipponen—has the major responsibility in domestic-policy areas as well as a growing role in foreign policy. The prime minister's "rainbow" coalition depends on the confidence of a parliamentary majority; otherwise, it must resign as the government. The Finnish Parliament (Eduskunta) is a unicameral chamber with 200 seats; in 2000, 37 percent of the seats were held by women, a gender proportion exceeded only in Sweden and Denmark. In recent years, the trend has been toward parliamentary majoritarianism; for example, one third of members can no longer delay passage of a law. The Parliament, which can override presidential vetoes, has significant legislative and oversight powers.

In Finland, party competition used to be very fierce. However, since World War II, this competition has been noticeably mitigated by two factors. First, as has been the case in other European political systems, coalitions are considerably less volatile than they used to be, due to growing convergence of views among party leaders. A second factor is that the former Soviet Union unambiguously indicated that it did not want Finland's Social Democratic Party ever to become a government party. Finland was not in a position to ignore this mandate, which infringed on its sovereignty. Thus, the Social Democrats were for many years relegated to the opposition.

In the March 1999 parliamentary elections, Finnish voters returned to power the same broad coalition of parties that had governed since 1995. Its core has been Prime Minister Lipponen's Social Democrats and the National Coalition Party (better known as the Conservatives), which have been joined by the ex-Communist Left Alliance, the Greens, and the Swedish People's Party. The Social Democrats lost heavily in 1999 but remained the largest winner with 22.9 percent of the votes, compared to 22.5 percent for the opposition Center Party (the former Farmers' Party) and 21.0 percent for the Conservatives, both of whom had scored gains. The Left Alliance slipped to 10.9 percent. The Greens won 7.3 percent of the votes, their best ever. Support for the Swedish People's Party held at 5.1 percent, and the opposition Christian League received 4.2 percent. Thus, the Finnish party system in recent years has been characterized by a lack of bipolar (left–right) dynamics; Lipponen's new government includes Conservatives and former Communists. In contrast to Denmark and Norway, Finland's party system has also not had an electorally significant xenophobic, right-wing party in recent years. On the other hand, the downward trend in electoral turnout continued in 1999: Only 68.3 percent of the Finns voted.

Finland declares its independence; civil war between the Whites and the Reds
A.D. **1917**

The Winter War between the Soviet Union and Finland
1939–1940

Finland decides to wage war on the Soviet Union
1942

Finland, suing for a separate armistice, is forced to expel all the German troops on its soil
1944

The Treaty of Paris imposes harsh peace terms on Finland
1947

Finland gains international prestige when the Olympic Games are held in its capital
1952

The Helsinki Accords are signed
1975

Finland's economic sights shift to Western Europe
1980s

Finland becomes a member of the European Union on January 1, 1995; in 1999, it eagerly joins the Euro Zone; the economy rebounds
1990s

2000s

Finland gains world prominence as a technological power

Tarja Halonen is elected president

While left–right differences are bridged by the "rainbow" coalition, its parties tend to be split within and among themselves over the issue of building a fifth nuclear-power plant in Finland. In 1993, parliament narrowly rejected expansion of nuclear power. Its renewed consideration revolves around Finnish competitiveness in a global economy, the creation of jobs, the needs of energy-intensive industries, and the country's treaty obligations to reduce the emission of climate-warming "greenhouse" gases. Conservatives and Social Democrats are mostly pronuclear, along with all of Finland's big manufacturers; while the Left Alliance, the Swedish People's Party, most of the leading opposition Center Party, and especially the Greens (who have threatened to leave the coalition) are antinuclear.

THE SHIFT TOWARD EUROPE

Finland, much more so than any other nation in Western Europe, was dumbfounded when its giant neighbor, the Soviet Union, started to experience overwhelming problems in the late 1980s. The Finnish government took a strictly neutral attitude when the nearby Baltic states, particularly Lithuania, began to agitate for independence, although it would have been possible for Finland to encourage Lithuania in its effort. The conditions of Finland and the Baltic states are very different—Finland broke away from Russia when the latter was about to become the Soviet Union, while the Baltic states had been independent nations from 1918 until 1940.

In an address that Esko Aho, then the prime minister of Finland, presented in 1992 at the National Press Club in Washington, D.C., he touched on two points of paramount importance. First, he stated, "The world around us has dramatically changed. The Soviet Union is no more, and gone are the last vestiges of the Cold War. The institutional framework of the old world order is gradually adjusting itself to new realities, the exact and final nature of which are not yet known. Even the cozy old world of domestic policies as usual in the Western industrial democracies seems to be on its way out. And here I do not speak only of my own experience and Finnish experience."

A little later the prime minister introduced a new reality, that is, the subject of Finland becoming part of the European Union (then referred to as the European Community) in the following words: "Finland is an integral part of Western Europe, and the well-being of the Finns hinges on trade with the rest of the pack and increasingly on our ability to keep Finland an attractive place to live and invest in. Less than two months ago Finland submitted her application to join the European Community. Negotiations are expected to start sometime next year, maybe the beginning of next year, and membership may be possible as early as the mid-1990s." Indeed, Finland became an EU member on January 1, 1995.

Of all the Nordic countries, Finland is the most enthusiastic supporter of the European Union and its economic and political goals. In January 1999, Finland was one of the 11 EU member nations to participate in the launch of the Euro. Fellow Nordic EU members Denmark and Sweden chose not to join the Euro Zone. Prime Minister Lipponen, like his Danish and Swedish counterparts, has favored speedy expansion of the European Union to include the Baltic and Eastern/Central European countries. However, he has parted company with them dramatically in other regards. In November 2000, speaking to the College of Europe in Bruges, Lipponen supported further integration via the "community method," which entails strong EU institutions (such as the Commission); and he criticized intergovernmentalism, which is favored by Britain and Denmark. In contrast to Swedish prime minister Goran Persson's moderate stance, Lipponen proposed a broadly representative "Constitutional Convention," which would participate in drafting a basic constitution for the European Union.

DEVELOPMENT

Against all odds, postwar Finland changed from a country based on primary industry (agriculture, forestry, and mining) into a semi-industrial country with modern manufacturing plants. Economic diversification in recent years has meant greater emphasis on high technology. The World Economic Forum's 2000 report rated Finland the world's second-most economically creative country.

FREEDOM

Finland rates high in the realm of human rights. Its citizens enjoy basic rights to the fullest extent. Finland also hosted the ceremonial signing of the Final Act, which led to the Helsinki Accords in 1975. Finnish troops have participated in many UN peacekeeping efforts.

HEALTH/WELFARE

Finland's health score is very solid (some argue that the Finnish sauna has contributed to high health levels). The social-welfare system is modeled on that of Sweden. Unemployment, however, has been higher than in other Scandinavian countries.

ACHIEVEMENTS

In the not-too-distant past, Finnish rated low as a language. People "of culture" in Finland spoke Swedish among themselves; books were always published in Swedish. After World War II, however, a linguistic battle concluded in the removal of the stigma that Finnish carried. Finland has been rated as the world's "least corrupt" country.

France (French Republic)

GEOGRAPHY

Area in Square Miles (Kilometers):
213,700 (547,026) (about twice the
size of Colorado)
Capital (Population): Paris
(2,152,000)
Environmental Concerns: acid rain;
air and water pollution
Geographical Features: mostly flat
plains or gently rolling hills in the
north and west; the remainder is
mountainous, especially the Pyrenees in the south and the Alps in
the east
Climate: temperate

PEOPLE

Population
Total: 59,330,000
Annual Growth Rate: 0.38%
Rural/Urban Population Ratio: 26/74
Major Languages: French, with rapidly
declining regional languages
Ethnic Makeup: Celtic and Latin with
Teutonic, Slavic, North African, In-
dochinese, and Basque minorities
Religions: 90% Roman Catholic; 2%
Protestant; 1% Jewish; 1% Muslim;
6% unaffiliated

Health
Life Expectancy at Birth: 75 years
(male); 83 years (female)
Infant Mortality Rate (Ratio):
4.5/1,000
Physicians Available (Ratio): 1/361

Education
Adult Literacy Rate: 99%
Compulsory (Ages): 6–16; free

COMMUNICATION
Telephones: 34,000,000 main lines
Daily Newspaper Circulation: 235
per 1,000 people
Televisions: 579 per 1,000 people
Internet Service Providers: 128 (1999)

TRANSPORTATION
Highways in Miles (Kilometers): 535,500
(892,500)
Railroads in Miles (Kilometers): 21,273
(32,027)
Usable Airfields: 474
Motor Vehicles in Use: 29,500,000

GOVERNMENT
Type: unitary republic
Independence: A.D. 486 (unified by Clovis)
Head of State/Government: President
Jacques Chirac; Prime Minister Lionel
Jospin
Political Parties: Rally for the Republic;
Union for French Democracy;

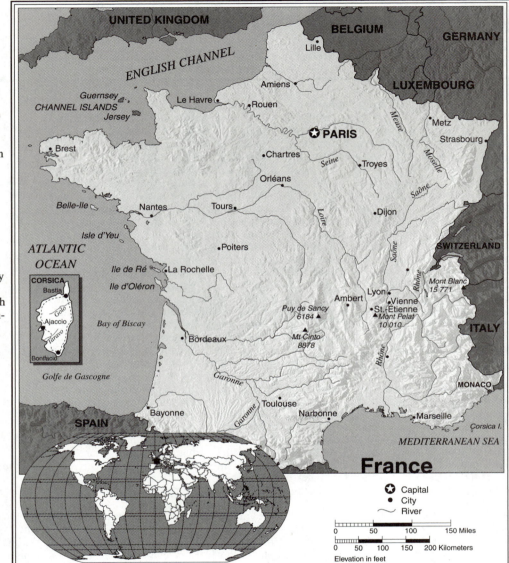

France

Republican Party; Democratic Force;
Radical Party; Socialist Party;
Communist Party; National Front;
Greens; others
Suffrage: universal at 18

MILITARY
*Military Expenditures (% of Central
Government Expenditures):* 2.5%
Current Disputes: minor territorial disputes

ECONOMY
Currency ($ U.S. Equivalent): 7.75 francs = $1
Per Capita Income/GDP: $23,300/$1.37
trillion
GDP Growth Rate: 2.7%
Inflation Rate: 0.5%
Unemployment Rate: 8.7%

Labor Force: 25,400,000
Natural Resources: coal; iron ore; bauxite;
fish; timber; zinc; potash; arable land
Agriculture: beef; dairy products; wheat;
cereals; sugar beets; potatoes; wine
grapes; fish
Industry: steel; machinery; textiles;
chemicals; automobiles; food processing;
metallurgy; aircraft; mining; electronics;
tourism
Exports: $304.7 billion (primary partners
EU, United States)
Imports: $280.8 billion (primary partners
EU, United States)

 http://www.odci.gov/cia/publications/
factbook/geos/fr.html

A LENGTHY HISTORY

Around 1200 B.C., The Gauls, a large tribe related to the Celts, settled in what is now called France. One of the oldest sources on Gaul, as the area came to be called, was a Roman general, Julius Caesar, who aimed at extending the Roman Empire to the north between 60 and 50 B.C. He recorded his forays and battles in Gaul in his book *de Bello Gallico*. The Gauls—if one goes by Caesar's account—were determined and ferocious fighters, but once they had been decisively beaten, they allowed their area to be incorporated into the Roman Empire. Only after the fall of the West Roman Empire, several centuries later, did the Gauls regain a measure of independence.

MONARCHISM

The history of France was for a long time the history of kingdoms. The French monarchy was, from the sixth century until its final demise in the nineteenth century, the equivalent of absolutism and despotism. The four best-known monarchs of that era are Charlemagne, Henri IV, Louis XIV, and Napoleon Bonaparte.

Charlemagne, solemnly crowned in a cathedral in A.D. 800, ruled an empire that was considerably larger than present-day France. He was, in fact, called the "Emperor of the Western World," a title that expresses the notion of the illusory Holy Roman Empire.

Like many other parts of Europe, France was rocked by religious strife in the sixteenth century. When the French throne was vacated and Henri of Navarre's Protestantism stood in the way of his succession, he simply announced, "*Paris vaut bien une messe,*" which implied, "Becoming king is well worth my conversion to Roman Catholicism." In spite of this crass opportunism, as Henri IV, he was one of France's most popular kings.

The seventeenth century witnessed the power and the glory of Louis XIV. The so-called "Sun King" was extremely fond of games. During one of the matches (which, remarkably, he always won), one of his ministers reminded him that they had still to discuss matters of state. Louis XIV, taken aback, retorted, *L'état, c'est moi* ("The state, it is I"), a phrase that has gone down in history as the most concise definition of absolutism.

And then there is Napoleon Bonaparte, who wrought havoc in all of Europe. Although he was an offshoot of the French Revolution, which attempted to do away with monarchy and aristocracy, he reestablished monarchical splendor. Indeed, even after Napoleon's defeat, monarchy and republicanism alternated. Monarchism

disappeared permanently only after Napoleon III, Bonaparte's nephew, had plunged France into despair and defeat at the hands of the Prussian armies in 1870. Even today, many Frenchmen and -women, while paying lip service to democratic principles, are ready to condone rather than condemn "Bonapartism," the tendency on the part of strong political leaders to be careless when it comes to observing democratic rules. Some consider such negligence a necessity since, in their opinion, it is, at least in France, the only alternative to chaos.

THE SUCCESSIVE REPUBLICS

French history since 1870, when the monarchy disappeared from the scene, has generally been divided into the Third Republic (1870–1940), the Fourth Republic (1946–1958), and the Fifth Republic (1958–present). These ordinal numbers correspond to the respective Constitutions in the periods concerned.

The Third Republic ended as it began: as a result of German aggression. It was marked by a high degree of partitocracy, which led to parliamentary supremacy. Nothing was achieved as a result of the balance between left and right and of the endless disputes between Catholics and anticlericals; its nickname "the stalemate society" appeared to be well deserved. In fact, the Third Republic exemplified both the need for and the distrust of the strong state that French people so often experience. The centralization that the absolute monarchy had manifested continued not only because the various *départements* ("departments") were governed by prefects who were all appointed by the Ministry of the Interior in the capital, but also because Paris became the the hub of the railway network.

During the interwar years (1920–1940), the French population declined, a fact due to the colossal losses suffered during World War I. While France recovered from the German onslaught in 1914, beat the Germans back, and emerged victorious from the Great War, the cost in young male lives had a prolonged effect.

World War II, which ominously loomed behind Adolf Hitler's expansionist tactics, found France poorly prepared and poorly equipped. A little more than a month after the German armies had forced their way through the Low Countries, France surrendered. A small band of French officers and soldiers, headed by General Charles de Gaulle, managed to escape to England, where the Free French Movement was founded. The vanquished France they left behind was divided into two parts. One part was occupied by the German armies,

and the other, named *État Français* ("French State," better known as *Vichy*), was rendered a German satellite. Resistance groups soon sprouted in both areas; these illegal forces gradually grew and greatly undermined the German war effort. On the other hand, many French people, mesmerized by the 1940 defeat, and some sympathetic with Nazi Germany's anti-Semitism, sought solace in cooperating with the Germans. This was particularly the case in Vichy. To France, the war ended when the Germans were expelled from its area in the latter half of 1944. A bitter debate remains to this day as to how much responsibility France should assume for the crimes and collaboration of the Vichy regime.

After World War II, a provisional government, semi-military in nature and often referred to as a "constitutional dictatorship," emerged, headed by de Gaulle. During the next two years, frequent clashes occurred between those who had helped to liberate France from outside (the so-called Free French) and those who had served the French cause in German-occupied or Nazi-oriented areas (the Resistance). All French people, however, were looking forward to the creation of the Fourth Republic. General de Gaulle expressed his distaste for Third Republic politics on numerous occasions, notably in his celebrated Bayeux Speech, in which he announced his ideas on how France was to be governed.

The Fourth Constitution, however, disappointed him, because it was almost a replica of the Third. And although it met with little enthusiasm on the part of the French public, it was nevertheless adopted in a referendum. De Gaulle decided to retire from political life. The period that followed turned out to be exactly as de Gaulle had predicted, characterized by prolonged party strife and a series of ineffectual governments.

The Fourth Republic was also plagued by great colonial crises. First, successive governments attempted for eight years (1946–1954) to restore French rule to what used to be French Indochina. In this first Vietnam war, France failed miserably. It ended its involvement after military defeat in 1954.

The French hoped to be more successful in Algeria, which was officially designated as an "overseas department" rather than as a colony. Algeria was the home not only of the native Arab population but also of hundreds of thousands of French settlers. As is the case in colonial societies, discrimination was rampant. Now that Algerian self-determination was in the offing, the French settlers feared that the former

(Photo courtesy Lisa Clyde)

The spires of the Cathedral of Notre-Dame at Chartres, located at the top of a hill in the ancient town, soar above the houses on the small streets. The main part of the cathedral was built in the mid-thirteenth century. This cathedral illustrates the all-encompassing presence of the Roman Catholic Church in the daily lives of citizens in medieval France.

colonial subjects would seek revenge. First there were skirmishes and then regular battles between the nationalists among the native Arab population and the French Army. The French settlers found their protection insufficient and established a secret army (the OAS).

The Algerian crisis brought about the demise of the Fourth Republic. De Gaulle, when called back from his self-imposed exile, indicated that he was willing to resolve this crisis but that he should be given powers greater than the Fourth Constitution yielded. In other words, the French political leadership of that time was blackmailed into agreeing to a new Constitution. Michel Debré, a lawyer and close friend of the general, was entrusted with the task of putting the Fifth Constitution together. Predictably, it closely followed de Gaulle's preferences, doing away with the power of political parties, their coalitions, and, above all, parliamen-

tary supremacy. It gave considerable powers to the executive branch of government, particularly to the office of the president. The Fifth Constitution was adopted by referendum, after which de Gaulle was, almost automatically, elected the first president of the Fifth Republic.

Government by Referendum
Since the new Constitution rendered the French Parliament very weak, there was no reason to bypass it. Still, de Gaulle developed a predilection for government by referendum. Rather than have the Parliament discuss major proposals, he would couch these in a referendum, which he would submit to the French people. He would explain his proposal to the nation in a televised speech, stressing the importance of voting "yes" to the referendum. Sometimes he would threaten to step down should the proposal not pass. A not-inconsiderable number of people would

vote in favor of the referendum, fearing that if they did not, the country might fall prey to chaos again.

The Evian Accords
One of the first points that de Gaulle addressed was the termination of colonialism and imperialism. The so-called Union Française was abolished, and a global referendum was designed that allowed all dependencies and former dependencies to become members of a new French Community, which resembled the British Commonwealth of Nations.

Contrary to what many people had been led to expect, de Gaulle did not use the sword in order to bring the Algerian rebellion to an end. He appeared to recognize the forces of history and started to negotiate with the nationalist leadership of the former overseas department. It took four years to arrive at some agreement, but finally, in 1962, the accords that declared

Algeria's independence were signed at Evian. A large proportion of the settlers in Algeria were disappointed with the outcome, in which they had had no voice; they left the country in droves and settled in southern France (most of them had never even been in France before).

De Gaulle contributed certain biases to the operation of the new Fifth Republic government that were distinctly anti-British and anti-American. If the former surfaced in his repeated refusal to allow Britain into the European Union (then known as the European Economic Community), the latter was clearly evidenced by his objection to and obstruction of American hegemony in the North Atlantic Treaty Organization. Some analysts attribute de Gaulle's hostility vis-à-vis Britain and the United States to the war years, when he felt slighted by Winston Churchill and Franklin Roosevelt.

In de Gaulle's view, France seemed destined to become the leader of the third force—that is, a Europe of the fatherlands that would not participate in the deadly confrontation between the United States and the Soviet Union. These notions caused the French president to order NATO's headquarters out of Paris. In 1966, de Gaulle withdrew all French troops from the integrated military command of NATO (although 60,000 remained stationed in West Germany). France continued to attend NATO's political meetings. In 1992, the French defense minister indicated that although there was no chance that France would return to NATO's integrated military command, his country would like to "take part more fully than previously in politico-military discussions." Later in the 1990s, President Jacques Chirac initiated discussions with NATO on possible French reintegration into NATO's military structure. France has actively participated in recent peacekeeping operations.

The French–German Rapprochement
As an actor on the world scene, de Gaulle committed gaffes that caused analysts to think that he lacked diplomatic skills. It was clear that the general was less interested in diplomatic niceties than in getting his message across. However, it must be noted that his monumental achievement—a French–German rapprochement—has certainly found its place in history. De Gaulle's opposite number in West Germany, Konrad Adenauer, belonged more or less to the same age group; both statesmen appeared to like each other very much. Without de Gaulle's initiative and without the leaders' mutual sympathy, this unprecedented rapprochement would never have been achieved between the two

nations, which had for a long time been hostile toward each other. The French–German rapprochement has been a major factor in the emergence of the "new Europe."

Social Problems
If most French people appeared to like France's increasingly independent stance in international politics, other issues produced dissent. Very little had been done to adapt university curricula to modern times. Most learning was still geared toward life as it had been half a century or a century before, and as a result, many university graduates remained unemployed for prolonged periods of time. Repeated petitions to the government had no effect, and, in 1968, large-scale student rebellions erupted, mainly in Paris but also in other parts of the country. Street fights were the order of the day, and the police were unable to arrest the rebels, who barricaded themselves on the Left Bank and in the heart of Paris. Although de Gaulle had been forewarned that trouble was brewing, he had left on a scheduled state visit to Romania. Georges Pompidou, the premier at the time, had great difficulty in coping with what came to be known as the "student revolt," especially once Renault automotive workers decided to support the students. De Gaulle thus had to be called back from Bucharest. Yet, even after his television address upon his return to Paris, the rebellious acts did not immediately cease. The combined student and worker revolt, which is often referred to as *les événements* ("the events"), made it clear that de Gaulle had lost part of his charisma, if not his grip. In 1969, de Gaulle called a referendum on a couple of minor constitutional changes, which the majority of voters rejected. Interpreting the outcome as a vote of no-confidence in his presidential leadership, de Gaulle resigned. He died the following year. De Gaulle's popularity has grown through the decades; polls have ranked him as the greatest leader in French history.

THE POST–DE GAULLE ERA
In the contest that developed as to who would be his successor, François Mitterrand, the leader of the French Socialist Party (PSF), competed for the first time. Interestingly, de Gaulle gave only lukewarm political support to Georges Pompidou, the man who had been his faithful "co-pilot." Pompidou, nevertheless, won the election, and as president never departed from *Gaullisme*. However, he died in office before his first term had run its course. New elections were held in 1974, and the Socialist Party again made Mitter-

rand its presidential candidate. The conservative contender now was Valéry Giscard d'Estaing, who had been part of de Gaulle's administration, though not a Gaullist himself. Giscard won. As a former minister of finance, he endeavored to find solutions to the financial and economic problems associated with the changing nature of the French economy. At the end of his term, however, the French economy was still behind those of other European Union (then known as the European Community) members.

A serious drawback to reelection, however, was Giscard's personality. He was by nature an elitist who made little effort to ingratiate himself with the public at large. In 1981, he again ran against Mitterrand, an outgoing man who radiated bonhomie, and he lost.

Mitterrand had intermittently aligned himself with the French Communist Party (PCF). In order to avoid fierce competition between the two leftist parties, the Union of the Left had put together a *Programme Commun*, which served as a joint platform. Mitterrand consequently had to share his victory with Communists. Four lesser ministerial posts were given to the PCF leadership. The new premier, Pierre Mauroy, belonged to the left wing of the Socialist Party.

The early part of the Mitterrand administration was marked by large-scale nationalization and increased government spending. It soon became obvious that these leftist measures were no remedies to the lingering economic ills and, after four years, the president appointed a young economist, Laurent Fabius, to be his premier. The latter was a technocrat, not given to ideological stands. The Communists strenuously objected to this move, correctly perceiving Fabius's appointment as a shift toward the center, departing from the Programme Commun. When Mitterrand did not yield, the Union of the Left collapsed and the Communists left their ministerial posts.

The government was thus entirely in Socialist hands; however, Fabius was given little time to implement his remedies. Parliamentary elections were held in early 1986. The major contender on the right was Jacques Chirac, the mayor of Paris, who called himself a "neo-Gaullist." Socialists and Communists saw their support in Parliament reduced to such an extent that a different cabinet, headed by another premier, had to be established.

Chirac, who (predictably) accepted the post, initially insisted that Mitterrand step down and that new elections be held for the office of president, as the government might otherwise be unworkable. France's

Fifth Constitution, however, lacks clarity on this point. Its founding fathers apparently did not foresee the anomalous situation emerging from elections scheduled at different points in time. As a result, intragovernmental backbiting played a large part in the final years of Mitterrand's first term. The Fifth Constitution was then severely tested, since political conflict was the order of the day in these two years of *cohabitation* (as the forced cooperation between two opposing blocs came to be called). So divided was the French government that both the Socialist president and the neo-Gaullist premier appeared at international conferences to which France had been invited, occasionally debating each other! And when the presidential elections of 1988 approached, Chirac announced that he would run.

The incumbent remained silent, indicating that he would be a candidate only a few months before the election was to take place. Since the Socialist candidate was reelected, one may assume that the period of cohabitation had worked to Mitterrand's advantage. More important, Mitterrand's record had overall been a fairly positive one, in spite of the strident claims by Chirac to the contrary. It is true that the economy left much to be desired, but in France, the adage prevails that "There is no final victory in economics."

Foreign Relations

On the international scorecard, the Socialist government had done well. The closer bonds with West Germany, established in de Gaulle's time, continued to be emphasized even though both countries had experienced an ideological reversal (France had moved to the left, West Germany to the right). And, most ironically, socialism proved to be less anti-American than might have been expected. The noticeable improvement of the French–American relationship coincided with France becoming more amenable in NATO matters.

On the other hand, France proved unwilling to conform to the Nuclear Test Ban Treaty, and its traditional nuclear testing in the South Pacific did incur the anger and annoyance of a great many nations, notably those in the Southern Hemisphere. In the port of Auckland, New Zealand, in July 1985, an explosion caused the sinking of the *Rainbow Warrior*, the flagship of the Greenpeace environmental group, which had been scheduled to obstruct French nuclear testing farther east. Suspicions that the French government might be involved were later confirmed, and the French were forced to pay a large compensation. Surprisingly, the incident never became an issue in the election campaign a few years later. More recently, the French government decided to sign the Nuclear Non-Proliferation Treaty.

After Mitterrand was reelected, he dissolved the National Assembly and ordered new elections to be held. These did not grant the president's party a majority, but the Socialists were now at least Parliament's largest group. The Socialists were short of a majority in the 577-seat National Assembly, but they could only be ousted from government if the Communists teamed up with the Conservatives, which appeared unlikely.

This time, Mitterrand chose Michel Rocard to be his premier. Rocard was a technocrat who preferred practicable policies. The appointment clearly separated the presidential and prime-ministerial fields of concern. Rocard was chiefly interested in domestic policies, whereas Mitterrand had started to move into foreign affairs. Both men worked closely together and did not attempt to override each other.

(French Consulate General)

François Mitterrand, leader of the French Socialist Party, was elected president in 1981 with the support of the Communist Party, which he then marginalized. He was reelected in 1988, and, due to failing health, Mitterrand retired in 1995, having served longer than any other French president. Well after his death in 1995, financial improprieties in his administration were revealed.

The Socialist president maintained an amicable relationship with West Germany's Christian Democratic chancellor, Helmut Kohl. He also paid a state visit to the United States, strengthening the bonds with the country that had been the target of de Gaulle's snide remarks. But during the second Mitterrand administration, new foreign policy facets developed. The president's interest in the European Union, which had grown by leaps and bounds, stabilized at a very high level, causing some analysts to believe that France was destined for leadership within the EU. However, the increased French interest should be attributed in part to the fact that the presidency of the European Commission was held by Jacques Delors, an extremely active Frenchman who sought to speed the development of a single European market.

To some extent Mitterrand also moderated the traditional pro-Arab policy that had characterized French attitudes and sentiments vis-à-vis the Middle East. It was not Iraq's 1990 invasion of Kuwait by itself but, rather, the massive hostage-taking and the way in which French citizens were treated in Baghdad and Kuwait that caused France to join those who had organized themselves militarily against Iraq.

Thus far Mitterrand has been the only president to serve two terms. When he died of cancer in 1995, he was eulogized by members of the political left and right alike. Only years later did his financial improprieties come to light.

THE EXECUTIVE
France, like many other Western European democracies, has a dual executive. The position of chief of state is taken by the president. However, departing from the tradition of the Third and Fourth Republics, the president under the Fifth Republic does not merely have a ceremonial role but actually exercises a great deal of power. Presidents have been elected for a term of seven years and, as the 1988 election exemplified, may be reelected. In a September 2000 referendum, voters overwhelmingly approved the shortening of the presidential term to five years. The primary agrument for the constitutional change was to make it more likely that the president and the Parliament would reflect the same political majority. The incumbent president and the leaders of the major parties all favored the change, so the positive outcome was hardly a surprise. Only 31 percent of potential voters bothered to participate, which is a rather low turnout by French electoral standards.

With the Fifth Republic, France has taken a big step in the direction of the presidential system. However, the relationship between president and premier is a delicate one, depending on the degree of forcefulness that emanates from the respective incumbents. This is evidenced by the responses of the public that the various premiers inspire. A great deal of criticism was directed at France's first female premier, Edith Cresson, whose tenure was brief. The succeeding premier, Pierre Bérégovoy, seemed lackluster. Ultimately, he proved no match for the withering attacks from the opposition; he resigned and committed suicide. The subsequent premier was neo-Gaullist Edouard Balladur, a former minister of finance and a more solid figure.

The president's executive powers are shared with the premier, who serves as the chair of the Council of Ministers (cabinet). The Fifth Constitution discusses the policy areas of these top executives in broad terms but establishes no clear demarcation. Yet it is obvious that the premier, who is supposed to "direct the operation of government" and to "ensure the execution of the laws," is in fact subordinate to the president. Thus, the premier's task in effect consists of particularizing the policies that the chief of state has outlined. De Gaulle actually claimed all policy-making powers for himself and allowed his premier only to implement them. And even during Giscard's presidency, a premier resigned, disgusted with the role of puppet.

The premier is appointed by the president. However, the latter is not completely free in his or her choice, since the premier and the Council of Ministers should reflect the strength of the various political parties in the National Assembly. When the opposition to the presidential party (or coalition) has firm control of the majority of seats, then the premier is hardly a "co-pilot."

THE LEGISLATURE
The French Parliament is bicameral. The more important chamber is the National Assembly, whose 577 members are directly elected. The candidates run in single-member districts, as in the United States and Britain (where a simple plurality elects). In France, however, the candidate must win an absolute majority. If not, there is a second ballot the next week, including only those candidates who received at least 12.5 percent of the votes. In practice, the parties of the left, and likewise the parties of the right, cut deals so that the second ballot typically includes a single rightist versus a single leftist. The

Senate, which has less important duties, has 274 members, who are indirectly elected for nine-year terms. In the past, it was possible for members of the government to be members of Parliament as well. The Fifth Constitution, however, rules this out, though it allows members of the government to have access to the two assemblies. Parliamentary powers have been greatly reduced in the Fifth Constitution. Gone are the days of parliamentary supremacy that haunted the Third and Fourth Republics. Not only is it no longer easy for the National Assembly to topple the government, but its tasks are strictly circumscribed. Thus, although Article 34 of the Constitution states blandly that "All laws shall be passed by the Parliament," it is, in fact, empowered only to tackle legislation within certain specified policy areas.

THE JUDICIARY
France has traditionally had a large and strong hierarchy of courts. However, due to the doctrine of parliamentary supremacy during the Third and Fourth Republics, the courts did not have the power of judicial review (which the U.S. federal courts have exercised since 1803 to declare laws unconstitutional). The Fifth Republic Constitution added a Constitutional Council, which has the power to supervise elections and to rule on the constitutionality of bills before they become law. The Council's nine members are appointed by (three each) the president of the republic, the president of the National Assembly, and the president of the Senate in staggered fashion for nine-year terms. At first the Council was somewhat of a joke, but since de Gaulle left the scene, it has become steadily more assertive. For example, in the early 1980s, it forced President Mitterrand to modify his nationalization program.

Since the mid-1990s, another part of the French judicial system has also become increasingly assertive: investigating magistrates. Hundreds of businesspeople and politicians have been placed under formal investigation for various financial improprieties, such as illegal party fundraising, kickbacks for contracts, and falsifying documents. In 2001, the magistrates' hit list included several former ministers, the mayor of Paris, officials of the Elf state oil company, the son of former president Mitterrand, the central bank's governor, the leader of the Communist Party, and the former president of the Constitutional Council. Financial scandals even brought into the picture President Chirac, who as mayor of Paris (1977–1995) is alleged to have known of kickback schemes that mainly benefited his party. As an incumbent president, Chirac is immune to prosecution (and French Presidents can be impeached only for treason), but the charges could be politically damaging. Although the French public traditionally has been cynical about politicians (viewing corruption as expected behavior), recent polls have indicated growing support for the magistrates' crusade, which is bringing members of the political elite to trial.

THE ECONOMY

France is among the top Western industrialized economies. Government policy stresses promotion of investment and maintenance of fiscal and monetary discipline. The French economy has traditionally been based on two major strands: *paysantisme* and *dirigisme*. The former concept constitutes the emotional relationship of the French with the soil, the source of all produce, a relationship that has persisted through times of industrialization. Thus President Giscard could, in a 1977 speech, refer to agriculture as *pétrole vert* ("green oil"), a national asset of supreme value. French farmers have been well organized and have maintained the most extensive legislative lobby. They have cultivated close ties with agricultural officials. Furthermore, farmers have a militant tradition of taking direct and sometimes dramatic action when they don't get what they want from government.

Dirigisme has in effect been a concurrent theme, instituted as early as the seventeenth century, when Jean-Baptiste Colbert introduced his system of mercantilism to the statecraft of King Louis XIV. Colbert hoped that government intervention, which gradually became entrapped in a host of regulations and bureaucratic devices, would ultimately benefit the country. A century later, mercantilism was attacked by the physiocrats, who combined governmental direction with an emphasis on nature.

In the mid-twentieth century, Keynesianism became a beacon to many French economists, even if in detail they departed from the original conclusions that its founder had established. Upon taking office in 1981, Mitterrand embarked upon more-than-moderate government spending in order to cure the sagging economy. That, too, may have been inspired by the lessons of John Maynard Keynes.

France also offers many contemporary examples of the practice of state involvement in industries. Although de Gaulle paid lip service to *laissez-faire* principles (it is ironic that this term originated in a country that has persistently preferred statism), he introduced heavy and basic industries that were to be controlled by the bureaucracy. In the context of contemporary examples of government interference, one may also point to nationalizations, which as a matter of course extend state control. Even leading companies such as Thomson and Rhone Poulenc were nationalized (although in the cases cited, the motivation may have differed from the traditional ones—infusions of public capital were made in industries weakened by the oil crises of the 1970s).

It may be possible to detect institutionalized statism in the establishment of the Economic and Social Council. It was intended as an advisory body, but its foundation is nevertheless symptomatic of governmental interest. A much stronger case may be made for the Planning Commission, which has churned out government plans since the mid-1950s. There is ample evidence that during the Fifth Republic, the French economy has been more thoroughly planned than those of its neighbors, even when a centrist or right-wing executive presides in the Elysée Palace.

France has coped remarkably well with the changes wrought on its economy in an extraordinarily brief period. It is also true, though, that, overall, the country has lagged behind other industrializing democracies in its march toward postindustrialism. Statistics reveal that only half a century ago, the country was still basically agrarian. The proportion of the workforce employed in the agricultural sector exceeded that of those working in industry. In 1939, these percentages were 37 and 30, respectively. Currently, the respective percentages are approximately 5 and 26. The rural exodus took place after World War II. At that time, farm mechanization made itself felt, and literally millions of people left the small rural communities in search of city jobs. The postindustrial era will also undoubtedly have sociopolitical implications. The numbers of workers in the old industries are bound to decline. There will be a rise of a white-collar, better-educated working class, a trend that will likely further hurt the Communist Party.

While France is currently the largest food producer and food exporter in Western Europe, it is also one of the larger industrial powers. French industry is advanced in computer science and in telecommunications. France also has the world record as far as nuclear energy consumption is concerned: More than 70 percent of all its energy derives from nuclear energy.

The overall picture of the French economy would be very rosy if it were not for the high unemployment rate. At the same time, like a number of other Western European countries, France has witnessed a veritable invasion of foreign workers. Most of them arrived in the 1960s, when there was a strong need for extra laborers. The majority of guest workers in France came from former French colonies, notably Morocco, Algeria, and Tunisia. To most of the foreign workers, the culture shock was somewhat mitigated by the fact that, as a rule, they had had a foretaste of French culture and often spoke the language. Nevertheless, the ultra right wing, notably the National Front (FN), has made political capital out of their presence. Arousing French ethnocentric and xenophobic sentiments in recent elections, it has argued that the foreign workers increase French unemployment and has advocated that they all be sent back to where they came from.

During the 1990s, the FN scored some local successes in economically weak areas of southern France. Then, in early 1999, the party split as a result of conflict (over means, not ends) between its founder Jean-Marie le Pen and his deputy Bruno Mégret. Furthermore, dissidents from mainstream right parties have set up new political formations that have attracted former FN voters.

A poll in 2000 found that 20 percent of the French people own up to having racist or xenophobic views, much higher than in Germany. However, violence against foreigners has occurred less frequently than in Germany. As Jon Henley reports in *The Weekly Guardian*, discrimination in France "emerges in small everyday incidents, in coy linguistic double-thinks . . . [and especially] in employment."

RECENT DEVELOPMENTS

The extent of presidential power in France made the outcome of the 1995 presidential election extremely important, not only to France, but also to the European Union. In the first round of the election process, on April 23, 1995, Socialist Lionel Jospin got the most votes, with 23.3 percent, a surprise even to his own pollsters. In a very close battle for second place, Paris mayor Jacques Chirac edged out Premier Edouard Balladur, with 20.8 percent and 18.6 percent respectively. Among right-wing candidates, National Front leader Jean-Marie Le Pen won 15 percent of the vote, and Philippe de Villiers won 4.7 percent. This showing was the highest for the far right in Europe since the end of World War II. On the left, Communist, Trotskyite, and environmentalist candidates won a total of 17.2 percent. The run-off election was held on May 7, between the two

Louis XIV becomes king of France A.D. **1651**	Outbreak of the French Revolution **1789**	Napoleon is defeated; the Congress of Vienna **1815**	The fall of the Second Empire; the founding of the Third Republic **1870**	The end of the Third Republic; France is defeated by Germany; World War II **1940–1945**	The founding of the Fourth Republic **1946**

leaders, Jospin and Chirac. As most polls had predicted, the run-off was won by Chirac.

Although most French people consider the spontaneous and charming Chirac quite likeable, the government has appeared rather directionless during his presidency. The economy has improved somewhat during his years in office, but there are many trouble spots. In March 2001, unemployment stood at 8.7 percent in France (slightly above the average for the Euro Zone). While the overall rate has moved downward in the last couple of years, the situation for many young people has remained bleak: Nearly a quarter of those under 25 have no job. Chirac has had some achievements, including implementing measures to end conscription and streamline the armed forces, overhaul the public-health system, and sell off or deregulate parts of the public sector.

During his first two years in office, President Chirac's prime minister was Alain Juppe, from Chirac's party, the neo-Gaullist Rally for the Republic. Chirac and Juppe benefited from a very large majority in the National Assembly (470 out of 577 seats). In order to ensure that France met the Maastricht criteria for the single European currency (to bolster the franc's stability and strength), and mindful that the government might have to make politically costly decisions in advance of the legislative elections planned for spring 1998, Chirac decided in April 1997 to call early elections. This impulse turned out to be rash. The left, led by Lionel Jospin, the Socialist Party's leader, unexpectedly won a solid National Assembly majority (319 seats, with 289 required for an absolute majority) in the two rounds of balloting, which took place May 25 and June 1, 1997. President Chirac named Jospin prime minister on June 2, and Jospin went on to form a government composed primarily of Socialist ministers, along with some ministers from allied parties of the left, such as the Communists and the Greens. In 1997, the Communists received about 10 percent of the (first-ballot) votes, about half what they could count on before the 1980s. Although there have been four Communist ministers in Jospin's government, the party organization has continued to sag, losing by its own statistics 90,000 members since 1997. Although the Greens won only about 7 percent of the

votes, they not only entered the National Assembly for the first time but also obtained the environmental ministry for Dominique Voynet, the leader of *les Verts* (the largest Green party). Voynet established a public profile in battling against nuclear power and for ecological tax reform. She was a key player at the November 2000 Hague climate change conference, where the EU and the United States deadlocked on how to reduce greenhouse gases.

The tradition in periods of cohabitation is for the president to exercise the primary role in foreign and security policy, with the dominant role in domestic policy falling to the prime minister and his or her government. Prime Minister Jospin has stated, however, that he will not leave any domain exclusively to the president, though President Chirac still claims to have the "final word" regarding foreign and security issues. However, both the president and the premier were in the limelight at European Council and international meetings as "duo" representatives of France.

Jospin, likely rival of Chirac in the 2002 presidential election, has pushed major domestic reforms. For example, in 2000 he crafted an autonomy deal for the French Mediterranean island of Corsica, which has been the site of a long-running separatist struggle against French authority. The Corsican regional assembly voted overwhelmingly to accept the peace accord that would grant it authority in numerous policy areas and allow it to "adapt" French legislation to local circumstances (after 2004 without the approval of the National Assembly). Furthermore, the Corsican language would become a part of the island's official curriculum. Jospin's initiative provoked the resignation of his interior minister, Jean-Pierre Chevènement, who was upset that the separatists were not required to renounce violence before negotiations could begin. Others, including President Chirac, articulated concerns about the implications for the unity of the French nation-state. In spring 2001, the autonomy plan moved to the agenda of the National Assembly, whose approval would set the stage for a possible challenge in the Constitutional Council.

In recent years, the French left has energetically endeavored to increase the role

of women in political and social life. France, where women received the right to vote only in 1944, has lagged behind other major Western democracies. Prior to the 1997 election, the Socialists adopted gender quotas to increase the number of women among their parliamentary candidates. Jospin appointed 10 women to his cabinet of 32 ministers. Still, only 9 percent of France's national legislators were women, compared to 13 percent in the United States and the 20 percent average in Europe. Jospin's majority in Parliament passed a new law that obliges all political parties to nominate equal numbers of men and women as candidates not only in parliamentary elections, but also in municipal elections. France has had among the most conservative laws granting legal abortions in Europe; in 2000 reforms were passed to extend the time limit for most abortions. A new law is also on the books allowing for legal unions for gays and lesbians.

In its rhetoric, the French government has expressed many of the same concerns of French intellectuals about negative effects of globalization. Jospin has declared, "The world is more than a market." Most politicians still regard neo-liberal "Anglo-Saxon" capitalism as a negative model. However, as *The Economist* observed in its 1999 French survey, "Over the past few years, their famously *dirigiste* economy has been liberalized beyond recognition: markets in electricity, telecommunications and gas have been opened to competition, and one-time icons of the French state. . . have been partially released into private hands." Furthermore, Jospin's finance minister has pushed significant tax reduction, and even the politically touchy issue of pension reform is on the table. On the other hand, his leftist coalition remains committed to the 35-hour work week and to social solidarity by raising the living standards of the working poor. Despite the record of accomplishments, Jospin's popularity plunged as a result of soaring fuel prices in autumn 2000. In the face of mounting protests by truckers, farmers, and fishermen, the government caved in and made retroactive concessions on fuel taxes, prompting the French "disease" to spread rapidly to neighboring countries, even those like Britain that lack a tradition of grassroots militance.

The Fifth Republic is founded by General de Gaulle 1958	Students and autoworkers revolt 1968	De Gaulle leaves office 1969	François Mitterrand is elected as the first Socialist president under the Fifth Constitution 1981	The Socialist cabinet is phased out and replaced by a right-wing cabinet; cohabitation 1986	Mitterrand is reelected; the right-wing cabinet is replaced by a center-left cabinet 1988	Jacques Chirac wins the presidency; Alain Juppé and then Lionel Jospin serve as prime minister 1990s	Protests mount: against EU agricultural policies, against the *lycée* system

2000s

Green parties gain power

President Chirac shares power with Prime Minister Jospin's left coalition

Corruption among high government officials is scrutinized

FRANCE AND THE EU

In 1992, the Maastricht Treaty (the Treaty on Political and Economic Union), endangered by a negative referendum outcome in Denmark, was rescued when a similar referendum in France resulted in 51 percent favoring the treaty, with 49 percent opposing it. Certainly this was a very small majority, but the majority in Denmark that rejected large parts of the treaty was even weaker. In subsequent years, France has continued to display ambivalence about the European Union and its goals, though it plays a central role in the organization.

Public-opinion polls have provided little evidence that the French, after a half-century, have developed strong emotional ties to the European Union. On the other hand, they have indicated that the French have had fewer misgivings in giving up the franc for the Euro single currency than the Germans have had in giving up the deutschmark. Furthermore, when asked in 1997 whether France's membership in the EU is "a good thing," French respondents were less likely to say yes (48 percent) than the Italians (69 percent), but more likely than the Germans (38 percent) and the British (36 percent). Yet compared to results in previous decades, the relative numbers of the French replying "a bad thing" or "neither good nor bad" has grown. When asked directly whether the country has benefited or not from EU membership, the French came down 44 percent versus 37 percent on the side of "benefited," which was close to the EU's average response in 1997. Certainly French farmers have received the largest share of the EU's Common Agricultural Program funds through the years. In 2000, when asked whether the EU should be given more powers to deal with the environment and food safety, the majority of French surveyed agreed. When the German foreign minister presented his vision of a "Euro-

pean Federation" in his May 2000 Humboldt University speech, he provoked negative responses from the French foreign and interior ministries; the latter suggested that the Germans still hadn't learned from their past attempts to dominate Europe (he later apologized).

President Chirac's turn to present his view of Europe's future came in his June 2000 speech in Berlin to the Bundestag. Although vague on details, he supported the idea of a written constitution for the EU. His vision was of a "united Europe of states rather than a United States of Europe." Chirac renewed calls for a "pioneering" group of EU members (to be sure, France and Germany, and the Benelux countries) which would move ahead toward closer integration than the others, which could later catch up. The British expressed misgivings about any scheme that sets up first-class and second-class membership. At the December 2000 Nice summit, which climaxed France's six-month EU presidency, a new draft treaty was agreed to by the 15 member states after prolonged, contentious negotiations, beyond the EU norm. Although Chirac declared it "one of the great successes" in the EU's history, most outside reaction was critical of his bullying of the small countries and his snubbing of the European Commission. From the viewpoint of French national interests, the summit had produced a treaty that would allow Eastern (and Southeastern) EU enlargement to go forward, reinforce the position of the big countries within the EU framework, and preserve France's parity with Germany in the Council.

The French leadership has supported further development of a "European Common Foreign and Security Policy," the so-called second (and intergovernmental) "pillar" of the EU. Accordingly, it has committed 12,000 troops to the Rapid Reaction Force that the EU is planning to

have in place by 2003 to deploy in crisis circumstances where NATO partners might decline to support intervention. The U.S. administrations of both Bill Clinton and George W. Bush have expressed misgivings about an independent European defense force, seeing it as a duplication and diversion of resources form NATO's proven collective-defense capability. The British government conceives of the new combined force as a supplement to NATO, which would remain the indispensable guarantor of European peace and security. On the other hand, the French government envisages an embryonic "European Army." And the German government's view is somewhere in between. French foreign minister Hubert Védrine has made clear his apprehension about a world dominated by a single "hyperpower" (the United States). Thus, from his view, with the EU already a major economic player in the global economy, a policy whose goal is an autonomous capability to intervene militarily in hot spots, like the Balkans, should now be pursued. The outcome would be a more equal partnership with the Americans, and ultimately a more multipolar and diverse world—from his view, a better world.

DEVELOPMENT

Until World War II, France was still largely agrarian, with little industry. This has changed, particularly under de Gaulle, who emphasized heavy and basic industries. France is entering the postindustrial era.

FREEDOM

France ranks high in Humana's human-rights rating. However, French government has become somewhat arbitrary in matters of asylum of refugees. The strength of several far-right groups that are venomously anti-Semitic and anti-immigrant is disturbing to many observers both within and outside of France.

HEALTH/WELFARE

France has a welfare system that includes health care. However, patients have to pay for health services first and subsequently have 80% of charges reimbursed. Under the welfare system, all French people have 5 weeks of paid vacation annually.

ACHIEVEMENTS

France has traditionally been known for its cultural sophistication. Several Nobel Prizes have been won by French scholars and authors. France has also become very advanced in computer, space, and telecommunications technology.

Germany (Federal Republic of Germany)

GEOGRAPHY

Area in Square Miles (Kilometers):
139,412 (356,853) (about the
size of Montana)

Capital (Population): Berlin
(3,459,000)

Environmental Concerns: air and
water pollution; acid rain;
hazardous-waste disposal

Geographical Features: lowlands
in the north; uplands in the
center; Bavarian Alps in the
south

Climate: temperate and marine

PEOPLE

Population

Total: 82,798,000
Annual Growth Rate: 0.29%
Rural/Urban Population Ratio:
13/87
Major Language: German
Ethnic Makeup: 92% German;
2% Turkish, 6% others
Religions: 38% Protestant;
34% Roman Catholic; 28%
unaffiliated or other

Health

Life Expectancy at Birth: 74 years
(male); 81 years (female)
Infant Mortality Rate (Ratio):
4.7/1,000
Physicians Available (Ratio): 1/293

Education

Adult Literacy Rate: 99%
Compulsory (Ages): 6–15

COMMUNICATION

Telephones: 46,500,000 main lines
Daily Newspaper Circulation: 375
per 1,000 people
Televisions: 551 per 1,000 people
Internet Service Providers: 625
(1999)

TRANSPORTATION

Highways in Miles (Kilometers): 403,087
(650,140)
Railroads in Miles (Kilometers): 25,312
(40,826)
Usable Airfields: 615
Motor Vehicles in Use: 44,500,000

GOVERNMENT

Type: federal republic
Independence Date: January 18, 1871
(German empire unification)
Head of State/Government: President
Johannes Rau; Chancellor Gerhard
Schroeder
Political Parties: Christian Democratic
Union; Christian Social Union; Free
Democratic Party; Social Democratic
Party; Alliance '90/Greens; Party of
Democratic Socialism; others
Suffrage: universal at 18

MILITARY

*Military Expenditures (% of Central
Government Expenditures):* 1.5%
Current Disputes: legal issues (restitution)
remaining from World War II and its
aftermath

ECONOMY

Currency ($ U.S. Equivalent): 2.31 marks = $1
Per Capita Income/GDP: $22,700/$1.86
trillion
GDP Growth Rate: 1.5%
Inflation Rate: 0.8%
Unemployment Rate: 10.5%

Labor Force: 40,500,000
Natural Resources: iron ore; coal; potash;
timber, lignite; uranium; natural gas;
salt; nickel; copper; arable land
Agriculture: grains; potatoes; sugar beets;
fruit; meat
Industry: iron; steel; coal; cement;
chemicals; machinery; shipbuilding;
motor vehicles; food and beverages;
machine tools; others
Exports: $610 billion (primary partners
EU, United States, Japan)
Imports: $587 billion (primary partners
EU, United States, Japan)

 http://www.cia.gov/cia/publications/
factbook/geos/gm.html

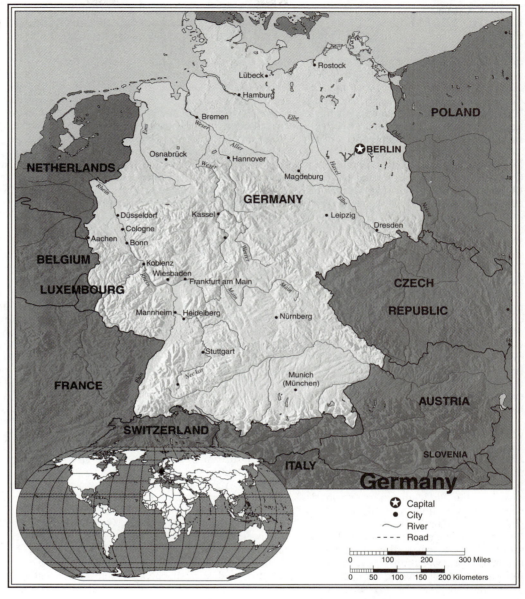

THE GERMAN IDENTITY PROBLEM

Of the larger and more important countries in Western Europe, Germany seems to be the one that is most afflicted by identity problems. On the one hand, the German "nation" is very old; on the other, its experience in the nation-state format has been relatively brief. Roman records mention the existence of Germanic tribes living in the general area between the Ems and what is currently known as the Oder–Neisse border. The Roman province of Gaul, created in 51 B.C., extended to the Rhine River, which was regarded as an ethnic frontier. Nevertheless, the general area was extremely turbulent since, in addition to the earlier *Völkerwanderung* ("great migration"), there were other—smaller—population movements as well. The Roman historian Tacitus, who lived from A.D. 55 to 120, ventured some scathing observations. He believed that the Germans had not progressed a great deal in civilization. By way of proof, he noted that they were dressed in skins and armed with clubs. He also noted that the Germans had a penchant for war as well as for gambling. Conversion to Christianity resulted from the unrelenting efforts of such missionaries as Boniface, an English monk (680–754).

After the death of Charlemagne and the partitioning of his empire in A.D. 843, Germany became part of the East Frankish Empire, which was later to become known as the Holy Roman Empire. The empire was officially to endure until 1806; however, after the thirteenth century, its political system was little more than a loose collection of hundreds of principalities (and free city-states) with no central institutions. The German portion of the multinational empire was further divided by the Reformation, as it became the battlefield in that prolonged and bloody struggle between Protestants and Catholics. (It is still possible centuries later to observe cultural differences between mostly Protestant northern Germany and mostly Catholic southern Germany.) Certainly the doctrine *cuius regio, eius religio,* which caused the religion of the ruler to be the prime determinant for the religion of the particular area, wrought havoc with whatever unity may have existed shortly after the Peace of Westphalia in 1648.

At this time, the nation-state format was emerging in Western Europe. While Germany, from medieval times, had rendered significant contributions to the intellectual, cultural, musical, and artistic enrichment of Europe, its contributions to political thought and practice were less "enriching." Many Germans were more attracted to strong centralized authority than to the liberal notions of restricting the role of the state. Psychologically, such an authoritarian penchant may well have been a backlash against the powerlessness that the small states experienced over many centuries of fragmentation.

But against the *Wille zur Macht,* as philosopher Friedrich Nietzsche called the impulse for power, one must also take into account the German definition of the *Rechtsstaat,* a state bound by laws, a concept that became the continental counterpart of the Anglo-American Rule of Law. It came to be formulated by Prussia, a unit east of the Elbe. Prussia had grown out of the Electorate of Brandenburg around the city of Berlin. Aided by its aristocracy, it became an important military power in the eighteenth century, excelling in such "Prussian virtues" as administrative efficiency.

By the end of the eighteenth century, Prussia had emerged as one of the great powers of Europe. After Napoleon's defeat, Prussia resumed its efforts to expand its control. Under the leadership of Otto von Bismarck, Prussia fought and won wars against Denmark, Austria, and France between 1864 and 1870. Finally, in 1871, Bismarck founded the Wilhelmine Empire, putting all the German states under one roof and under one ruler. Bismarck, who held the post of chancellor (the German term for prime minister) of Prussia, became chancellor of the new construction as well. By taking control of France's Alsace-Lorraine and the Saarland, Bismarck set the stage for nearly 80 years of nationalist rivalry and two devastating wars between France and Germany. (Only in the 1950s did a Franco–German alliance come into being.)

In World War I, which France hoped would yield a revanche, England and eventually the United States became involved. On November 11, 1918, an armistice was proclaimed. Germany had lost the war. (Refer to the regional essay for more information on World War I.)

The German Empire had come down with a bang, and what came in its stead was the Weimar Republic, founded in 1919. (It lasted until 1933.) Germany had been forced to cede territory and had to pay formidable amounts in reparations. As a result, its economy deteriorated even further. Against this backdrop, a little-known Austrian agitator named Adolf Hitler, who in 1923 had attempted a coup that had failed miserably (and for which he spent time in prison, during which he wrote his anti-Semitic manifesto *Mein Kampf*), started his steady ascent to power in Germany. He used to the utmost the bad economic conditions, which in his view derived from the Versailles *Diktat* (the dictated peace of 1919) and from alleged Jewish conspiracies. Gradually his National Socialist (Nazi) Party gained power. It never gained a majority during the Weimar years; in elections in 1933, it received 34 percent of the vote. With the Communists (the Nazis' arch-enemy) it held a "negative majority" that could torpedo any policy designed by the moderate parties of the government. Finally, President Paul von Hindenburg, then 86 years old, was convinced by advisers to appoint Hitler as chancellor.

THE NAZI YEARS

Hitler was sworn in as chancellor on January 30, 1933, commencing 12 years that constituted a horrifying period for Germany and for the world. Believing that his aims would be better served through a highly centralized government apparatus, Hitler immediately put an end to the federal form of government adopted by the Weimar Republic. On the pretext that Communists had set fire to the Parliament building, Hitler declared a state of emergency only two months after having assumed office. The Weimar Constitution allowed him in that case to rule through decrees, giving him extraordinary powers. This "emergency" was maintained until the very end of Hitler's reign (May 1945).

In foreign-policy matters, Hitler was drastic: He almost immediately tested the Allies' will to enforce the conditions of the Versailles Treaty. It soon became clear to him that France and Great Britain were war-weary and reluctant to risk any conflict. While German strength was still below par, Hitler launched his march into the Rhineland, something the Versailles Treaty had specifically prohibited. Gradually the chancellor (or *Führer,* "leader," as his title had become) embarked upon implementation of two major policies: 1) the formation of a large and powerful army; and 2) the wholesale, state-sponsored elimination of the Jews of Europe. The first policy led to World War II. The second led to the Holocaust, the horrific period when the Nazis murdered 6 million European Jews and 5 million other people. (This period is discussed in greater detail in the regional essay.)

Shortly before World War II began, Hitler concluded a nonaggression treaty with the Soviet Union. He feared that he would be preoccupied in the west and did not want to fight a war on two fronts. The areas that were incorporated into Germany before the war were Austria (which Hitler regarded as an *Anschluss*—an annexation, or voluntary merger), Sudetenland, and, subsequently, Czechoslovakia in its entirety. After that, a number of countries were invaded and occupied by Germany: Poland (1939, setting off World War II), Denmark and Norway (April 1940), the Low Countries (Belgium, the Netherlands, and Luxembourg, May 1940), and France (June 1940). Then came a pause, since the German armies did not succeed in invading England. Nevertheless, in April 1941, Hitler felt strong enough to fight on another front: He directed his attention eastward and invaded the Balkans, notably Yugoslavia and Greece. These countries too were conquered. Finally, Operation Barbarossa (the code name for the German invasion of the Soviet Union) was launched, a little later than originally scheduled, due to unanticipated Greek resistance, in June 1941. This move proved to be Hitler's downfall, since what he had feared most—to have to fight on two fronts—now became a reality.

Now two new factors became of prime importance. First, the Russian winter turned out to be considerably fiercer than the German Army leaders had anticipated; it forced the war machine to stop. Second, Japan had launched a war against the United States in December 1941, causing the European war to become a world war. Thus, the United States, which had refused to be dragged into the European theater, now found itself at war first with Japan but by implication also with its allies, Germany and Italy. The Nazi German empire had been at its largest when Japan entered the war, and although the hostilities were still to last more than three years, that period was marked by gradual diminution—the Soviet armies driving back the Germans from the east, and the Anglo–American forces, after an invasion in Normandy had been staged, from the west. For Germany, the drama finally came to an end in May 1945: Hitler had committed suicide in a bunker, besieged from all sides, having made not entirely successful attempts to transfer government to others outside Berlin. His ravaged armies fought until May 7, 1945, when Germany officially surrendered unconditionally.

RENEWED DIVISION

Once the European war had become a world war, the Allied leaders were in frequent contact with one another, notably at historic "summit" meetings (the term had yet to be coined) held in Cairo, Teheran, Yalta, and Potsdam. These conferences were intended to map out the course of the war. In addition, these "war councils" fully considered the future—how the postwar world would be constructed. It was decided to establish zones of occupation for the time being. The Soviet armies coming from the east would be responsible for those parts of Germany and Austria that would come under their control when the Third Reich surrendered. Similarly, the armies coming from the west would institute American, British, and French zones of occupation. Berlin, the German capital, which would fall in the Soviet-occupied area, would become a miniature replica of Germany, in that it too would have four occupation zones. Naturally, these divisions were believed to be only temporary.

However, within two years of the end of the war, the friction between the Soviet Union and the United States had degenerated into what came to be known as the "Cold War," a bitter enmity to the brink of armed hostilities. This Cold War diverted some of the vengeance and vindictiveness on the part of the victors vis-à-vis their former enemies. Thus, in 1947, the United States decided to include Germany (or at least the parts that were under the control of the United States, Great Britain, and France) among the recipients of Marshall Plan aid. This assistance scheme (its official name was the European Recovery Plan), provided goods and money to Western European countries devastated by the war, facilitating the recovery of their economies. U.S. Secretary of State George Marshall, feared that these countries would otherwise be a ready prey for communism. The aid had initially been offered to some countries behind the Iron Curtain as well. But Moscow did not want any area that fell under its control to participate in the plan.

(Photo credit National Archives)

On January 30, 1933, following a steady rise to power in a time of economic turmoil, Adolf Hitler was sworn in as German chancellor. He immediately set about implementing radical domestic and foreign policies that eventually would fuel a world war and bring about the Holocaust, a genocide in which literally millions of Jews and other peoples were murdered.

The Marshall Plan and the intensifying Cold War caused growing disaffection between Central/Eastern and Western Europe, and thus between the two parts of Germany. The western part recovered economically in a remarkably short time. By contrast, the eastern part was plundered, its industry shipped to the Soviet Union as "reparations."

The Cold War also caused the two Germanys to develop extremely hostile attitudes toward each other. This was particularly the case once separate states had started to take the place of the occupation zones. In 1949, the United States wanted to include the western part of Germany in an alliance that was to prevent the Soviet Union from controlling all of Europe. Since only an independent country could be the buffer that the United States wanted, the U.S., British, and French occupation zones were united and converted into a new West German state, to be named the Federal Republic of Germany. Moscow followed suit with its eastern zone, and soon the western and eastern parts of Germany had become member states of rival collective-defense organizations (respectively, NATO and the Warsaw Pact).

East Germany—that is, the part that became the German Democratic Republic—was considerably smaller in area and in population than its Western counterpart. In addition, the Soviet Union annexed a small part of East Germany, and a larger slice of eastern Germany was given to Poland in order to compensate that country for a slice of eastern Poland that the Soviet Union had incorporated into its body politic. As a consequence, hundreds of thousands of Germans born and raised in the eastern prewar part of their country found themselves living under the Polish or Soviet government.

THE GENESIS OF MINORITY PATTERNS IN GERMANY

The Federal Republic of Germany not only exceeded the German Democratic Republic in prosperity, it soon also outstripped most Western European countries. West Germany's economic strength was such that it was invited to be a party to the Treaty of Rome in 1957, a pact that was to lead to the European Union. In this period—the late 1950s and early 1960s—West Germany experienced its "economic miracle." No wonder that it developed into a magnet for the entire Eastern bloc, but particularly for the German Democratic Republic.

Some 3 million people fled from East Germany before the government of that country prevented further mass flights by strengthening its border controls and shooting at persons who were attempting to flee. One of the favorite escape routes at the time was through Berlin. East Germans went to East Berlin (which had been declared the capital of East Germany); it was very easy to get from West Berlin to West Germany at that time. The population movement from East Germany to West Germany had been fairly gradual, which allowed the West German economy to absorb the newcomers.

There was still a great demand for labor, and West German industries started to attract workers from the Mediterranean basin, notably from Turkey. These *Gastarbeiter* ("guest workers") were hired on a short-term basis, usually for two or three years. However, when their contracts expired, they were normally renewed. Many of them did not return to Turkey at all or did so only for a brief leave or vacation. The guest workers started to constitute minorities in West Germany who often failed to integrate into the German culture. The Turks were Muslims in the overwhelmingly Christian host country. Today there is a Turkish minority population of some 2.1 million, who live in parts of Berlin, Frankfurt, and other big cities. Friction between the guest-worker populations and Germans historically has become more acute in times of economic stress.

These tensions, however, appear minor when compared to the social and economic problems associated with asylum-seekers beginning in the 1980s. The "plane people," as they came to be called, hailed from countries in South Asia. Interflug, a Central/Eastern European airline with a dubious reputation, sold individuals one-way tickets to East Germany, ostensibly for the purpose of vacationing. However, that was not the final destination. The "plane people" proceeded to East Berlin, which yielded easy access to West Berlin for non-Germans. As soon as they found themselves on West German soil, they claimed political persecution in their home countries, trying to achieve refugee status and eventually asylum.

It was up to the German authorities to find out whether these asylum-seekers were covered by Article 16 of the Basic Law (the German Constitution), which declares, "Persons persecuted on political grounds shall enjoy the right of asylum." That article had left the asylum question simple because, at the time it was written, only persons fleeing from Communist regimes east of Germany were expected to qualify for asylum. (Germans fleeing Communist East Germany did not require asylum, since they already had German citizenship.) It usually took West German authorities two to three years to find out whether or not the Basic Law had been abused. In the meantime, those aspiring to asylum were housed and fed at German taxpayers' expense. In the vast majority of cases, the claims of political persecution appeared groundless. But even if refugee status was denied, the "fake asylum-seekers" would rarely be forcibly expelled. During 1992–1993, the major German parties reached a consensus about restricting the constitutional right to asylum. (The new law does not apply to ethnic-German resettlers.) Asylum-seekers reaching Germany through "safe" third countries and those coming from countries judged "democratic" can now be sent back immediately. Furthermore, Germany has provided aid to Poland and the Czech Republic to allow them to better control their borders and care for refugees. The ethnic cleansing and armed conflict in the former Yugoslavia was treated by the German government as an humanitarian emergency. Thus, during the mid- and late 1990s, Germany accepted hundreds of thousands of these refugees (more than any other country) until conditions would allow their safe return home.

Early in the 1990s, the German government announced that all residents of Central/Eastern European countries, including the successor states of the former Soviet Union who were able to trace their ancestry to Germany, could demand to be brought back—and fed and sheltered—at the Federal Republic's expense. Thus far, the call "home" has been answered by more than half a million Germans or former Germans resident in the ex-Soviet Union alone. There are still close to 3 million ethnic Germans in Central/Eastern and Southeastern Europe, many of whom do not even speak German. In combination with the costs of upgrading the economy of the former German Democratic Republic (now referred to as "eastern Germany"), these repatriation plans have proved to be so colossal that the German government has been forced to modify them. Germany has begun sponsoring German-language courses and cultural activities in other European countries and simultaneously encouraging ethnic Germans to stay where they are.

OSTPOLITIK AND DEUTSCHLANDPOLITIK

In the early 1970s, West German chancellor Willy Brandt embarked upon what he called *Ostpolitik* ("Eastern Policy"). In fact, the term was originally part of the German lexicon in the days of Otto von Bismarck. Bismarck, "the Iron Chancel-

(Photo courtesy Cornelia Warmenhoven)

In 1937, the Nazis created the Buchenwald concentration camp in Germany. Today the site is a national memorial, helping to teach people about what happened to the millions of Jews and others who were murdered in the Holocaust.

lor," appeared dead-set against letting Germany have "her place in the sun" (the standard expression for a colonial empire). Instead, he cast a favorable eye on countries and areas east of Germany that could easily be annexed as *Lebensraum* (living space). Bismarck's Ostpolitik was never implemented, however, since the chancellor finally gave in and allowed Germany to have overseas colonies.

In Brandt's time, the term referred to West Germany's attempts to arrive at some reconciliation with those eastern countries that had been plundered and pillaged by the German armies. In the 1970s, Ostpolitik consisted of three major gestures: Brandt, on behalf of his nation, apologized for the atrocities committed; he came up with a compensation program to persons who had particularly suffered at the hands of the Nazis; and he proposed mutual rec-

ognition to be followed by the establishment of diplomatic relations.

In 1974, it was found that an East German spy had been able to penetrate Brandt's inner circle, thereby endangering vital NATO secrets. The chancellor immediately assumed responsibility and resigned. Brandt was succeeded by Helmut Schmidt, who, in the late 1970s and early 1980s, developed what came to be called the *Deutschlandpolitik*. This policy addressed the relationship between the two Germanys, which had frequently been overtly hostile, especially whenever the Cold War flared up. While it did not create a climate of reconciliation, the Deutschlandpolitik increased communications between the Germanys. Still, it is difficult to see any direct correlation between the budding but often intermittent dialogue that West Germany and East Germany en-

tertained and the events that started to unfold in the latter half of 1989. Both Germanys were then 40 years old and more often than not victims of, as well as participants in, the Cold War. Those four decades had witnessed a host of Cold War incidents such as the Berlin Airlift, the Hungarian Revolution, the Prague Spring and its bitter aftermath, and repeated crises in Poland. If these events punctuated the period, there was no accumulation of hostility; however, as the years went by, the hopes for a reunification faded. Back in 1949, when the Federal Republic of Germany was established, its Basic Law had indicated that "The entire German nation remains invited to complete the unity and freedom of Germany."

REUNIFICATION

It is difficult to point to a single factor or, for that matter, to a combination of factors that were to lead to the reunification. Possibly the armaments race, with its emphasis on high technology, proved too costly to the Soviet Union, causing its economy to become overburdened. Perhaps the changing direction of the Soviet Union through Mikhail Gorbachev's glasnost and perestroika policies produced totally unexpected side effects. However that may be, in early 1989 Gorbachev made the rounds in Central/Eastern Europe, telling its Communist rulers that they would be on their own. The era of the Brezhnev Doctrine was over. The factor of relative deprivation may also have played a role. The affluent lifestyle of their Western neighbors was something East Germans were thoroughly aware of: They heard it on radio broadcasts, they watched it on television. It was a lifestyle that they would have preferred for themselves; in comparison, their economic lives continued to be grim.

With remarkable candor, Central/Eastern Europeans had once made clear that they wanted "socialism with a human face." That had been the popular slogan during the days of the Prague Spring in 1968. In the aftermath of the clampdown on Czechoslovakia by Warsaw Pact troops, communism (socialism and communism were identical in Central/Eastern Europe) was put on the defensive. The region's economies lagged further behind the West. Corruption became widespread, especially among those who held positions of power.

In a way, it is somewhat surprising that it took 21 years for dissent to ferment into protests, but it should be remembered that until early 1989, the Soviet presence and pressure were intense. All over Central/Eastern Europe, Soviet troops were in place; and memories of bloody episodes that had cost many lives (especially in Budapest in 1956) made it difficult to start thinking about new protests.

But 1989 was the year of the explosion. Shortly after Gorbachev had paid his respects to the heads of state or government in the Eastern bloc, the increasing disenchantment revealed itself in two ways.

East Germans started to take to the streets, day after day, in ever-larger numbers, in Leipzig and later in other large cities. While these demonstrations, which involved hundreds of thousands of people, were peaceful, they must have appeared threatening to East German authorities, since many demonstrators had started displaying antigovernment slogans. Others again made attempts to flee East Germany by asking for asylum at foreign embassies, and occasionally remaining their guests for weeks on end. The Hungarian Embassy was often selected, because Hungary had become a maverick in the Eastern bloc. But Hungary too was only used as a transit country, the goal being to enter West Germany. This massive "voting with their feet" was extremely embarrassing to the East German government, which made statements, announcements, and promises that things would soon get better.

The Berlin Wall, which bisected the city, had since its erection in 1961 been a symbol of the separation in which East Berliners lived. Suddenly, in November 1989, it proved possible to get across this monument; indeed, people started to stand on the wall, to walk on it, to hack away at it. Thus, freedom of access was achieved by the collective will of hundreds of thousands of people. East Germans could now enter the promised land directly; there was no longer any need to make use of detours through Austria and Hungary.

The East German regime tried to save itself by changing party leadership and by opening up to the democratic opposition via "roundtable" talks. However, the verdict of the March 1990 free elections in East Germany was clear: The parties and movements that favored rapid unification with the West prevailed over those that favored a more cautious approach. In July, the two economies were unified as the East and West German currencies were equalized (though the eastmark had been considerably lower in value than the deutschmark). In September, Germany officially accepted its eastern boundaries as redrawn after World War II. In October, East and West Germany were united under Article 23 of the Basic Law, which allowed for the rapid incorporation of the

East as five new eastern states; otherwise, reunification under Article 146 would have required much more time, with the drafting of a new constitution and its approval in a ˙referendum. In December 1990, all-German parliamentary elections returned to power Helmut Kohl, "the Unification Chancellor," and his coalition government of Christian Democrats and Free Democrats. In 1991, a parliamentary majority voted to return the federal capital to Berlin. A decade later, the costly transition has been accomplished.

Since reunification in October 1990, the social and economic difficulties associated with merging the two states and their peoples have dampened the initial euphoria of the government and their citizens. Satisfactory resolution of these tremendous problems will likely take many more years to achieve.

THE NEW GERMANY

Officially, the name of the reunified political system is the Federal Republic of Germany. Unofficially, some have attempted to label it the "Berlin Republic" to differentiate it from the "Bonn Republic" of 1949–1990. However, except for the expansion of the system to include the new eastern states (and East Berlin), its institutional framework remains the same. As the official name indicates, Germany is a federal system, one of only four in Western Europe. The 16 states have their own policy-making domains, governments, parliaments, and courts. Furthermore, the state governments, participate in federal policy-making by sending delegates to the upper house of Parliament, the *Bundesrat*. By European standards, the Bundesrat is a powerful upper house, since it can veto about 60 percent of the legislation passed by the popularly elected lower house of the Parliament, the *Bundestag*. Because different parties or coalitions may control the majorities of the two houses, the upper house can force compromises by the federal government.

The federal government is headed by the chancellor, who is the most significant political figure in Germany. The chancellor is selected by the majority of members of the Bundestag and is removable only by a "constructive" vote of no-confidence, meaning that the opposition must agree on a replacement by a majority vote. In a constitutional sense, the German chancellor is stronger than the British prime minister, in that only the chancellor is given the power "to determine the guidelines of policy" (and is not just first-among-equals). However, in a political sense, the German chancellor is weaker than a British prime minister in that the chancellor

shares power with coalition partners, due to Germany's proportional representation election laws and misgivings about any party having an absolute majority. The symbolic leader as head of state is the federal president, who is typically a well-respected senior politician. The President is elected by the Federal Assembly of state and federal parliamentarians for a once-renewable five-year-term. There is no limit on the chancellor's tenure; Helmut Kohl was chancellor from 1982 to 1998. To date, all federal chancellors and presidents have been men.

The Bundestag is an active legislative chamber. Despite party discipline, it enjoys institutional autonomy from the executive, electing its own leadership and organizing its own agenda. Although most bills originate with the government, the Bundestag has a powerful set of committees, compared to the British House of Commons, which may amend legislation and scrutinize administration. Currently Germany ranks seventh in the world in its parliamentary representation of women; 31 percent of the members elected to the Bundestag in 1998 were women.

Half of the 16 judges of Federal Constitutional Court are chosen by the Bundestag, and half by the Bundesrat. In contrast to the lifetime appointments of U.S. Supreme Court justices, they serve nonrenewable 12-year terms. The Federal Constitutional Court has evolved as an active, well-respected court with the power of judicial review; that is, it can (and does) declare federal and state laws unconstitutional. Its current presiding judge is a woman.

At the outset of the Federal Republic, the party system somewhat resembled that of the old Weimar Republic, when 10 parties (and independents) won Bundestag seats in the 1949 election. But over the next decade, almost all of the small parties disappeared from the Bundestag, leaving West Germany with two large "catch-all" (or *Volksparteien*) parties that sought to appeal to a broad spectrum of voters: the right-of-center Christian Democrats (CDU/CSU), and the left-of-center Social Democrats (SPD). The small, liberal Free Democratic party (FDP) has played the role of the "king maker," forming coalitions with the CDU/CSU or (after 19690 with the SPD. In 1983, an environmentalist/antiwar party, the Greens, entered the Bundestag after winning seats in a number of state parliaments. Initially the Greens maintained a strategy of fundamental opposition: no coalitions with other parties. Yet within a decade, their realist wing had prevailed and they were sharing power with the SPD at the state level. Reunifi-

Beginning of the Protestant Reformation A.D. 1517–1521	The Thirty Years' War ravages most of Germany 1618–1648	The Congress of Vienna 1814–1815	Otto von Bismarck pushes forward the unification of Germany under Prussian leadership 1860s	The unification of Germany; Bismarck becomes chancellor of the German Empire (Second Reich) 1871	Bismarck is dismissed as chancellor by the new emperor, William II 1890	World War I ends in the defeat of Wilhelmine Germany; the Weimar Republic is established 1914–1918	The Weimar Republic 1919–1933	Adolf Hitler is appointed chancellor and proceeds to consolidate power 1933	Annexation of Austria and Sudetenland; increased terror against Jews in Germany 1938
●	●	●	●	●	●	●	●	●	●

cation brought another party into the Bundestag: the ex-Communist Party of Democratic Socialism (PDS), which has managed to survive by becoming the advocate of eastern interests. Through the years, there have also been various right-wing extremist parties, which have on occasion won seats at the state level. However, none has managed to win any seats in the Bundestag.

In the election of September 27, 1998, Chancellor Kohl's coalition lost its parliamentary majority. His CDU/CSU won 245 seats, with 35.1 percent of the second votes. (German electors cast a first vote for a district representative and a second vote for a party list; the latter is more important in that it more or less determines the share of seats that a party wins.) This outcome represented not only a significant loss of 49 seats for Kohl's party, but also its lowest share of the votes in a Bundestag election since 1949. His coalition partner, the FDP, won 6.2 percent and 42 seats. Gerhard Schröder's SPD finished first with 40.9 percent and 298 seats. By forming a coalition with the Greens (who won 6.7 percent and 47 seats), the SPD had the majority to elect Schröder as chancellor. He formed a new government, whose 17 ministers included three Greens. This development was significant because it was the first time in the history of the Federal Republic that an election had produced a complete turnover of governing parties, and a "bourgeois" party was not sharing power in the federal government and the (once "anti-establishment") Greens were. Nationally, the PDS won 5.1 percent of the votes and 35 seats; in eastern Germany, it won 21.6 percent. The combined support for the three right-wing extremists parties came only to 3.3 percent—much lower support than the far right enjoys in France, Belgium, and Austria. Turnout in German elections has traditionally been high; in the 1998 Bundestag election, 83 percent of Germans voted, compared to 49 percent of Americans in the 1996 presidential election.

Many factors were involved in this historic alternation of governing parties produced by the voters. First, many were reacting against unemployment, which had reached a postwar high of 12.6 percent in January 1998. The rate of unemployment in the east was nearly double that in the west, and the CDU in relative terms lost twice as many votes in the east.

Second, the unpopularity of the Kohl government's reduction in social-entitlement programs, coupled with tax cuts for the wealthy, was a factor. Third, Schröder projected the image of a new, modernizing leader, which contrasted with that of an out-of-touch, exhausted Kohl "clinging to power."

During the campaign, Schröder had sought to appeal to the *Neue Mitte* (the "new middle") beyond the traditional SPD electorate by promising "economic stability, domestic security, and continuity in foreign policy." But his early performance as chancellor was not so reassuring; as some wags observed, "he hit the ground stumbling." Most notable among the new government's problems was the power struggle between the moderate chancellor and his more leftist finance minister, Oskar Lafontaine (who was also the SPD party chair), over economic policy; this struggle was resolved by Lafontaine's resignation. Second, there were public quarrels between Schröder and his environment minister, Jürgen Trittin (a prominent Green), regarding the new atomic law; it was put on hold. In its first year, Schröder's government could point to only a couple of major policy successes. It passed the first new citizenship law (though watered down to obtain FDP backing in the Bundesrat) in 83 years. Now children of foreigners born in Germany are allowed to have two countries' passports until the age of 23, when they must decide which to retain. Furthermore, the naturalization period for adult foreign residents has been shortened. In addition the government began a phased-in program of ecological taxes on energy to encourage conservation, as well as to generate revenues that would allow reductions in social-insurance contributions of individuals and employers.

In its second year, Schröder's "Red-Green" government (as it was commonly known) began getting its act together; meanwhile, the CDU/CSU struggled to put a series of scandals, especially former chancellor Kohl's illegal handling of secret political contributions, behind them. The *Spendenaffäre* ("Donation Affairs") not only forced Kohl to resign as CDU honorary chair, but it also stigmatized his successor as party chairman, Wolfgang Schäuble, who eventually resigned. To symbolically "turn the page," CDU

elected its first-ever chairwoman, Angela Merkel (a relatively young, moderate easterner). Whether she will be the CDU/CSU's candidate for chancellor in 2002 will depend on party leaders' and activists' assessment of her performance and prospects. (Germany does not hold primary elections to nominate candidates.)

Alliance for Work, Schröder's cooperative job-creating program with the unions and employer associations (both of which are well organized in Germany), has not appeared to be particularly productive. Nevertheless, the German economy has recently seemed to be turning the corner, which the government attributes to its sound policies (with the weak Euro helping German exports). In November 2000, unemployment was down to 9.3 percent, gross domestic product growth was projected at 3.1 percent, and consumer prices were up only 2.4 percent over the previous year. The chancellor's biggest coup in 2000 was getting the votes of state governments, where the SPD shares power with the CDU or FDP, so that his program of individual and corporate income-tax cuts passed through the Bundesrat despite the opposition of the national CDU and FDP leaders. His next goal is to pass pension reform, which includes governmental incentives for private investment.

Germany continues to be at the forefront of environmentally friendly policy. In 2000, the much-delayed new atomic law passed when negotiations produced a consensus with the electrical-power industry on closing nuclear-power plants after an operating life of 32 years. The Greens had long favored a shorter period, but they compromised. Thus, Germany has become the first major industrial power to begin the phase-out of nuclear-power plants. Furthermore, the Schröder government has remained committed to the next stages of ecological tax reform, despite mass protests about high fuel prices in the autumn of 2000. Environment Minister Trittin has presented an action program by which Germany can meet its 2012 goal for reduced greenhouse-gas emissions set by the Kyoto Protocol, something the U.S. government has not done.

One of the negative points in any evaluation of the new Germany is the hostility that has been experienced by members of minority groups. Many assaults have been carried out on them by individuals who

| World War II begins in Europe with the German attack on Poland **1939** | The defeat of Germany ends World War II in Europe; shortly before the end, Hitler commits suicide in Berlin **1945** | Germany is under Allied military occupation **1945–1949** | The founding of the Federal Republic of Germany and the German Democratic Republic **1949** | West Germany becomes a member of NATO; East Germany joins the Warsaw Pact **1955** | West Germany is one of the charter members of the European Union **1957** | The Berlin Wall is constructed **1961** | Willy Brandt becomes West Germany's first Social Democratic chancellor and initiates Ostpolitik **1969** | Helmut Schmidt becomes the second Social Democratic chancellor after Brandt steps down **1974** | Helmut Kohl, is chancellor; the Berlin Wall falls **1980s** | Reunification; Gerhard Schröder unseats Kohl **1990s** |

2000s

| A decade after reunification, eastern reconstruction continues | A Red–Green coalition; environmentally friendly legislation is passed | Victims of forced labor in the Nazi era are to be compensated |

call themselves "neo-Nazis," and on occasion firebombs have been thrown into the apartments of immigrants and asylum-seekers. Violent incidents of right-wing extremism have occurred in both eastern and western Germany but are more frequent in the east, especially in areas of high youth unemployment. Although brutal incidents have brought thousands of anti-Nazi marchers into the streets, Foreign Minister Fischer in mid-2000 lectured the general public about the risks of being silent in the face of racist behavior. The government has begun the legal process of banning the most extreme of the right-wing parties, the small National Democratic Party (NPD). Yet it is recognized that though this may send a message, it is hardly a quick-fix solution to the hatred.

Six decades after World War II, Germany still is coming to terms with the legacy of the old Nazism. In June 2000, the Bundestag passed a bill, supported by all parliamentary parties, to set up a fund of $5 billion (half paid by the government and half paid by German industry) to compensate victims of Nazi forced labor.

Schröder's government is a watershed in that (with one exception) all of its cabinet ministers are too young to have personal memories of World War II. Thus, foreign observers have been alert for any signs of changes in direction, now that Germany is in the hands of the "successor generation." Except for an early comment that NATO might rethink its policy of first use of nuclear weapons, Foreign Minister Fischer of the Greens has communicated the new government's support of its predecessor's international commitments. However, NATO's air war against Serbia in the spring of 1999 threatened the Red–Green coalition. Antiwar opposition at the grassroots level forced a special conference of the Greens to decide the party's stance on

the issue. In the end, a compromise motion passed, which did not undermine Fischer, who supported the air war as the only remaining way to stop genocide in Kosovo. Subsequently, Germany contributed aircraft to the war effort and 8,500 troops to the NATO peacekeeping force. Both coalition parties favor streamlining the army, although the Greens favor deeper personnel cuts. Both Schröder and Fischer favor Germany's moving toward an international political role more commensurate with its position in the global economy. For example, they have made the case for a permanent seat for Germany on the Security Council of the United Nations, a body that grew out of the wartime alliance to defeat Nazi Germany.

GERMANY AND EUROPE
During the 1950s, the Federal Republic of Germany become one of the six charter members of the European Coal and Steel Community and the European Economic Community (since 1993 incorporated under the umbrella of the European Union). Chancellor Konrad Adenauer participated fully in European integrative efforts with more than economic gains in mind; he saw the "Europeanization" of Germany as the only way to rehabilitate it in the eyes of its neighbors. Adenauer and his successors accepted de facto French leadership of Europe and generally deferred to French interests. Over the years the German economy became the largest in Europe, and Germany dutifully paid the largest share of the EC budget.

When the process of European integration speeded up in the mid-1980s, the German government supported both the completion of the "Common Market" and the strengthening of European institutions. In 1990, Helmut Kohl viewed "United Germany" as one side of the coin and "United Europe" as the other side. In De-

cember 1992, the Basic Law was amended to make explicit Germany's commitment to European unity through its engagement in the European Union. Despite public opposition to giving up the deutschmark for the common Euro currency and the economic costs involved in meeting the convergence criteria set by the Maastricht Treaty (ratified in 1993), Kohl pushed European Monetary Union so that it could come into existence on schedule (January 1, 1999). The EMU had not been a major campaign issue in 1998, but more of the public was beginning to feel that Germany is paying an unfair share of the EU budget. In a late 1999 survey, Germans were evenly split over the question of whether their country had benefited from European Union membership.

In the 1990s, Schröder made comments indicating that he may be less enthusiastic than Kohl about "deepening" the European Union. However, as chancellor he has shared his predecessor's commitment to the "widening" of the EU to include countries of Central/Eastern Europe. Foreign Minister Fischer's vision of Europe goes beyond Kohl. In May 2000, to the dismay of his French counterpart, he spoke out for a federal Europe with a written constitution, a bicameral legislature, and a directly elected president. In December 2000, however, the outcome of the EU summit in Nice indicated that a federal Europe is still a long way off and that a more assertive Germany is still prepared to compromise to accommodate French interests.

DEVELOPMENT

Germany has the leading industrial economy in Western Europe. It ranks second in the world in international trade and fourth in economic output. A strong base in some traditional industries, such as heavy goods, autos, and chemical products, is now being supplemented by high-tech development. In 2000, Germany became the first major industrial country to decide to phase out nuclear power because of its risks.

FREEDOM

Germany is a representative democracy with civil rights and liberties guaranteed by the Basic Law. There is a vigorous press, and many checks and balances are built into the governmental system. German electoral law has been copied by many post-Communist political systems in Central/Eastern Europe. Extensive formal study of the Holocaust is required of all German students.

HEALTH/WELFARE

Germany's social-market economy combines an emphasis on private enterprise with an extensive social network of protection and welfare benefits for the citizenry. A conservative statesman, Otto von Bismarck, introduced the rudiments of the welfare state more than 100 years ago. Despite the economic and social difficulties of the east, reunified Germany still ranked 14th best in the world in the UNDP's 2000 quality-of-life report (HDI).

ACHIEVEMENTS

The emergence of a well-functioning, stable West German democracy after 1949 must be regarded as a political miracle that outranks even the economic miracle of rebuilding a ruined country and becoming one of the world's most prosperous states. The reunification of Germany in 1990 was achieved nonviolently. Germany is the biggest net contributor to the EU budget, and the European Central Bank is based in Frankfurt.

Greece (Hellenic Republic)

GEOGRAPHY
Area in Square Miles (Kilometers):
51,146 (131,940) (about the size of Alabama)
Capital (Population): Athens (3,093,000)
Environmental Concerns: air and water pollution
Geographical Features: mostly mountains, with ranges extending into the sea as peninsulas or chains of islands
Climate: temperate

PEOPLE

Population
Total: 10,601,600
Annual Growth Rate: 0.21%
Rural/Urban Population Ratio: 41/59
Major Languages: Greek; English; French
Ethnic Makeup: 98% Greek; 2% Turkish, Vlach, Slav, and others
Religions: 98% Greek Orthodox; 2% Muslim and others

Health
Life Expectancy at Birth: 76 years (male); 81 years (female)
Infant Mortality Rate (Ratio): 6.5/1,000
Physicians Available (Ratio): 1/259

Education
Adult Literacy Rate: 95%
Compulsory (Ages): 6–15; free

COMMUNICATION
Telephones: 5,536,000 main lines
Daily Newspaper Circulation: 135 per 1,000 people
Televisions: 442 per 1,000
Internet Service Providers: 23 (1999)

TRANSPORTATION
Highways in Miles (Kilometers): 70,200 (117,000)
Railroads in Miles (Kilometers): 1,484 (2,474)
Usable Airfields: 80
Motor Vehicles in Use: 3,101,000

GOVERNMENT
Type: presidential parliamentary republic
Independence Date: 1829 (from the Ottoman Empire)
Head of State/Government: President Konstandinos Stephanopoulis; Prime Minister Konstandinos Simitis
Political Parties: Panhellenic Socialist Movement; New Democracy; Democratic Social Movement; Political Spring; Coalition of the Left and Progress; Communist Party; Rainbow Coalition; others
Suffrage: universal and compulsory at 18

MILITARY
Current Disputes: constant friction with neighboring Turkey; tensions with Macedonia over name; Cyprus question with Turkey

ECONOMY
Currency ($ U.S. Equivalent): 402.9 drachmas = $1
Per Capita Income/GDP: $13,900/$149.2 billion
GDP Growth Rate: 3%
Inflation Rate: 2.6%
Unemployment Rate: 10%
Labor Force: 4,320,000
Natural Resources: bauxite; lignite; magnesite; petroleum; marble; hydropower
Agriculture: wheat; olives; tobacco; corn; cotton; fruit; olive oil; wine; meat; dairy products
Industry: food and tobacco processing; textiles; chemicals; metal products; tourism; mining; petroleum
Exports: $9.8 billion (primary partners EU, United States)
Imports: $27 billion (primary partners EU, United States)

http://www.greekembassy.org
http://www.odci.gov/cia/publications/factbook/country-frame.html

THE HELLENIC REPUBLIC

Ancient civilizations have always fascinated and intrigued historians, politicians, and the public in general, particularly those civilizations that left a clearly visible heritage of monuments. It is tempting to assume the existence of direct links between societies currently resident in the same general area and the peoples of past glory. The interval between classical Greece and modern Greece was lengthy— 2,000 years between "the cloud rising in the West," a signal that spelled the Roman conquest of ancient Greece, and the time that modern Greece arose. In the interim, the country was part of the Roman Empire, the Byzantine Empire, and the Ottoman Empire. In the early nineteenth century, the Greeks were at last able to liquidate the hated *tourkokratia* (rule by Turks) in areas that were to become the new nation-state of Greece. Modern Greeks nevertheless point with pride to their classical ancestry beyond the chasm of foreign rule, finding the discontinuity to be of less importance than the large number of historic and linguistic data supporting the existence of a direct link.

Greece lies on the southern tip of the Balkan Peninsula, flanked on the west by Italy, across the Ionian Sea, and on the east by Turkey, across the Aegean Sea. It also shares a common land border with Turkey in Thrace. Many of the country's several hundred islands are within sight of Turkish Anatolia. Crete, another center of an ancient civilization, is the largest of the Greek islands. The country—both the mainland and the islands—is mostly mountainous and dry. Only a quarter of the land is arable.

HISTORY

Only in the early nineteenth century did Greeks regain a measure of independence, after having been under Turkish rule for centuries. Yet even a decade later, they still did not have an exclusive say in such matters as their independence, the boundaries of the new state, and other issues that attend self-determination. A conference held in London in 1827 established all this, although it was at least left to the Greeks to convene a National Assembly in order to draw up the Constitution.

When apprised of the size of their new home, many Greeks felt acutely unhappy, since one out of every five was destined to live in areas that were not included. Whatever the glorious past of its ancient ancestry, modern Greece started out with a yearning for *enosis*, that typically Greek version of irredentism, which would not be satisfied until all Greek communities had become part of Greece. The fulfillment of these enosis ideals met with occasional disappointments, but it was nevertheless achieved (with a few exceptions) within 1½ centuries. The turmoil that more recently descended on Cyprus and that nearly triggered a war in the Eastern Mediterranean was in effect provoked by a resurgence of the enosis trauma.

The greatest expansion of the Greek national borders took place during the Balkan Wars (1912–1913), fought under the charismatic leader Eleftherios Venizelos. After these wars were concluded, shortly before World War I, Greece's national territory appeared to have expanded by 70 percent and its population by 75 percent. Some of these gains were offset by massacres and murders that Turks perpetrated in Anatolia in 1922 (where, it is believed, some 600,000 ethnic Greeks were killed). In addition, 1½ million Greeks were forcibly expelled from their ancestral lands.

POLITICAL CHARACTERISTICS

Since monarchies were the order of the day in Western Europe until well into the twentieth century, Greece was destined to become a kingdom upon achieving independence. This at least was the conclusion of the big powers that convened in London in 1827. But, as Greece had been part of foreign empires for more than 20 centuries, no aristocracy had been able to develop. This proved no obstacle: Monarchs were, after all, often recruited from the aristocratic elites of other countries. Louis I of Bavaria, who was party to the London convention, accepted the crown on behalf of his 17-year-old son, Otto. Surrounded by Bavarian advisers and troops, Otto

(Photo courtesy Lisa Clyde)

The Acropolis is a distinct and prominent part of Athens, Greece. The most famous Acropolis temple is the Parthenon. Close by is the temple Erechtheion, built in the fifth century B.C. On the southeast corner of this temple is the "Porch of the Caryatids," pictured above. Athens stretches off into the distance.

made his debut in Nauplia, the provisional capital.

No attempts were made by what was basically a foreign court to assimilate or to bridge the differences in culture, and it consequently did not take long for Bavarian elitism to become an irritant. When Greece seemed on the road to becoming a Bavarian colony, a backlash forced Otto to step down, in 1862. George I, the second son of the heir to the throne of Denmark, was then invited to rule Greece; he managed to stay in power for half a century. During this period, Greece was both a monarchy and a democracy (like the remaining monarchies in Western Europe today). George I was assassinated during the turmoil that the Balkan Wars created, and he was succeeded by Constantine (George II).

The period that followed was one of utter confusion and instability. Kings were on occasion set aside, exiled, and subsequently returned to power. However, this monarchical instability was only a reflection of the overall turbulence that has been Greece's fate since its birth as a modern nation-state.

Since 1831, Greece has experienced seven changes of constitutions, five removals of kings, three republics, seven military dictatorships, and more than a dozen revolutions or attempted coups. More than 150 cabinets have governed Greece since then, one third of these since 1945. Assessing Greek politics with these statistics in mind, one cannot escape the impression that the country is in perpetual turmoil. Yet the overall record is not entirely negative. Greece has enjoyed some form of parliamentary rule with adequate guarantees of individual rights for about 80 percent of the time. One should also take into account that for many years during the period at issue, Greece was at war, a circumstance that is bound to weaken democracy. Indeed, it is possible to conclude that modern Greeks appear to have inherited a natural preference for democracy.

POST–WORLD WAR II DEVELOPMENTS

Greece has experienced war not only on its eastern frontier but also to its west. Shortly after Italy joined World War II on the side of Germany, in June 1940, it sought to expand across the Adriatic Sea. To that end, it attacked first Albania and then Greece. In the latter operation, the Italians met with stiff resistance; in fact, the Greek Army proved in many ways superior to the invading Italians. The situation became embarrassing—so much so that, in April 1941, Adolf Hitler decided

to come to the rescue of Italy, its Axis partner, and "clean up the Balkans" before launching "Operation Barbarossa"—the invasion of the Soviet Union—on June 22, 1941. As a result of this delay, Operation Barbarossa had to forfeit summer months that were essential to its drive. (Operation Barbarossa eventually led to Hitler's defeat.) The Greeks were unable to hold out against the Germans for very long, and their government as well as a significant portion of their army were relocated to the Middle East. (While numerous Western European governments fled to London, Greece instituted a government-in-exile to its east.)

The German occupation provoked massive armed resistance and shaped Greek political attitudes for the next four decades. Greek resistance mainly operated in the mountains, and it was there that it became increasingly infiltrated by Communist partisans. It was only natural that the resistance troops would attempt to fill the power vacuum once the German armies retreated. Thus, when the government-in-exile returned to Greece, now headed by George Papandreou, it found a country dominated by Communists. However, it was obvious that obstacles to the return to normalcy had been foreseen, since the returning government was accompanied by no fewer than 15,000 British troops. In December 1944, their leader was forced to put down a revolt in Athens that had been instigated by the Communist-led National Liberation Army (ELAS). This military group, which had been extremely courageous in its resistance against the Germans, now attempted to capitalize on its popularity by seizing power, as its counterparts in such neighboring countries as Yugoslavia, Albania, and Bulgaria had done.

The monarch still had not returned, and Archbishop Damaskinos, who had been appointed regent, organized the first postwar elections, to which some 2,000 foreign observers were invited through the intermediary of the United Nations. The elections strongly favored bourgeois conservative parties, which received more than 70 percent of the popular vote. The Greek Communist Party (KKE), still very strong in the mountains, boycotted the process.

Having more or less settled what government would rule Greece, the time had come to make a decision regarding the monarchy. A referendum revealed that 64 percent of the participants wanted the return of the king.

By that time, however, the KKE had commenced a civil war, which was to last for four years. The Greek Civil War was

one of the specific situations to which the Truman Doctrine referred, and the United States offered a great deal of material assistance to the Greek government. The war was won by the nationalist forces, but it left scars on Greek society. The KKE was outlawed between 1946 and 1974. The war influenced the political orientation of Greek governments for 35 years. Between 1946 and 1981, Greece was under conservative government for a total of 28 years, whereas liberal–center parties were in power for less than seven years.

In practice, some of the conservative governments turned out to be far less than democratic; the "regime of the colonels," as the (conservative) military dictatorship has often been called, represented a clear departure from democracy. One of the most surprising facts of modern Greek political history is that this highly authoritarian phase ended not with a bang but with a whimper. In the seventh year of their rule, the military dictators (who were not completely united) were preoccupied with the possibility of the long-awaited enosis, or integration, of Cyprus. To render Cyprus, which had been independent for 14 years, a part of Greece would naturally have amounted to a gross violation of international law. The general ineffectiveness of the United Nations with respect to sudden changes in the status quo, as well as the supposed silent support of the United States (which had continued its full recognition of the junta), may have caused the regime to assume that the politics of *fait accompli* would work as it so often does. Domestically, of course, the enosis of Cyprus would have put a feather in an otherwise featherless cap.

The Greek government, however, had apparently not anticipated a Turkish response. The Turkish government tried to safeguard itself against encirclement by preempting enosis. Its troops landed on the north side of Cyprus and soon held more than one third of the island. It was obvious that an attempt to remove Turkish troops would cause a full-scale war between the two neighbors, who had been mutually hostile for a long time. Not only did the gamble not work, but it actually proved counterproductive. The public turned even more against the military dictatorship, which, without further ado, abandoned power.

There was no immediate backlash; the first post-junta government proved conservative and centrist (but democratic!). The royal house, having left the country at the military takeover, returned. However, there was some uncertainty about their political future, and it was decided to let the people determine that in a referendum. A strong preference for a republican form of government resulted.

VAGARIES OF A SYSTEM

But it was not only the somewhat anticlimactic eclipse of the monarchy, which had headed Greece for 1½ centuries without ever being very popular, that generated the urge for a new constitution. The military junta, coming into power in 1967, had never paid any attention to the existing Constitution, a charter that dated from 1952. Only after it had been dusted off and briefly revived did the new civilian government realize that it had become outdated and that Greece needed a completely new one. Put together in a surprisingly brief time and promulgated in June 1975, the new Constitution contained a number of innovations. It identified the Greek political system as a "presidential parliamentary system"—an unusual designation, since democracies are customarily classified as either presidential *or* parliamentary. The new Constitution's founders wanted the world to know that Greece had become a *republic*.

The office of the chief of state now fell to a president, to be elected for five-year terms. If the old charter had granted the king little more than ceremonial powers, the new Constitution gave the president a balancing role in politics. The president would thus moderate, and occasionally reconcile, extreme views. Such a task could only be performed if the office were granted some power.

Meanwhile, a leftist party, the Panhellenic Socialist Party (PASOK), which had been created by Andreas Papandreou shortly after the demise of the rule of the colonels, gained in strength. Papandreou had lived in the United States for a number of years and had even become an American citizen. Yet his dislike of American imperialism was obvious. In 1981, PASOK won the elections and Papandreou became prime minister. He immediately made clear that he did not like the power-sharing that the new Constitution implied. He managed to get adopted a constitutional revision that took away the large majority of presidential powers, granting them to the prime minister. The designation "presidential parliamentary system" as a consequence became less appropriate.

The new Constitution also converted the bicameral Legislature into a unicameral Parliament with 300 members. If nominally there was a separation of power among the three branches of government, in practice the executive branch of government (headed by the prime minister, who was also the leader of the majority party in Parliament) had the lion's share

of the power, completely overshadowing the two other branches. In some ways, this system may be compared to Great Britain's prime-ministerial government.

When Papandreou rose to power after the 1981 general elections, he was supported by 48 percent of the popular vote. Through a system of weighted proportional representation, his party received a majority of seats in Parliament.

Papandreou, although a democrat by inclination, could be somewhat autocratic. Generally the leader of PASOK chose to ignore public opinion that expressed disapproval of his basic policies. In the latter half of the 1980s, Papandreou was besieged by spectacular allegations of moral and political corruption, which greatly weakened the respect accorded him as well as his popularity. The fact that, after numerous escapades, he left his wife and married a flight attendant young enough to be his daughter did not help.

MAJOR POLITICAL ISSUES

The first seven years after the junta's demise proved a testing ground for the new Panhellenic Socialist Party. During the election campaign of 1981, it made foreign policy a central issue, claiming that all Greek postwar governments had in effect been "servants of foreign interests." Papandreou committed himself to radical changes and made *Allaghi* ("Change") his party's chief theme. The following sweeping changes were envisaged: 1) a withdrawal of Greece from the North Atlantic Treaty Organization; 2) the removal of U.S. military bases from Greek soil; and 3) a referendum on Greece's accession to the European Community (today known as the European Union) of which Greece had just become a full member.

However, virtually nothing was done with respect to these strident claims once PASOK had achieved victory and Papandreou had become prime minister. Greece remained a NATO member (for "national reasons"); the status of the American bases was renegotiated behind closed doors, and the promised referendum was never held. Nevertheless, the stark discrepancy between campaign promises and postelection performance did not cause the prime minister to tone down his anti–NATO and anti–U.S. rhetoric.

The difference between anti–NATO rhetoric and the implications of withdrawing from NATO had become painfully obvious to Greek governments. Once Greece withdrew somewhat halfheartedly—that is, only from the NATO command structure—steps were taken to put more emphasis on Turkish membership. The threat

that Turkey might benefit from Greece's withdrawal provided a strong incentive for Greece to retain its membership.

The status of the U.S. bases in Greece (which naturally relates closely to the NATO issue, if only because NATO has always been identified with the United States) will surely remain on the governmental agenda for a long time to come. It involves specific terms, or periods: If an agreement is about to expire, a new one will have to be negotiated or the bases removed. As for EU membership, that is no longer a serious issue, as it has become an asset rather than a liability. Indeed, Greece became a staunch supporter of the Treaty of Maastricht. The "Eurobarometer" poll in 2000 indicated that a majority of Greeks want the EU to play a greater role in their daily lives.

Greek foreign policy is greatly influenced by Turkey's immediate presence and the apprehensions that this presence evokes. Some of these sentiments may be traced back to centuries of cultural conflict, punctuated by a prolonged period of oppression. But even after the liquidation of the Ottoman Empire and the emergence of Greece as a modern nation-state, the adversarial relationship between Greece and Turkey lingered. Turkey was particularly aggrieved when a string of islands close to its coast was ceded to Greece after the Balkan Wars (1912–1913). And the Greeks still feel the slight of the Turkish invasion of Cyprus. Shortly after the invasion, Turkey added insult to injury by occupying 38 percent of the island. A great many Turks were then transported from Turkey into the occupied part of Cyprus. Undoubtedly this was done to strengthen any Turkish claim at future negotiations. It was a transparent ploy, but nothing was undertaken internationally to prevent it. And suddenly, in 1983, nine years after the Turkish invasion, the Republic of Northern Cyprus was proclaimed, with much fanfare. Thus far, the United Nations has refused to consider its admission, and Turkey is the only country to have recognized it. To most countries—and the UN—it is obvious that such recognition would amount to taking sides in the Greco–Turkish conflict.

Another bitter dispute developed with respect to oil drilling in the Aegean Sea. Shortly after oil was discovered off the Greek island of Thasos, Turkey claimed the continental shelf east of the median line between the Greek and Turkish mainlands. Such a claim would have been in compliance with recent rules of international law, more specifically with the UN Conference on the Law of the Sea, were it not for the Greek islands that lie a stone's throw from the Turkish coastline. Future exploration and exploitation will thus be conducted from the Greek mainland and will be of no concern to Turkey.

Popular sentiments and governmental attitudes vis-à-vis Turkey tend to remain the same, regardless of what government happens to be in office. Since the fall of the first PASOK government, frictions between Greece and Turkey, particularly with respect to the Cyprus issue and the Aegean oil, have continued. And in August 1990, when Turkey developed into a staunch ally of the United States in the latter's conflict with Iraq, it seemed foreordained that Greece would reluctantly grant lukewarm support, in the context of the EU and NATO. On the other hand, it is entirely possible that Turkey might have refused to shut off the oil pipeline coming from Iraq had Greece immediately thrust itself into the fray.

CHURCH–STATE RELATIONS
The Greek Orthodox Church, to which 98 percent of the Greek people belong, has been an important social institution since the Byzantine era. Through centuries of subservience, Greece had lost its political institutions; but in those dark days, the Church played a critical role in the survival of the national spirit. Under Ottoman rule, the Church performed numerous quasi-political functions that customarily are undertaken by state agencies.

The Orthodox Church has deep roots in Greek society. It is also quite wealthy; the Socialist government asserted the right to expropriate Church property and to turn it over to "cooperative farming." The announcement of such measures evoked strong reactions on the part of the Church and its membership, endangering the traditional church–state harmony. The Orthodox hierarchy threatened to place itself within the fold of the Ecumenical Patriarchate of Istanbul, which would embarrass the PASOK administration. The strained relations between the Church and the government contributed to domestic tensions.

THE GREEK ECONOMY
Many scholarly and technical terms in our lexicon can be traced back to classical Greek. *Economy* is such a term, meaning "law of the house." Freely translated, *oikonomos* means "the way in which the house is customarily run," a phrase that perfectly describes the economies of extended families or of the smaller city-states that were politically independent and economically self-sufficient. The city-states came to be absorbed by the Macedonian Empire, after

which Greece remained submerged in foreign empires for many centuries. A fully integrated, national Greek economy, therefore, is of relatively recent origin.

Another factor that may have hampered economic coordination and integration is the country's geography: the Hellenic Republic includes some 1,400 islands, about 170 of which are inhabited. Regular shipping services are maintained among them and with the mainland but, in spite of this network of connections, some island economies have remained largely separate.

The poor soil, on the mainland as well as on most of the islands, is another handicap. Greece nevertheless produces important crops, such as grains, rice, corn, cotton, tobacco, citrus fruits, figs, olives, and raisins. Although most of these are exported, the tendency toward a subsistence economy has persisted in some parts.

An industrial base was introduced into Greece in the mid-1950s. When PASOK gained office in 1981, industrialization was intensified. Attempts to shift the country's livelihood away from agriculture have met with some success; the Organization for European Cooperation and Development has designated Greece a "newly industrialized country"(NIC).

Although Greece has been an intermittent source of labor for other countries, it has also found itself in the position of having to import foreign laborers. Sometimes these labor migrations have even proceeded simultaneously. However, unemployment is currently a troubling 10 percent.

Generally, the Greek economy is stagnant. The stagnation may be attributed in part to external factors: Greece shares in the European Union's overproduction of wine (a number of other Mediterranean members of the EU also produce wine). In addition, revenues from tourism have fluctuated in recent years, partly because of occasional fear of terrorism.

In contrast to general expectations, Papandreou refrained from massive nationalization of private enterprise. The Greek economy remains a mixture of privately owned and state-controlled enterprises. All major economic enterprises—banks, surface and air transportation, telecommunications, and so on—are state-owned. Shipping, an important source of hard currency, remained in private hands during the Papandreou administration of the 1980s. By the same token, PASOK's extensive nationalization record caused alarm among Greek ship owners, many of whom removed their vessels from Greek registry. Also, the Socialist government was blamed for creating a climate that discouraged foreign investment, thereby

causing an outflow rather than an inflow of capital.

Whenever and wherever elections are held, the state of the economy is bound to be a factor of considerable moment. During the PASOK years in the 1980s, economic conditions fluctuated greatly, an urbanization-related phenomenon. To illustrate this, one may argue that had the election been held in June 1988, PASOK might have won a third term. The countryside, where the Socialists were strongest, enjoyed unprecedented prosperity, if only because of the generous injection of EU funds. It had also been politically freed from right-wing hegemony, which often had proved stifling. The two years of stringent wage restraints were over, and real income was on the rise again. Finally, Prime Minister Papandreou himself had not yet been subject to smears and allegations. By June 1989, however, renewed stagnation and an increase in unemployment had weakened the PASOK position so much that Papandreou could no longer point to it with pride. He was forced to concentrate on noneconomic concerns, such as narrow nationalism and foreign issues.

THE RIGHT STRIKES BACK
Nevertheless, the result of the June 1989 election was difficult to predict. Papandreou turned out to be a shrewd strategist, able to render the sound and fury of his opponents to nothingness or even to make them counterproductive. Still, if eight years earlier his campaign slogan had been "Change," the 1989 election was fought (and won) on the opponents' slogan of *Katharsis* ("Purge," or "Cleanup"). Although the charges may have been exaggerated, Papandreou's PASOK administration was marred by a degree of corruption and fraud.

Spirited campaigning—evading areas of public discontent—enabled the PASOK leader to capture 39 percent of the vote. The conservative New Democracy Party, led by Constantine Mitsotakis, won 44 percent. The balance was held by Harilaos Florakis, whose Communist-dominated Coalition of the Left won 13 percent of the vote. Florakis, who had called Papandreou an "untrustworthy partner" and PASOK a party that harbored "neo-Fascist tendencies," was unwilling to consider a union of the left. The alternative was a temporary government alliance of the major opposition parties, which, in spite of ideological differences, were prepared to collaborate in what they called a *katharsis* administration—what Papandreou called an "unholy alliance of nongovernment." However that may be, executive immunity no longer protected

A Greek revolution commences against the Ottoman Empire A.D. 1821	The boundaries of the modern Greek state are established (excepting the islands) 1827	King George I ascends to the throne and begins his 50-year rule of Greece 1863	Greece joins its Balkan neighbors in a war against the Ottoman Empire 1912	Greek expeditionary forces are defeated by Turks in Anatolia; all Greek residents in Turkey are resettled in Greece 1922	Germany occupies Greece; Greek resistance troops operate in the mountains 1941–1944	The returning Greek government, aided by British troops, faces Communists in powerful positions 1944

(Photo Courtesy of Lisa Clyde)

This is the town square of Arakhova, a small ski resort just outside of Delphi, Greece. A ski resort seems incongruous for Greece, but that is, indeed, snow on the slope in the distance.

PASOK leadership; and Papandreou plus a handful of former PASOK ministers were indicted on charges of accepting bribes, involvement in a fraud against the European Union, and widespread telephone tapping.

Nevertheless, the interim government could not agree on sensitive issues. As soon as the worst excesses of PASOK rule had been investigated, a second election was held, on November 5, 1989. The major protagonists both increased their shares of the vote: The New Democracy Party got 46 percent and PASOK 40 percent. The loser in this election was the Left Coalition, whose share dropped to 11 percent.

After weeks of bargaining, the three party leaders decided to put aside their differences and to agree on a national-unity government until April 1990. Curiously, the negotiations stopped then and there, and no attempt was made to resolve the gridlock in the intervening period. When the April deadline arrived, the only decision made by the national-unity government was to replace the aged Xenophon Zolotas by another old hand, Tzanne Tzannetakis, who in turn yielded the prime minister's office to Constantine Mitsotakis.

The latter felt (wrongly, as it turned out) confident enough to engage in nepotism. His daughter, Dora Bakoyannis, was made a junior cabinet minister in charge of co-ordinating government activities. It did not take Greeks long to complain that Prime Minister Mitsotakis's family was in charge. With his New Democracy Party holding but a two-seat majority in the 300-member Parliament, Mitsotakis could not afford to offend too many conservatives. Many ministers tendered their resignations. Only Antonis Samaras, the sole conservative with a continuously high approval rating, seemed destined to remain in charge of foreign affairs.

Finally, however, friction erupted between Samaras and Mitsotakis over the Macedonian question. Yugoslavia, while crumbling, had yielded a new independent

| Civil war erupts between Communist guerrillas and the Greek government **1946** | Greece embarks on the road to industrialization **1950s** | The Coup of the Colonels establishes a military regime **1967** | Civilian rule is restored; the monarchy is abolished by a referendum **1974** | Andreas Papandreou is prime minister **1980s** | PASOK is voted out of power **1989** | PASOK and the once-discredited Papandreou are elected again to head the government **1990s** | **2000s** |

Papandreou's successor, Costas Simitis, wins reelection

Greece becomes the 12th member of the Euro Zone in 2001

state that called itself "Macedonia." This also happens to be the name of a northern Greek province. Such a quasi-duplication would normally be taken in stride. (Belgium, for example, has a Province of Luxembourg, located quite close to the country of the same name.) But it became a major issue in Greece. While Prime Minister Mitsotakis was prepared to compromise and have the new state call itself "Slav Macedonia," Foreign Minister Samaras argued that the fledgling republic should choose a completely different name. Samaras was fired, but Mitsotakis soon adopted his stance in an attempt to appease public opinion. Eventually, an uneasy compromise was struck in that the area came to be called, the "Former Yugoslav Republic of Macedonia."

Meanwhile, Greece's relations with Albania, another country that had rid itself of a rigid Marxist regime and self-imposed isolation, degenerated into shooting incidents. Albania, fearing that Greece still had designs on the region known as Northern Epirus, refused to allow an ethnic Greek political party to take part in elections on March 22, 1991.

RECENT CONTROVERSIES

It is not uncommon for political parties of the left and right to alternate in office, but recent events in Greece are highly unusual. After the PASOK defeat in 1989, and due to the poor health of its leader, most analysts believed that it would take a long time for the Greek socialist party to get on its feet again. But PASOK and Papandreou were voted back into power in October 1993, possibly because voters were impatient at the slow pace of the changes promised by Mitsotakis and the New Democracy Party. In January 1996,

a gravely ill Papandreou finally resigned; PASOK picked Costas Simitis, a more pro–EU politician, to replace him. Simitis immediately called for, and won, new elections to confirm his mandate. Then, despite some major foreign-policy troubles, he also narrowly won reelection in 2000.

Greece suffered the embarrassment of being the only EU country not invited to join the new currency union, which introduced the new monetary unit known as the Euro in 1999. The government's austerity budgets, which had precipitated a series of long and painful agricultural strikes and transportation blockages, failed to convince the EU of Greece's financial health. The NATO allies also expressed doubts about Athens' ability to control terrorists operating on its territory, especially after the assassination of the British military attache in June 2000. A 1996 military dispute with Turkey over the tiny Aegean island of Imia (Kardak in Turkish) ended with President Simitis bowing to NATO pressure to pull back Greek forces, a compromise for which some nationalists labeled him a traitor. Then the mysterious Greek role in the 1999 capture of the fugitive Kurdish leader Abdullah Ocalan in Kenya caused public soul-searching about why Simitis's government seemed to be cooperating so much with the Turks.

Domestic issues gave Simitis little relief. Over the vociferous objections of the Orthodox Church, he had religious information removed from citizens' identity cards; he also approved the construction of the first mosque in Athens since Ottoman times. The state visit of U.S. president Bill Clinton in late 1999 was accompanied by tremendous anti–U.S. demonstrations and even some car bombs.

Many Greeks were still angry over the perceived U.S. support for the junta between 1967 and 1974 and over the recent NATO bombing of Serbia. Indeed, Greek relations with Serbia, and even with Milošević personally, remained close into 2000.

Simitis's main goal now is to curb the country's inflation and budget deficit and to update labor laws so that Greece can integrate more closely with the other EU countries and attract foreign investment. Observers note, however, that Greece as a society is still of two minds about "joining Europe": While many Greeks are attracted to the image and prosperity of Western Europe, they also bristle—perhaps in the long tradition of Greek freedom fighters against various historical foes—at the thought of taking orders from other countries in the EU or NATO.

DEVELOPMENT

Greek merchant marines constitute more than 20% of the European Union's shipping capacity. While Greece's economy has improved over the past several years, its people still have one of the lowest per capita incomes in the EU, and aid from the EU accounts for about 4.5% of GDP.

FREEDOM

Greeks have defended their freedom of expression with a passion. About 70 dailies of various sizes of circulation are published, 16 of them in Athens alone. All major Western and Eastern newspapers are readily available.

HEALTH/WELFARE

All Greeks are covered by some form of medical insurance. Although the quality of health care is not comparable to that of many other Western European countries, it is at least readily available.

ACHIEVEMENTS

Greece boasts many outstanding literary figures, some of whom have been internationally recognized. George Seferis, Yannis Ritsos, Odyseus Elytis, Kostis Palamas, and Nikos Kazantzakis are considered giants of modern literature. Seferis and Elytis are recipients of the Nobel Prize in Literature.

Iceland (Republic of Iceland)

GEOGRAPHY
Area in Square Miles (Kilometers):
39,758 (103,000) (about the
size of Kentucky)
Capital (Population): Reykjavik
(105,500)
Environmental Concerns: water
pollution; inadequate wastewater treatment
Geographical Features: mostly
plateau interspersed with
mountain peaks, icefields; the
coast is deeply indented by
bays and fiords
Climate: temperate; moderated
by North Atlantic Current

PEOPLE

Population
Total: 277,000
Annual Growth Rate: 0.57%
Rural/Urban Population Ratio: 8/92
Major Language: Icelandic
Ethnic Makeup: a homogeneous
blend of people of Nordic and
Celtic origin
Religions: 91% Evangelical
Lutheran; 9% others or no
affiliation

Health
Life Expectancy at Birth: 77
years (male); 82 years (female)
Infant Mortality Rate (Ratio):
5.3/1,000
Physicians Available (Ratio): 1/335

Education
Adult Literacy Rate: 100%
Compulsory (Ages): 6–16; free

COMMUNICATION
Telephones: 168,000 main lines
Daily Newspaper Circulation:
515 per 1,000 people
Televisions: 285 per 1,000 people
Internet Service Providers: 14 (1999)

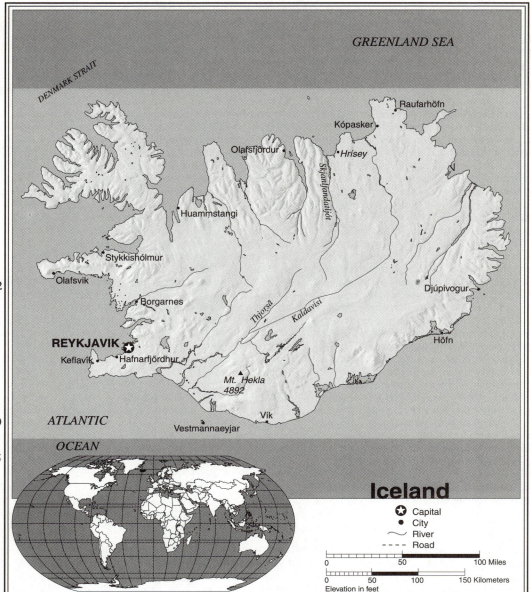

Iceland

- ★ Capital
- • City
- River
- - - - Road

0   50   100 Miles
0   50   100   150 Kilometers
Elevation in feet

TRANSPORTATION
Highways in Miles (Kilometers): 7,406 (12,341)
Railroads in Miles (Kilometers): none
Usable Airfields: 86
Motor Vehicles in Use: 150,000

GOVERNMENT
Type: constitutional republic
Independence Date: June 17, 1944 (from
Denmark)
Head of State/Government: President
Olafur Ragmar Grimsson; Prime
Minister David Oddsson
Political Parties: Independence Party;
Progressive Party; Social Alliance;
Left-Greens; Liberals
Suffrage: universal at 18

MILITARY
Military Expenditures (% of GDP): none
(Iceland's defense is provided by the
U.S.–Icelandic Defense Force)
Current Disputes: Rockall continental
shelf dispute with Denmark, Ireland,
and the United Kingdom

ECONOMY
Currency ($ U.S. Equivalent): 104.41 kronur =
$1
Per Capita Income/GDP: $23,500/$6.42
billion
GDP Growth Rate: 4.5%
Inflation Rate: 1.9%
Unemployment Rate: 2.4%
Labor Force: 131,000

Natural Resources: fish; hydropower;
geothermal power; diatomite
Agriculture: cattle; sheep; potatoes;
turnips; fish
Industry: fish processing; aluminum smelting;
ferrosilicon production; geothermal
power; tourism
Exports: $1.9 billion (primary partners
EU, United States, Japan)
Imports: $2.4 billion (primary partners
EU, United States)

http://www.iceland.org/
http://www.odci.gov/cia/publications/
factbook/geos/ic.html

LAND OF ICE AND FIRE

Iceland, a lonely outpost on the northwestern fringe of Europe, halfway between the European and North American continents, is located just south of the Arctic Circle. It is actively volcanic, and both lava flows and glaciers meander through its landscape, which explains the graphic name bestowed on Iceland: "land of ice and fire." Nearly 10 percent of the land area is covered by glacial ice. This terrain appears so barren and desolate that those arriving by air may get the impression of landing on the moon.

One would not expect much in the way of human habitat in Iceland, and indeed the island appears not to have been inhabited before medieval times. For a considerable period, the unprepossessing area served merely as an orientation point and sometimes as a temporary shelter to fishermen braving the winds and waves of the North Atlantic. The first sparse settlements were established by Irish monks intent on dissociating themselves from the world. It is not known what happened to them. They disappeared without a trace.

One of the first settlers was Ingolfur Arnarson, a chieftain from West Norway who wanted to break away from royal authority. In A.D. 874, he, his family, and his dependents built settlements on the site of what is now the capital city of Reykjavik. Apparently this move stimulated a flow of settlers to the area that continued and increased in volume toward the end of the ninth century, probably because Harald Fairhair had become king of Norway around the year 885—the conquest of Norway by this king had been at the expense of many noblemen, who thereupon decided to leave Norway. Some of them left for the British Isles; others proceeded to Iceland and started a new life in these harsh surroundings. It is possible that Harald Fairhair's ascent to power served as a catalyst for the Age of Discovery, especially since Iceland was also used as a base for the discovery and subsequent settlement of Greenland. This achievement is largely attributed to Eric the Red, a Norwegian Viking. His son, Leifur Eriksson, one of the first persons to have been born and raised in Iceland, in turn led an expedition from Greenland to Vinland, which must have been the northeastern part of North America. Over time, numerous items have been discovered that help to document the Viking Age there.

Conditions remained punishing in Iceland itself, its people eking out a living through fishing and farming. Sheep farming became an important livelihood. But whether the settlers had come as refugees or voluntarily, the laws that they took with them were fragmentary and often difficult to interpret and enforce. A new legal system that would reflect the situation seemed urgently needed. To that end, the various settlements started to institute local assemblies, which in turn led to the establishment of a Parliament (in Old Icelandic, the *Althing,* or "Whole Assembly"—the term indicates that all people were to participate in its proceedings; if the British Parliament is often nicknamed "the mother of parliaments," the Althing may be viewed as the grandmother of parliaments).

The Althing operated as a rule-making institution for three centuries. Iceland was free from outside interference during that period, and since monarchical rule was absent, it could be classified as an independent republic of sorts. But in the late thirteenth century, the island lost its independence to Norway, not quite by force but, rather, as the result of a treaty. And a century later, both Iceland and Norway came under Danish rule.

The population on the island was fairly stable until the fifteenth century, when the plague arrived. Living conditions in Iceland generally remained extraordinarily severe until the middle of the eighteenth century, past the time when life in other European countries had started to evidence some comfort. Icelandic health care was inferior, and recurrent pestilence took its toll, decimating the populace. The question of whether or not to resettle the survivors on the mainland was raised from time to time.

INDEPENDENCE

But even if the social and health conditions left much to be desired, a great deal of progress was made in terms of political emancipation. Denmark granted home rule (independent self-government in internal affairs) to Iceland in 1903. And during World War I, Denmark became serious about shedding its overseas holdings. It sold its Virgin Islands in the Caribbean to the United States, and it concluded a treaty with Iceland that recognized the island as an independent state under the king of Denmark. That treaty, which led to the so-called Act of Union, was concluded for a definite term—25 years—and Denmark was occupied by Nazi German forces before it expired. The British, apprehensive that the fall of Denmark in April 1940 would be interpreted as implying the surrender of Iceland as well, took immediate steps to prevent a German occupation of the strategically located island.

The treaty with Denmark as well as the Act of Union may be said to have been canceled through circumstance, but during the British occupation (which subsequently became an Anglo–American occupation effort), the Althing passed legislation to terminate the relationship that had existed with Denmark. Formally, Iceland became an independent republic on June 17, 1944, although it remained occupied, in view of the ongoing war.

In return for the wartime cooperation, Iceland received a share of U.S. Marshall Plan aid, even though it had not been damaged by the war. And a couple of years later, the island government was invited to help found the North Atlantic Treaty Organization. The United States by that time had become convinced of the strategic value of the small nation-state, which had no military. Its major contribution thus became the air base at Keflavik.

Iceland's membership has occasionally caused the NATO leadership to resort to subterfuge. For example, when the Icelandic government included Communists or persons of a radical leftist persuasion, NATO excluded it from sharing its classified information and from deliberations concerning top-secret plans. The Icelandic government, however, has never taken offense, and Iceland has continued to be a member under these somewhat humiliating circumstances.

THE ECONOMY

Iceland was hit in the early 1990s by a severe recession. The adverse circumstances may have been due in part to Iceland's gradually changing economy. Iceland's National Research Council concluded that the country was about to enter the postindustrial age, a time when skills, ability, and innovation are seen as more important than equipment and structures. In more recent years, the economy has rebounded impressively, with low inflation and unemployment and much improved per capita GDP incomes and growth rates. Iceland has made strides toward liberalizing its economy. And thanks to its membership in the European Economic Area, Iceland can participate in the European Union's free-trade market (except in fish and food.)

In 1880, more than 73 percent of the working population were engaged in farming; by 1950, this proportion had declined to 30 percent. Currently, agriculture employs only about 4 percent of the workforce. These figures reflect the growing urbanization of the island, but it should also be taken into account that farmers are largely occupied with cattle and sheep breeding. (The amount of arable land is only 0.5 percent.)

Occupationally, agriculture and fishing, which both are forms of primary industry, have an interesting inverse relationship in Iceland, in that fishing increases as agriculture declines. So great has Iceland's dependence on fish and fish products become that its government, beginning in the mid-1900s, placed severe restrictions on foreign fishing vessels, which could deplete the waters that surround the island and thereby ruin its economy. A free-market economy predominated in Iceland, though government ownership of businesses expanded, particularly in the fishing industry.

In 1952, Iceland extended the limits of its fisheries outward to a perimeter four miles from lines drawn between its longest promontories and between islets and rocks offshore, enclosing many fishing grounds previously open to foreign fishermen. Some fishing boats in violation of the new rule were detained. In retaliation, British trawler owners at Hull and Grimsby refused facilities to Icelandic vessels. However, this was only the beginning of what later came to be called the "Cod War," one of the larger intra-NATO conflicts. One of the main reasons for extending Iceland's fishing rights in the North Atlantic was the modern fish-harvesting technique being practiced, especially by the British fishing vessels. Gradually, Iceland extended its own zone of fishing rights, by implication prohibiting foreign vessels from fishing in those waters. The International Court of Justice ruled in 1974 that the 50-mile limit that Iceland established in 1972 could not be imposed unilaterally. Iceland, however, rejected the ruling. Iceland's fishing rights eventually reached the 200-mile limit, which has now been universally accepted as the legacy of the United Nations Conference on the Law of the Sea.

The most recent partition of fishing grounds has not removed the danger that some fish species may become extinct. It means only that the predator role has become monopolized. The Icelandic press reported that the exports of frozen seafood from Icelandic Freezing Plants Corporation reached a record high of 91,500 tons, worth some $300 million, in 1993.

Some 67 percent of all Iceland's exports are fish-related. Cod has traditionally been the most important species of the nation's fishing industry. The Marine Research Institute reported that by mid-1992, the lowest cod catch in half a century had been reached. It is a matter of course that fluctuations in fish catches will cause ups and downs in Iceland's economy. In addition, marketing and delivery have become momentous factors, and one wonders whether Icelandic fish-auction markets can compete with their foreign counterparts. Mod-

(Iceland Tourist Board photo by Johan Henrik Piepgrass)

Fish and fish products constitute an important part—67 percent—of Iceland's total merchandise exports and involve employment of about 11 percent of the workforce. These fishing boats and trawlers are at port in Reykjavik.

ern telecommunications enable buyers to bid on fresh, unprocessed fish from virtually anywhere. As a result, the competition has become enormous.

Whaling is one of the more contentious subjects that a foreigner should avoid bringing up while visiting Iceland. Most Icelanders feel that whales are just fish, only bigger. Early Icelanders are said to have harpooned small whales from yawls, and Icelanders have always been quick to utilize beached whales. Icelanders argue that their whaling record has undeservedly rendered them into pariahs in the international community. They believe that Iceland has done as much as, if not more than, other countries to protect the whale from extinction.

Whaling reached a peak at the turn of the century; by 1902, it had become clear that world whale stocks were on the decline. The Althing banned whaling in 1915, with the moratorium in place until 1935. However, the Althing's power extended only to Icelandic whaling, and many foreign whaling operations continued. In 1935, Icelandic whaling was resumed, but it stopped again when World War II started in 1939. After the war, Icelanders resumed whaling. In 1986, an international whaling ban was issued. Iceland, although a member of the International Whaling Commission (IWC), had never endorsed the international agitation on the ethical level, but it pledged that it would fully abide by the new mandates. To enforce the IWC ban, in 1986 the Sea Shepherd Conservation Society's "eco-saboteurs" sank two of Iceland's four whaling ships and destroyed a whale-processing plant. (Iceland declined to prosecute Sea Shepherd leader Paul Wat-

son.) In 1992, after numerous recriminations, Iceland resigned from the IWC. It joined with other critics—Faeroe Islands, Greenland, and Norway—to form the North Atlantic Marine Mammal Committee (NAMMCO), which favors the harvesting of these species. In March 1999, the Icelandic Parliament voted to resume whaling. Polls indicate that 75 percent of the adult population agreed with the decision. However, fearing negative effects on tourism and fish exports, the government has followed a go-slowly approach.

No discussion of the Icelandic economy would be complete without mention of the extensive use that Iceland makes of its natural resources. It derives a large portion of its energy supply from hydropower and from thermal springs heated by molten lava. Thus, although the winters can be grueling, nature seems to compensate by providing thermal waters that can be used for heating systems. This low-cost energy is also used for the development of the aluminum-smelting industry. For its transportation needs, the country has been largely dependent on oil deliveries from the former Soviet Union and Britain. As a result of the economic prosperity of recent years, Iceland has nearly as many motor vehicles per capita as the United States. To reduce its dependence on foreign oil, the government favors the development of hydrogen fuel cells to replace gasoline and diesel for vehicles.

SOCIAL POLICY

Iceland, in common with other Scandinavian countries, provides a comprehensive system of social welfare. More than half of the national budget goes to education, public-assistance programs, and health.

The earliest Norse settlement in Iceland A.D. **874**	The Althing meets for the first time **930**	Christianity is introduced in Iceland **1000**	Union with Norway **1264**	Union with Denmark **1380**	Recurrent pestilence **1400–1800**	The Althing is abolished during the period of Danish absolutism **1800**

Cost-free medical care supports one physician for every 335 people in the country. Iceland's rates of life expectancy for men and women are among the highest in the world. According to UN statistics, 92 percent of Icelanders live over age 60, the highest figure in the world. The good news, however, has serious implications for the future. With people living so long, Iceland will not be able to go on providing such a generous welfare system. The government has begun trimming its largesse. Indeed, Iceland now spends less than half of what Denmark does on social welfare, as a percentage of GDP.

GOVERNMENT
Iceland is a constitutional republic. Like most Western European nations, it has a dual executive—a head of state, the president, who serves terms of four years; and a head of government, the prime minister. The president, in his or her largely ceremonial role, must appoint the prime minister, who presides over a government responsible to Parliament. Iceland's Parliament, which is still called the Althing, has 63 members, popularly elected by proportional representation. The indirectly elected upper house was abolished in 1991.

There are half a dozen major political parties. They tend to fall into two groups: the Socialist bloc and the non-Socialist bloc. Since none of the political parties receives a majority vote, a government has to rely on coalitions. Interestingly, these coalitions often include political parties from both blocs. The conservative Independence Party invariably constitutes the bulk of such coalitions. During 1995–1999, Prime Minister David Oddsson headed a majority coalition comprised of his Independence Party and the (centrist) Progressive Party. This government actively sought to balance the budget, and to privatize and liberalize the economy.

Historically, the Socialist bloc has included the Social Democratic Party and the radical-left People's Movement (including the former Communist Party of Iceland). The 1999 campaign saw a restructuring of the opposition. The Social Democrats (including the People's Movement) formed a "Social Alliance" with the People's Alliance and the Women's Alli-

ance. The end of the Cold War has made cooperation easier among moderate and radical leftists. However, unity still was not achieved in 1999. Rising concerns about the impact of power-intensive industry on the natural environment and lingering skepticism about the market economy translated into the emergence of a new party, the Left-Greens.

The Independence Party reaped most of the benefits from recent years of prosperity and tranquility; it won 40.7 percent of the parliamentary votes in May 1999 (its best performance since 1974). Although its partner, the Progressive Party, slipped to 18.4 percent, the coalition was renewed under Oddsson's leadership. The Social Alliance's campaign turned out to be not very effective; it won fewer votes than its constituent parties did in 1995. The two new parties, Left-Greens and the Liberals (who emerged as a protest against the government's quota system of fisheries management), both won parliamentary seats, with 9.1 percent and 4.2 percent of the votes, respectively.

WOMEN
The women's-liberation movement made vast strides in Iceland. The status of women started at a low point in Iceland in contrast to the other Nordic countries and has always been inferior to that of men in sociopolitical respects. In 1976, Iceland enacted its Equal Rights Amendment. Until that time, 80 percent of Icelandic women employed outside the house had earned on the average about 40 percent less than their male counterparts.

Then, in 1980, Vigdis Finnbogadottir, director of the Reykjavik Theater, ran for the office of president. She was elected, a feat that provided an enormous stimulus to the women's movement. Encouraged, it put forward its first candidates in local elections in 1982. When they too were successful, the movement changed from a pressure group into a political party, the Women's Alliance, which in 1983 participated in national elections.

In the 1987 election, the Women's Alliance won 10.1 percent of the vote, which, under the proportional-representation system, translated into six seats. More important, the parliamentary role of the Women's Alliance became pivotal, since neither of the two blocs won a majority.

In 1984, the Women's Alliance ordered a women's general strike, which was honored by the president (who normally may not be identified with any party); she left her office for the day!

A presidential election had been scheduled for June 26, 1992. However, the incumbent was the only person to register as a candidate. The election was therefore canceled, and President Finnbogadottir embarked upon her fourth term of office on August 1. She subsequently announced that she would not seek a fifth term. In 1996, conservative Olafur Ragnar Grimsson took office as president.

The Women's Alliance fared poorly in the 1995 parliamentary elections. Its support dropped by half, to 4.9 percent, and it won only three seats. In 1999, the party submerged itself in the Social Alliance. Though fading as a party, its emphasis on women's representation has had a historic influence on other parties. In 1983, there were only three women in the Parliament and none in the cabinet. After the 1999 elections, there were 22 in Parliament and four in the cabinet.

The inferior position that women traditionally held in Iceland is all the more remarkable because Iceland has followed Scandinavian social mores in many other respects. One may point to the patronymic tradition, for instance. Iceland has an unusual concept of family names; it pays more attention to first names. Traditionally, the first name of the father becomes the last name of the son or daughter. Thus, Jon's daughter Ragnihildur will be named, in full, Ragnihildur Jonsdottir. However, in ancient sagas, the *Edda* (either of two Old Icelandic volumes of myths and poems compiled during the thirteenth century), and other poetic literature, the position of women is never portrayed as subordinate; rather, women are equal to men in valiance, courage, and other virtues.

LANGUAGE
A factor that looms large in the popularity of ancient literature in Iceland is that its language has been preserved much better than in many continental European countries. Over many centuries, the language did not change to a great extent. Indeed, a study of source material over various centuries reveals a rigidity that most prob-

Timeline:

The Althing is restored as a consultative body
1843

Home rule
1904

Act of Union; Iceland becomes independent in a personal union with the Danish crown
1918

British and U.S. occupation
1940–1945

Iceland is proclaimed a republic; the last ties with Denmark are severed
1944

The Cod War
1970s

Vigdis Finnbogadottir becomes the first female president of Iceland
1980

Recession and labor tensions ease by the late
1990s

2000s

Icelanders debate joining the European Union and the Euro—and decline

Iceland's quality of life is rated among the best in the world

ably results from Iceland's isolation. Like the United States, Iceland started off as a colony of settlement; but, in sharp contrast to the United States, it was never a melting pot in any sense. Basically, all its immigrants arrived over a strictly limited period from Scandinavia—thus the same source in linguistic respect. (Danish, Norwegian, and Swedish differ only dialectically.) There has been an absence of other outside influences. Thus, the fact that twentieth-century Icelanders are still able to enjoy the *Edda* and other sources of ancient literature is due to this linguistic introversion.

THE INTERNATIONAL ARENA
Although culturally introverted and geographically insular, Iceland has hardly been isolationist. In 1949, it became a charter member of NATO. Iceland's contribution is its strategic North Atlantic location and in particular its radar sites. In addition, Iceland and the United States have maintained a special bilateral relationship. As of the late 1990s, there was only a small contingent of U.S. military personnel deployed at Keflavik, in the southwestern part of the country. Today, most Icelanders seem largely content with this low-profile arrangement. Since 1992, Iceland has also been an associate member of the Western European Union, the slowly emerging military arm of the European Union's common foreign and security policy.

Since the European economy in the early 1990s had started to absorb 80 percent of Iceland's exports, it was only natural that the idea to join the European Union started to take hold. However, resistance also gathered strength. By mid-1992, the press reported that 48.3 percent of Icelanders opposed the idea of Iceland's becoming a member of the Union, 15.7 percent

supported it, and 34.5 percent had yet to make up their minds. "The classical stumbling block to Icelandic membership," Foreign Minister Jon Baldvin Hannibalsson stated in June 1992, "has been the [EU] Common Fisheries Policy, which would oblige Iceland to open its fishing grounds—the nation's only immediate resource—to the oversized European fleet."

For three decades, Iceland has been a member of the European Free Trade Association. In the early 1990s, the European Economic Area developed as a linkage between EFTA and EU countries (except for EFTA member Switzerland, which declined to join the EEA). Subsequently, three EFTA countries became members of the EU. Basically, the EEA confers on Iceland virtually all trading opportunities and obligations that go with full EU membership, except in farming and fisheries. In 2000, 69 percent of Iceland's exports went to countries within the EEA.

Outside the all-important economic field, Iceland also succeeded in gaining international recognition for its ecological initiatives. During preparations for the 1992 Rio Summit, a world conference on issues pertaining to the environment and energy, Iceland's delegation in New York managed to achieve a reclassification of energy sources, to highlight which are environmentally safe and which are destructive. At the Rio meeting itself, Iceland became a distinctive force among Western nations for its unequivocal stand on environmental questions.

While some of the industrialized nations, particularly Great Britain and the United States, have consistently attempted to block effective restrictions on discharging and dumping at sea chemicals and even nuclear waste, and restrictions on nuclear submarine activity, Iceland has spoken out against them. Iceland argued that

pollution of the seas by persistent organic substances posed a far more serious threat to whale and seal populations than a regulated hunt ever could.

Iceland's government has received international attention by granting a license to a private Reykjavik-based company (deCode Genetics) to create a comprehensive, computerized database that cross-references medical, genealogical, and genetic information of Icelanders. The goal is to profit from the country's relatively homogeneous gene pool as biotech researchers seek to understand the relationships between genetic traits and certain diseases. Although the project has the support of over 75 percent of the people, members of the medical community have expressed opposition due to possible violations of the right to privacy. As one American medical authority observed, "Iceland gives the whole world an opportunity to think through issues that nobody has been able to yet."

As a small country on the geographical fringe of Europe, Iceland entered the new millennium with a sense of national pride over its accomplishments. In 1999, Foreign Minister Halldor Asgrimsson reported to Parliament, "We have greater security, greater welfare, less pollution, better education than almost any other country in the world."

DEVELOPMENT

Icelandic society has become considerably more urbanized since World War II. The industrial base continues to be small; fishery remains strong. Farming has declined, while tourism has increased tremendously. Forty percent of Icelanders use the Internet, the highest percentage in the world in 2000.

FREEDOM

Plans to terminate the centuries-long rule by Denmark materialized at the end of World War I. Independence came during World War II, when Iceland was cut off from Denmark, which was occupied by the Germans.

HEALTH/WELFARE

Iceland has an excellent social-welfare system. Benefits cover maternity and child care, health care, education, unemployment compensation, and pensions. Iceland's quality of life is considered by the UN Development Program to be the 5th highest in the world.

ACHIEVEMENTS

Iceland's unilateral extension of the limits of its fishing rights was accepted by other nations and eventually became universalized by the UN Conference on the Law of the Sea. Literacy is virtually total. Icelanders read more books per capita than any other population. Even the ancient *Edda* is widely read. In 2000, Iceland had the lowest unemployment rate among the 29 OECD countries.

Ireland (Eire)

GEOGRAPHY

Area in Square Miles (Kilometers):
27,135 (70,280) (about the size
of West Virginia)
Capital (Population): Dublin
(481,000)
Environmental Concerns: water
pollution from agricultural
runoff
Geographical Features: mostly
level to rolling interior plain
surrounded by rugged hills
and low mountains; sea cliffs
on west coast
Climate: temperate maritime

PEOPLE

Population
Total: 3,798,000
Annual Growth Rate: 1.16%
Rural/Urban Population Ratio: 42/58
Major Languages: English; Gaelic
Ethnic Makeup: Celtic; English
minority
Religions: 92% Roman Catholic;
3% Anglican; 5% other

Health
Life Expectancy at Birth: 73 years
(male); 79 years (female)
Infant Mortality Rate (Ratio):
5.6/1,000
Physicians Available (Ratio): 1/681

Education
Adult Literacy Rate: 100%
Compulsory (Ages): 6–15

COMMUNICATION
Telephones: 1,600,000 main lines
Daily Newspaper Circulation: 151
per 1,000 people
Televisions: 279 per 1,000
Internet Service Providers:
14 (1999)

TRANSPORTATION
Highways in Miles (Kilometers): 55,458
(92,430)
Railroads in Miles (Kilometers): 1,206
(1,947)
Usable Airfields: 44
Motor Vehicles in Use: 1,230,000

GOVERNMENT
Type: parliamentary republic
Independence Date: December 6, 1921
(from the United Kingdom)
Head of State/Government: President
Mary McAleese; Prime Minister Bertie
Ahern
Political Parties: Fianna Fáil; Democratic
Left Party; Labour Party; Fine Gael;
Communist Party; Democratic Left;
Sinn Fein; Progressive Democratic
Party; Green Party
Suffrage: universal at 18

MILITARY
Military Expenditures (% of GDP): 0.9%
Current Disputes: Northern Ireland issue
with Britain; Rockall continental-shelf
dispute concerning Denmark, Iceland,
and Britain

ECONOMY
Currency ($ U.S. Equivalent): 0.93 punt
(Irish pound) = $1
Per Capita Income/GDP: $20,300/$73.7
billion
GDP Growth Rate: 9%
Inflation Rate: 7%
Unemployment Rate: 3.8%
Labor Force: 1,770,000
Natural Resources: zinc; lead; natural gas;
barite; copper; gypsum; limestone;
dolomite; peat; silver
Agriculture: livestock and dairy products;
turnips; barley; potatoes; sugar beets;
wheat
Industry: food products; brewing; textiles
and clothing; chemicals; machinery;
pharmaceuticals; transportation
equipment; tourism; glass and crystal;
software
Exports: $66 billion (primary partners
EU, United States)
Imports: $44.9 billion (primary partners
EU countries, United States, Japan)

http://www.irlgov.ie/frmain.htm
http://www.odci.gov/cia/publications/
factbook/geos/ei.html

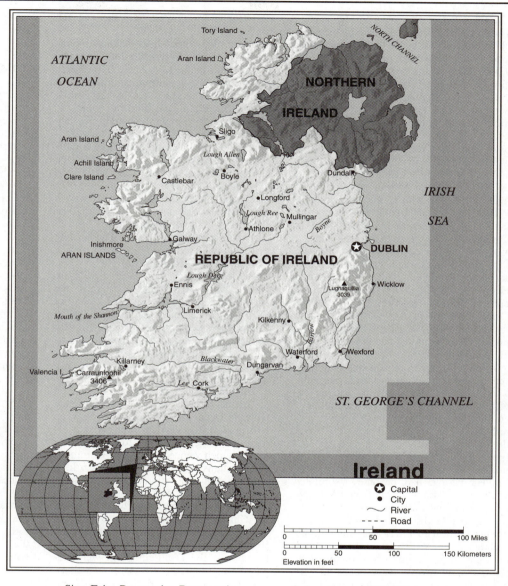

IRELAND (EIRE)

The large island immediately to the west of Britain, which came to be named Ireland, or Eire, has been inhabited for about 9,000 years. At the beginning of recorded history, a Celtic language was spoken in the area, a language closely related to Scottish, Welsh, Gaelic, and Breton. Celtic tribes were not confined to Ireland but extended over parts of Britain and the European continent as well. The Celts had well-developed social and legal systems and a rich, orally transmitted culture.

Christianity was introduced on the island by the man who is now known as St. Patrick. He is still greatly revered by the Irish. The conversion from Druidism to Christianity took place in the fifth century A.D., and in the centuries that followed, Irish monks established centers of learning at home, in Britain, and in continental Europe. From 700 to 900, Vikings invaded parts of Britain, which caused some of the Celtic residents to flee westward and to seek refuge on the island. However, the raiders also conquered and colonized parts of Ireland itself.

Although English expansions may have started earlier, a massive transmigration took place in the sixteenth and seventeenth centuries, a time when religious conflict had started to make Britain very turbulent. The English swarmed all over the island, rendering it a colony, and brought over their aristocracy as landowners. Earlier, and smaller, groups of English people had become assimilated and hardly distinguishable from the native inhabitants; but this had now become impossible, since the English took over government and governmental institutions. Another group settled in northern Ireland, different from the English in many ways: the Scots. They were very poor and very religious, adhering to the Presbyterian denomination. They refused to mix with the native Irish, because the latter were universally Roman Catholic.

These invasions, which took away their land and government, caused resentment among the Irish. In 1798, Wolf Tone and the United Irishmen, inspired by the American and French Revolutions, started a rebellion to terminate British rule. The rebellion, which was more an emotional reaction than a carefully calculated enterprise, was suppressed. The nineteenth century witnessed sporadic rebellions, but there was no sustained revolt, since the Irish people at that time lacked the wherewithal to rise against an infinitely stronger Britain. The potato famine of the 1840s proved devastating; more than 4 million Irish people died as a result of it. A great many others who survived this economic catastrophe emigrated to the United States.

At the time when Great Britain was immersed in World War I, the long-simmering conflict came to a head. The Easter Rising afforded a glimpse of Irish armed potential. It was put down by the British troops, the Black and Tan, but what the rebels failed to gain on the battlefield, they secured in the British parliamentary elections held in 1918. The Sinn Fein, a radical Irish party, won 73 seats, whereas the Unionists/Conservatives in the north won only 26 seats. After this victory, Sinn Fein proceeded to set up an Irish Parliament, with Eamon deValera as its head. The members of the new Parliament refused to take their seats in the British Parliament.

THE ANGLO–IRISH WAR

Such defiance was not taken lightly by Britain, and after some skirmishes, the Anglo–Irish War broke out. The conflict (1919–1921) served as another catalyst to Irish independence. Some British (mostly mercenary) troops committed such atrocities on the Irish that public opinion in the United States and Great Britain was very adversely affected.

The changing climate of opinion forced the British to a truce and to the conference table. In December 1921, a treaty between Irish and British leaders brought into effect the Government of Ireland Act of 1920, which conceded British Commonwealth status and a parliament to 26 Irish counties (which jointly assumed the name Irish Free State) and kept six northern counties with a separate parliament within the United Kingdom.

The treaty and its main product, the Irish Free State, caused bitter disputes in the ranks of the Sinn Fein. Some were prepared to accept the British concessions, believing that they were to lead to the ultimate ideal of complete independence of the entire island. Others, such as deValera, adamantly opposed what was being offered, certain that Great Britain was shortchanging them. After losing the internal fight, deValera founded a new organization, the Fianna Fáil (which means "Warriors of Destiny"). By 1932, it was the dominant political party in Ireland; its leader then formed a government and embarked upon his long tenure as prime minister (1932–1959).

Although the terms of Irish independence were gradually strengthened, the British refused to make further concessions in regard to the six counties in the north. DeValera had made a turnabout by endorsing a system that he had initially rejected. Seeming prepared to wait for whatever time that could aid the Irish cause, deValera turned a turbulent situation into a stable one.

IRELAND'S NEW IDENTITY, HOME RULE, AND INDEPENDENCE

Since 1921, the Irish Republic has endeavored to pursue two broad goals simultaneously: On the one hand, it has wanted to convey the image of an independent nation; on the other, it has fervently aspired to resolve the border question and related issues. For eight decades, Eire has been enmeshed in a de facto triangular relationship with Northern Ireland and Britain.

This triangle has served as a persistent reminder that Ireland's jurisdiction is limited—that it does not cover the entire island, but only 26 out of 32 counties. The British presence, from which many generations have wanted to disentangle themselves, is right on Ireland's doorstep. One cannot study a map of the island without looking at a part of the United Kingdom (Northern Ireland).

Ireland has built up a degree of hostility against the English. It has differed from Britain on four communal dimensions: language, ethnic origin, culture and tradition, and religion. And then there is the moral consideration: During the heyday of imperialism, Britain planted Scottish colonies in Northern Ireland. Here, too, the typical colonial picture prevailed; the Scottish vanguard became landowners at the expense of the Irish. Later tenants came from Scotland to strengthen the Protestant cause. Thus, a whole new society was created in Ulster, completely different in character from every other part of Ireland. The Presbyterian Scots constituted a majority in the six northern counties; but if these counties would also be ceded to Ireland, their numbers would be so diluted that they would constitute a minority. They would be a small Protestant fragment in a population that was almost universally Roman Catholic. For the British Conservative Party and some Liberals, giving Ireland any independence would not only engulf Protestants in a Catholic state but also destroy the Constitution as defined by the empire that was built around Britain and Ireland.

The Irish were convinced that the British concern with the Protestant cause in Northern Ireland was nothing but a hypocritical pretense and that the British wanted only to keep a foothold on the island. For the same reason, the Irish Free State was inhibited by numerous restrictions after officially gaining independence. As a part of the Commonwealth, it was obliged to recognize the British Crown and a governor-general. It also had to subscribe to an oath of loyalty.

In 1937, the "colonial" Constitution was swept away by deValera's government. But only when World War II broke out and the Irish government announced that Ireland would remain neutral did the Irish realize that they had at last attained a sovereignty that extended even to matters of foreign policy. Imperial powers have historically been little aware of injustices they have perpetrated. In Ireland's case, Irish neutrality was puzzling if not totally incomprehensible to the British. It did not take long for rumors to circulate that German submarines found refuge, and were even refueled and resupplied, in Irish ports. These rumors may have been groundless—it seems unlikely that Eire would have risked getting into a war with the United Kingdom—but they were nonetheless indicative of the British inability to appreciate Irish neutrality.

It is possible that the United Kingdom had a hand in the delay that the Irish application for admission to the United Nations experienced. Ireland had been excluded from charter membership in large part because the UN was in effect an extension of the wartime alliance into peace, and since Ireland had emphasized its neutrality throughout the war, it did not qualify for charter membership. It finally became a member in 1955.

So strained was the relationship between Ireland and Britain that the former abandoned the Commonwealth of Nations, a global organization that comprises Britain and all its dependencies and former dependencies. DeValera and his government found even this nonpolitical bond too restrictive, and as soon as this last tie with the United Kingdom was severed, in 1948, Ireland proclaimed itself a republic.

By that time, the Cold War had started. As a result, the North Atlantic Treaty Organization was established. This collective-defense pact, sponsored by the United States, included most of Western Europe. Ireland nevertheless refused to join NATO.

By the 1970s, some of these attitudes gave way to a more cooperative spirit, and in 1973, Ireland joined the European Union (then known as the European Economic Community) at the same time as the United Kingdom.

GOVERNMENT

Ireland has adopted a system of parliamentary democracy that in many ways resembles the original model of parliamentary government created by Britain. Thus there is a dual executive; but since Ireland is a republic, its head of state is not a monarch but a president who is elected by direct vote of the people. The

president has a fixed term of six years, and is a largely ceremonial head of state. The person who is pivotal in all other respects is the prime minister (*taoiseach*). The prime minister heads the party (or coalition of parties) that has the support of a majority of the members of the lower house (*Dail*) of the Parliament (*Oireachtas*). As in the British parliamentary system, elections must be held every five years; however, prime ministers can call for early elections to improve their majority or if they lose majority support for their government's programs.

The Dail is easily the more important of the two houses of the Parliament. It has 166 members. They are elected by a complex system of proportional representation, known as " single transferable vote," in multimember constituencies. Basically, voters rank-order their preferences for candidates; votes not needed to elect candidates (once they have obtained a majority) or wasted on hopeless candidates are transferred down the list to others. Proponents argue that the positive effect of these complicated and slow tabulations is a democratic outcome that more accurately reflects voters' preferences than does the single-member simple plurality (first-past-the-post) system, used in American legislative elections and British parliamentary elections.

The upper house (*Seanad*) of the Parliament has 60 members; 11 are appointed by the prime minister, six by the universities, and 43 by panels that represent various economic and cultural interests. The Seanad can amend or delay legislation passed by the Dail; however, in the end, it can be overridden by the majority of the Dail.

The five main political parties represented in the Dail have been Fianna Fáil, Fine Gael, the Labour Party, the Progressive Democrats, and the Workers' Party (now the Democratic Left). Fianna Fáil, the same party that deValera established about 65 years ago, has been in power the longest, although it has often depended on coalitions. It used to be a radical party but has over the course of time moderated its stand.

The current government is another coalition led by the Fianna Fáil, which tends to attract support from rural westerners, businesspeople, professionals, and urban workers. In the 1997 elections, it won 39.3 percent of the first votes and, with transferable votes, ended up with 77 seats in the Dail. Its junior partner, the right-of-center Progressive Democratic Party (which originated in 1985 as a splinter from the Fianna Fáil) won 4.7 percent of first votes and ended up with only four

seats. The government has survived with the support of four of the independents in the Dail. The second party in 1997 was the Fine Gael, which historically has been the more "establishment" party, drawing disproportionate support from upper and middle classes; it received 27.9 percent of first votes and 54 seats. The Labour Party was third, with 10.4 percent and 17 seats. Minor parties also picked up seats: Democratic Left (four), Green Alliance (two), and Sinn Fein (one). The latter is noteworthy because it was the first parliamentary seat won in the Irish Republic by the nationalist left party (known as the political arm of the Irish Republican Army in Northern Ireland).

The office of president was created by the first fully republican Constitution (of 1937). It is possible for the president and the prime minister to represent different parties, especially since the election dates can also differ. Currently both the president (Mary McAleese) and the prime minister (Bertie Ahern), who has been in office since 1997, are from the Fianna Fáil.

Bertie Ahern has been termed the "Teflon *taoiseach*" because of his ability to survive a series of scandals associated with his mentor, Charles Haughey, three times the prime minister. The latter's "high-living shenanigans" and serious allegations of sleaze involving party campaign solications and favors have been an embarrassment for the Fianna Fáil. The personally modest Ahern has been establishing a statesmanlike aura, through his working relationship with British prime minister Tony Blair in common pursuit of peace in Northern Ireland. Even more, Ahern and his party stand to benefit from Ireland's booming economy; but given Ireland's electoral system, power hinges on coalition politics and thus the electoral fortunes of the small parties in the upcoming 2002 elections.

Irish law is based on common law and legislation enacted by Parliament under the Constitution. The Constitution guarantees freedom of conscience and the free profession and practice of religion to all citizens. The Supreme Court can declare legislative acts unconstitutional. Regulations of the European Union have the force of law in Eire, as they do in all member states.

THE ECONOMY

Generally speaking, the late 1970s and early to mid-1980s were economically calamitous to Ireland. Then–prime minister Charles Haughey declared that "deficit budgeting on the current side since 1973 has brought disaster." (Deficit budgeting means that in the budget, the government expenditures exceed the revenues.) Haughey continued, "We intend to draw a line under that unhappy 20-year period and get back to balanced annual budgets."

By 1991, the Irish economy had experienced a complete turnaround, so much so that *Time* magazine praised the country as the "showpiece of the European Community [Union]." In part this was due to an economic pact, the "Program for National Recovery," which brought government spending under control and led to a lowering of inflation and interest rates. Average economic growth rates have steadily risen over the past decade and now stand at about 9 percent per year. In its trade with the rest of the European Union, Ireland is running a healthy surplus. Foreign companies (predominantly American), taking advantage of free access to the EU market, have also contributed to this picture of economic health. Economists see Ireland as benefiting from its well-educated, English-speaking workforce. The government has actively courted foreign investors and, in the mid-1990s, created an industrial-development agency to assist small indigenous firms.

Ireland has an open economy that depends to a large extent on international trade. Industry now accounts for more than 80 percent of Ireland's total exports, 24 percent of Ireland's gross domestic product, and 27 percent of its total employment. The Irish industrial sector has developed rapidly over the last 20 years. Foreign investment, stimulated by tax exemptions, capital and training grants, as well as other means, has greatly contributed to the growth of the Irish economy. The main components of the manufacturing industry in Ireland are the food-and-drink and the mechanical-engineering sectors. The industrial sectors showing most rapid growth are electronics and pharmaceuticals, both of which have been aided by substantial foreign investment, especially on the part of the United States. Ireland now rivals the United States as the world's biggest exporter of computer software.

Historically, the weakest spot of Ireland's economy has been unemployment. In 1993, the official unemployment rate stood at 15.9 percent. It has steadily fallen, to 3.7 percent in late 2000. In the early 1990s, the state training and employment agency (FAS) traveled abroad seeking jobs for Irish youth. Now it seeks to lure Irish home to work. After generations of emigration, Ireland has suddenly become a land of immigration. Half of recent arrivals were returning Irish, 19 percent were British, and 13 percent were from other EU countries. *The Economist*

estimates that to keep the economy booming, Ireland will need 200,000 additional workers within five years. Restrictions on work permits, not only for skilled jobs, are being loosened by the government for non-EU residents. However, immigrants confront one of the downsides of the Irish economic miracle: a housing crisis. In recent years, house prices have more than doubled, while rents have been going up 30 percent per year. Workers' wages have not been able to keep up with the spiraling prices.

Labor unions have always been important elements in the Irish economy, and a factor that is bound to contribute to the upsurge or downfall of the economy is the attitude of the unions. In 1987, unions were persuaded to endorse the "Program for National Recovery," which played a major role in Ireland's recovery by keeping wage increases and inflation below average EU levels. It has now been found that most of the sacrifices were in effect borne by the lower-paid and higher-taxed blue-collar workers. Subsequent signs of budding affluence have made the unions more determined to exact a tougher deal for the Program for National Recovery than the one that was negotiated in 1987, when the economy was in deep recession.

In 1999, the government, employers, and unions agreed upon a "Program for Prosperity and Fairness," which traded tax cuts for nominal pay increases. However, energy-price increases in 2000 caused inflation to soar to 7 percent (almost three times the EU average) and eroded take-home pay. Ignoring the advice of the president of the European Central Bank, the Irish government sought to counter labor unrest with a new expansionary budget of significant tax cuts and welfare spending, to renew the "social partnership."

Agriculture is a very important sector of the Irish economy. Approximately 10 percent of workers are employed directly in agriculture, which yields about 6 percent of the gross domestic product and about 15 percent of total exports. Of the total land area of approximately 17 million acres, some 14 million acres are agricultural land (more than 82 percent of the total). Ireland's membership in the European Union has benefited the country's agriculture. The EU provides a continental market and subsidies for Irish farmers' products through the Common Agricultural Program (CAP). Furthermore, Irish farms have received substantial developmental aid to modernize their operations. Although the European Commission has encouraged environmentally friendly farming practices, the Irish Envi-

ronmental Protection Agency's (EPA) 1996 Report identified agricultural run-off as the single greatest threat to water quality. Environmentalists have been critical of the Irish EPA's preference for voluntary compliance to deal with pollution problems.

Fishing has gradually become an important part of the economy. Among the fish found in Irish waters are herring, cod, whiting, mackerel, plaice, and shellfish; Irish salmon is exported to many countries. Investment in fish processing is greatly encouraged by the government.

The Irish economy also benefits from mining operations. The country is the leading producer of zinc ores, and one of the world's largest zinc-lead deposits is located at Navan, County Meath. The principal mining activity is the quarrying of sand, gravel, and stone for the construction industry. There is also coal mining, though on a more limited scale. And in the Celtic Sea off the County Cork coast, a gas-production platform has been built.

Ireland's tourist industry, which has received EU assistance, has grown rapidly since the mid-1990s. The restoration of normal political conditions in Irish border areas would mean further increases in tourism as well as an extension of a network of cross-border cooperation in economic and cultural sectors.

Ireland has adopted, like most industrialized democracies, a comprehensive welfare system. A complete range of medical services is provided free of charge to lower-income groups. In addition, there is a social-insurance program, which entitles insured workers to free medical services. Generally, however, there has been some retrenchment in the benefits that the Irish welfare system provides.

SEARCH FOR COMPROMISE

Relentless terrorism has generated considerable concern on both sides of the Ireland–Northern Ireland border. In 1983, a "New Ireland Forum" was established after political parties in both Eire and Northern Ireland had agreed on consultations about the manner in which lasting peace and stability can be achieved in a new Ireland through the democratic process. Participation was open to all democratic parties that had members elected or appointed to the Irish Parliament or the Northern Ireland Assembly.

Party leaders convened for the first time in Dublin in May 1983. The report of this first session, in which the four main nationalist parties, North and South, participated, was made public on May 2, 1984. The central operational conclusion of the report was that "the validity of both the

St. Patrick arrives to convert the Irish from Druidism to Christianity **A.D. 432**	Ireland is colonized by Viking raiders, who mostly operate from Wales **700–900**	British groups enter Ireland in small colonies but soon assimilate into the main body politic **1200s**	Large-scale transmigration from England, which soon assumes imperialistic overtones **1500s–1600s**	The potato famine; more than 4 million die; many more emigrate **1840s**	"The Troubles" start with the Easter Rebellion; violence and strife **1916–1919**	Anglo–Irish War **1919–1921**	The Government of Ireland Act is adopted, creating a northern and southern Ireland **1920**

(UN/P.Slaughter)

Agriculture is very important to the economy of Ireland. More than 82 percent of Ireland's total land area is agricultural land, and about 10 percent of workers are employed directly in agriculture. In this photo, it is harvesting time at an Irish farming community located in a rich, green valley.

Nationalist and Unionist identities in Ireland and the democratic rights of every citizen of this island must be accepted; both of these identities have equally satisfactory, secure and durable political, administrative and symbolic expression and protection."

If the establishment of the New Ireland Forum signaled a degree of progress in the meeting of minds, the Anglo–Irish Agreement of 1985 went even further. It had its roots in a meeting of the leaders of the Anglo–Irish Intergovernmental Council, which enabled both parties to give institutional expression to the unique relationship between the two countries. On November 15, 1985, the historic meeting at Hillsborough was held (at the level of heads of government). Here the prime minister of Ireland and the British prime minister signed the Anglo–Irish Agreement. It established an Intergovernmental Conference, which enabled the Irish government to express views and put forward proposals on various aspects of Northern Ireland affairs. Thus, for the first time, the Republic of Ireland would exercise some influence in matters relating strictly to Northern Ireland, the area that was withheld from Ireland when it embarked on the road to independence—an area that, in the words of one authority, is a "factory of grievances."

Eamon deValera is prime minister of the Irish Free State 1932–1959

The new Constitution is ratified, changing the nation's name from Irish Free State to Eire 1937

Ireland declares itself an independent republic, no longer part of the British Commonwealth 1939

Ireland is neutral in World War II 1939–1945

The Irish become progressively more resentful of the British presence in Northern Ireland 1950–1980

Ireland becomes a member of the European Union 1973

The New Ireland Forum; the Anglo–Irish Agreement 1980s

Irish voters favor the "closer union" of the Maastricht Treaty; Ireland signs the Northern Ireland Peace (Good Friday) Agreement 1990s

2000s

Ireland's stunning economic upswing continues; its economy grew by a remarkable 11% in 2000

In referendum results, Irish voters reject the Nice Treaty, dealing a blow to European enlargement

In 1998, history was made when Ireland, Northern Ireland, and Britain signed the Northern Ireland Peace Agreement, after years of discussion and negotiation. This "Multi Party Agreement" is designed to bring about a lasting peace on the island. In February 2000, the power-sharing agreement between Protestant Unionists and Catholic Republicans was suspended due to conflicts over decommissioning Irish Republican Army arms. Compromises by the Ulster Unionist party leader and first minister of Northern Ireland, David Tremble, allowed his regional government to be reinstated. However, serious conflicts about implementation have continued, particularly within the Protestant camp. Only time will tell if this power-sharing experiment will succeed. Optimists believe that although there will be relapses, the corner was turned with the "Good Friday" Agreement of 1998: For the majorities on both sides, there is no going back to the bad old days. On November 26, 1998, Tony Blair became the first British prime minister to address the Irish Parliament since the country's independence in 1921. Further evidence of new Anglo–Irish relations was provided by Irish prime minister Ahern, who remarked that Ireland might consider rejoining the (British) Commonwealth.

IRELAND AND THE EU
June 2, 1992, was a dismal day for the European Union: Contrary to general expectations, a Danish referendum rejected the Treaty on European Union, better known as the Maastricht Treaty. Shortly afterward, it was Ireland's turn to tell the world whether or not it accepted the treaty.

If the question had been put before a panel of authoritative analysts which country, Denmark or Ireland, would conduct a referendum that was to have a negative outcome, most would immediately have pointed to Ireland. It was obvious that the Irish referendum was dogged by emotional issues such as abortion and nationalism. In this overwhelmingly Roman Catholic society, abortion is viewed as a deeply religious issue. Less than a year before the referendum, the Irish Supreme Court had stirred public sentiments by decreeing that abortion could be legally sought in foreign countries; anti-abortion activists now feared that a "yes" vote could reinforce that secular view.

The nationalists, on the other hand, rejected the pact because it was bound to compromise Ireland's independence and neutrality. What must have counted was the monetary argument: The government indicated that although some sovereignty might have to be conceded under the treaty, Ireland would gain financially. According to government leaders, the Union would yield about $10 billion over the next five years.

The turnout for the Irish referendum was 57.3 percent, which is not very high, considering that after the Danish referendum, the eyes of Europe were focused on this important contest. Another negative vote would almost certainly have jeopardized the Maastricht Treaty. However, in spite of intense discussions and debates that manifested a vociferous opposition to the ratification of the treaty, 69 percent voted in favor. Once the results were in, then–prime minister Reynolds stated, "There is a lot of relief in the cabinets of Europe today." He added, "It is a great day for Ireland and a

great day for Europe.... Our decision will give renewed momentum to the ratification of the treaty across Europe.... We have succeeded in putting European union back on the rails."

A survey in 1999 indicated that nearly 90 percent of the Irish people felt that their country had benefited from membership in the European Union; this was the highest level of all 15 EU-member national populations. Indeed, Ireland has received some $28 billion of assistance from the European Union since it joined in 1973. Although there are several factors that have contributed to its booming economy, generous EU aid has certainly been a plus. The expansion of the European Union in the early years of the new century to include the poorer (and larger) ex-Communist countries of Central/Eastern Europe will change the relative position of Ireland. The Irish may become less enthusiastic about the European Union when they become a net contributor to its budget after 2006.

DEVELOPMENT

Great steps have been taken to overcome underdevelopment. In becoming an industrialized nation, Ireland has especially been aided by its inclusion in the European Union. In 1999, Ireland's GDP per capita was higher than that of Britain.

FREEDOM

Ireland is highly democratic. Although the country is overwhelmingly Roman Catholic, there is freedom of religion. While abortion is still illegal in Ireland, the Supreme Court has ruled that it is constitutional for physicians and clinics to provide women with addresses of foreign abortion clinics.

HEALTH/WELFARE

Ireland has an extensive welfare system, whose major defect has been that it does not have adequate mechanisms to check whether or not recipients are really eligible for the benefits they receive.

ACHIEVEMENTS

Ireland's rich cultural heritage in literature, music, and dance has enjoyed a phenomenal resurgence of interest around the world in recent decades. Dublin has become one of the world's "happening" cities.

Italy (Italian Republic)

GEOGRAPHY

Area in Square Miles (Kilometers):
116,303 (301,255) (about the size of Arizona)

Capital (Population): Rome (2,645,000)

Environmental Concerns: air and water pollution; acid rain; waste treatment and disposal

Geographical Features: mostly rugged and mountainous; some plains and coastal lowlands

Climate: Mediterranean except north of the Po River, where it is temperate or alpine

PEOPLE

Population

Total: 57,634,400

Annual Growth Rate: 0.09%

Rural/Urban Population Ratio: 33/67

Major Languages: Italian; French; German

Ethnic Makeup: Italian; with small German, French, Slovene, and Albanian minorities

Religions: predominantly Roman Catholic

Health

Life Expectancy at Birth: 76 years (male); 82 years (female)

Infant Mortality Rate (Ratio): 5.9/1,000

Physicians Available (Ratio): 1/193

Education

Adult Literacy Rate: 98%

Compulsory (Ages): 6–13; free

COMMUNICATION

Telephones: 26,000,000 main lines

Daily Newspaper Circulation: 126 per 1,000 people

Televisions: 436 per 1,000

Internet Service Providers: 219 (1999)

TRANSPORTATION

Highways in Miles (Kilometers): 190,562 (305,388)

Railroads in Miles (Kilometers): 11,376 (18,961)

Usable Airfields: 136

Motor Vehicles in Use: 34,000,000

GOVERNMENT

Type: republic

Independence Date: March 17, 1861

Head of State/Government: President Carlo Azeglio Ciampi; Prime Minister Silvio Berlusconi

Political Parties: Christian Democrats; Democratic Party of the Left; Italian Socialist Party; Forza Italia; Social Democratic Party; Italian Renewal; Refounded Communist Party; National Alliance Party; Greens; others

Suffrage: universal at 18 (except in senatorial elections—minimum age 25)

MILITARY

Military Expenditures (% of GDP): 1.7%

Current Disputes: property and minority-rights issues with Slovenia and Croatia

ECONOMY

Currency ($ U.S. Equivalent): 2,289 lire = $1

Per Capita Income/GDP: $21,400/$1.21 trillion

GDP Growth Rate: 1.3%

Inflation Rate: 1.7%

Unemployment Rate: 10%

Labor Force: 23,193,000

Natural Resources: mercury; potash; marble; sulfur; dwindling natural-gas and crude-oil reserves; fish; coal; arable land

Agriculture: fruits; vegetables; meat and dairy products; fish

Industry: tourism; machinery; motor vehicles; iron and steel; chemicals; food processing; textiles; ceramics; footwear

Exports: $242.6 billion (primary partners EU, United States)

Imports: $206.9 billion (primary partners EU, United States)

 http://www.cia.gov/cia/publications/factbook/geos/it.html

THE REPUBLIC OF ITALY

Italy has from time to time been the target of highly unflattering characterizations. While it is true that the country did not exist yet when the Vienna Congress aimed at restoring pre-Napoleonic Europe, presiding Prince Metternich is reported to have exclaimed: "Italy is not a country; it is a geographical concept!"

In contemporary times, phrases such as "a difficult democracy," "a republic without a government," and "a stalemate society" have been used to describe Italy. Another stigma, referring generally to the character of Italian society, is described by "amoral familism." This particular term points to excessive emphasis on family relations, which to the Italian will always take precedence over public interests. Indeed, the interest in government and in governmental processes is notably low among Italians. But it is not the real government that solves problems and gets itself out of dilemmas. For such eventualities, the *sottogoverno* (literally, "undergovernment") exists. All sorts of unofficial wheelings and dealings take place, and the persons who provide these functions often are not even in government positions. Sottogoverno may be compared to a spare tire that comes into operation when all else fails—and such emergencies are not at all rare in Italy.

THE PROBLEM OF THE STATE

Italy's main handicap may be summarized as follows: It has neither a strong government nor an efficient civil service, nor, for that matter, a tradition of centralization. A number of historical factors accounts for these characteristics. First, unification and independence were late in coming. Also, the prolonged mix and overlap of church and state has to be taken into account—particularly in the 1870s, when the Roman Catholic Church subverted plans and processes for unification. In addition, regional disparities, which in Italy appear to be very pronounced, have, as a matter of course, been utterly dysfunctional to integrative policies.

The Resurgence Movement

As a Western European nation-state, Italy was a very late starter. For much of the period between the fall of the West Roman Empire in the fifth century A.D. and the *Risorgimento* (a movement toward liberation and unification in the nineteenth century), Italy was internally fragmented, consisting of numerous sovereign city-states and principalities of varying size. In addition, parts of Italy were dominated by foreign powers, particularly Austria.

The Risorgimento was narrowly based; it did not involve the bulk of Italian peoples. Indeed, the population at large was sometimes hostile, and at best indifferent, toward what poets and politicians perceived as the second birth of the Roman Empire, or the supreme achievement of Italy's destiny. Instead of a grass-roots movement, the Risorgimento was in essence an elitist endeavor, largely initiated by the Piedmontese monarchy and supported by the nobility elsewhere. Nor could a common means of communication help to resolve the problem of Italy's divisiveness. A variety of languages was spoken throughout nineteenth-century Italy; French was the language that served the Piedmontese court.

As was mentioned earlier, the Roman Catholic Church frustrated attempts at unification and independence. The Church's power had often been based on divide-and-rule precepts, and conditions in Italy facilitated the implementation of such directives. Since the variety of regimes prevailing in Italy were usually very weak, the Church had often assumed extra-ecclesiastical duties. Indeed, it had in effect become a quasi-governmental institution, taking care of people's needs in worldly matters as well. Now that a real government loomed with the Risorgimento movement, the papacy naturally feared that the competing system would take away tribute as well as loyalty normally owed to the Church. When in 1870 Italian troops entered Rome, the Vatican denounced the new state and called on Catholics not to participate in its politics. The Risorgimento—like the French Revolution a century or so earlier—thus abounded with anticlerical sentiments, its statesmen having become convinced the Vatican interference would always be the albatross around Italy's neck. They therefore proclaimed the separation of church and state as their central precept. But while in France the state proved sufficiently strong to dissociate itself from the Church and to become the dominant power in the land, for a long time the Church remained a rival force of legitimacy in Italy.

From Papal State to Vatican

The Roman Catholic Church has owned land since medieval times, exercising in effect sovereignty over extensive areas of Italy. These lands came to be called the papal states. Their area shrank considerably in the nineteenth century, particularly under the impact of Italy's gradual unification. Their size today is little more than the papal mansions—the living quarters of various cardinals and the Vatican offices

in the heart of Rome, plus a summer palace in Castel Gandolfo, some 15 miles away.

The Holy See, or Vatican, is now a state by itself and is as such internationally recognized. In addition, there are millions of Roman Catholics all over the world who support the construction of a Vatican sovereignty—that is, a Church having secular sovereignty and worldly powers as well. The fascist Italian dictator Benito Mussolini bought Vatican support by reaffirming this quintessence in the Lateran Treaties of 1929. These treaties between the Kingdom of Italy and the Vatican recognize the Holy See as a sovereign state. Mussolini may have regretted this concession, since it failed to end the political and secular influence that the Vatican enjoyed outside its own territory.

One may argue that after World War II, the strife between Church and state started to become less serious—certainly less intense—than it had been in the past. After all, the first four postwar decades witnessed a great deal of cooperation, if not symbiosis, between the Roman Catholic Church and the Christian Democrats (DC). Nevertheless, the DC lost some important battles that had the full backing of the Vatican, such as a referendum that concerned divorce and legislation regarding abortion.

A radical change came in 1994. It was then that the Christian Democrats—who, in spite of ups and downs, had been Italy's major power brokers for nearly half a century—lost a national election. This also meant that the Vatican lost much of its power to influence, let alone to manipulate, secular politics.

REGIONAL DISPARITIES

Regional disparities—that is, the divergence in cultures among the various parts of the country—have continued to be vast in Italy, much more so than in France or Britain. When territorial unification had finally been achieved to the extent that it invalidated Prince Metternich's cynical observation that Italy is a geographical concept, one of the Piedmontese unifiers reportedly exclaimed, "We have made Italy, now we must make Italians!"

Language, normally a major integrative factor, in Italy's case turned out to be a major obstacle. Italian, once spoken by only 2 percent of the population, became the country's *lingua franca* (common language), an imposition that forced a great many Italians to become bilingual. They now speak Italian in school, at work, and in other public places, but often use their own language or dialect at home or among themselves. Valle d'Aosta, a region in northwestern Italy close to the borders of France and Switzerland, is French-speaking. A much larger area in the northeast that includes South Tirol, ceded to Italy by Austria after World War I, is German-speaking. (For some time it remained an Austrian irredenta, and even now, after having been part of Italy for several generations of rule, a resentful population often refuses to communicate in Italian.)

Although regionalism as a rule causes cultural and lingustic diversity, in Italy it appears to include developmental aspects as well. A north–south dichotomy, which some other countries also display, is very pronounced in Italy. The north is a well-developed, industrial area, inhabited by people who do not differ significantly in lifestyle from most other Western Europeans. The south is relatively undeveloped and rural, by contrast, inhabited by people almost universally involved in agricultural pursuits. So uneven has development been that some analysts consider Italy to consist of two countries. It is difficult to explain this disparity satisfactorily. One may assume that northern Italy moved along with the rest of Western Europe, sharing its economic development and its cultural heritage. And since Rome has traditionally been the capital and trendsetter, a demarcation line north of the so-called Eternal City would have been impossible.

In the past few decades, the many Italian governments have made it a point to try to equalize the development of the two parts of the country by encouraging industry in the south. However, most governmental efforts have failed to produce the desired effect. To be sure, a number of industries moved south, but usually they left after a brief period of time. Their abandonment of the ill-fated industrial sites in the south produced what came to be called "cathedrals in the desert": large, vacant buildings in the dismal context of rural poverty. Some economists have argued that the infrastructure of southern Italy is simply not there, or that it has at least not reached the take-off point that would render injections of money and capital equipment profitable.

A great many southern Italian villages and small towns have been demographically affected, in that their young people have often had to leave the area in search of work elsewhere. First they found employment in the north. When that area was saturated they went farther afield, settling in countries of the European Union. It is only a matter of course that these migrations have adversely affected the prospects of the south.

The weakness of the state, which in Italy manifests itself in regional disparities as well as in developmental discrepancies, is to some extent offset by a strong civil culture. This is evidenced by the lack of success that terrorist groups (on both the left and right) have had. It is clear that the Italian people have been firm in their condemnation of violence.

WORLD WAR II AND THE NEW ITALY

While in World War I Italy joined the Allies (a policy for which the country subsequently believed itself to have been poorly rewarded), in World War II it chose Adolf Hitler's side, becoming an Axis partner of Nazi Germany in June 1940. However, Italy fared poorly, soon losing *mare nostrum* (Latin for "our sea," meaning the Mediterranean) to the British, and fighting the ground war so dismally, first in North Africa and subsequently in Sicily, that its Axis partner had to come to the rescue. Massive German aid notwithstanding, Italy lost these areas. Once Sicily, the largest Mediterranean island, had been taken, the Anglo–American forces started their invasion of the Italian peninsula, a much more arduous task. This was, in effect, the "first second front." Here, they met with considerably more resistance than they had expected. Nevertheless, the Italian fascist regime collapsed in 1943.

Although Italy then joined the Allies, officially declaring war on its former Axis partner, it could not immediately formalize the turnabout. In fact, many of the major changes had to await the end of the war. Since the king had compromised the monarchy by allowing Benito Mussolini and his fascism to gain a solid political foothold after the March on Rome in 1922, a referendum was conducted on the question of whether Italy should become a republic. By a majority of 54 percent, Italy rejected the continuance of the monarchy, and the Constituent Assembly that convened in 1948 embarked upon producing a constitution with a republican design. In this Constitution, two principles prevail that may help to explain why Italy has a "blocked" democracy: legislative supremacy and the multiparty system. The Constitution introduced legislative supremacy (which still prevails today). This allows the Legislature (the Chamber of Deputies or the Senate) to oust a government through a vote of no-confidence. On important issues, various political parties in opposition team up against the government. Invariably, a few members of the government coalition abstain or join the vote of no-confidence. When a new government has been established, as a rule made up of members of the victorious parties, new confrontations may soon be in-

itiated. This has been an ongoing process in Italy for half a century.

It would be wrong to think that the new cabinets are always made up of completely new ministers. Often there is a great deal of what in the political trade is called "replastering"—that is, the new ministers are recruited from the same set of politicians. When Aldo Moro was kidnapped and subsequently killed by the Red Brigades in 1978, he had been prime minister five times. Giulio Andreotti has been in positions of power over a span of four decades, sometimes as defense minister, at other times as foreign minister or prime minister. On June 28, 1992, party leaders again compromised by electing a relatively young professor of constitutional law, Giuliano Amato, to the post of president of the Council of Ministers (prime minister).

Amato had been a deputy leader of the Socialist Party, and his name came up when the candidacy of Bettino Craxi, the party's leader, was rejected (the latter, who once had been prime minister, was implicated in a scandal). At that time, the Christian Democrats appeared to be riddled with corruption. That party has traditionally been the dominant force in Italian politics; dogma; its close association with the Church sometimes made it seem that the Vatican was participating in Italian politics through the DC. (Another intermediary is afforded by Catholic Action, or AC, a large pressure group closely associated with both the Church and the DC.) Amato's reign did not last long, however; in April 1993, he was replaced as prime minister by Carlo Ciampi, who did not belong to any political party.

The multiparty system has contributed to the never-ending chaos in Italian party politics. The number of political parties in Italy competing for seats in the Chamber of Deputies or in the Senate invariably hovers around 20. It is a matter of course that none of them will receive a majority, which, in turn, implies that coalitions, often fragile, will have to be formed. That situation was still in full force during the 1994 elections.

By early 1994, Italy had used up more than 50 governments. A new election loomed in the spring. But now the "politics as usual" seemed to have come to an end. The man who wrought this change was not an average politician; indeed, he was not even a politician to begin with. Silvio Berlusconi started out as a businessman who, over the course of decades, had amassed a truly formidable fortune. Like Ross Perot, to whom he was often compared, Berlusconi was a billionaire who was very unhappy about the way

(Courtesy United Nations photo/S. Jackson)

The canals of Venice are famous the world over. Boats like these have been used to transport people for hundreds of years.

things were going in Italy. He thus thrust himself into the political arena, spending huge amounts of money in founding a new party, the Forza Italia, and in buying up the media to trumpet what the party stood for. The Forza Italia tilted strongly to the right.

In May 1994, Berlusconi and 25 ministers from his right-wing coalition came to power, amid fanfare of a new political era for Italy. Soon there were tensions within the coalition, which led the Northern League, the regional separatist party of Umberto Bossi, to withdraw its support of Berlusconi. On December 22, besieged by charges of corruption, Berlusconi submitted his resignation. The following month, Lamberto Dini, the treasury minister in the previous government, was named prime minister by president Oscar Luigi Scalfaro.

In subsequent years, Berlusconi has remained politically active despite a number of convictions for bribery and other wrongdoing (he contends that the charges are politically motivated).

ITALIAN PARTY POLITICS
The DC was always bound to garner the largest number of votes in an election, but only once did it avail itself of a majority.

On all other occasions, it was necessary to enter into coalitions. These coalitions never included the second-largest party, the Italian Communist Party (PCI, which is now known as the Democratic Party on the Left, or PDS), although the PCI expressed a willingness in the late 1960s to cooperate with a "bourgeois" party. The "historic compromise," as the Communists labeled their olive branch, was never taken seriously by the Christian Democrats, although informal consultations between the DC and the PCI may well have amounted to the cooperation that the DC officially wanted to avoid. Since the DC was badly in need of coalition partners, it regularly included the Socialists.

The Socialist Party of Italy (PSI) had until 1956 been closely aligned with the Communists. From 1956 to 1963, the PSI was in a period of reassessment, and after 1963 it became receptive to the overtures of the Christian Democrats; this tendency on the part of the DC was nicknamed "the opening to the left." Cooperation between the DC and the PSI has in general been harmonious. On one occasion, the Christian Democrats allowed a Socialist to assume the prime ministership. That cabinet consisted largely of DC members, but it did make a difference in that Bettino

Craxi, the Socialist prime minister, remained in office considerably longer (1983–1987) than most of his predecessors. However, it cannot be denied that the Christian Democrats had a great deal of influence as the major coalition party.

The Communists and Socialists
The old Communist Party reached the peak of its popularity when led by Enrico Berlinguer, who was a populist as well as an ideologue. When Berlinguer died, in 1984, the party started to suffer from a steady attrition, which accelerated after the collapse of the Soviet Union. At that point, the name change to the Democratic Party on the Left was effected.

The old PCI was an offshoot of the Socialist Party. The PCI had been banned in the mid-1920s, when Benito Mussolini's fascism became well established. One of the Communists' earliest leaders was an ideologue and writer named Antonio Gramsci. Although his works differ considerably from those of Marx, he has often been called the "Karl Marx of Italy." Mussolini, who had initially also been on the left end of the political spectrum, had Gramsci thrown in jail, where he eventually died. The PCI then moved underground and became extremely active in

Italian troops enter Rome; the Risorgimento and the new Italian state are denounced by the papacy A.D. **1870**	Unification is achieved by the Risorgimento **1871**	The new Italian state becomes expansionist; attacks and conquers Somaliland and Eritrea **1889–1890**	Italy joins Great Britain and France in World War I **1914**	March on Rome; King Victor Umberto appoints Benito Mussolini as prime minister **1922**	Lateran Treaties; Mussolini buys off the Vatican **1929**	Mussolini attacks and conquers Ethiopia **1935**	Italy enters World War II as an Axis partner **1940**

the Resistance. The enormous popularity that the party enjoyed immediately after World War II may in part be traced to the heroic conduct of Communists in the *Resistenza* ("Resistance").

While in prison, Gramsci laid the theoretical foundations for an international Communist system differing sharply from prevailing patterns. One of his major theses was that all nations had different cultures and that it would thus be dysfunctional to impose the same communism on all of them. While Gramsci remained a Marxist all his life, he steadily alienated himself from the Soviet Union. Palmiro Togliatti, who as one of Italy's Communist leaders had fled abroad when Mussolini started to persecute Communists, adopted some of Gramsci's earlier ideas upon his return to Italy, particularly those associated with the "Italian road to socialism."

Generations of political observers and analysts have argued that the DC and PCI (now the PDS) are in fact religious parties that have their roots in old Italy. The argument runs that as modernization of the economy and society continues, the Catholic and Marxist "faiths" will wither away.

This view was shared by Bettino Craxi. Italy's economic performance during his years as prime minister was very solid— no small accomplishment—but Craxi failed to create a third political bloc. The PSI was weak, with only around 12 percent of the vote, and it had a reputation for corruption at the local level. It could not claim with any plausibility to be the dominant force in a coalition that included the DC.

Political Parties and Factions

One of the more fascinating aspects of Italian politics is the kaleidoscopic nature of its political parties. Before the 1994 elections, there was a large array of political parties, headed by the Christian Democrats and, in second place, the Party of the Democratic Left. Under the circumstances, the DC had to rely on a coalition that involved two, three, and sometimes four other parties. If this external division was not enough, most political parties, but particularly the DC, were also highly multifactional—that is, they contained many factions that had a great deal of autonomy. To illustrate this, in Italy, factions have

their own offices and their own newsletters or newspapers.

THE ECONOMY

The Italian economy expanded dramatically in the 1950s, when big companies such as Fiat and Olivetti drew thousands of southern workers to the northern cities of Turin and Milan. Growth was rapid but lopsided; labor relations were bitter. The 1960s were years of varying expansion. The early part represented an economic miracle almost on the scale of the West German experience; in the later part there was the "hot autumn" (1969), during which a wave of strikes and protests marred economic achievements. Then the dynamism of the economy was slowed by the international oil crises of the early 1970s.

The strengths and weaknesses of Italy's economy are peculiar. The private sector has too few large companies, but these few are perfectly capable of competing in world markets. The state sector has fallen victim to Italian forms of bureaucracy, especially since the Christian Democrats expanded their clientelism. Nationalization is no longer viewed as a panacea, since corporations often deteriorate when managed by the government. It is for that very reason that the PDS voted against the nationalization of the Fiat autoworks. (A curious phenomenon—a Communist party voting against nationalization!)

Government spending in general is far too high, a factor that causes budget deficits and is bound to cut investments. (Italy's out-of-date banking system has greatly restrained investments anyway.) And if the bureaucracy can be burdensome, the legislation is certainly stifling. A maze of government regulations has descended on matters like hiring and firing, and this too will increase the inflexibility of the labor market. The black-market economy nevertheless flourishes, not merely in areas such as textiles and shoes but even in technologically advanced sectors like electronics.

During the 1980s, the south continued to lag behind, but the north witnessed a revival of its economy. Today, although national unemployment remains high— more than 10 percent—exports have boomed, and Italy has matched Great Brit-

ain in gross domestic product. (To some extent, the Craxi government may take credit for that. It modified the system of wage indexation, which reduced inflation and labor costs.) However, a greater impetus has come from the private sector. Small companies are thriving, which has led experts to suggest that the weakness of the Italian state may be an advantage for the private sector. It can—so the argument goes—adapt and modernize, free from burdensome governmental regulations which the state can enact but not enforce. There could be some truth in this, but Italian industrialists generally argue that what they need is fewer but more sensible regulations. In this sense, the problem of the state remains in spite of Italy's economic revival.

Throughout 1997 and 1998, the government struggled to meet the criteria for inclusion in the upcoming currency union within the European Treaty. The Maastricht Treaty, for instance, allows a country's budget deficit to be only 2.7 percent of its annual gross domestic product if it wants the right to use the new Euro; similarly, the accumulated public debt may total only 60 percent of the annual GDP. Even though Italy met these requirements, and does now use the Euro in addition to the lira, its fellow EU members expressed concern that the compliance might only be short-lived, since some of it came from a one-time special tax and since the country's leftist parties bitterly resent the curtailing of government services.

As the rural Italian south drops in population, many areas of the north are undergoing an industrial and high-tech boom; this contrast is deepening the secessionist sentiments among some northerners, especially those in the populist right-wing Northern League, which would like to set up a separate country called "Padania," or at least see Italy become a federation.

RECENT CONDITIONS

The recent recession that has plagued the industrialized nations of Western Europe has had its effects on the Italian economy. Unemployment is a particular concern. But Italy does not face the huge economic problems alone: It is an enthusiastic member of the European Union, which endeav-

Italy drops out
of the war after
heavy losses in
manpower and
territory
1943

A referendum on
the monarchy: a
majority favor a
republic
1946

The first
republican
Constitution of
Italy becomes
effective
1948

Italy experiences
an "economic
miracle"
1958–1963

The "hot autumn"
of strikes
1969

The divorce
referendum
1974

The abortion
referendum
1980s

Despite financial
problems, Italy
manages to join
the countries
launching the
Euro in 1999
1990s

2000s

The economy of
the north
flourishes while
the south lags

Illegal
immigration
remains a
problem

Berlusconi's
center-right
coalition handily
wins the 2001
parliamentary
elections

ors to find joint solutions, and is also one of the prestigious Group of Seven (G-7).

Another problem is the mafia. On July 19, 1992, a car bomb exploded in central Palermo, killing the most prominent anti-mafia judge, Paolo Borsellino. The blast had widespread ramifications. In Rome, it hindered the Amato government in its twin tasks of fighting the mafia and restoring the battered economy. The government finally decided to move mafia prisoners to an island off the coast of Tuscany. In Palermo, seven out of 16 deputy prosecutors resigned, one of them commenting that the job was "like standing in front of a firing squad."

Finally, Italy, which has for many years been a country of emigration, has become a target of immigration. Nearby Albania is in a pitifully poor economic and political state; many Albanians are fleeing or trying to cross into Italy. Those who left Albania first were lucky: Their numbers were small, and the Italian government was prepared to allow them in. But they were followed by larger groups, literally thousands and thousands of people, and soon Italy made it known that economic refugees would no longer be welcome. The Italian Coast Guard was strengthened, and other measures were taken to prevent Albanians from crossing the Adriatic Sea. The continuing crises in the former Yugoslavia also remain of serious concern to Italy.

However, in the face of abundant problems, the Italian party system has collapsed, and it is now up to a few leaders to make things work. This was demonstrated when, during the elections of April 1996, Romano Prodi, certainly not a political veteran, came up with his center-left ULIVO ("Olive Tree") Alliance. The new alliance won, and Prodi became the 55th Italian prime minister since the end of World War II.

Prime Minister Prodi's avowed goal was to make Italy a "normal democracy." While he did not achieve that—to no one's surprise, as could Italian government ever be normal?—he did manage to cut public spending, to move ahead with privatization, and to get Italy included in the slate of European Union countries that launched the Euro. Perhaps most important, Italians began to get a sense of continuity in government in the course of his 2½-year stewardship.

However, the able manager's political steam ran out in the fall of 1998, and Prodi finally fell victim to a vigorous opposition (including the redoubtable Silvio Berlusconi) in October, when he narrowly lost his parliamentary majority in a vote of confidence. The prime ministership was then handed over to parliamentary leader Massimo D'Alema, an ex-Communist (a choice that was sharply criticized by the Vatican). While D'Alema's stated goal was to provide more stability, his government was soon characterized by political maneuvering and bickering—not a promising sign for longevity.

D'Alema, whose more moderate party, called Democrats of the Left, has split from the ideologically minded Refounded Communist Party, shared the enthusiasm of most Italians for remaining a vigorous and full member of the European Union. He met with the pope to try to improve relations. In what has become a familiar tactic in postwar Italian politics, D'Alema resigned in late 1999, then reshuffled his cabinet and resumed his role as prime minister. He resigned for good in 2000, after his party lost ground in a series of regional elections. Italy's next post–World War II government was then formed by Giuliano Amato.

In the 2001 parliamentary elections, Silvio Berlusconi's center-right coalition (House of Freedoms) won solid majorities in both houses of the Parliament over the center-left coalition (Olive Tree), headed by Francesco Rutelli. Berlusconi's own party, Forza Italia, won 30 percent of the votes (up nine points since 1996), while his right-wing allies the postfascist National Alliance (12 percent) and the once-separatist Northern League (4 percent) saw their votes decline. The Olive Tree coalition won 39 percent of votes (down only one point since 1996), but was bested in many seats awarded according to first-past-the-post rules. As *The Economist* reported, "Unbothered by international warnings about his past business methods, current legal tangles and continuing spectacular conflict of interest, Italian voters gave Silvio Berlusconi the green light." In his "Contract with Italians," Berlusconi promised *radical* solutions to the country's many problems.

Besides endless government corruption scandals and the "maxi-trials" of large numbers of members of organized-crime families, recent Italian governments have undertaken little in domestic policy outside of economic reforms.

In significant foreign-policy moves, recent Italian governments have sent troops to participate in stabilization forces in Albania, Bosnia, and Serbia (Kosovo); and Italy joined the other G-7 countries in announcing $65 billion of debt relief for developing countries in 2000, money that, it is hoped, those countries will use for health and education programs.

DEVELOPMENT

Northern Italy (upward from Rome) is well developed, on a par with most Western European nations. The southern part is still very much uninindustrialized, despite governmental efforts. Italy is a member of the Group of Seven, the prestigious global committee that reviews economic conditions on an annual basis.

FREEDOM

A 1984 antiterrorist law introduced search warrants by telephone. However, since the elimination of fascist repression, Italy has bent over backward to maintain democracy. Certain areas that are culturally different have high degrees of autonomy.

HEALTH/WELFARE

Both men and women in Italy enjoy high life expectancy rates, and infant mortality is low. However, the country has thus far failed to develop a universal health-care system.

ACHIEVEMENTS

Italy's cultural brilliance has long been acknowledged. An achievement test that is not so well known is that fewer terrorists appear to be operating in contemporary Italy. The country has successfully walked the thin line between the strict surveillance that antiterrorism necessitates and a respect for people's freedoms and rights.

Luxembourg (Grand Duchy of Luxembourg)

GEOGRAPHY

Area in Square Miles (Kilometers):
1,034 (2,586) (about the size of Rhode Island)

Capital (Population): Luxembourg (76,400)

Environmental Concerns: urban air and water pollution

Geographical Features: mostly gently rolling uplands with broad shallow valleys; uplands to slightly mountainous in the north; steep slope down to moselle floodplain in the south; landlocked

Climate: modified continental with mild winters, cool summers

PEOPLE

Population

Total: 437,400

Annual Growth Rate: 1.27%

Rural/Urban Population Ratio: 10/90

Major Languages: Lëtzbuergesch; German; French; English

Ethnic Makeup: Celtic base, with French and German blend; about 25% foreigners (guest workers)

Religions: 97% Roman Catholic; 3% Protestant and Jewish

Health

Life Expectancy at Birth: 74 years (male); 81 years (female)

Infant Mortality Rate (Ratio): 4.8/1,000

Physicians Available (Ratio): 1/454

Education

Adult Literacy Rate: 100%

Education Compulsory (Ages): 6–15

COMMUNICATION

Telephones: 293,000 main lines

Daily Newspaper Circulation: 381 per 1,000 people

Televisions: 384 per 1,000 people

Internet Service Providers: 13 (1999)

TRANSPORTATION

Highways in Miles (Kilometers): 5,108 (3,167)

Railroads in Miles (Kilometers): 165 (275)

Usable Airfields: 2

Motor Vehicles in Use: 248,000

GOVERNMENT

Type: constitutional monarchy

Independence Date: April 17, 1839

Head of State/Government: Grand Duke Jean; Prime Minister Jean-Claude Juncker

Political Parties: Christian Social People's Party; Luxembourg Socialist Workers' Party; Action Committee for Democracy and Pension Rights; Democratic Party; Green Alternative

Suffrage: universal and compulsory at 18

MILITARY

Military Expenditures (% of GDP): 1%

Current Disputes: none

ECONOMY

Currency ($ U.S. Equivalent): 47.6 Luxembourg francs = $1

Per Capita Income/GDP: $34,200/$14.7 billion

GDP Growth Rate: 4.2%

Inflation Rate: 1.1%

Unemployment Rate: 2.7%

Labor Force: 236,400 (1/3 foreigners)

Natural Resources: iron ore; arable land

Agriculture: barley; oats; potatoes; wheat; fruit; wine grapes; livestock

Industry: banking; iron and steel; food processing; chemicals; metal products; engineering; tires; tourism

Exports: $7.5 billion (primary partners Germany, France, Belgium)

Imports: $9.6 billion (primary partners Belgium, Germany, France)

 http://www.cia.gov/cia/publications/factbook/geos/lu.html

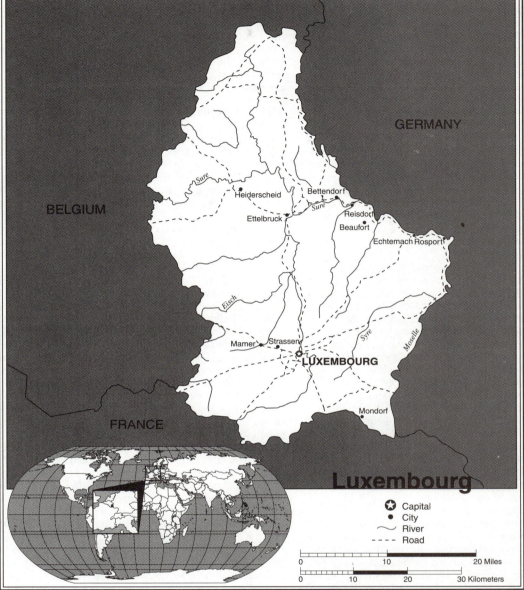

Capital
City
River
Road

| 0 | 10 | 20 Miles |

| 0 | 10 | 20 | 30 Kilometers |

LUXEMBOURG

In recent decades, the world has witnessed the birth of numerous countries characterized by populations of only a few thousand to a few hundred thousand. Most of them are small in area as well as geographically remote, economically weak, and/or of limited international significance. As a result, they have customarily been ignored as significant political actors.

The Luxembourg case is altogether different. An ancient state in the heart of Western Europe, it has, to a surprising degree, earned much favorable attention in the recent past for the successes of its economy; for its roles in European Union institutions; and, perhaps most dramatically, for the growth of a large international banking sector since the 1970s.

At the center of a highly industrialized region where France, Germany, and Belgium meet, Luxembourg "sits on a mountain of iron," as a German industrialist, trying to persuade his government to annex the country a century or so ago, reported. Iron and steel have constituted the basis of Luxembourg's modern development, though the economy has become increasingly diversified. Of late, the country has been wooing computer and other high-tech companies to establish themselves within its borders. Nevertheless, agricultural production has also been important for generations, and Luxembourg wines have become known all over the world.

Iron-ore resources have become depleted over the years, but the country's international visibility and the substantial improvements in its standard of living have spurred progress in Luxembourg. National pride, revitalized in recent years, plays a large role. One may cite as an example the increased use of Lëtzbuergesch, an ancient West Frankish dialect more closely related to German than to French. (Luxembourg has two other officially recognized languages—French and German—while English is widely understood.) Lëtzbuergesch is first and foremost a spoken language; there are hardly any written records in Lëtzbuergesch, and only in recent decades have linguists developed methods of transcription. Other sources of pride that deserve mention are the hard bargains that the country drove in the EU, the renegotiation of its treaty of economic unity with Belgium, and the fact that Jacques Santer, at that time the prime minister of Luxembourg, was selected to become the president of the European Commission in 1994.

Such developments have helped to sustain a positive momentum in the face of a rapidly aging population and acute labor shortages. This is no small achievement, given that in recent years, the surrounding larger states in Western Europe have experienced severe deficits, foreign debts, high unemployment, cutbacks in social programs, and, in general, the politically troublesome consequences of global forces.

THE ROAD TO SOVEREIGNTY

Luxembourg was founded in A.D. 963 by Siegfried, the count of Ardennes, who made it into a fortress. For half a millennium, Luxembourg was a major actor in Central Europe and in the Holy Roman Empire. However, its defeat by Burgundy in 1443 signaled a decline; and from then onward, it was controlled by one or another of its larger neighbors, notably by Burgundy, Spain (which still had possessions in the Southern Netherlands), Austria, and France (during Napoleonic times). The 1815 Congress of Vienna, fearful that France might again wreak havoc in Western Europe, ordained that Holland, Belgium, and Luxembourg were to constitute one country that could act as a buffer against such an eventuality. The measure revealed gross ignorance of cultural differences, when Belgium seceded from the Netherlands in 1830, Luxembourg eventually followed suit. However, it did not want to be part of Belgium either, and the Treaty of London, while granting independence to Belgium, provided some separate status to Luxembourg. The latter drifted into the so-called Germanic Confederation, which it left in 1866. The foundation of Germany in the next decade solidified Luxembourg's independence, since the two rivals, France and Germany, denied each other the possession of Luxembourg. When the Low Countries were overrun by the German war machine in 1940, Adolf Hitler made a distinction between Belgium and the Netherlands (which were viewed as occupied countries) and Luxembourg (which was re-integrated into Germany).

The Dutch flag and the Luxembourg flag are very similar, for good reason: For some time there was a personal union between the monarchies of the two countries. However, when King William III of the Netherlands (who was also king of Luxembourg) died in 1890 without a son, the union dissolved, since the Luxembourg Constitution did not allow females on the throne. Luxembourg then became a grand duchy, and all connection with the Netherlands was severed.

EUROPEAN CONNECTIONS

Although Luxembourg has always prided itself on its independence, it realized that its interests would be promoted if it became a member of an economic union. As soon as the German *Zollverein* (customs union) was disbanded, Luxembourgers voted for an economic union with France. The latter held a dim view of Luxembourg's potential and rejected the offer. A union with the voters' distant second choice, Belgium, was concluded; in 1922, after prolonged negotiations, the Belgo–Luxembourgeois Economic Union (BLEU) was signed for a term of 50 years. The arrangement included linkage of Luxembourg's currency to the Belgian franc, which in practice implies that one may pay in Belgian currency in Luxembourg, but not the other way around. BLEU had little regard for the fact that Luxembourg's economy may be smaller than that of Belgium, but it is in fact superior, in that when Belgium would decide to devalue its franc, Luxembourg would have to follow suit. Although a new BLEU surfaced in the early 1970s, this particular wrinkle had not been removed. When in the early 1980s Belgium decided to devalue its currency, Luxembourg reluctantly devalued the Luxembourg franc.

Key members of the governments of Belgium, the Netherlands, and Luxembourg fled into exile in London when Germany invaded their countries in 1940. The respective governments-in-exile agreed to form an economic union, *Benelux,* once the war would be over. That union was formally launched in 1948, but the tasks of postwar recovery as well as the politicization of their economies kept the three countries' policies little affected by the Benelux pact until the entire area had become submerged in what is now called the European Union. Paradoxically, it was only then that the Benelux started to operate: The three small countries saw an important advantage in a closer coordination of their policies within the new setting, in that it allowed them to obtain a measure of strength against larger members.

Luxembourg sought advantages through Benelux for its rapidly growing international monetary operation. Luxembourg now qualifies as one of Europe's few banking and tax havens, and as of 1999, more than 220 banks had set up shop in the small country. The EU's push toward harmonization of its member states' banking and tax laws could threaten economic catastrophe for Luxembourg. However, Luxembourg has been diversifying into other areas. High technology may one day replace banking as the country's key industry. For example, the country has been at the forefront of satellite-television broadcasting in Europe.

| The founding of Luxembourg by Siegfried, Count of Ardennes A.D. **963** | The conquest by the Duchy of Burgundy spells the beginning of more than 400 years of foreign domination **1443** | The Congress of Vienna makes Belgium and Luxembourg parts of the Netherlands **1815** | Independence as a result of the Belgian Revolution **1839** | Separation from the Netherlands crown **1890** | The end of the Zollverein; the beginning of the Belgo-Luxembourgeois Economic Union **1919–1922** | Luxembourg becomes an international financial and banking center **1980s** | Luxembourg is one of the first countries to enter the postindustrial era **1990s** | **2000s** |

Luxembourg's voice in the European Union remains strong

Luxembourg has fared well as the smallest member of the European Union. It is much more influential than its share of the population or the economy of the EU would seem to warrant. Since the 1950s, the leaders of Luxembourg have actively supported a more supranational ("federal"), less intergovernmental path for European integration. Several EU institutions have long been based in Luxembourg City, most notably the European Court of Justice, the European Investment Bank, and the European Court of Auditors. EU employees provide a significant contribution to the local economy. A EUROSTAT survey in 1999 indicated that more than 70 percent of Luxembourgers felt that the country had benefited from EU membership. During its six months tenure in 1997, Luxembourg ably occupied the Presidency of the Council of the EU.

Traumatic violations of its neutrality in both world wars made Luxembourg an eager charter member of the North Atlantic Treaty Organization. Although its military includes only 800 troops, the country has contributed to collective action. NATO's early warning aircraft (AWACS) are registered under the Luxembourg flag. Lacking its own air force, Luxembourg financially supported the Belgian Air Force during NATO's air war against Serbia in 1999.

In the aftermath of World War I, Luxembourg's constitutional monarchy moved rapidly toward modern parliamentary democracy, in which the grand duke's role in politics is largely symbolic, with the government responsible to the directly elected Parliament. The royal family has remained popular, but the political culture of the country is also thoroughly democratic. Voting is a legally compelled duty. The complex electoral system allows voters to vote not only for candidates of their choice, but also for parties of their choice. In recent elections, about half a dozen political parties have competed for parliamentary seats. The Social Christians (PCS) have tended to be the strongest party but have needed a coalition partner to form a government. Due to the swing of voters from the Socialists and the Greens, the Democrats (right-wing liberals by European standards) finished ahead of the Socialists as the second-strongest party in the June 1999 election and joined a new governing coalition, headed by Prime Minister Jean-Claude Juncker of the Social Christians.

Coalitions tend to be moderate, slightly left or slightly right of center, and committed to the maintenance of a comprehensive social welfare system. The scope of coverage and the level of benefits of social programs in Luxembourg rank near the top among all nations. For example, the World Health Organization 2000 survey rated Luxembourg's health-system performance the fifth best in the world. Effective industrial policies, a smooth transition from manufacturing to services in large portions of the economy, and limited pressures for new jobs have made looking out for the public welfare easier for the Luxembourg government than it has been in many other industrialized countries.

One of the main difficulties is that the markedly aging population (Luxembourg has one of the world's lowest birth rates) has created a chronic shortage of workers. In part this has been remedied by the steady introduction of "guest workers" and the use of automation. Luxembourg has had the lowest unemployment rate within the EU; only 2.2 percent of its workers were unemployed in mid-2000. Luxembourg also stands out as having the highest proportion of workers who are non-nationals—over a third, mostly from Southern Europe. Despite their growing presence, to date there has been no serious xenophobic reaction to fuel right-wing protest parties, as has occurred in neighboring countries.

THE FUTURE
As is the case with most Western industrialized democracies, the main challenge that Luxembourg faces today is to maintain the social and economic gains that it has achieved since World War II. Political stability and healthy industrial and service sectors, particularly in international financial transactions, bode well in this regard. Growth rates have slowed markedly since the early 1980s in Luxembourg, but no signs of economic danger are evident. The economy's openness to extraneous forces does make it subject, if not vulnerable, to international trends; but it is better positioned than most national economies to withstand the shocks and to minimize their effects.

DEVELOPMENT

From primary industries (agriculture and mining), Luxembourg's economy developed to secondary industry (manufacturing) and has now become a postindustrial society in which banking, finance, and other services play a predominant role in the economy. Luxembourg has become a financial center that emulates Switzerland. Banks maintain a high degree of secrecy.

FREEDOM

Luxembourg has a very favorable human-rights rating. There is a great deal of tolerance for foreigners, who comprise substantial proportions of the workforce and the population in general.

HEALTH/WELFARE

Luxembourg, a very prosperous country, has high life expectancy rates, and a relatively low natural population growth. The state offers a comprehensive social-welfare system. According to the UN's Human Development Index, Luxembourg's quality of life is the 17th best in the world.

ACHIEVEMENTS

The University of Luxembourg, operating only during the summer, invites participants from all over the world to join in its Summer Seminars. These always focus on a particular theme, and the speakers are all experts in the fields selected.

Malta (Republic of Malta)

GEOGRAPHY

Area in Square Miles (Kilometers):
123 (320) (about twice the
size of Washington, D.C.)
Capital (Population): Valletta
(7,200)
Environmental Concerns: increasing
reliance on desalination; recycling of solid waste
Geographical Features: mostly
low, rocky, flat to dissected
plains; many coastal cliffs;
archipelago; many good harbors
Climate: Mediterranean

PEOPLE

Population
Total: 391,700
Annual Growth Rate: 0.74%
Rural/Urban Population Ratio:
10/90
Major Languages: Maltese;
English; an Arabic dialect
Ethnic Makeup: Maltese (descen-
dants of ancient Carthaginians
and Phoenicians, with strong
elements of Italian and other
Mediterranean stock), Italian
Religions: 91% Roman Catholic;
9% others

Health
Life Expectancy at Birth: 75
years (male); 81 years (female)
Infant Mortality Rate (Ratio):
5.9/1,000
Physicians Available (Ratio):
1/403

Education
Adult Literacy Rate: 88%
Compulsory (Ages): 5–16; free

COMMUNICATION
Telephones: 191,500 main lines
Daily Newspaper Circulation:
145 per 1,000 people
Televisions: 739 per 1,000 people
Internet Service Providers: 4 (1999)

TRANSPORTATION
Highways in Miles (Kilometers): 1,080
(1,742)
Railroads in Miles (Kilometers): none
Usable Airfields: 1
Motor Vehicles in Use: 141,000

GOVERNMENT
Type: parliamentary democracy
Independence Date: September 21, 1964
(from the United Kingdom)
Head of State/Government: President
Guido de Marco; Prime Minister Ed-
ward Fenech Adami

Political Parties: Nationalist Party; Malta
Labour Party; Alternativa Demokra-
tika/Alliance for Social Justice
Suffrage: universal at 18

MILITARY
Military Expenditures (% of GDP): 5.5%
Current Disputes: discussions with
Tunisia regarding the commercial
exploitation of the continental shelf
between their countries

ECONOMY
Currency ($ U.S. Equivalent): 0.466
Maltese lira = $1
Per Capita Income/GDP: $13,800/$5.3 billion
GDP Growth Rate: 4%
Inflation Rate: 1.8%

Unemployment Rate: 5.5%
Labor Force: 144,000
Natural Resources: limestone; salt; arable
land
Agriculture: fodder crops; fruits and
vegetables; flowers; livestock
Industry: tourism; electronics; shipbuilding
and repair; clothing; construction; food
and beverages; textiles; footwear; tobacco
Exports: $1.8 billion (primary partners
France, United States, Germany)
Imports: $2.7 billion (primary partners
Italy, France, United Kingdom)

 http://searchmalta.com/
http://www.cia.gov/cia/publications/
factbook/geos/mt.html

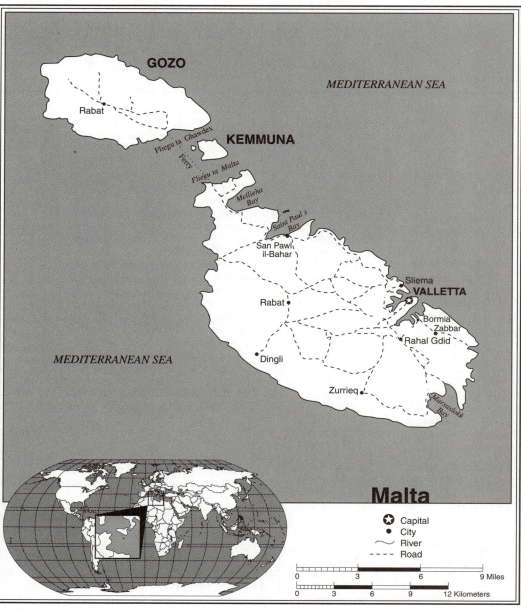

Malta

- ✪ Capital
- ● City
- 〜 River
- --- Road

THE REPUBLIC OF MALTA

Malta, a group of rocky, low-lying islands in the Mediterranean Sea, is poised at the crossroads of ancient seafaring nations. Not surprisingly, this exceptional location has made it a magnet to peoples for many centuries. The name Malta derives from a Phoenician term meaning "hiding place." Apostle Paul is said to have been shipwrecked there on his way to Rome in A.D. 60; after swimming ashore safely, he remained for a time to convert the inhabitants to Christianity.

Over the centuries, Malta has been conquered, colonized, occupied, or protected by Phoenicians, Carthaginians, Romans, Byzantine Greeks, Arabs, Normans, and a succession of modern peoples. In 1530, the Knights Hospitalers (also known as the Knights of St. John, a Roman Catholic religious and military order) made Malta their base, having been driven away from Rhodes by Turks. One of their most prominent leaders, Grandmaster Jean de la Valette, organized the Knights to resist the long siege by Turkish land and naval forces in 1565, thereby depriving the Turks of access to the western Mediterranean. La Valette supervised the construction of new fortifications, around which the capital city that now bears his name gradually grew.

In the 1790s, Malta was occupied for brief periods first by the Russians and subsequently by the French. Napoleon Bonaparte's preoccupation with large imperial conquests caused the French occupation to languish; and, in order to frustrate the proposed reinstatement of the Knights by the French emperor, the Maltese requested protection by the British in 1802. Malta thus became a British Crown colony, and for the next 150 years, British naval and military forces made it impossible for emerging Italy to follow the Romans in speaking of the Mediterranean as *mare nostrum* ("our sea"). Britain had finally started to realize the strategic importance of the three islands that jointly constitute Malta. Thus Lord Horatio Nelson, a British naval hero in the late eighteenth century, commented: "Malta is always in my thoughts."

The isles are located some 60 miles south of Sicily and 200 miles north of the coast of Libya. Thus, Malta's political, economic, and psychological orientation can be traced to Western Europe on the one hand and to the Arab world on the other. This mixture has strengthened over many centuries and remains part of contemporary Maltese politics and its people's worldview. The two worlds—the Arab and the Western European—seem to join in Malta's Semitic language, which is related to Arabic but written in Latin characters and is strongly influenced by the Sicilian dialect of Italian.

During World War II, the Maltese experienced great stress; they were mercilessly bombed by the Germans and Italians. They endured the pressure of a siege that lasted for nearly three years (1940–1943) and resulted in 5,000 casualties. For their courage and endurance, the people of Malta were collectively awarded the George Cross, Great Britain's highest civilian decoration. This was the first time that the distinction was conferred on any part of the British Commonwealth. The emblem is displayed today on the Maltese flag.

INDEPENDENCE

The British colony became self-governing in 1947. The debate on Malta's eventual emancipation found its small size and population to be major stumbling blocks. However, these criteria lost their validity in the decade that followed, and a referendum held in 1956 made clear that complete independence was preferred by a wide margin. Nevertheless, the Maltese political parties could not agree upon the method to achieve that goal. Finally, in 1964, the impasse was overcome, and a Conservative government led the island nation to independence.

Like most former British colonies, Malta opted to remain in the Commonwealth of Nations. In this political construction, Malta continued to recognize the sovereignty of the British Crown. In 1974, a Socialist government took the initiative to amend the Maltese Constitution so as to make Malta a republic, but without surrendering its membership in the Commonwealth.

THE SYSTEM OF GOVERNMENT

The more than 160 years of British overlordship left the political heritage of a parliamentary form of government. Executive authority is exercised by a prime minister and a cabinet, who are responsible to the unicameral Legislature in Malta, the 75-member House of Representatives. Since independence, the two principal parties have been the (Conservative) Nationalist Party and the (Socialist) Labour Party.

The Nationalist Party is supported by business interests and a middle-class constituency, although it naturally also tries to appeal to broader cross-sections of the electorate. The party belongs to the European Christian democratic tradition and relies on the cooperation of the Roman Catholic hierarchy in Malta. Its foreign-policy orientation is decidedly pro-Western and European.

In contrast, the Labour Party, which regained power in October 1996, with Alfred Sant as prime minister, is closely associated with Malta's large and powerful trade-union movement. While in office from 1971 to 1987, the Labour government attempted to identify Malta's interests more closely with the developing world. Ideological links were forged between this island nation of devoutly Roman Catholic people and Libya (200 miles across the Mediterranean Sea to the south), North Korea, and other anti-Western states. The Labour Party's pronounced anticlericalism brought it into conflict with the Roman Catholic Church, which, in the 1960s, threatened supporters of the party with excommunication. The Labour Party nevertheless managed to gain office. Under Prime Minister Dom Mintoff, however, mutual hostility soon escalated, particularly when schools and hospitals that had been operated by religious orders were nationalized.

Mintoff, who remained prime minister until 1984, abrogated Malta's mutual-assistance security agreement with Great Britain almost immediately after assuming office. This action caused the departure of the Royal Navy, and the consequent loss of revenue could not easily be remedied. The Labour prime minister flirted with the Soviet Union, which had gained in prominence in the Mediterranean. At one point, he offered docking facilities for Soviet vessels in return for political and economic support.

The volatile prime minister then turned to an anti-Western and anticapitalist country closer to home: Libya. Opponents of the Labour Party feared that Malta would become a satellite of the Muammar Qadhafi regime, and they also pointed with alarm to the authoritarian streak in Mintoff's political temperament. He had promoted his loyal party followers in the civil service, police, and state enterprises; and he might well have "tweaked" election results in 1981 in order to remain in office. Yet, he was also a skillful politician in a more traditional way, managing to stay in power with a one-vote majority in Parliament during his first term. Mintoff made Malta very visible on the international scene by its participation in the UN Conference on the Law of the Sea.

Six years later, in the 1987 election, the Nationalist Party finally succeeded in terminating Labour control. Shortly after taking office, the new Nationalist Party prime minister, Edward Fenech Adami, pleaded with the Maltese people "to feel as one people" regardless of political party. The defeated Labour supporters, however, were reluctant to accept the decision of the voters, taking their frustrations into the streets. The postelection rioting underscored the deep cleavages of class, ideology, and political traditions that divided the country.

Adami, who had won the election by a wafer-thin majority, pledged rapid shifts in

Malta becomes a British Crown colony
A.D. 1802

The Maltese are besieged by the Germans and Italians during World War II
1940–1943

Malta becomes self-governing
1947

Independence within the Commonwealth of Nations
1964

The Labour Party and Prime Minister Dom Mintoff take office
1971

179 years of British military presence in Malta ends
1979

Prime Minister Edward Fenech Adami and his Nationalist Party come to power
1987

Power fluctuates between Labour and the Nationalists
1990s

2000s

Malta's full candidacy for membership in the EU continues to be a topic of conflict

Adami is prime minister again, after a time out of power

foreign policy that would underscore Malta's solidarity with Western Europe. His government, for instance, wanted to apply for European Union membership. Indeed, this was one of the major planks in the Nationalist platform (the Labour Party saw fit to warn the electorate against integration into Europe because of that continent's AIDS cases!). However, with regard to Malta's position vis-à-vis the North Atlantic Treaty Organization, the two parties concluded a preelection agreement: They reached a consensus about Malta's status as a neutral country—a neutrality that specifically ruled out the long-term use of repair facilities in Maltese dockyards by warships from NATO fleets.

MALTA AND THE EU

On July 16, 1990, Malta formally applied to join the EU, an initiative that was largely launched by the Nationalist Party. The opposition Labour Party would have preferred to maintain Malta's associate membership in the organization. Further steps toward full membership should, according to the opposition, be a matter of "watching and waiting." The Nationalists, however, argued that such a hesitant skepticism was bound to dissipate, as events in the British Labour Party and PASOK (the Greek Socialist party) have shown. These two Socialist parties had initially opposed EU membership but had subsequently come to endorse the mergers, more or less enthusiastically.

In 1996, Alfred Sant was again elected prime minister, and he immediately pulled the plug on accession to the EU. When he was voted out, however (once again losing to Adami), in late 1998, the path to negotiations with Brussels was again clear. Malta, like Slovakia, Romania, Bulgaria, Lithuania, and Latvia, belongs to what is called the "second tier" of applicants to the European Union. Although their prospects for admission are not as good as

those of Poland, Hungary, the Czech Republic, Estonia, and Slovenia, the EU determined in 1995 that it would eventually begin talks with these states. These discussions finally started in February 2000. EU requirements include economic reform but also issues of security and political stability. Malta's Nationalist Party hopes that the country can meet the criteria by 2010.

Security issues might be Malta's greatest stumbling block. Malta is not accused of sponsoring international terrorism, despite its often close relationship with Libya. But it has fairly often been the site of terrorist attacks. There were bloody hijackings in 1985 and 1997, and in 1995 a leader of the movement known as Islamic Jihad was gunned down on the street; many Arabs blamed Israel for the murder. There were also widespread accusations that Maltese banks and businesspeople helped the Yugoslav government of Slobodan Milošević violate sanctions on oil and financial transactions.

THE ECONOMY

Economically, Malta is closely linked to the European Union; indeed, EU countries already account for about three quarters of Malta's imports and exports. Yet many people in Malta are ambivalent with respect to full EU membership. The island's employers' federation, for instance, fears that becoming a full member will aggravate rather than solve Malta's economic problems. One should also bear in mind a visit by one of the EU's high officials who somewhat tactlessly argued that the membership of small states could ravage democratic aspects in the European Union's decision-making process unless such members would resign themselves to allowing others to conduct policy. On the other hand, Malta has more inhabitants than Luxembourg, and nobody can deny that the latter is disproportionately part of the EU's decision making. The recent em-

phasis on subsidiarity, moreover, could appear to reduce the dangers of what extreme nationalists might qualify as "internal colonialism." (*Subsidiarity* means that the EU institutions will take care of policy matters only if the member states and their subunits are unable to deal with them.)

Although 38 percent of Malta's land surface is arable, most of its food supplies have to be imported. Malta's chief crops are potatoes, onions, and beans. The biggest sources of income are the dockyards and tourism; nearly all of the latter comes from EU countries. Tourism declined by 28 percent in 1990, however, because of the Perisan Gulf conflict and the expense of flights from Europe. Since then, tourism has rebounded. There are few sightseeing opportunities on the island, which makes the country a haven for tourists seeking peace and quiet. Recent prosperity among EU countries prompted a rise in visitors in the 1990s, with a record 1 million going to Malta in 1992. The tourism infrastructure is improving also, with the opening to the public of new archaeological sites. Three new major tourist resorts have also been opened by Western corporations.

DEVELOPMENT

Most of Malta's trade is with EU countries. Malta is lacking in raw materials and must import its entire energy requirement. Unemployment—currently about 5.5%—is believed to have occasionally soared to more than 20%—far more than the typical official estimate.

FREEDOM

Class divisions are quite sharp in Malta, as is the cleavage between devout Catholics and anticlerical elements identified with the Labour Party. Despite its rough-and-tumble history as an independent country, Malta has remained committed to political liberty and the protection of civil rights.

HEALTH/WELFARE

The Labour government's nationalization of Catholic schools and hospitals was an important issue in the 1987 election, bringing about its downfall. Malta has a comprehensive state-supported welfare system. Improved health care has reduced infant mortality to a rate that is typical of the more developed countries.

ACHIEVEMENTS

Malta is a popular travel destination for Western Europeans, who go to the island to take advantage of its mild climate, beaches, and attractive scenery. The government endeavors to make tourism less volatile as a source of revenue.

The Netherlands (Kingdom of the Netherlands)

GEOGRAPHY
Area in Square Miles (Kilometers):
15,865 (41,532) (about twice the
size of New Jersey)
Capital (Population): Amsterdam
(717,000); The Hague (seat of
government) (443,000)
Environmental Concerns: water
and air pollution; acid rain
Geographical Features: mostly
coastal lowland and reclaimed
land; some hills in southeast
Climate: temperate; marine

PEOPLE

Population
Total: 15,893,000
Annual Growth Rate: 0.5%
Rural/Urban Population Ratio:
11/89
Major Language: Dutch
Ethnic Makeup: 91% Dutch; 9%
Moroccans, Turks, and others
Religions: 34% Roman Catholic;
25% Protestant; 3% Muslim;
2% others; 36% unaffiliated

Health
Life Expectancy at Birth: 75 years
(male); 81 years (female)
Infant Mortality Rate (Ratio):
4.4/1,000
Physicians Available (Ratio): 1/412

Education
Adult Literacy Rate: 100%
Compulsory (Ages): 5–18

COMMUNICATION
Telephones: 9,337,000 main lines
Daily Newspaper Circulation: 299
per 1,000 people
Televisions: 495 per 1,000 people
Internet Service Providers:
70 (1999)

TRANSPORTATION
Highways in Miles (Kilometers):
76,200 (127,000)
Railroads in Miles (Kilometers): 1,675
(2,791)
Usable Airfields: 28
Motor Vehicles in Use: 6,525,000

GOVERNMENT
Type: constitutional monarchy
Independence Date: 1579 (from Spain)
Head of State/Government: Queen
Beatrix; Prime Minister Wim Kok
Political Parties: Labor Party; Christian
Democratic Appeal; Liberal; Democrats
'66 (D'66); Green Left; Socialist
Party; others
Suffrage: universal at 18

MILITARY
Military Expenditures (% of GDP): 2.1%
Current Disputes: none

ECONOMY
Currency ($ U.S. Equivalent): 2.60
guilders = $1
Per Capita Income/GDP: $23,100/$365.1
billion
GDP Growth Rate: 3.4%
Inflation Rate: 2.2%
Unemployment Rate: 3.5% (but generous
welfare benefits prompt many to drop
out of the labor market)
Labor Force: 7,000,000
Natural Resources: natural gas; petroleum;
arable land

Agriculture: grains; potatoes; sugar beets;
fruits; vegetables; livestock
Industry: agroindustries; metal and
engineering products; electrical and
electronic machinery and equipment;
chemicals; petroleum; construction;
fishing
Exports: $169 billion (primary partners
EU, Central/Eastern Europe)
Imports: $152 billion (primary partners
EU, United States, Central/Eastern
Europe)

http://www.odci.gov/cia/publications/
factbook/geos/nl.html
http://www.travel.org/nether.html

THE NETHERLANDS

The Netherlands (frequently referred to as "Holland," although that term in fact applies to the two oldest and most important provinces) is a small country with a glorious past. Once a European power ruling a large colonial empire, it still plays a greater role in international affairs than could be expected of a nation of its size and population.

About one third of the country's land area is actually *below* sea level, reclaimed by a complex system of dikes, polders, and drainage canals whose construction started as early as the thirteenth century and culminated in the execution of the so-called Delta Plan. Its execution, which took more than a quarter of a century, was a reaction to a huge flood in 1953 that claimed nearly 2,000 lives. The twelfth province, which the Dutch recently added to their kingdom, is entirely made up of reclaimed land.

The Netherlands straddles the area where three large rivers—the Rhine, the Maas, and the Scheldt—flow into the North Sea. The fact that Rotterdam is now the world's busiest port is a twentieth-century reminder of a maritime tradition that began when William Beukelszoon discovered in 1384 how to cure herring with salt. This invention initiated a lively trade in cured fish, a trade that stretched from the Baltic to the Mediterranean Seas. In turn, this experience may have prepared Dutch seafarers for the lucrative trade in spices that they took over from the Portuguese two centuries later.

The Portuguese were the first Europeans to sail east to the Spice Islands (the Moluccas, which are now part of Indonesia). Upon their return to Europe, they deposited their precious products in Lisbon, and the Dutch then distributed them all over Europe. However, the king of Spain became king of Portugal as well in 1580; since Spain was at war with the rising Netherlands, Dutch traders were no longer allowed to come to Lisbon. The Dutch then started to contemplate seriously getting the spices first-hand. Initially they believed that sailing through arctic waters to the north of Scandinavia, Russia, and Siberia would yield a significant short cut. But a Dutch ship got stuck at Nova Zembla, where the crew had to pass an unbearably harsh winter before returning to Holland the next spring.

The Dutch then decided to benefit from Portuguese explorations. However, they varied the customary Portuguese route considerably, no doubt because they wanted to minimize contacts with the people who had become their enemies. They thus first sailed to the Cape of Good Hope but did not go around it or go north along the African coast, as the Portuguese did. Instead, the Dutch went straight east, which carried them, after weeks and weeks, to western Australia. They then went north, finding their way to the Spice Islands. There they soon realized that although it was possible to avoid other European explorers en route, it was impossible to ignore the Portuguese and Spaniards once they had arrived.

The beginning of the sixteenth century witnessed the establishment of the United East India Company (better known by its Dutch initials VOC), which was granted a monopoly of the spice trade in the Moluccan Islands. Eventually, Dutch maritime activity also extended to the West; tobacco and sugar were carried from the New World to Europe, and Dutch seafarers took slaves from Africa to the British colonies in North America. By the mid-1600s, the Netherlands had developed into a global commercial power whose trade and prestige were protected by a powerful navy, the match of those of England and Spain. Amsterdam, the capital of the Netherlands, had grown into Europe's richest city, a center of commerce and finance. The great prosperity was enriched by a culture that highlighted art. Indeed, the Dutch refer to the seventeenth century as their "Golden Age."

It would, however, be wrong to think of the seventeenth century as a period in which peace prevailed in the Netherlands. In the decades preceding the Golden Age, the Reformation—a religious movement that led to the rejection of some Roman Catholic doctrine and to the establishment of Protestant churches—had spread like wildfire. The Catholic king of Spain, to which the Netherlands belonged at that time, felt aggrieved about this and attempted to reverse the tide. King Philip II thus dispatched his armies to the Northern Netherlands, which had embraced the Calvinist faith. The year 1568 marked the beginning of the Dutch War of Independence, which was to last 80 years. It was for all intents and purposes a Protestant war against King Philip II.

By the time the Eighty Years' War finally came to an end, at the Peace of Westphalia in 1648, the Netherlands had become a haven for Protestants and numerous other "dissidents." Jews fleeing from the Inquisition in Spain and Portugal, and Huguenots fleeing from France after the revocation of the Edict of Nantes, sought and found asylum in the Netherlands. Even the Pilgrims stayed in Leiden for some years before finding their way to the New World. The Eighty Years' War, which often epitomized religious intolerance, may have had an unexpected benefit, since it caused future generations in Holland to have a greater appreciation for religious and political freedoms.

THE RELIGIOUS DIMENSION

Religion has for a considerable time been a strong factor in Dutch politics and culture. Although the country has gone through a period of secularization, it is still possible to find evidence of the Calvinist religion, which reigned supreme during the Eighty Years' War and afterward. Even the monarchy, which can be traced back only to 1813, offers an example, as the Constitution decrees that members of the royal house may adhere only to the Dutch Reformed Church. When it appeared that one of the princesses had become Roman Catholic in order to marry a Spanish nobleman, she was immediately scratched as a member of the royal house.

Nowadays, however, Dutch society is highly diverse, at least in the religious respect. The Roman Catholic Church has the largest following, though many are only nominally Catholic. Also, the upper clergy of the Dutch Catholic Church during the 1960s and 1970s openly rebelled against the Vatican, which strongly disapproved of the liberalizing tendency of the Dutch bishops and cardinals, fearing that it might erode established doctrine.

Although the Roman Catholics are now the largest religious denomination in the Netherlands, they were for a long time less accepted in Dutch society. Indeed, as indicated above, the tolerance that was so generously extended to religious refugees from other countries was not accorded to Catholic compatriots. They were viewed as persons who had been, at best, indifferent to King Philip II's fight against the Protestants. They had never been driven into churches that were then set on fire, as happened to the Protestants. The Catholics' lack of involvement caused them to be denied full civil rights, a condition that was to persist for two centuries. Until the mid-nineteenth century, Dutch Catholics were not even allowed to enter the public bureaucracy. The Constitution of 1848 started to change all this, but it was not until 1917 that all Dutch males, regardless of their religious beliefs, were enfranchised.

Protestantism in Holland may be less nominal than Catholicism, but it is certainly very diverse, ranging from free-thinkers and unitarians on the one hand to rigid fundamentalists on the other. About one third of the Dutch population are unaffiliated with any religion, and this proportion is steadily growing.

The year 1917 was also significant in that it ended what the Dutch called the

"School Struggle." All private schools (in large part denominational) were henceforth to receive governmental subsidies whose amounts closely correspond to the schools' enrollment sizes. Arend Lijphart, a well-known Dutch scholar, considers 1917 to be the beginning of what he has termed "consociationalism," a political feature typical of the Netherlands. Dutch society, according to Lijphart, was in effect divided into three segments: the Protestants, the Roman Catholics, and those unaffiliated with any religion (who were often referred to as "humanists"). So strong was the religious or humanistic orientation of these subsocietal groups that all their activities and associations revolved around their specific belief system. Educational facilities from kindergarten through university catered to members of the respective groups. There also were associations whose activities had no bearing on religion, such as stamp collecting and chess, that were either Protestant, Roman Catholic, or unaffiliated. It was theoretically unnecessary to mix or even to meet with persons from the other groups.

The complete insulation of these three groups throughout one's entire life would naturally be utterly dysfunctional in a political system if it were not for their political elites. The elites of the segments communicated and negotiated with one another. They made political compromises. As had been the case with the schools, the compromises were in large part based on numbers or, rather, proportions.

Lijphart asserts that the consociational system in the Netherlands came to an end in 1967. By then, people had defected from their respective churches in large numbers and refused to be categorized religiously when it came to associations. Only politico–religious associations like political parties and trade unions remained faithful to consociational patterns for another decade.

THE GOVERNMENT

The Netherlands is headed by a constitutional monarchy, which implies that the king or queen does not have any political power and is in effect "above" politics. The monarchy has been retained for ceremonial and symbolic purposes only. The king or queen (for more than a century, the Dutch have had a queen as head of state) plays a formal role in the process of government formation. Opinion surveys have indicated that the monarch enjoys strong support among the Dutch people (in 1999, 73 percent favored the status quo). Nevertheless, a leading figure of the Democrats '66 stirred debate by proposing

(Credit: U.N. Photo 186853/A. Brizzi)

The International Court of Justice is the principal judicial organ of the United Nations. Its seat is at the Peace Palace in The Hague, the Netherlands.

its "modernization" by trimming the queen's residual powers. Queen Beatrix, who has reigned since 1980, is reportedly one of the richest women in the world.

The Netherlands has a multiparty system, and it frequently happens that 10 or more parties compete for seats in the Second Chamber, the lower house that functions as the main political arena. None of these parties receives majority support; as a result, coalitions have to be formed in order to establish a government. Frequently three or more parties participate in such a coalition; although parties that jointly constitute a cabinet are usually more or less like-minded, such governments tend to be fragile.

The less important First Chamber consists of 75 representatives elected indirectly for four-year terms by the provincial parliaments. It has a noteworthy role in the amending of the Constitution. By its refusal (by one vote) in May 1999 to approve an amendment to add a "corrective" popular referendum to the Constitution, it caused the temporary breakdown of the ruling three-party "purple" coalition. The crisis was overcome by the party leaders' agreeing on an alternative reform (advisory referendum) that would not necessitate changing the Constitution.

During the 1970s, the large religious parties merged into the Christian Democratic Appeal, an ecumenical power bloc that led the government for two decades in conjunction with the Labor Party. The most formidable Socialist party was the Labor Party, which was established shortly after World War II and has often been part of the government. On the left, one also finds the D'66, a social-democratic party founded in 1966, which today represents a left-liberalism. The Communist Party, which shortly after World War II secured more than 10 percent of the vote, is no longer represented in Parliament.

In the Netherlands, voter support is translated into parliamentary seats through the proportional representation system. A party will generally be allocated the same proportion of seats in the lower house that it has secured of the total vote. Thus, while the Dutch system does not have a standard minimum limit expressed in percentages, there is an automatic threshold, in that for each seat, a party has to gain the quotient of the total number of votes divided by 150 (the total number of seats in the Second Chamber).

The outcome of the 1994 parliamentary elections was historic. The Christian Democrats (CDA) lost heavily (down 20 seats), as did their coalition partners, Labor (down 12 seats). Labor's leader, Wim Kok, was commissioned by Queen Beatrix to form a majority coalition. For the first time, Christian Democrats were left out of power. A new configuration emerged— a "purple" coalition of Labor, the Liberals (VVD), and the Democrats '66, who bridged the gap in the economic views of the two larger coalition partners. Kok has enjoyed personal popularity as prime minister, and in the May 1998 parliamentary elections, the electorate rewarded his governing coalition for the remarkable performance of the Dutch political economy. Labor and the Liberals both enjoyed big electoral gains, with the latter reaching their best result ever. Although the smaller partner, D'66, lost 10 of its 24 seats, Kok formed a second "purple" coalition. In the opposition, the CDA lost five more seats, the Green Left party saw its support double and ended up with 11 seats, and the small Socialist Party won five seats. An extreme right-wing party, the Center Democrats, lost its three seats. Voter turnout was 73 percent, the lowest since 1970, when compulsory voting was repealed.

POLARIZATION VS. PILLARIZATION

One may argue that the disintegration of the *verzuiling* (the term means "pillarization" and refers to the system of religious compartmentalization discussed earlier) was brought about by a strong decline in religious orientation during the late 1960s. Simultaneously, many young persons adopted a counterculture lifestyle; it did not take long for the city of Amsterdam to develop into the capital of the international hippie movement.

In matters of morality and sex, the Dutch witnessed the swing of the pendulum as well, having been very staid in the past. In the 1960s and 1970s, drugs started to be introduced into this part of Europe. Amsterdam came to succeed Marseilles as the drug capital of Western Europe. Some attempts were made to legalize less harmful drugs. The experiment failed, however, because the governments of neighboring countries complained, finding that their young people had flocked to the Dutch distribution centers. Indeed, the drug traffic has become the crux in the negotiations concerning the opening of borders for the single market of the European Union.

THE WELFARE SYSTEM

A popular theory holds that attachment to religious values and economic prosperity relate inversely. In the Dutch case, it was certainly true that the decline of religious values was attended by an unprecedented prosperity.

The improved economy resulted in part from the discovery of natural gas in the northern province of Groningen. The subterranean gas bubble turned out to be so large that great quantities of natural gas were sold to neighboring countries. This steady income was a windfall for the Netherlands. Then, in the early 1970s, the world was struck by an energy crisis. The domestic use of natural gas, as well as the revenues from its sale to other countries, expanded enormously as a consequence of the three- to fourfold increase in energy prices in the period 1973–1976. However, the income that these new ventures generated was not used for capital investments but for consumption—that is, for the proliferation and enlargement of the already generous system of welfare benefits. The system eased the financial burdens resulting from unemployment, sickness, and disability.

In the Netherlands, welfare benefits are paid out for as long as they are needed by an individual. Until recently, they were even indexed so that they automatically matched the increases in wages in the private sector. The upgrading of the welfare system resulting from the economic prosperity of the 1970s caused more people to be eligible for benefits, while the established benefits grew in size. At one point, more than 25 percent of the workforce received some income from the government without working! The systemic generosity also extended to migratory workers, refugees, and the like.

The governmental service sector had already become top-heavy when, in the 1980s, energy prices started to plummet. This, of course, had a disastrous effect on the revenues that the Dutch government received from its sale of natural gas. The shrinking budget forced the authorities into a retrenchment, and subsequently into a reversal, of the welfare system. The retrenchment, coming at a time when the Dutch workforce appeared to include 900,000 disabled persons, nearly caused a government crisis in 1993. The Dutch welfare system has since been trimmed. Many benefits that resulted from excessive paternalism have been revoked.

ETHNICITY AND EQUALITY

The urgency to come to grips with the economy has in recent years raised questions about the extent to which social solidarity should apply to non-Dutch residents. A majority among them are Indonesians and Surinamese who came to the Netherlands as political and economic refugees when their homelands, former Dutch colonial possessions, gained independence. Among the Indonesians are the South Moluccans, who fought on the Dutch side during the Indonesian Revolution. During the late 1970s, they briefly engaged in domestic terrorism.

In addition to the immigration waves clearly associated with its loss of empire, the Netherlands invited "guest workers" from the Mediterranean basin during periods of full employment. These people were scheduled to work in the manufacturing industry for two years. Yet their terms were renewed again and again—until unemployment made itself felt. They are now often called the "guest unemployed." Having paid into the unemployment and sickness funds while they were working, the foreign laborers are entitled to the same benefits in matters of unemployment or health as Dutch citizens. Although it is less prosperous than neighboring Germany, the Netherlands has also been a target for political or economic asylum-seekers, most of whom are not refugees in the usual sense of the word.

The Dutch tradition of tolerance has generally held up quite well in dealing with ethnic diversity. Indeed, the government has allowed the families of guest workers to come and join their relatives. Foreigners have even been permitted to participate in municipal elections and to run for local office. The great cultural differences have nevertheless imposed strains on Dutch society, strains that occasionally culminate in offensive graffiti or riots. The Dutch government has instituted programs to help these non-European immigrants assimilate. But then, not all appear to be interested in becoming assimilated.

The Netherlands has traditionally been a land of immigration. But with 10 percent of the current population foreign-born, and with the greatest national density of population in Europe, pressures have mounted for action to hinder the waves of immigrants. Recently the Dutch government initiated a campaign of mobile response units to stem the flux of illegal immigrants.

PROGRESSIVE POLICIES

In spite of its rigid adherence to certain principles (such as free trade, political liberty, human rights, and, in a sense, what remains of the Calvinist morality), Dutch society has manifested a remarkable flexibility, a willingness "to go with the flow." It has adapted well to changing circumstances.

The loss of empire was soon followed by the country's integration into the North Atlantic Treaty Organization and the European Union. In both regional organizations, the Netherlands is small but not insignificant, loyal but vocal.

The Netherlands was one of the charter members of both the European Steel and Coal Community and the European Economic Community, the precursors of the European Union. Dutch governments have long favored the strengthening of the supranational (federal) aspects of the European institutional framework. The

The new Constitution eliminates the political role of the monarch and introduces direct elections to Parliament **1848**	The consociational approach in government, resulting from pillarization **1917–1967**	Five years of Nazi occupation leave the Dutch economy in ruins **1940–1945**	The Marshall Plan helps the Netherlands to get back on its feet **1947–1949**	Transfer of sovereignty to Indonesia; the Netherlands loses the cornerstone of its empire and joins NATO **1949**	The Netherlands joins the European Coal and Steel Community, a forerunner of the European Union **1958**	Queen Beatrix succeeds her mother, Juliana; a center-left coalition returns to office Prime Minister Ruud Lubbers **1980s**	Prime Minister Wim Kok and his Labor Party meet with voters' approval in two parliamentary elections; The Netherland participates in the launch of the Euro **1990s**

2000s

Homosexual couples gain full legal rights to marriage, divorce, and adoption

Euthanasia is legalized

Netherlands is one of the net contributors to the EU budget, after Germany. The historic Maastricht Treaty (Treaty of European Union) bears the name of the Dutch city where it was signed by the member states. Dutch banker in Wim Duisenberg became the first president of the new European Central Bank in Frankfurt. Polls have long indicated that the Dutch people strongly support European economic and political integration. In 1997, only 20 percent of the Dutch surveyed felt that their country had not benefited from membership in the European Union.

The Netherlands' influence in NATO was demonstrated when the United States insisted on the deployment of cruise missiles in the Netherlands. The majority of the Dutch initially disagreed. There were vociferous protests, demonstrations, and rallies. The Netherlands government indicated that it needed time to convince its opponents that the only alternative to accepting missile deployment would be to get out of NATO. Only a minority of people appeared to favor that solution. Finally, on November 1, 1985, Prime Minister Ruud Lubbers announced that the Netherlands had accepted deployment of the U.S. missiles. The Dutch government also demonstrated its loyalty to NATO in spring 1999 by participating in the air war against Serbia to stop massive abuses of human rights in Kosovo.

In domestic matters, too, this mixture of principle and pragmatism has prevailed. When it was found that the 1980s decrease in energy prices had created huge liabilities for future generations, since a large part of the welfare system had been financed by the profits on natural gas, steps were taken to trim the welfare budget and to integrate it into the overall budget. The effective political response to this economic crisis demonstrated again that the

Dutch give their government authority to change policies that no longer work without abandoning the main premises that Dutch society holds dear.

Matters of family planning and birth control have also been tackled forthrightly in this country with a high population density. The prolonged debates on abortion have resulted in its conditional legalization.

In recent decades, the Netherlands has been a pioneer in many areas of social policy. For a considerable time, it was the only country to discuss on the national level the legalization of euthanasia for the terminally ill. In 1994, Parliament did not legalize euthansia, but it did pass careful guidelines for physicians so that they might avoid criminal prosecution in participating in such. In 2001, Parliament fully legalized the practice for Dutch citizens. After earlier permitting the registration of homosexual partnerships, in 2000, by an overwhelming majority, Parliament granted full-fledged rights to homosexuals to marry, divorce, and adopt children. Dutch society has also long tolerated prostitution (in special districts) and the use of soft drugs (in "coffee shops"). Authorities have preferred to allocate resources toward cracking down on serious crime and hard drugs, rather than "recreations." The government is moving to license brothels and regulate them for health and safety reasons. There have been experiments with giving free heroin to addicts to lower crime, as well as efforts to combat large-scale drug dealers. The Dutch drug policies have long been criticized by neighboring countries.

The Netherlands is also known for its "progressive" criminal-justice and corrections policies. Prisoners do not lose their right to vote and are not cut off from family and community, which they will reenter after serving (by U.S. standards) relatively short times behind bars.

During the late 1980s and early 1990s, the Dutch economy spiralled downward. While opening Parliament in September 1993, Queen Beatrix, in her Speech from the Throne, highlighted the "alarming" rate of unemployment and indicated that there was no light at the end of the tunnel. In April 2001, the Dutch unemployment rate was one of the lowest in Europe, healthy growth was forecasted, and inflation was only slightly above the average rate of the Euro Zone.

The Netherlands has embraced work flexibility and economic incentives without tearing its social net. Kok's government is promising to lower taxes and state indebtedness in the near future. Critics see a disturbing trend away from the consensus society to the market society, where "consumers" replace "citizens." Nevertheless, the Dutch model continues to be a reference point for other European governments struggling to modernize their countries' economies and still retain some sense of solidarity.

DEVELOPMENT

The Netherlands is a highly developed country that has in effect entered the postindustrial era. About 89% of the total area is considered urban. The people have become very sensitive to concerns related to the environment. The Dutch have been leaders in national environmental plans to combat pollution, regulate land use, and move toward sustainability.

FREEDOM

The Netherlands ranks very high among the world's countries in terms of freedom. Foreigners are allowed to vote or run for office in municipal elections. In 2000, the Netherlands granted homosexuals full equality in marriage, divorce, and adoption. In 2001, euthanasia was legalized by the Parliament.

HEALTH/WELFARE

Health indicators are generally good, though the social-welfare system has been trimmed. The UN Development Program (2000) rated the Netherlands' quality of life as the 8th best in the world.

ACHIEVEMENTS

One of the greatest achievements of recent times is the completion of the Delta Works, which protects the Netherlands from devastating floods, such as the ones that took place in 1953 and 1995. Dutch economic reform has produced hundreds of thousands of new jobs while retaining an ample social-welfare net.

Norway (Kingdom of Norway)

GEOGRAPHY

Area in Square Miles (Kilometers):
125,149 (324,220) (about the
size of New Mexico)
Capital (Population): Oslo (492,000)
Environmental Concerns: air and
water pollution; acid rain
Geographical Features: glaciated;
mostly high plateaus and rugged
mountains broken by fertile
valleys; small, scattered plains;
coastline deeply indented by
fjords; arctic tundra in north;
about 2/3 mountains
Climate: temperate along coast;
colder interior

PEOPLE

Population

Total: 4,482,000
Annual Growth Rate: 0.5%
Rural/Urban Population Ratio: 27/73
Major Language: Norwegian
Ethnic Makeup: Norwegian; a
minority of Sami (Lapp)
Religions: 86% Evangelical
Lutheran (state church); 3% other
Protestant and Roman Catholic;
11% others, none, or unknown

Health

Life Expectancy at Birth: 75 years
(male); 81 years (female)
Infant Mortality Rate (Ratio):
3.9/1,000
Physicians Available (Ratio): 1/285

Education

Adult Literacy Rate: 100%
Compulsory (Ages): 6–16

COMMUNICATION

Telephones: 2,736,000 main lines
Daily Newspaper Circulation: 498
per 1,000 people
Televisions: 459 per 1,000 people
Internet Service Providers:
21 (1999)

TRANSPORTATION

Highways in Miles (Kilometers): 54,157
(90,261)
Railroads in Miles (Kilometers): 2,512
(4,027)
Usable Airfields: 103
Motor Vehicles in Use: 2,172,000

GOVERNMENT

Type: constitutional monarchy
Independence Date: October 26, 1905
(union with Sweden dissolved)
Head of State/Government: King Harald
V; Prime Minister Jens Stoltenberg

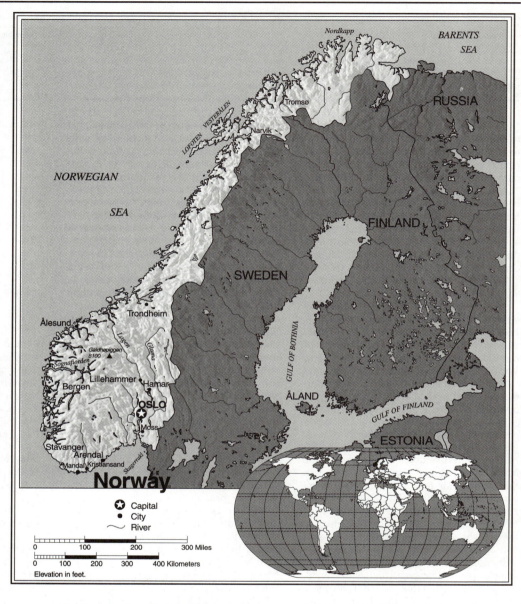

Political Parties: Christian Democrats;
Labor Party; Conservative Party; Center
Party; Socialist Left Party; Norwegian
Communist Party; Progress Party;
Liberal Party; Red Electoral Alliance
Suffrage: universal at 18

MILITARY

Military Expenditures (% of GDP): 2.1%
Current Disputes: territorial claim in
Antarctica

ECONOMY

Currency ($ U.S. Equivalent): 9.43 kroner
(crowns) = $1
Per Capita Income/GDP: $25,100/$111.3
billion
GDP Growth Rate: 2.7%

Inflation Rate: 3%
Unemployment Rate: 3.2%
Labor Force: 2,700,000
Natural Resources: petroleum; copper;
natural gas; pyrites; nickel; iron ore;
zinc; lead; fish; timber; hydropower
Agriculture: animal husbandry; feed
grains; potatoes; fruits; vegetables
Industry: oil and gas; food processing;
shipbuilding; pulp and paper products;
metals; chemicals; textiles, fishing
Exports: $47.3 billion (primary partners
EU, United States)
Imports: $38.6 billion (primary partners
EU, United States, Japan)

 http://www.odci.gov/cia/publications/
factbook/geos/no.html

LAND OF THE MIDNIGHT SUN

The Kingdom of Norway, often called the "Land of the Midnight Sun," is not a major world power. But it is affluent, its social problems are relatively few, and the quality of life is high for almost all of its residents.

Norway lies on the Scandinavian peninsula and borders Sweden, Finland, and Russia. The distance from one end of the country to the other is about 1,100 miles—the same distance as from Oslo to Marseilles, France. There is a great deal of variation in Norway's climate, a result of the different degrees of latitude, differences in altitude, varying topography, the effectiveness of the prevailing westerly winds, and the effects of the Gulf Stream. Thus, even though Norway stretches as far north as Alaska and Siberia, its entire coast is open for shipping throughout the winter. Hay and potatoes grow in even the northernmost region of the country.

About half the land area is mountainous, and more than a quarter is covered with forest. Only 3 percent of the land is arable. Glaciers created many sharp mountain peaks and dug out numerous long valleys. The Norwegian coastline is punctuated by long, narrow fjords and thousands of islets. Most Norwegians live by the fjords or along the coast.

A QUEST FOR SOVEREIGNTY

The first recorded period in Norwegian history was the Viking Age, from approximately A.D. 800 to the mid-1000s. The unification of Norway took place during this era. This integrative phase culminated in the adoption of Europe's first national code of common law, in the 1270s. After that period, the Norse kings started to expand their control over the country, tightening their rule.

However, the so-called Black Death, moving across Europe, reached Norway in 1349 and killed off at least half the population. This catastrophe may well have caused the rapid loss of Norse influence. In 1397, Norway came under Danish rule; it remained a dependency of Denmark for more than 400 years.

In 1814, Denmark, which had followed Napoleon Bonaparte to defeat, tried to cede Norway to Sweden. The Norwegians protested, and a constitutional convention at Eidsvoll endeavored to reestablish Norway as an independent country. Negotiations with the Swedes resulted in a compromise among the neighboring countries. The Norwegians accepted the Swedish king as their own, but Norway was recognized as a separate country, and the democratic provisions of the Eidsvoll Constitution were preserved.

However, in the course of the nineteenth century, Norway lost some of the independence it had been given in the compromise following the Eidsvoll convention. Norwegian resentment toward the union grew, and the government asserted greater autonomy. In 1905, under pressure from the Parliament, the Norwegian government abolished the union with the Swedish crown.

As soon as Norway became independent, a Danish prince was invited to become king of Norway, taking the name Haakon VII. The current monarch, King Harald V, is his grandson. He was the first Norwegian prince to be born on Norwegian soil in 567 years.

THE SYSTEM OF GOVERNMENT

Norway is a constitutional and hereditary monarchy. Norway's founders had originally—that is, on May 17, 1814—decreed that royal succession would be possible only through male heirs. But this provision was changed as a result of subsequent amendments in 1990.

The Constitution vests legislative power in the *Storting* (the Norwegian Legislature, which has 165 seats). The king, as head of state, enjoys the so-called suspensive veto (a type of veto that can only delay but not annul legislation). Thus, the royal veto may be exercised; but if the same bill is passed by two successive Stortings (established by subsequent elections), it will become law without the assent of the sovereign.

Formally, the king has the command of the land, sea, and air forces; and he also heads the government administration, which means that all appointments are made in his name. In times of dire crisis, the king may have more power.

In 1991, King Olav V, who had been a symbol of resistance to the German occupation of Norway during World War II, died, at age 83. He was succeeded by his son, who became the first King Harald since the Viking Age.

The Norwegian parliamentary system differs from the British system in that the Storting is elected for a full term of four years; thus, it cannot be dissolved early in favor of new elections. If the government should lose a vote of confidence, a new governing coalition or minority government (tolerated by other parties in the current Parliament) would have to be formed. Upon election, several parliamentary members are assigned to the Upper Chamber (*Lagting*) while most are assigned to be Lower Chamber (*Odelsting*); the two chambers function separately only in dealing with constitutional issues. A special committee is chosen by the Storting to award the Nobel Peace Prize each year. Like Denmark and Sweden, Norway has a number of ombudsmen, who are selected by the Storting, to protect the rights of citizens, consumers, and military personnel. As in the other Nordic countries, women are well represented in politics, holding 36 percent of the parliamentary seats in 2000.

Norway's electoral system is based on proportional representation, which means that political parties receive the proportion of seats in Parliament that they have obtained of the votes in the election to the Storting. This encourages a multiparty system as well as splinter parties. Six parties, for example, won seats in the 1985 election. All Norwegian citizens have the right to vote at age 18 and about 80 percent of those eligible to vote do so at each national election.

Norwegian politics is generally stable. The Labor Party was in power virtually without interruption from 1935 to 1965, and it still remains the largest party. In the parliamentary elections of September 1997, Labor won 35 percent of the vote and garnered 65 of the 165 seats. Closely aligned to trade unions, labor adheres to social-democratic principles.

The period following 1965 witnessed strong power shifts between the Labor Party and a coalition of non-Socialist parties. In 1985, the non-Socialist parties received a slight majority, but their coalition collapsed the very next year. Gro Harlem Brundtland, who in 1981 had become Norway's first female prime minister, was returned to office in 1986 as the head of a minority Labor government. The legislative elections that were held in 1989 again produced ambiguous results. Jan Peder Syse became prime minister of the government that then took office. Syse was forced to step down, and Gro Harlem Brundtland again became prime minister. Although left and right appear to match each other in Norway, most voters have a slight preference for a center-left government.

In the fall of 1996, Brundtland resigned as prime minister, and Thorbjørn Jagland was appointed to replace her. Jagland and the Labor Party retained power through the parliamentary 1997 elections. He was succeeded by Kvell Magne Bondevik, a Lutheran priest, of the Christian Democrat Party. His minority government, holding only 42 seats, was based on a coalition of the Christian Democrats (who in 1997 won 13.7 percent of the vote), the rural Center Party (7.9 percent), and the free-market Liberals (4.5 percent). Their goal was to strengthen the social-welfare state, including more redistribution of wealth

(UN/DPI Photo/E. Schneider)

Gro Harlem Brundtland was first elected prime minister in 1981. She was returned to that post in 1986, in 1990, and again in 1993. She resigned in 1996. A medical doctor and master of public health, Dr. Brundtland became director-general of the World Health Organization (WHO) in 1998.

and a reduction in private consumption. To pass a budget or any law, the government would have to negotiate with other parties. From the outset, the coalition appeared fragile, especially since its parties between and within themselves were divided over the questions of joining the European Union and participating in the Euro Zone. However, Bondevik and his cabinet ultimately fell over an environmental dispute. The parliamentary majority, led by Labor and the Conservatives (who had received 14.3 percent of the 1997 vote), was pushing for natural-gas power plants, even if this would require amending Norway's strict antipollution laws. The government favored waiting until new technology would be available to cut emissions. A vote of confidence was called, and the government lost, 81 to 71 (with 13 absent).

In March 2000, at age 41, Jens Stoltenberg, a protégé of Gro Harlem Brundtland and son of a former foreign minister, became Norway's youngest prime minister. Former prime minister and current Labor Party leader Thorbjörn Jagland took over as the foreign minister. Not only does the new cabinet stand out for its youthfulness (average age 43), but also because 42 percent of the ministers are women. While Labor's seats number 23 more than the old coalition's, the new government will also require the support of other parties to get its program passed in Parliament. The political honeymoon turned out to be brief: Complaints about the government's economic and social policies correlated with the Labor Party's decline in the polls in 2000.

An interesting character to have emerged in Norwegian politics in recent years is Carl I. Hagen of the right-wing, populist Progress Party. Hagen is a maverick who defies the Norwegian tradition of consensus (though he says he wants his party to be formally included in a ruling coalition). He has been snubbed by the country's mainstream parties because of his perceived hostility to immigrants; but in the last few years, he and his party have become more prominent and popular. Hagen has toned down his party's xenophobic rhetoric, but all Norwegians know that he stands for a more restrictive immigration policy. Foreigners make up only about 5 percent of the population, yet Hagen favors a referendum to shut the door. The Progress Party also promises to cut taxes, to spend more on the welfare state, and to abolish development aid to poor countries. Hagen has been able to exploit citizens' frustrations over declining standards in public services and high prices at the gasoline pump. While recent governments have set aside oil revenues in a fund to guarantee Norway's prosperity after the oil boom (and to keep the economy from overheating in the current period), Hagen and a large majority of the public favor spending more now. Late 2000 polls indicated that the Progress Party had surpassed the long-leading Labor Party in popularity.

THE WELFARE STATE
In Norway, taxes pay for a national insurance system, which tries to preserve the standard of living of those who cannot work because of old age, disability, illness, unemployment, or other reasons. The welfare system pays the full cost of hospitalization and most doctors' fees. Workers are paid regular wages when they are sick and receive four weeks of paid vacation each year. The government also subsidizes housing costs and provides for education. University students pay no tuition, and they may receive government loans for their living expenses.

Most newspapers get subsidies from the government, which also provides assistance to artists and writers as well as to most theaters and museums. Nevertheless, artists and the mass media do not feel inhibited about criticizing the government. A government corporation is responsible for radio and television broadcasts; no commercials are allowed.

Norwegian taxes are among the highest in the world, constituting almost half of the country's gross domestic product. Norwegians complain about the tax burden, but so far they have not tolerated major cutbacks in services.

A HOMOGENEOUS POPULATION
The Norwegian population is very homogeneous. The vast majority are Nordic and adhere to the Lutheran faith. The main minorities are the Sami (Lapps)—the 20,000 indigenous people of northern Scandinavia—and about 15,000 recent immigrants, mainly from developing countries. These peoples sometimes encounter a degree of hostility from the mainstream Norwegians.

Norwegians value their leisure time and spend much of it in the forests and mountains and along the fjords and coast. By international standards, the crime rate is low and drug problems are minimal, except for alcohol abuse. In Norway, driving under the influence invariably results in the immediate and permanent revocation of one's license.

THE ECONOMY
The main goals of Norwegian economic policy have been to maximize employment and production value, preserve rural settlements, and maintain Norwegian con-

Many Norwegians live near the sea. These fishing boats are docked in the harbor at Lofoten, Norway.

trol of natural resources and an equitable distribution of wealth and income. The government has been very successful in achieving these aims. (The unemployment rate reached 7.5 percent in 1993; but in 1999, it was only about 3.2 percent.) Only half the population live in cities of 10,000 or more people, and few countries have greater equality of income.

The traditional industries—farming, fishing, and forestry—are still important stimuli to boost economic patterns and self-sufficiency. However, they no longer employ many people or contribute much to the gross national product. Service industries, especially social and educational institutions, have become the main employers. The trade balance is very dependent on manufacturing, shipping, and petroleum exports. Most factories process Norwegian raw materials or exploit Norwegian energy sources. Norway primarily trades with European Union countries—although it still is not an EU member itself—and the United States.

Electric consumption per capita in Norway is the highest in the world, but hydropower from waterfalls and mountain lakes provides inexpensive energy from perpetually renewable sources for aluminum production and other manufacturing. Since 1975, Norway has also exported large amounts of oil and natural gas found in the North Sea. Norway shares these resources with Great Britain. Although certainly an asset, the exploitation renders the Norwegian economy dangerously dependent on the world market price of oil. Before oil prices suddenly plummeted in 1986, revenues from the "offshore" sector (that is, petroleum and natural-gas production) constituted more than 20 percent of all government income. The sharp reduction in these revenues forced adjustments in the state budget and the introduction of an austerity program that could affect social services. In addition, inflation and high wages are serious threats to the competitiveness of traditional Norwegian industries, such as shipbuilding and textiles. Inflation in consumer prices was estimated at 3 percent and economic growth at 2.7 percent in 2000.

The Norwegian government is heavily involved in the economy, since it owns corporations, extends subsidies to industry and agriculture, and regulates private firms. But the economy is in no sense socialist, and most Norwegian businesses are privately owned and subject to market forces.

INTERNATIONAL RELATIONS

Although its landmass is considerable (Norway is larger than Italy), the country has a relatively small population. That condition causes its foreign policy to be somewhat subdued; apart from maintaining Norwegian sovereignty, it confines itself mainly to abstract targets such as promoting peace and reducing tensions. Norwegians are strong supporters of the principles embedded in the United Nations Charter. (The first secretary-general of that organization, Trygve Lie, was a Norwegian.) Norway spends a higher percentage of its gross domestic product on aid to poor countries than any other state in the world.

In 1993, promoting peace and reducing tensions materialized in an exceptionally important initiative. Secret meetings were convened in a farmhouse near Oslo to which delegates of Israel and the Palestine

First unification of Norway A.D. 885	Height of the Norwegian Empire: first attempts at the codification of national law ca. 1275	The Black Death kills half the population 1350	Union with Denmark 1397	End of Danish rule; Norwegian Constitution; union with Sweden 1814	Norwegian independence 1905	German occupation 1940–1945	Norway becomes a charter member of NATO 1949	Oil is discovered in the North Sea 1968	Norway hosts the 1994 Winter Olympic Games; voters reject the Maastricht Treaty and opt to stay out of the European Union 1990s

2000s

Kvell Magne Bondevik is succeeded by Jens Stoltenberg as prime minister

Liberation Organization had been invited. There, outside the glare of the media, they met to work out a declaration of principles in September 1993—a document that was to lead to a peace agreement between the two parties, which had never ceased hostilities since the birth of the State of Israel in 1948. Norway has also been involved in low-profile peacemaking efforts in Myanmar (Burma), Colombia, and Sri Lanka.

To assure the Russians that no attack will come from Norwegian territory, Norway has allowed no foreign troops and no nuclear weapons to be based within its borders, and no military training exercises are held close to the Russian border.

Military service is universal and compulsory. In peacetime, one may be drafted from ages 19 to 44. The training period in the Norwegian Army, Coastal Artillery, and Anti-Air Artillery is 12 months, and in the Air Force 15 months. Relative to its size, Norway has been a noteworthy contributor to NATO, of which it has been a member since 1949. For example, significant numbers of Norwegians have been deployed to Kosovo in the aftermath of NATO's 1999 air war against Serbia. Norway has also played an active role, building bridges between East and West, in the Organization for Security and Cooperation in Europe. Prime Minister Stoltenberg boasts that, per capita, his country has provided more UN peacekeepers than any other.

In the early 1970s, Norway and Denmark were scheduled to be admitted to the European Union (then called the European Economic Community). However, a public referendum held in Norway in 1972 made clear that a majority of Norwegians did not want admission to the organization. Denmark, with which Norway had very intricate economic relations, did become a member on January 1, 1973. Norway has been a member of the European Free Trade Association since 1960. Nor-

way joined the European Economic Area in 1992. The EEA emerged out of the mutual desire of EFTA countries and the European Union to have closer links. While EFTA members Sweden, Austria, and Finland soon became full-fledged EU members, 52.3 percent of Norwegians again balked on EU membership in a November 1994 referendum. Although the government favored membership, voters saw a viable, prosperous future outside of the EU for their small country with its huge oil riches. What is often missed is that the EEA Treaty binds Norway (and Iceland) tightly to the European Union in terms of the flow of goods, services, capital, and even people, under rules determined in Brussels.

There have been recent signs of movement toward a common EU foreign and security policy, which could ultimately leave Norway exposed. Although Prime Minister Stoltenberg declared that "Europe's future is Norway's future," the Danes' rejection of the Euro in September 2000 is sure to delay further any new EU referendum in Norway.

In 1987, Prime Minister Brundtland had assisted in putting together the UN Report on the Environment. She firmly believed that economic development and environmental protection should be linked, her argument being that the rehabilitation of the European environment could be a test case of whether "we are capable of dealing with our common responsibilities." Norway is one of Europe's largest producers of oil and gas, both hydrocarbon fuels that, when burned, produce carbon dioxide, a "greenhouse" gas that many scientists believe will cause a calamitous rise in global temperatures.

Other important foreign-policy issues of concern to Brundtland and her successors include expansion of Nordic cooperation, control of the Barents Sea around Svalbard, and protection of the North Sea

petroleum installations. Somewhat uncharacteristically, in 1993, Brundtland announced that Norway would resume commercial catching of Minke whales. This decision provoked heated criticism from environmentalists around the world, followed by threats to boycott Norwegian exports. Norway has been unmoved. In fact, in 2000, Norway (and Japan) began courting support from developing-world countries so that Minke whale meat (already sold on the local market) could be sold on the international market. Environmentalists oppose such legal trade, contending that it makes whale poaching easier and paves the way for a wholesale return to commercial whaling.

THE FUTURE

There is fear that Norway's social harmony might dissipate, in that its people's consensus about values and goals could disintegrate. Norwegians are no longer used to suffering; a few serious mistakes by political or economic leaders could have devastating effects. In many respects, however, Norway can count on natural advantages. The country has numerous natural resources, and its people have accumulated considerable expertise in using them. There is a national tradition of honest and effective political leadership, of humanitarian solidarity, and of acceptance of the need for strong effort. Their high level of education is also bound to help the Norwegians to deal with the problems that may come their way.

DEVELOPMENT

Thanks to its vast oil and natural-gas reserves, Norway has become one of the wealthiest countries per capita in the world.

FREEDOM

Norway is a parliamentary democracy. Civil liberties are guaranteed in the Constitution, and the press enjoys considerable freedom. Norway attempts to further democracy, peace, and human rights around the world.

HEALTH/WELFARE

Norway's welfare state is a costly, but comprehensive, cradle-to-grave system. The country has one of the highest life expectancy rates in the world as well as a high per capita income. The UN Human Development Index (2000) ranks Norway second in the world in terms of quality of life.

ACHIEVEMENTS

Norway has more than 300 art galleries, most of which are supported by public funds. A cultural fund guarantees the purchase of at least 1,000 copies of every work of Norwegian literature for circulation in the public library system. Nearly 1% of Norway's GDP goes toward helping developing countries, more than any other nation.

Portugal (Portuguese Republic)

GEOGRAPHY
Area in Square Miles (Kilometers): 35,672 (92,391) (about the size of Indiana)
Capital (Population): Lisbon (582,000)
Environmental Concerns: soil erosion; air and water pollution
Geographical Features: mountainous north of the Tagus; rolling plains in the south
Climate: maritime temperate; cool and rainy in the north; warmer and drier in the south

PEOPLE

Population
Total: 10,048,300
Annual Growth Rate: 0.18%
Rural/Urban Population Ratio: 64/36
Major Language: Portuguese
Ethnic Makeup: homogeneous Mediterranean stock with a small black African minority
Religions: 94% Roman Catholic; 6% others

Health
Life Expectancy at Birth: 72 years (male); 79 years (female)
Infant Mortality Rate (Ratio): 6.0/1,000
Physicians Available (Ratio): 1/332

Education
Adult Literacy Rate: 87%
Compulsory (Ages): 6–15; free

COMMUNICATION
Telephones: 4,117,000 main lines
Daily Newspaper Circulation: 47 per 1,000 people
Televisions: 333 per 1,000 people
Internet Service Providers: 20 (1999)

TRANSPORTATION
Highways in Miles (Kilometers): 41,239 (68,732)
Railroads in Miles (Kilometers): 1,914 (3,068)
Usable Airfields: 66
Motor Vehicles in Use: 3,480,000

GOVERNMENT
Type: parliamentary democracy
Independence Date: 1140; republic proclaimed October 5, 1910
Head of State/Government: President Jorge Sampaio; Prime Minister Antonio Manuel de Oliviera Guterres

Political Parties: Social Democratic Party; Portuguese Socialist Party; Portuguese Communist Party; Popular Party; National Solidarity Party/United Democratic Coalition; The Left Bloc
Suffrage: universal at 18

MILITARY
Military Expenditures (% of GDP): 2.6%
Current Disputes: none

ECONOMY
Currency ($ U.S. Equivalent): 237 escudos = $1
Per Capita Income/GDP: $15,300/$151.4 billion
GDP Growth Rate: 3.2%
Inflation Rate: 2.4%
Unemployment Rate: 4.6%.
Labor Force: 4,750,000
Natural Resources: fish; forests; cork; tungsten; iron ore; uranium ore; marble; arable land; hydropower
Agriculture: grains; potatoes; olives; grapes; other fruits
Industry: textiles; footwear; wood pulp; paper; cork; metalworking; oil refining; chemicals; fish canning; wine; tourism
Exports: $25 billion (primary partners EU, United States)
Imports: $34.9 billion (primary partners EU, United States, Japan)

Portugal

PORTUGAL:
WHERE CULTURES MEET

Modern Portuguese culture is a patchwork reflecting the influences of many different peoples, regions, and historical eras. Prehistoric cave painters left their decorations in Escural. The Romans left their mark in the township of Conimbriga and in the temple of Diana in Évora. Such southern towns as Olhão and Tavira exhibit the Moorish influence in their architecture. The Flemish, French, and Italian cultures have also enriched Portugal, and the voyages of the Portuguese explorers opened the country to Asian influences—The opulence of the Baroque period owed much to the gold and diamonds of colonial Brazil. To visit Portugal is to travel backward in time and around the world within a single country.

Stretched along the westernmost part of the Iberian Peninsula, Portugal has played an important part in world history as well as in European history. Unlike Spain, the country that it borders on two sides, Portugal today has a homogeneous population—or, at least, no ethnic or linguistic minorities of any consequence. In the mid-1970s, when the Portuguese colonies in Africa attained independence, a large number of Portuguese whose forefathers had settled in Angola, Mozambique, and Guinea-Bissau in the early part of the century repatriated to Portugal. Surprisingly, however, these 700,000 "returnees" caused only temporary dislocations in a Portuguese society of barely 9 million. Among these *retornados* were many persons of mixed blood and about 100,000 blacks. A fairly large proportion of the returnees left for Brazil, a former Portuguese colony of great consequence.

HISTORY

The Iberian Peninsula was part of the Roman Empire for centuries, and the Roman presence left an indelible mark on the Portuguese culture and language. After the fall of the West Roman Empire in the fifth century A.D., Portugal was left to itself for a time (although it was, as were other parts of the Roman Empire, subject to interference on the part of the Visigoths). Little is known about this period; in the eighth century, however, Muslim invaders entered from the south. Coming from North Africa, they crossed the narrow body of water that came to be called the Strait of Gibraltar and brought Spain and Portugal under Muslim rule. Their occupation of the entire Iberian Peninsula lasted many centuries; the *Reconquista* ("Reconquest"), or reclamation, of the western (Portuguese) region proved to be a particularly arduous and intermittent process.

When the Reconquest of Portugal was finally completed, shortly before the fifteenth century, the country appeared to have assumed an identity of its own, one that differed markedly from that of its larger neighbor, Spain. It was only natural for the two countries to become rivals, since both excelled in seafaring and trade, exploits that almost inevitably led to commercial expansionism and a global presence. But whereas Portugal ventured south, Spain went west (severely underestimating the size of the globe!). The Portuguese, at least initially, viewed moving into Africa as a logical extension of the Reconquest: The Muslims had been driven back and expelled and now were to be fought on their own soil. It should be remembered that the Caliphate, expanding from the Arabian peninsula, had conquered the northern belt of Africa. Christianity, freed from its siege in Southern Europe, was to be carried into Africa.

Thus, at the end of the fifteenth century, Portugal became the first nation since the Roman Empire to "discover" and explore the African continent. Skilled Portuguese navigators sought to reduce the risk in seafaring by keeping the coastline in view; unlike the Spanish and, later, the Dutch and English, they did not venture onto the open sea. When the elements conspired against them, they hurried toward some shelter along the coast, a bay or another inlet that would provide refuge. In this fashion they ventured farther and farther south, until, in 1488, Bartholomeu Diaz de Novaes rounded the Cape of Good Hope, eventually reaching the East African coast and the Indian Ocean.

This remarkable feat was a milestone soon to be overshadowed in importance by the news that Christopher Columbus, an Italian sailing for Spain, had (he assumed) found the way to the Indies by crossing the Atlantic in a westward direction. The serious rivalry that ensued between Europe's two top seafaring nations was somewhat reduced by the Treaty of Tordesillas (1494), which divided the world into Spanish and Portuguese areas of expansionism. (This treaty, sponsored by the pope, was not recognized by any nation other than Spain and Portugal.)

In 1580, however, the Portuguese throne was left without a homegrown successor, and Spain's King Philip II, whose wife was related to the Portuguese dynasty, "inherited" Portugal. The merger, which caused Portugal to be dominated by Spain, was to last some six decades.

Empire

Only in 1640 did Portugal regain its independence, under the leadership of the House of Braganca. However, the memories of the Spanish domination rendered Portugal, the smaller and less populous of the two nations, perpetually apprehensive of its neighbor's overtures. As a result, the country developed close commercial and political ties with England, which by that time had replaced Spain as an imperial rival.

The enormous empire that the Portuguese built and maintained for nearly half a millennium resulted from two different impulses. The first wave of expansionism was based on the spices that the almost legendary islands in the East produced—as well as on the gold that Brazil, far to the west of Portugal, was assumed to have. Some of these typically seventeenth-century operations had been either relinquished by Portugal or taken over by rival empires. The Dutch, for example, had done a great deal to destroy or at least reduce the Portuguese presence in the East Indies. That first wave included the conquest of Goa on the Indian subcontinent and of half of Timor, where Portuguese rule was maintained until 1975.

The second wave of Portuguese expansionism took place in the era of modern imperialism, a late-nineteenth-century phenomenon. Imperial powers, suddenly realizing that the world was finite and that this would be their last chance to enlarge their territories, divided all that was left. Modern imperialism initiated the scramble for Africa, since little had been done to that continent in terms of commercial colonialism. Traditions that dated back to early discoveries and explorations may well have revived Portuguese interests in Africa. It is surprising that small Portugal, a country by then much in decline, participated in this scramble on such a large scale. The three colonies in Africa that it managed to claim had a total area about 22 times the size of Portugal itself. To the west, Brazil, to which ancient Portuguese claims were reaffirmed, added a territory about 92 times as large as the mother country!

Brazil also provided a haven of refuge to the king of Portugal when Napoleon Bonaparte's armies invaded the Iberian Peninsula in 1808. During the occupation of Portugal, Brazil was the official seat of government. Upon its return to Portugal, the monarchy introduced various liberal innovations, which, however, did little to prevent political instability. The general restlessness increased steadily throughout the nineteenth century and finally came to a head in 1910, when a military revolt ousted King Manuel II. The Portuguese monarch managed to escape to Gibraltar. From there he went to England, where he died in exile many years later.

Political Disorder

Portugal's first experiments with a republican form of government were equally

Lisbon has a number of beautiful squares that echo their long history. The statue of Sedubal is located in the large, open plaza pictured above.

dismal. The parliamentary system that was adopted soon yielded the same legislative supremacy that democratic polities elsewhere in Western Europe experienced at different points in time. The parties not in power would team up against the government coalition, causing its downfall, and then would establish a new coalition government, which in turn would be subject to attacks.

During the endless succession of governments, the Portuguese economy steadily deteriorated. At one point, the military decided to appoint Antonio Oliveira de Salazar, a well-known economics professor, as minister of finance. He introduced a variety of austerity measures that enabled Portugal to get on its feet again. When elections were held a few years later, the post of prime minister went to Salazar, who embarked upon constitutional reform and a complete overhaul of the country's political institutions. The Constitution of 1933 outlined the *Estado Novo* ("New State"), a construction entirely based on Salazar's ultra-right-wing philosophy, which bore a remarkable resemblance to other Mediterranean varieties of fascism, such as those in Spain and Italy.

Salazar ruled with an iron hand for more than three decades. In 1968, he was felled by a stroke, and Marcello Caetano took over. Caetano seemed willing to introduce gradual changes into a system that had become ossified, but ultimately, every step forward was matched by a step backward. He thus proved unable, on the one hand, to maintain the structure that he had inherited and, on the other, to conceive meaningful reform. The sense of drift during these post-Salazar years provided the climate for the Revolution of 1974.

A major problem was Portuguese colonialism. Salazar's rigidity with respect to colonialism was what ultimately proved to be his regime's undoing. By the 1970s, all other colonial empires had been liquidated; yet Portugal's colonial empire was still intact, with the exception of the Portuguese possessions in India, which the Indians had retaken in 1961 after 14 years of fruitless negotiations.

As a result, during the 1960s, the United Nations largely concentrated its anticolonial rhetoric on Portugal. The Portuguese government, however, stubbornly insisted that there was no question of colonialism and that various amendments to the Constitution of 1933 clearly indicated that Portugal did not have "colonies" but only "overseas provinces." As could be expected, the subterfuge did not work, and the UN pressures continued. In fact, Por-

tugal became an outcast in the family of nations. But its extremely rigid system of government failed to seal off the colonies, and finally the winds of change started to affect the peoples of Angola, Mozambique, and Guinea-Bissau. Rural rebellions proliferated, and urban uprisings were the order of the day. The Portuguese government, predictably, responded with force. Larger and larger troop contingents were sent to quell the insurrections.

The End at Last

These reactionary policies did not work either, and the situation in the colonies deteriorated rapidly. The African possessions, which had in the past been reasonably productive in Portugal's service, became grave liabilities. In the early 1970s, Angola, Mozambique, and Guinea-Bissau constituted enormous drains on the metropolitan budget. This was particularly the case because the mere maintenance of the colonial system demanded prohibitive defense expenditures, which a small country lacking in resources could not possibly afford, as well as a rising toll in lives. Also, Portugal, which had become an international pariah because of its stubborn colonialist attitudes, was denied developmental aid from UN agencies and other

| The Treaty of Tordesillas establishes the demarkation between Portuguese and Spanish expansion A.D. 1494 | Portugal comes under the Spanish Crown 1580 | Portugal breaks away from Spain; the House of Braganca becomes Portugal's ruling monarchy 1640 | Portugal moves its seat of government to Brazil 1808 | Portugal becomes a constitutional monarchy 1882 |

organizations. Military morale plummeted, and most Portuguese became convinced that even simple containment would be a hopeless task that could lead only to economic ruin.

In late 1973, a group of younger officers came together and established the Armed Forces Movement (MFA), initially intended as a pressure group. When conditions continued to worsen, some senior officers, including General Antonion de Spinola, joined the movement and, in April 1974, launched the coup that brought the Estado Novo down. De Spinola became interim president. Ultimately, the brutal jungle wars in Angola, Mozambique, and Guinea-Bissau had determined the fate of the New State.

RETURN TO DEMOCRACY

The Estado Novo ended not with a bang but with a whimper. Considering that the die-hard regimes of Salazar and Caetano had been thoroughly entrenched for half a century, one might have expected a full-fledged revolution that would entail prolonged fighting. But there was no revolution, although many of the coup participants were inclined to view it as such. Some of the authorities of the *ancien regime* were put in protective custody; others were persuaded, if needed, to leave the country. A number went into exile once Brazil offered political asylum.

The army had come into power, for after the coup, the MFA could be seen as representative of the entire armed forces. However, it lacked ideological underpinnings, and at this point in time, political parties, which had been banned in Portugal for more than 50 years, resurfaced. Among them, the Portuguese Communist Party (PCP) was particularly active in competing for government power. Its initial success may well have been aided by the backlash that the PCP provided against the ultra-right positions and policies that had pervaded Portuguese politics for decades. But the possibility of prolonging the uneasy coalition between the military and the Communists receded quickly after a transitional phase.

Parliamentary elections, held in April 1975, must have given the Communists the message that they were not viewed as the wave of the future: Non-Communist parties received more than 80 percent of the vote. Also, during the summer that followed, massive anti-Communist demonstrations were the order of the day, in cities as well as in rural areas. In August, another blow was delivered to the PCP, when non-Communist elements within the MFA forced the resignation of the pro-Communist prime minister, Colonel Vasco Goncalves.

In November 1975, Communist Party leaders, realizing that their power and popularity were slipping, staged a coup in a desperate attempt to reverse the tide. The coup was foiled, and its failure spelled the end of meaningful competition on the part of the PCP. It was then that massive purges were initiated within the armed forces. The removal of the Communists must have provided relief to the member states of what is today known as the European Union (then named the European Economic Community) and other Portugal-watchers, such as the United States. Still, the possibility that Portugal would experience a backlash to the right-wing dictatorship and turn Communist was very real for more than a year.

After 1975, Portuguese politics started to be characterized by competition among parties that were in varying degrees to the left of center but not Communist. The Socialist Party (PS) remained very important, although it failed to secure outright majorities in elections. Its main rival, the combination of the Popular Democratic and Social Democratic Parties (PPD/PSD) progressively increased its support. But if the late 1970s were still marked by a measure of unpredictability, the decade of the 1980s revealed that the country had become fully committed to democratic principles, and that it had, as *The Economist* phrased it, "found its political center of gravity." The early 1980s also witnessed the preparation for the Treaty of Accession—that is, Portugal's admission to the European Union, which would never have been granted had any doubt lingered regarding Portuguese democracy. Portugal's membership in that important organization became effective on January 1, 1986.

By that time, a relatively young man named Anibal Cavaco Silva had become Portugal's man of the hour. His confidence, enthusiasm, and powers of persuasion aided him in building up the Social Democrats, at the expense of the orthodox Socialists, and in gaining the prime ministership.

THE ECONOMY

In the Portuguese economy, too, the legacy inherited from the Estado Novo was dismal. The authoritarian regime created by Salazar did not seek rapid modernization and industrialization for Portugal; indeed, it aimed at keeping Portugal self-sufficient, quiescent, and immune to pressures from outside. The results were mixed. On the one hand, Portugal remained a rural and basically underdeveloped country. On the other, since it could rely on its African colonies (particularly Angola) for raw materials and oil, it was not subject to the vagaries of the international economy. At the same time, Portugal expanded its trade relations with other European countries during the 1950s and 1960s. Portugal was a charter member of the European Free Trade Association. In 1972, it signed a preferential trade agreement with what would later become known as the European Union.

As soon as the repressive structure of the Estado Novo had been removed, the new leadership realized how much the economy had lagged behind the rest of Europe. The first step taken to modernize the economy—the constitutionally irrevocable nationalization of some 800 firms—was carried out when the PCP was still in power. Then, in 1977–1978, negotiations were begun with the International Monetary Fund, which demanded that the government adopt an austerity program before it would issue loans. This shift in course to a more centrist economic policy worked well.

In the early 1980s, the momentous decision was made to include Portugal in the European Union, although both the EU and Portugal realized that a number of stumbling blocks presented themselves. In the first place, Portugal requested free circulation of labor, in accordance with the terms in the Treaty of Rome. This would mean that Portuguese workers could settle and work wherever they wished within the EU. But the poor economic conditions in the Salazar period had caused a great many Portuguese workers to emigrate, not only to the African colonies but also to more economically advanced countries in Western Europe. In fact, 1 million to 2 million Portuguese laborers were working elsewhere in Europe, and the countries concerned did not want this number to increase. Another stumbling block toward accession was the huge discrepancy be-

King Manuel II is ousted; Portugal becomes a republic
1910

Antonio Salazar becomes prime minister and establishes the Estado Novo
1932

A new Constitution is adopted
1933

Salazar, incapacitated, is succeeded by Marcello Caetano
1968

The "Revolution of the Carnations" demolishes the Estado Novo; political parties resurface
1974

Portugal becomes a member of the European Union
1986

Portugal and China reach agreement over Macau's future
1987

Portugal moves toward a free-market economy
1990s

2000s

Portugal joins the Euro Zone and holds the EU Council presidency during the first half of 2000

The government continues to explore ways to diversify and strengthen the economy

tween Portugal's economy and the economies of the EU member states. The Portuguese economy in fact resembled that of a developing nation.

Specific problems that needed to be worked out concerned Portugal's inefficient farming methods and its antiquated fishing fleet. Little mechanization had been introduced on either front; two thirds of the fishing vessels even lacked engines. Generally, Portugal absorbed all the new measures very well. Indeed, when it came to becoming an EU member, a measure of impatience could be noted.

On June 1, 1989, the Portuguese Parliament approved a package of reforms that did away with the socialist economy and endeavored to reprivatize a large number of the firms that had been nationalized in 1975. This proved a giant step toward a free-market economy. However, a great deal of obstruction arose on the part of the former owners, who wished to be compensated. Compensation claims had been ignored at the time of the nationalization; small amounts had been paid many years later. It was now, according to the original owners, that the day of reckoning had come, especially since in most cases the firms had not been returned to them.

Throughout the 1990s, Portugal and Greece remained the poorest members of the European Union. Their gross domestic products per capita were not much higher than that of the Central/Eastern European country of Slovenia, which will likely be among the first of the former Communist countries to enter the EU. Still (unlike Greece), Portugal met the criteria for the introduction of the EU's new international currency, the Euro, in 1999. Other than an outbreak of mad cow disease in 1998, which led to many countries banning its beef, Portugal has had smooth economic relations with the rest of Europe.

Among Portugal's chief industries are textiles, footwear, wine, chemicals, fish

canning, paper, and cork (Portugal is one of the world's leading cork producers); its chief crops include grains, potatoes, rice, olives, grapes, and other fruits. The country also produces such minerals as tungsten, uranium, and iron. Its forests constitute a valuable resource as well. Although these resources appear very promising, the lack of capital may well be the greatest single obstacle to development in Portugal.

The Portuguese economy has made some strides since the turbulent days of the Revolution in the mid-1970s. This progress became more pronounced after the country became a member of the European Union. However, its recent economic history bristles with instances of social backslides. On August 4, 1993, for example, the minimum legal age of employment was moved back to 14. Although there was strong protest on the part of labor unions, which claimed that the measure violated the recommendations of the International Labour Organization, and public response in general was hostile, the Portuguese government did not reverse its decision.

RECENT DEVELOPMENTS
France is not as unique as far as the political phenomenon known as *cohabitation* (the sharing of power between a president and a prime minister of different parties) is concerned. Portugal, like several other countries, has developed its own version of cohabitation. Whereas it is considered dysfunctional in France, in Portugal neither the voters nor the two main parties appear to have any objection to the possibility of a deadlock in decision making. The current administration, at any rate, does not need to worry. Prime Minister Antonio Manuel Guterres, elected in 1995 and reelected in 1999, is of the Socialist Party, as is Jorge Sampaio, who won the presidency in 1996.

In 1999, the after-effects of Portuguese imperialism returned to the news. Macau, a Portuguese colony since 1557, reverted to Chinese rule, two years after the British handed back Hong Kong. Also in 1999, the former Portuguese colony of East Timor was consumed by bloodshed. It had been annexed by Indonesia in 1976 and was now attempting to gain independence. The Indonesian government responded with indiscriminate violence, killing thousands of civilians and causing mass flight. The United Nations sent peacekeepers to help the transition to statehood. In the 1970s, two other Portuguese colonies, the African countries of Mozambique and Angola, were also torn by bloody civil wars; to historians, this suggests a general pattern of Portugal's negligence or ineptitude in preparing its former colonies for their independence.

The burden of recent history was also felt in 1997, as investigations into Portugal's economic relations with Nazi Germany during World War II gained momentum. Like Turkey, Spain, Argentina, and other countries, mostly acting through Swiss intermediaries, Portugal sold raw materials to Adolf Hitler in exchange for gold plundered from individual victims of the Nazi terror and from German-conquered countries. Only a small portion of this gold was ever returned.

DEVELOPMENT

In June 1989, the Portuguese Parliament approved reforms that did away with a socialist economy. Since then, Portugal's economic development has been uneven, but the country did qualify to participate in the launch of the Euro.

FREEDOM

Since the harsh Salazar days, Portugal has made enormous progress in matters of human rights. While there is still a degree of child labor and one may be imprisoned for "insulting civil or military bodies," Portugal has nonetheless adapted itself in record time to democratic standards.

HEALTH/WELFARE

Health conditions are mixed in Portugal. Life expectancy is reasonable, but infant mortality rates are among the highest in Europe. Although Portugal has had Socialist governments since 1974, no major health-assistance scheme has been developed. Some 16% of Portuguese over 15 years of age are illiterate.

ACHIEVEMENTS

The country is well known for its *fados*—melancholic songs, usually about love. On several occasions, Portuguese fado singers have won prizes at European song festivals.

Spain (Kingdom of Spain)

GEOGRAPHY

Area in Square Miles (Kilometers):
195,988 (504,750) (about twice the size of Oregon)
Capital (Population): Madrid (2,867,000)
Environmental Concerns: water quality and quantity; air and water pollution; deforestation; desertification
Geographical Features: large, flat to dissected plateau surrounded by rugged hills; Pyrenees in the north
Climate: temperate; more moderate along the coast

PEOPLE

Population
Total: 39,997,000
Annual Growth Rate: 0.1%
Rural/Urban Population Ratio: 23/77
Major Languages: Castilian Spanish; Catalán; Galician; Basque
Ethnic Makeup: Mediterranean and Nordic composite; Romani (Gypsies)
Religions: 99% Roman Catholic; 1% others

Health
Life Expectancy at Birth: 75 years (male); 82 years (female)
Infant Mortality Rate (Ratio): 4.9/1,000
Physicians Available (Ratio): 1/241

Education
Adult Literacy Rate: 97%
Compulsory (Ages): 6–16; free

COMMUNICATION
Telephones: 16,290,000 main lines
Daily Newspaper Circulation: 104 per 1,000 people
Televisions: 490 per 1,000 people
Internet Service Providers: 49 (1999)

TRANSPORTATION
Highways in Miles (Kilometers): 205,918 (343,197)
Railroads in Miles (Kilometers): 8,986 (14,400)
Usable Airfields: 105
Motor Vehicles in Use: 18,660,000

GOVERNMENT
Type: parliamentary monarchy
Independence Date: 1492 (expulsion of the Moors and unification)
Head of State/Government: King Juan Carlos I; President José Maria Aznar
Political Parties: Popular Party; Spanish Socialist Workers Party; Spanish Communist Party; United Left; regional parties and others
Suffrage: universal at 18

MILITARY
Military Expenditures (% of GDP): 1.1%
Current Disputes: dispute with Britain over Gibraltar; territorial disputes with Morocco over places of sovereignty

ECONOMY
Currency ($ U.S. Equivalent): 196 pesetas = $1
Per Capita Income/GDP: $17,300/$677.5 billion
GDP Growth Rate: 3.7%
Inflation Rate: 4%
Unemployment Rate: 14%
Labor Force: 16,200,000
Natural Resources: coal; lignite; iron ore; uranium; mercury; pyrites; fluorspar; gypsum; zinc; lead; tungsten; copper; kaolin; hydropower; forests (cork)
Agriculture: grains; citrus fruits; vegetables; fish
Industry: textiles and apparel; footwear; food and beverages; metals and metal manufactures; chemicals; shipbuilding; automobiles; machine tools; tourism
Exports: $112.3 billion (primary partners EU, Latin America, United States)
Imports: $137.5 billion (primary partners EU, United States, OPEC)

Spain

⭐ Capital
● City
〜 River

Elevation in feet.

THE KINGDOM OF SPAIN

From a European perspective, Spain is a large country, comprising as it does 85 percent of the Iberian Peninsula, the Balearic Islands in the Mediterranean, the Canary Islands in the Atlantic Ocean, the enclave of Llivia in the Pyrenees, the Northern African towns of Ceuta and Melilla, and several islands off the coast of Morocco. In terms of surface area, it is one of the largest countries on the European continent, and it ranks fifth in population within the EU.

HISTORY

Spain is also one of the oldest nation-states in Europe. In its earliest history, successive waves of different peoples spread over the Iberian Peninsula, occasionally merging and mixing with previous settlers, but more often driving them into less habitable places. Early records mention the Iberians, whose presence (particularly in the eastern parts of the peninsula) was noted by colonizing Greeks and the omnipresent Romans. Celts, Phoenicians and their "subsidiaries," Carthagenians, Romans, Visigoths, and finally Arabs—all occupied Spanish territory at one time or another.

Spanish history became less blurred, or at least better recorded, when the West Roman Empire collapsed in the fifth century A.D. and the Visigoths began to drift into the peninsula, not massively but piecemeal over the course of a century.

In the eighth century, Muslims arrived from the increasingly Islamic lands of North Africa, crossing the Strait of Gibraltar. Their forceful presence left an indelible mark on Spain. Even the derogatory name that they were given by the Spanish—Moors—entered the international vocabulary, as Shakespeare's "Moor of Venice" and the "Moros," a religious minority in the Philippines, will attest. The Muslims ruled Spain for many centuries. The hostility between Muslim and Christian peoples (as well as Jews, for that matter) during a large part of that period has been consistently overstated. Modern historians tend to consider the Muslim presence ultimately a beneficial influence on the development of the Iberian Peninsula.

However, the arrival and departure of the invaders, the periods of conquest and reconquest respectively, were naturally marked by hostility and armed conflict. The *Reconquista* ("Reconquest") took an extraordinarily long time; it was finally completed when Granada fell in 1492—coincidentally, the same year in which Christopher Columbus, funded by Spain, sailed to what he believed to be the Indies. Spain, by then a monarchy, thus emerged as a seafaring and early imperial nation as soon as Muslim rule came to an end. It was then that conquests were made, colonies founded; the conquistadores ("conquerors"), with a sword in one hand and the Bible in the other, embarked upon what in those days was viewed as a holy mission. Colonialism, particularly the Spanish and Portuguese variants, had strong missionary (and authoritarian) streaks.

Spanish Imperialism

Initially, Spain, like Portugal, appeared primarily interested in Southeast Asia, but its influence in those parts rapidly declined and ultimately remained confined to the Philippines. It is possible that Spain found the overheads too high, the voyages too time-consuming and costly. Such a conclusion is also warranted by the fact that the Philippines were administered not from Spain but from Mexico. Spanish interests thereupon shifted to Latin America and the Caribbean. In due time, Spain became dominant in South America (with the notable exception of Brazil), in Central America, and on many Caribbean islands, as well as in parts of North America. Even after Spain declined in importance as a European power, a process introduced by the defeat of the Spanish Armada in 1588 and highlighted by the Peace of Westphalia in 1648, it managed to retain a formidable imperial presence in the lands across the Atlantic.

However, the nineteenth century proved disastrous to the Spanish Empire: The Napoleonic Wars (a primarily European series of events) generated the beginnings of the decolonization process. If most of Latin America had been anxious to follow the example of independence set by the United States, the turmoil in Europe afforded the opportunity.

While Spain's holdings in Central and South America attained independence in the first half of the nineteenth century, armed conflict with the United States took care of numerous colonies elsewhere. A war of three months' duration triggered by the sinking of the *Maine* forced Spain, the loser, to cede the Philippines, Guam, Cuba, and Puerto Rico to the United States. On the other hand, Spain did gain a number of colonies in the latter part of the nineteenth century, the era of modern imperialism. Generally, however, it failed to acquire possessions of great value in the scramble for Africa. Also, when the time had come to liquidate its empire, Spain turned out to be considerably less reluctant than Portugal to grant its colonies independence. Indeed, in the case of the Spanish Sahara (a stretch of desert), Madrid decided simply to abandon the territory without further formality, its argument being that the nomadic peoples in the area did not qualify as a nationalist group to whom sovereignty might be transferred. However, since the Western Sahara is rich in phosphate, the Spanish

(Photo courtesy Lisa Clyde)

Families display colorful, symbolic banners from the sides of buildings during a religious festival in Toledo.

withdrawal caused an armed conflict among various neighbors and claimants, such as Morocco, Mauritania, Algeria, and a group of guerrilla fighters that called themselves the Polisario. By the mid-1990s, Morocco had prevailed in the military struggle.

Gibraltar

Another anomaly is provided by Gibraltar, a spit of land that is little more than a rock at the tip of the Iberian Peninsula. It has been a British possession since the Treaty of Utrecht in 1713. This colonial leftover is a thorn in Spain's side. While Britain is not a Mediterranean country, it has always had an acute interest in the Mediterranean Sea (which in the past was also affirmed by its possession of Malta, Cyprus, and the Suez Canal). This special interest derived in large part from the much shorter voyage that the Mediterranean route afforded to the part of the empire generally called "East of Suez." And even now that the empire has been restructured into the Commonwealth, Great Britain is reluctant to cede Gibraltar to Spain.

In 1966, Spain pleaded with Britain to grant it "substantial sovereignty" of Gibraltar. When the plea fell on deaf ears, Spain instituted a partial blockade—that is, it cut Gibraltar off from its natural hinterland. The blockade was later lifted, but Anglo–Spanish relations remained very strained. At one point, the British proposed that a plebiscite decide the fate of Gibraltar. But to Spain's regret, the outcome of the plebiscite among the 30,000-odd residents of Gibraltar revealed a strong preference for the status quo, though the number of Spanish natives far exceeded those from Great Britain. In the early 1990s, Spain finally seemed resigned to reality, and the government decided to take away the Spanish guards who more or less surrounded Gibraltar. The fundamental dispute, however, still has not been resolved.

Ceuta and Melilla

If Spain views the Gibraltar situation clearly in terms of irredentism, it appears considerably less concerned about the continuance of colonialism when the shoe is on the other foot. Ceuta and Melilla are two towns in Morocco that had their origins in medieval times. They have been Spanish enclaves since 1580 and 1470, respectively. When France gave independence to Morocco, Spain might have surrendered these two tiny possessions, which had little strategic or economic value. But rather than granting them to Morocco, Spain chose to keep these towns under its rule, its main argument being

that the colonies had been in Spanish hands for so many centuries that they differed significantly from the Moroccan ambience. Spain has occasionally tightened its control, particularly when nonresident Moroccans have attempted to settle in Ceuta and Melilla, which have a lower unemployment rate and a higher standard of living than Morocco.

Monarchy and Republicanism

Although Spain has strong monarchical traditions, it has not been without republican intervals. And remarkably, the transition from monarchy to republicanism was more often than not achieved without force or violence. Amadeo of Savoy ruled briefly in the third quarter of the nineteenth century and then decided to step down. Things were unsettled for a short period, but the military then determined that a monarchy was preferable to a republic and proclaimed Alfonso XII king of Spain.

His son and successor, Alfonso XIII, abdicated voluntarily, because in the municipal elections of 1931 the republican votes outnumbered those of monarchists. However, this time it took longer for the monarchy to be restored, if only because Alfonso XIII's abdication augured the prelude to the Spanish Civil War (1936–1939).

Franco

General Francisco Franco, upon becoming the head of state and the supreme commander of the armed forces in 1936, remained the undisputed central figure in the Civil War. The strict dictatorship that he created was fascist in nature. Franco ruled with an iron hand for 40 years, during which time all political parties were banned except for his own, the Falange. Freedoms and democratic rights also fell by the wayside. Only during the last five years of his control did new trends manifest themselves.

Franco's approach toward the monarchy was somewhat ambivalent. He believed that, in general, Spanish monarchs had, as political symbols, helped to unify the nation. However, he certainly did not want to have a monarch as a top executive as long as *he* was head of state, since that would imply a competition for popular favor. His solution was simple: As long as he ruled, the institution of monarchy would continue to exist but would be, so to speak, on the back burner. Franco was well aware that his control would one day end, and although he wanted it to be followed by the reinstatement of the monarchy, he trusted that the authoritarian political system would survive. As to the question of who would be the future monarch, he made his choice early: It would

not be Alfonso XIII's son, who had offended him by his frequent criticism, but, rather, Alfonso's grandson, Juan Carlos.

The old dictator died on November 20, 1975. Two days later, Spain had a monarch at its head: King Juan Carlos I.

A NEW ERA

Franco's death proved a watershed such as Spain had never known. Few people, least of all Franco himself, could have foreseen the profound and rapid changes that were to come. Juan Carlos had been born when the Civil War was about to end. He had been groomed in the dictator's shadow, which implied that his style had been authoritarian. Juan Carlos had been fed the daily diet of dictatorship; democracy was little more than an alien, theoretical concept to him.

Yet after Franco's death, King Juan Carlos wasted little time in rendering Spain a constitutional democracy, or, as the Spanish liked to call it, a "parliamentary democracy." Referenda were instituted as instruments of change; only a little more than a year after the dictator's demise, a referendum on political reform was held. Of course, being a constitutional monarch, Juan Carlos did not have a great deal of power, but he applied whatever influence he could to arrive at democratic structures. Demonstrating rare political skill and good judgment, he named a young reformist politician, Adolfo Suárez, to the premiership in 1976. Both men cherished similar ideals and took British democracy as a model.

Suárez was mainly responsible for the execution of what has been called the "Democratic Transition." Trade-union organizations, other pressure groups, political parties—in short, all the linkage mechanisms in a democracy—were allowed to come out of the woodwork. They had been banned and, more often than not, ruthlessly persecuted for nearly four decades. But now, even party leader Santiago Carrillo, who had been in exile for that period, was allowed to reorganize the Communist Party of Spain (PCE), which had been anathema in Franco's days. (This party and its leader, who subsequently designed the concept of "Eurocommunism," were singularly unsuccessful in capturing public support.)

Basic liberties were recognized, and amnesty was granted for political offenses. The climax of these preparations for democracy was the parliamentary elections of June 15, 1977, the first elections held in Spain in 41 years. The *Cortés* (Parliament) was a necessary result of these elections. Shortly after its formal investiture, the Parliament started to work on a new constitution. After vigorous debate, it was approved in a referendum on December 6, 1978. Thus, only three years after Franco's death, a democratic Constitution became Spain's official guiding instrument. Its first article would make Franco turn in his grave; it reads:

> Spain is hereby established as a social democratic State, subject to the rule of law, and advocating as higher values of its legal order, liberty, justice, equality and political pluralism. National sovereignty is vested in the Spanish people, from whom emanate the powers of the State. . . . The political form of the Spanish State is that of a Parliamentary Monarchy.

The Constitution proceeds to outline the Legislature (*Cortés Generales*) as consisting of two houses, a Congress of Deputies and a Senate. The government, headed by a prime minister and a deputy prime minister, is collectively accountable for its political management to the Congress of Deputies.

As in all parliamentary systems, the Congress of Deputies exercises control over the executive by its ability to withhold confidence from the government. In addition, it has the option of adopting a motion of censure, which has a similar effect. This lower house thus represents the Spanish people and exercises the legislative power of the state, approving budgets and supervising government actions in other respects as well. If the government is not made up of party leaders whose party enjoys a majority in the Congress of Deputies, coalitions will have to be established with other, preferably like-minded, parties.

The Senate is the house of territorial representation. Each province has four representatives. Autonomous communities also nominate one senator each, plus another for every 1 million inhabitants in their respective territories. Both senators and deputies have four-year terms.

As compared to Franco's time, when only one party was allowed to operate, currently more than a dozen parties vie for seats in Parliament, including regional ones. Among them are extremely small splinter parties. Union organizations are considerably more limited in number, while pressure groups in general appear to be growing in numbers as well as in membership.

SOCIALIST RULE
Adolfo Suárez, having skillfully presided over the transition to democracy, stepped down in 1981. A brief interregnum by Leopoldo Calvo Sotelo followed, after which Felipe González Márquez was elected to the prime ministership. González led the Spanish Socialist Workers Party (PSOE), which in the parliamentary election of 1986 received a majority of seats in the Congress of Deputies. However, the late 1980s witnessed a drop in support for the Socialists. This soon turned out to be irreversible. At the general election in 1989, the PSOE lost its absolute majority, after which it declined further. In part this trend may be attributed to the fact that González deliberately freed the party from its doctrinaire positions. The prime minister was basically a technocrat and as such preferred the party to adopt some kind of technocratic neutrality, which did not sit well with the Socialists, still steeped in ideological causes. The real blow came in 1992 in Seville, the birthplace of the prime minister and a city that long had seemed the bedrock of socialism. Since the Expo '92 was being held in that city, González granted it a great deal of extra money. However, the municipal elections, to everyone's surprise, turned the Socialist mayor out and voted a non-Socialist party leader in.

Nevertheless, the fact that 1992 happened to be Spain's big year may have slowed down the decline of the PSOE. The summer Olympic Games were held in Barcelona. Madrid was proclaimed the "Cultural Capital of Europe," and throughout Spain celebrations commemorated the 500th anniversary of Christopher Columbus's first voyage to the Americas.

Gradually Prime Minister González became more interested in the emerging Europe. He concentrated his attention on the European Union, particularly after Spain became an official member in 1986. (The "Eurobarometer" 2000 poll indicated that a majority of Spanish citizens favor the EU's playing a greater role in their daily lives.)

The domestic political initiatives pursued by the Socialist government were as a rule moderate in character. Legislation concerning abortion and divorce, as well as educational reform, was approved by the Parliament, but the government was careful not to risk too open a confrontation with the Roman Catholic Church. Spanish legislation on abortion and divorce is among the most restrictive in Western Europe. And in pursuing educational reform, the Socialist government did not question the right of private (primarily Catholic) schools to continue receiving state funds.

DEVOLUTION
The post-Franco government soon realized that political pluralism (as furthered by the Constitution) and ethnic diversity are different matters. Other referenda were conducted during González's tenure, such as the so-called autonomy referenda, which were to decide whether or not certain regions populated by ethnic minorities would be allowed to have some type of minor constitution. Spain's trend toward devolution—decentralization of power—appears to have muted minority-group dissatisfaction somewhat.

The Basques' Nationalist Quest: Elusive But Ending
The Basques are a very ancient people straddled across the Pyrenees, the population cut in two by the French–Spanish border. The Basques on French soil, numbering only about 200,000, have had quarrels with the French government, but there has been very little violence. Spain, on the other hand, has about 2 million Basques, and while they possess the same culture, the same traditions, and the same language as their French counterparts, Spain has always experienced considerable Basque restlessness and violence. Indeed, it was here that the organization Basque Fatherland and Liberty (ETA) was established. Inspired by radical nationalism, the ETA soon developed into a terrorist organization that has been responsible for a great many killings. A few years ago, French and Spanish authorities agreed that Spanish Basques may be extradited to Spain; they hoped that this measure would weaken the ETA, since the group could no longer count on using France as a safe haven.

Other Groups
Basque separatism may have stimulated a similar stridency on the part of other ethnic minorities in Spain, such as the Cataláns, the Galicians, and the Andalusians. The Cataláns in particular are very proud of their heritage and language, which they do not wish to be confused with, much less identified as, those of Spain. Of the various regions, Catalonia has the second-largest population (approximately 6 million, as compared to Andalusia's 7 million).

Spain thus provides an interesting political paradox that is also being seen in many other European countries, such as Germany, France, and Britain. While it has come to terms with the drives that seek integration into a larger context, that is, the European Union, it is internally challenged by the forces represented by various minority-group demands. The Spanish government's decision to use devolution to give its citizens a greater voice is proving to have resulted in a greater desire among the various regional groups to stick together rather than to fragment.

The Reconquista is complete; Spain embarks on the road to discovery and empire A.D. 1492	The Armada is defeated; Spain loses its place as the world's main imperial power 1588	The Treaty of Westphalia 1648	The Treaty of Utrecht formalizes the British occupation of the Rock of Gibraltar 1713	The Spanish Civil War, won by the extreme right (the Franco fascists) 1936–1939	Franco dies; a new era in Spanish history begins 1975	The text of the new (post-Franco) Constitution is adopted by universal suffrage 1978	Spain joins NATO; Spain joins the European Union as a full member 1980s	Spain hosts Expo '92 and the 1992 Summer Olympic Games; Spain participates in the launch of the Euro 1990s

2000s

Economic adjustments to the EU and the increasingly global economy

Despite the ETA cease-fire, Basque separatist violence continues

THE ECONOMY

More than 40 percent of the land in Spain is arable, and the climate is benign. For a long time Spain was predominantly agricultural, with its products geared largely toward the domestic market. After World War II, Spain's position as an outcast among the European nations enhanced the tendency of economic introversion—of having as little trade and commerce with nations in the region as possible. For some time the Franco regime labeled this "self-reliance," but the situation suddenly changed in the 1960s, when Franco decided to diversify the economy. Apparently he had determined that the time had come to industrialize, or at least to expand Spain's very limited industrial base. The diversification was a remarkable success, in that the economy responded very favorably. In the economic history of Spain, the 1960s have come to be identified as an economic miracle—not perhaps of the same scope as that which took place in West Germany, but remarkable nevertheless. This feat helped Spain's image; it was no longer dismissed as a "backward" Southern European country.

In the late 1970s and early 1980s, exports started to increase, and, since the dictator had died and the new government appeared democratically inclined, the time had come to consider Spain's admission to the European Union. Both Iberian countries, Spain and Portugal, still had low levels of per capita income, but the Union was no longer so concerned about structural imbalances, believing that these wrinkles might be ironed out in time.

In the 1990s, on the economic front, the government concentrated on bringing inflation under control. The rate in April 2001 was 4 percent for consumer prices. In matters of unemployment, the news is still discouraging—around 14 percent. And during the European currency crisis in 1993, the Spanish peseta had to be devalued by 5 percent. It has been devalued several times since—a boon to foreign tourists. However, in joining the Euro Zone in 1999, Madrid turned future currency management over to the European Central Bank in Frankfurt, Germany.

RECENT DEVELOPMENTS

The Spanish economy remained strong in the 1990s, growing at an annual rate of more than 3 percent for much of the decade. The annual GDP growth rate in March 2001 stood at 3.7 percent. Traditionally regarded as a rather spendthrift country, Spain got its budget deficit under control and was able to participate in the launch of the Euro in January 1999. In 2000, Prime Minister José Maria Aznar of the Popular Party, despite the unease of some Spaniards with his conservative policies, easily won reelection, partly on the basis of the good economic performance. Another contributing factor was the continued disarray of the Socialist Party, which emerged after corruption scandals implicated long-time prime minister Felipe Gonzalez. The most serious allegations were that the Gonzalez government had run or condoned military "death squads" in the mid-1980s that killed more than two dozen suspected Basque terrorists.

On the international scene, Spain made headlines because of the investigative activities of one of its judges (who in the Spanish system also indict and prosecute), Baltasar Garzon. In 1998, he indicted former Chilean dictator Antonio Pinochet for crimes against Spanish citizens during the antileftist "Dirty War" of the 1970s. Pinochet was in England, where he was placed under house arrest while Garzon continued to lodge more and more charges against him. International human-rights groups enthusiastically supported Garzon and urged the British government to extradite Pinochet to Madrid for trial. After a great deal of hesitation, angry exchanges with Chilean military leaders, and the grudging U.S. release of controversial documents, the British government dodged the bullet by announcing in January 2000 that Pinochet was too ill to stand trial. He quickly flew home to Chile.

The Aznar government has faced two grave domestic problems. The most dangerous is Basque terrorism, which proceeded at full tilt in 1997 and into 1998. Plans to assassinate King Juan Carlos and to blow up the famous new Guggenheim art museum in Bilbao were discovered. Suddenly, in 1998, the ETA announced an "indefinite" cease-fire, which the government accepted. Optimism ran high that the avowedly separatist Basque political parties might now be content to pursue their nationalistic policies peacefully. But the cease-fire lasted only until December 1999; the bombings and shootings resumed and quickly reached a fever pitch. Arrests and convictions of dozens of Basque terrorists in both Spain and France have done little to stem the bloodshed, which continued with a series of car bombs in 2001. In the May regional elections, Basque voters voiced their rejection of terrorism by dealing the ETA's political wing a major setback.

Another, newer domestic concern is violence against foreigners and immigrants. Spain has recently seen the growth of far-right parties known, among other things, for their xenophobia. There were a number of attacks on North Africans, especially Moroccan "guest workers," of the type of anti-immigrant sentiment that has also been seen in recent years in France, Austria, and Germany.

DEVELOPMENT

Spain's transition from empire to nation has been remarkably smooth. However, it did take the country a long time to move away from its agricultural and rural base. Industrialization has progressed unevenly, with unemployment a persistent and serious problem.

FREEDOM

Freedom has increased progressively since Spain embarked upon the road to democracy after General Francisco Franco's death in 1975. Madrid has been the venue of a prolonged follow-up conference regarding human rights as outlined in the Helsinki Accords.

HEALTH/WELFARE

Health has greatly improved since Spain ceased to be an outcast in Europe and has begun participating in European medical conferences and other exchanges of medical science. Some cautious steps have been taken toward a welfare state.

ACHIEVEMENTS

On September 10, 1981, Picasso's *Guernica* arrived for permanent exhibition in Madrid. This famous painting, depicting the dismal destruction of a northern Spanish town by fascist troops and planes, had been banned by the Franco regime.

Sweden (Kingdom of Sweden)

GEOGRAPHY
Area in Square Miles (Kilometers):
179,986 (449,964) (about the size of California)
Capital (Population): Stockholm (718,500)
Environmental Concerns: acid rain; pollution of the North Sea and the Baltic Sea
Geographical Features: mostly flat or gently rolling lowlands; mountains in the west
Climate: temperate in south; subarctic in north

PEOPLE

Population
Total: 8,873,000
Annual Growth Rate: 0.02%
Rural/Urban Population Ratio: 17/83
Major Language: Swedish
Ethnic Makeup: a homogeneous Caucasian population; a small Sami (Lapp) minority; an estimated 12% are foreign-born or first-generation immigrants
Religions: 94% Evangelical Lutheran; 1% Roman Catholic; 5% others

Health
Life Expectancy at Birth: 77 years (male); 82 years (female)
Infant Mortality Rate (Ratio): 3.4/1,000
Physicians Available (Ratio): 1/384

Education
Adult Literacy Rate: 100%
Compulsory (Ages): 6–15

COMMUNICATION
Telephones: 6,010,000 main lines
Daily Newspaper Circulation: 515 per 1,000 people
Televisions: 476 per 1,000
Internet Service Providers: 29 (1999)

TRANSPORTATION
Highways in Miles (Kilometers): 130,758 (210,900)
Railroads in Miles (Kilometers): 7,574 (12,624)
Usable Airfields: 256
Motor Vehicles in Use: 4,006,500

GOVERNMENT
Type: constitutional monarchy
Independence Date: June 6, 1523
Head of State/Government: King Carl XVI Gustaf; Prime Minister Goran Persson
Political Parties: New Democracy Party; Social Democratic Party; Moderate Party; Liberal Party; Center Party; Christian Democratic Party; Left Party; Green Party
Suffrage: universal at 18

MILITARY
Military Expenditures (% of GDP): 2.5%
Current Disputes: none

ECONOMY
Currency ($ U.S. Equivalent): 10.91 kronor = $1
Per Capita Income/GDP: $20,700/$184 billion
GDP Growth Rate: 3.8%
Inflation Rate: 0.4%
Labor Force: 4,300,000
Unemployment Rate: 4%, plus about 5% in training programs
Natural Resources: zinc; iron ore; lead; copper; silver; timber; uranium; hydropower
Agriculture: animal husbandry; grains; sugar beets; potatoes; fish
Industry: iron and steel; precision equipment; wood pulp and paper products; processed foods; motor vehicles; tourism
Exports: $85.7 billion (primary partners EU, United States)
Imports: $67.9 billion (primary partners EU, United States)

http://www.luth.se/luth/present/sweden/
http://www.swedentrade.com/usa/

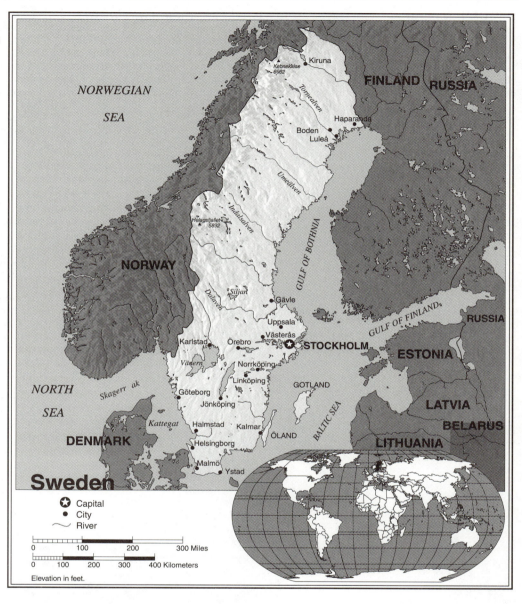

Sweden

- ⭐ Capital
- ● City
- 〜 River

0   100   200   300 Miles
0   100   200   300   400 Kilometers

Elevation in feet.

SWEDEN

Sweden is one of the oldest kingdoms in the world, and it has been one of the more significant countries in European history. With the largest population among the Nordic states, Sweden is a long country—indeed, the distance from Malmö, Sweden's most southern city, to the extreme north of Sweden equals that of Malmö to Rome, Italy!

Nordic history plays an important part in the Swedish world view. For some centuries Sweden was highly assertive in diplomacy and military matters. The beginning of modern Swedish history is usually associated with Gustav Vasa, who was elected king in A.D. 1523 after leading a revolt that terminated Sweden's century-long link with Denmark. It steadily rose in importance, becoming one of Europe's great powers during the seventeenth and eighteenth centuries. For a long time it waged war after war, usually in the role of aggressor. One of Sweden's greatest kings, Gustav Adolph II, was killed in a battle in southern Germany in 1632 while leading his army in the Thirty Years' War. Another king, Charles XII, was out of the country for more than 14 years in the early eighteenth century, warring in Poland and Russia.

But since the end of the Napoleonic era in 1814, a sharp reduction in Sweden's armed forces, its adoption of a policy of military/political neutrality, and its relative geographic isolation have all contributed to keeping the country out of war. Swedes pride themselves on having made that transition from a country frequently at war to a peaceful society, and they believe that they can help other nations, particularly those in the developing world, to do the same.

Sweden was one of the very few countries in Europe not to be engulfed by World War II. All postwar Swedish governments have studiously avoided becoming part of a collective defense pact. Thus, when the North Atlantic Treaty Organization emerged, the Swedish government indicated immediately that under no circumstances would it join.

In 1992, the question of Swedish neutrality came up again, possibly because the end of the Cold War demanded some reassessment. But Margaret af-Ugglas, Sweden's foreign minister, simply reaffirmed that the policy would remain "nonparticipation in alliances in time of peace, aiming at neutrality in time of war." Sweden is a member of the Nordic Council, an organization that facilitates cooperation and consultation in many matters that have no bearing on war.

RAPID DEVELOPMENT

Today, Sweden is one of the most prosperous, democratic, and highly developed nations in the world. It has certainly come a long way since the 1800s, when hundreds of thousands of Swedes emigrated to the United States, leaving behind them chronic poverty and shortages of economic opportunities. There were other reasons for leaving the country. Some simply wanted a new start in life; others believed that they had too little religious freedom in Sweden. Swedes continued to emigrate to the United States in large numbers well into the twentieth century. Nearly every Swedish family now has a living relative in the United States; at one time, in fact, the second-largest community of Swedes in the world was located in Chicago.

Sweden entered the industrial age behind Great Britain and Germany. The early 1900s marked tremendous changes in the country. Industries started to mushroom, natural resources were utilized as never before, and many people left the land for the cities.

Most Swedes are only one or two generations removed from rural life. The relatively recent urbanization may in part explain the typical Swede's dream of owning a cabin in a little meadow by a lake. This bucolic ideal often conflicts with reality, which for many is a high-rise apartment in the suburbs. While there are no slums or ghettos, it is now recognized that the rapid buildup of concrete structures

(Photo courtesy Lisa Clyde)

Stockholm, Sweden's largest city, was founded in the middle of the thirteenth century; it officially became the capital in 1436. The city is an attractive blend of old and new, with architecture represented from medieval times to the present. As pictured above, waterways and bridges are a prominent feature in the capital.

Carl XVI Gustaf became king of Sweden in September 1973. Today, the role of the Swedish monarch is largely ceremonial. Members of the royalty are quite accessible in this highly egalitarian country, as can be seen in the above photograph, taken as the king and his wife, Queen Silvia, left a parliamentary function at Stockholm Cathedral in 1994.

largely making up the "bedroom cities" is not without social consequences.

The "model" Swedish cities built in the 1960s are left as dreary monuments of a bygone era. Experiments on this scale were possible in part because Sweden's entire population numbers less than that of New York City, spread out in an area as large as California, and with a high per capita income. Sweden also has been able to organize and regulate its society in ways that more populous federated countries have not been able or willing to do.

THE SWEDISH POLITICAL SYSTEM

In the nineteenth century, Sweden made the transition from authoritarian monarchy to parliamentary democracy, achieving this without revolution. Politics in Sweden has always been pragmatic and methodical—mob rule, corruption, and extremism have played little part in the country's history since the days of King Gustav III, whose assassination in 1792 is portrayed in Verdi's opera *Masked Ball*. Nothing similar occurred until 1986, when Prime Minister Olaf Palme was murdered as he walked home from a movie theater in downtown Stockholm. Swedes were stunned; their vision of Swedish insulation from the turmoil and violence of the outside world was shattered.

Ingvar Carlsson, who became prime minister after Palme's murder, was relatively unknown outside political circles, because he had always worked in Palme's shadow. A professional politician who rose from union organizer and youth leader to ministerial rank, Carlsson quickly established a profile of a quietly pragmatic leader. His style contrasted sharply with Palme's flamboyant image. Carlsson directed his attention more to social and economic problems than to foreign policy.

Sweden's parliamentary democracy is headed by a constitutional monarch. Thus the king, as the head of state, has an entirely ceremonial role. The Swedish monarch, who never wears a crown, has even lost to the speaker of the Parliament (MIRiksdag) the formal power of selecting the prime minister, a power that other European monarchs still enjoy. The prime minister, as the head of government, exercises executive authority. The Riksdag, which since 1971 has consisted of a single chamber, has 349 seats, which are distributed in accordance with the proportion of votes that each individual party receives in the national elections. However, in order to prevent the fragmentation of the party system, a party must receive a minimum of 4 percent of the votes nationwide; or, alternatively, 12 percent of the votes in any of the 28 voting districts. A party that

does not meet either requirement will not be seated in the Riksdag. Voters cast their ballots for party lists, rather than for an individual, and it is the party leadership that determines the candidates and their position on the lists.

Political power in Sweden rests with parties and other organizations, primarily labor unions, employee associations, farmer groups, and popular movements such as single-issue social groups. There is a high density of interest-group organization in Sweden. For example, the Trade Union Federation (LO) organizes in excess of 85 percent of Swedish workers. The LO is closely allied with the Social Democratic Party (SAP). Employers are also well organized by the Confederation of Employers (SAF), which supports the Moderate Party (better known as the Conservatives). Similarly, a high percentage of farmers are members of the Farmer's Organization (LRF), which backs the Center Party (once named the Farmer's Party). The Swedish political norm is that these peak organizations and smaller ones will be consulted by the government when legislation is being considered, in the interest of consensus.

For most of the past century, the largest political party in Sweden has been the Social Democratic Party. It has often had to rely on the tacit support of a smaller party, such as the Left Party—formerly the Communist Party—in order to remain in power. The ex-Communists, who generally receive a small percentage of the vote, do not vote against the SAP on most issues. This may explain the political longevity of various Social Democratic governments.

Sweden has four major non-Socialist parties: the Conservative Party; the Center Party; the Liberal Party; and, during the 1990s, the New Democracy Party. The Center and Liberal Parties coordinated their policies in an arrangement known as the "union of the middle." (It should be noted that the term *liberal* has no leftist connotation in Europe, as it does in the United States. Indeed, more often than not liberal parties in Europe are protagonists of big business; although somewhat centrist in orientation, the Liberal Party in Sweden is an example of just that.)

For the non-Socialists to form a government, they not only must win more votes but must also agree on a common purpose. This is difficult, as the right and center in Sweden are typically much more divided among themselves than is the left, a phenomenon that was clearly evidenced in the years between 1976 and 1982.

Voter turnout in Sweden is normally very high—over 85 percent. Swedish elections are usually decided by narrow margins. The Social Democrats sometimes have no more than a plurality in Parliament, amounting to just a few more seats than the next-largest party.

The parliamentary elections of September 1991 witnessed the defeat of Ingvar Carlsson and his Social Democratic government. This was a strong indication of popular disenchantment with the welfare state—or, rather, with the high taxes that the welfare state demands. In addition, the worldwide recession of the early 1990s was taking its toll on Sweden. Maintaining a high employment rate was considered the most important method of keeping the economy strong, so what to other countries may have seemed a relatively low unemployment rate of 3 percent was viewed in Sweden as unacceptable. Then there was the country's high inflation rate. Although it seemed to be improving, it contributed to the temporary downfall of the Social Democratic Party.

Elections in the fall of 1991 brought Carl Bildt to power. Bildt was a non-Socialist whose views and policies were very much right-of-center. Still, his four-party coalition was unable to claim a majority in the Riksdag. Prime Minister Bildt thus needed to take into account, if not to con-

sult with, the opposition parties on any great project.

Upon taking office, Bildt announced that the new government would work to complete four major tasks during its term: negotiating to bring Sweden into the European Union; terminating the country's economic stagnation and reestablishing Sweden as a nation of growth and enterprise; initiating freedom of choice in welfare policy, thus enhancing social care; and shaping long-term environmental policies. Most of these programs were continued after the Bildt administration was voted out of office in 1994. The public brought Ingvar Carlsson back to power. Meanwhile, the two party platforms had grown together. In 1996, after Carlsson resigned for personal reasons, Finance Minister Goran Persson, of the Social Democratic Party, won the prime ministership.

The Swedish parliamentary elections of September 1998 were historic. Although the economy was improving, unemployment remained high by Swedish standards, and the pains of the Social Democratic–Center government's austerity measures lingered. Both parties lost significantly. The Social Democrats, while remaining the largest parliamentary party, won only 36.4 percent of the votes—their worst electoral results since 1921. The Center Party lost more than half of its votes (ending up with 5 percent), which lengthened its series of electoral setbacks to seven. The major opposition party, the Conservatives, was led by former prime minister Carl Bildt, who had gained a statesman's aura as the European Union's special envoy to Bosnia. His party, which campaigned for tax cuts and overhauling the welfare system and labor regulations, won 22.9 percent, but failed to gain any new seats. The big winners, with record highs, were the Christian Democrats, who, at the expense of the Conservatives, won 11.8% of the votes; and the Left Party, which, at the expense of the Social Democrats, won 12 percent. The People's Party Liberals barely made it back into the Riksdag, while right-wing populist parties did not.

Despite the Social Democrats' electoral disaster, Goran Persson was able to form a minority government with the parliamentary support of the Left Party and the Greens (who won 4.5 percent). The three parties could find common grounds on economic policies; however, in contrast to the Social Democrats, both the Left and the Green parties favor Sweden's withdrawl from the EU.

A Unique Governmental Check

One finds in Sweden that Scandinavian innovation, the ombudsman, a parliamentary

commissioner who exercises a check on the administration. The office of ombudsman was created to right wrongs that may have been perpetuated on average citizens by the government. In Sweden, the *Justitie-ombudsman* exerts supervision over all the courts of law, the civil service, and the military. There are also special ombudsmen for press complaints, antitrust (fair business competition), consumer affairs, and women's equality.

FOREIGN POLICY

The Swedes make an interesting distinction between nonalignment (which they wish to pursue in peacetime) and neutrality (which is to prevail in a war in which they themselves are not involved). But the Swedes reject the Danish solution of the pre–World War II era, when the Danes abolished their army: The Swedes favor armed neutrality. They believe that a small, nonaligned nation can help to bridge differences and improve the dialogue in world politics. The Swedish political elite also assume that their country will not be attacked unless there is a general state of war in the world.

These assumptions made it difficult for Sweden to come to grips with submarine violations of Swedish territorial waters. Soviet submarines were discovered near sensitive areas, such as Sweden's naval bases. The Swedish government preferred to deal with these violations quietly, through diplomatic channels.

Occasionally, however, its diplomacy vis-à-vis the Soviet Union was not so quiet. Sweden was the first country in 1986 to conclude that a major accident had taken place in Chernobyl's nuclear power installation. The Swedish government immediately alerted other European countries to the grave dangers of exposure to such high levels of radiation.

Gradually in 1993, support for Sweden's membership in the European Union grew. By the end of the year, polls found that all of the major political parties as well as the vast majority of business leaders backed the government in this endeavor. Sweden had already signed up for the preparatory stage, the European Economic Area—that is, the free-trade agreement between the EU and the seven-nation European Free Trade Association, which was scheduled to take effect in 1994. In the same year, 52.2 percent of Swedish voters supported in a referendum Sweden's entry into the EU. EU membership would follow on January 1, 1995. A difficult point that has recently loomed is to what extent it will be possible for Sweden to pursue nonalignment as a member of the Euro-

Kalmar Union with Denmark and Norway A.D. 1397	First Parliament (Riksdag) at Arboga 1435	Gustav Vasa is elected king 1523	A coup takes place and King Gustav IV is deposed 1809	Norway is in a union with Sweden 1814	Jean Baptiste Bernadotte becomes king and assumes the name Karl Johan 1818
●	●	●	●	●	●

pean Union. Or, conversely, will the EU also have defense implications?

Although Sweden could have qualified to join the Euro Zone on January 1, 1999, its government chose to delay membership in the monetary union. A survey in the fall of 1999 indicated that half of the Swedish public felt that their country had not benefited from EU membership; this negativism was the highest found in any of the 15 EU countries. The Conservatives are more pro-Euro than the Social Democrats, whose leaders are inclined to bide their time until political conditions would be more favorable for a referendum on the issue. On the other hand, their allies, the Left and the Green Parties, favor an early referendum (to defeat the Euro). The Danes' rejection of Euro Zone membership in the September 2000 referendum has caused Prime Minister Persson to be more cautious about timing. Recent polls have indicated that a majority of Social Democratic voters oppose joining the monetary union. They and many other Swedes see the country's economy as doing well outside of the Euro Zone, are suspicious of further political integration, and are concerned about the harmonizing of welfare benefits (downward). Persson's government favors EU enlargement, prioritizing unemployment as an EU concern, and stronger EU policies to protect the environment.

CULTURE AND LIFESTYLE
Sweden has a rich cultural and intellectual history. August Strindberg, the playwright; Carl Larsson, the artist; and Ingmar Bergman, the film and stage director, are just a few of the Swedes known around the world. Students of botany learn Carl Linnaeus's classification system, and Sweden has also become known for the Nobel Prizes, annually awarded to top authors and scientists.

In the world of sport, too, Sweden has made its mark. Skiers watched Ingemar Stenmark dominate slalom events for nearly a decade. Bjorn Borg, Mats Wilander, and Stefan Edberg became tennis kings; and in 1987, Sweden's national hockey team won the World Cup.

Culture and quality of life go hand-in-hand with health and longevity. Life expectancy in Sweden is higher than in most countries. The World Health Organization's 2000 global survey rated Sweden's health system as the fourth best in the world (the United States rated 15th best). According the the UN's Human Development Index 2000 (which combines educational, health, and income statistics), Sweden's quality of life was the sixth best out of 174 countries.

Some stereotypes of Swedes, however, present less flattering images. Swedes are sometimes thought by foreigners to be aloof, cold, preachy technocrats—a view that others vigorously refute. Swedish tourists are sometimes thought to overindulge in alcohol when abroad, perhaps because it is heavily taxed at home and thus expensive. Some critics have sought to portray Sweden's social-welfare system as a causal factor in the country's high suicide rate. However, Sweden does not rank first in Western Europe; five other countries have higher suicide rates.

The truth may lie somewhere in between the wholly positive and wholly negative images. The average Swede is a conformist who raises continuity to a virtue. Swedes generally follow the rules and expect others to do so as well—including foreigners. They believe that society should be fair and that, by and large, personal freedom and success should have limits. Swedish citizens are fond of their country and their flag, but many do not understand American-style public patriotism.

Sweden has one of the lowest birth rates in the world. Its population growth in recent decades has primarily been due to the influx of immigrants. Beginning in the 1960s, non-Nordic immigrants helped to meet the labor shortages in industry. Later, others came as political refugees. Social Democratic governments sought to conduct a moral foreign policy whose corollary was tolerance and support for immigrants at home. Integration of non-Nordic residents (who made up around 10 percent of the population in 2000) has not had much success; one finds segregated communities in the suburbs of Stockholm.

In recent years, the government has responded to growing public concern about the number of foreigners by implementing more restrictive regulations on immigration. A wave of death threats, bombs, and the murder of a trade-union official shocked normally nonviolent Sweden in 1999 into recognizing its growing subculture of xenophobic skinheads, neo-Nazi rockers, and Aryan power groups, which provide support for an estimated 200 right-wing terrorists. Critics have viewed the authorities as being too complacent, for too long. Swedish ministers have opposed banning neo-Nazi groups and parties, however, arguing that this would contradict the country's liberal freedom of speech laws and create martyrs for the extreme right.

Swedish society is among the most egalitarian on Earth when it comes to the role of women. Seventy percent of women hold full- or part-time jobs outside of the family, and the family has lost its patriarchal character. Many young people choose to cohabit rather than marry; more than half of Swedish babies are born out of wedlock.

The Equality Ombudsman helps to deal with charges of discrimination in the workplace. Women have been moving into many traditional male roles in Swedish society. In fact, a proposal to conscript women as well as men into the Swedish military is currently under active consideration. Royal succession has been changed so that the monarch's first child, regardless of sex, is the heir-apparent (as Crown Princess Victoria now is). There has been no female prime minister to date. However, 38 percent of ministerial posts in 1999 were held by women. In the Inter-Parliamentary Union's 2000 survey of the percentage of women in the lower (or single) house of parliaments in 177 countries, Sweden ranked *first,* with women holding 42.7 percent of Riksdag seats.

THE ECONOMY
Since Sweden has had Social Democratic governments for most of the past century, one would expect its economy to be based on government-run or government-owned enterprises. However, the percentage of *private* enterprise has been over 90 percent during decades of Social Democratic administration. In the mid-1970s, it was reduced slightly, mainly because many of the private companies appeared to be languishing. Ironically, nationalizations took place during a six-year period in which a non-Socialist government held office.

In the late 1970s, shortly after the nationalizations took place, the Swedish economy slumped. In fact, it eventually became so dismal that analysts started to refer to Sweden as the "sick man of Europe." Some attributed the poor per-

1.25 million
Swedes emigrate
to the United
States
1820–1930

The union with
Norway is
dissolved
1905

Universal
suffrage and a
democratic
political system
are introduced
1918–1921

The Social
Democrats come
to office;
introduction of
the "Swedish
model" welfare
system
1936

Prime Minister
Olaf Palme is
murdered
1986

Sweden joins the
European Union in
1995 but declines to
join the launch of the
Euro in 1999; Sweden
confronts challenges
to its neutrality during
World War II
1990s

2000s

Sweden hosts
an important
international
conference on
Holocaust
education

An anti-EU
demonstration
in Malmo turns
violent

Sweden holds
the EU Council
presidency
during the first
half of 2001

formance at that time to the high ratio of public spending (which reached 67 percent of the gross domestic product by 1982). Naturally, these critics were surprised to see Sweden recover within a decade, even though the role of the state in the economy was still very much larger than in other industrial economies.

Sweden established a huge "State Company" (*Statsföretag*), which serves as a type of umbrella organization for all the nationalized firms. Some 10 percent of Swedish workers are employed by state-run companies, which include post and telephone services, railroads, the iron-ore mining in Lapland, and a large part of the domestic energy production (mainly hydropower).

The Social Democratic Party prefers to leave the private character of the economy alone. The least that one can say is that there is no direct influence on the part of the government. Nevertheless, the SAP exerts tremendous influence on the economy, if only by taxing profits.

One of the major aspirations of SAP governments has been full employment. This ideal has not always been fulfilled in practice; most economies will, even in the best of times, be marked by pockets of unemployment, made up of people between jobs and people with no inclination to work. In the case of Sweden, the ambition to reach full employment has caused some cheating, in that the government has concealed unemployment by excluding those who registered for retraining programs as well as those who volunteered for emergency work (invariably on a temporary basis). It is obvious that these forms of "employment" actually are little more than waystations. But there may be a psychological gain: Laid-off workers continue to be in contact with the labor market.

Curiously, inflation has never loomed large as a government concern. Occasionally the inflation rate has soared while the government has done its utmost to fight unemployment. However, in general, the Swedish economy has surprised analysts, since there has been no inverse relation between inflation and unemployment.

In recent years, the Swedish economy has experienced annual economic growth of about 4 percent (well above the average for the 11 Euro Zone countries). In 2000, Sweden also enjoyed very low inflation in consumer prices: 1 percent. Unemployment, according to official statistics, has fallen to 4 percent, less than half of the average for the Euro Zone. Thus, counter to conventional neo-liberal wisdom, Sweden, where taxes are the highest in Western Europe and governmental expenditures represent 62 percent of the GDP, has reinvigorated its economy while retaining a comprehensive social-welfare system.

Sweden has its primarily agricultural past long behind it. It endorsed industrialization wholeheartedly at the beginning of the Industrial Revolution and is now predominantly an industrial and service economy. Only 2 percent of the workforce engage in agricultural pursuits; but, since the farming equipment and methods are extremely modern, these individuals are able to supply the vast majority of the country's food needs.

Traditionally, the Swedes have not been reluctant about embracing new technology. Certainly this has been the case with computers and the new information technologies. The government has energetically sought to extend the Internet "revolution" to the entire society, not just its highly educated professional strata. Surveys indicate that half of all Swedes are connected to the Internet. Stockholm boasts the highest number of Internet firms per capita of any place in the world. The Swedish cellular phone company, Ericsson, is a global trailblazer in wireless technology. In 2000, International Data Corporation ranked Sweden as the world's leading information society (the United States ranked second).

ENERGY AND THE ENVIRONMENT

On the receiving end of acid rain stemming from British smokestacks and other sources, Sweden became not only the site of the Stockholm Conference of 1972 (the historic first global environmental conference) but also fiercely protective of the environment at home and abroad. In 1980, in the aftermath of the Three Mile Island nuclear accident in the United States, Swedish voters approved a referendum to shut down all of the country's nuclear reactors by 2010. Sweden has depended on nuclear energy for around half of its electricity. The Conservative-led government in 1991 favored closing nuclear plants only after their operational lifetimes—a view that has found support among many industrialists and trade unionists worried about the availability of cheap alternative fuels. However, Prime Minister Persson, backed by the antinuclear Center and Green Parties, moved to shut down the first plant in 1999 *and* to reduce greenhouse gases as required by the 1997 Kyoto Protocol. In addition, the government has required "eco-labeling" of products, moved toward banning the landfill disposal of combustible wastes, protected old-growth forests, and passed a unified environmental code that rolls together 15 major laws. Persson's overall goal is to achieve sustainable development in Sweden within one generation.

DEVELOPMENT

The Swedish economy is characterized by extensive cooperation among the state, private enterprises, and trade unions. About 90% of industrial output is accounted for by private enterprises. Sweden is heavily dependent on foreign trade for its prosperity. Sweden is one of the world's leading exporters of iron ore. It is also one of the world's leading information societies.

FREEDOM

Sweden is a constitutional monarchy and a parliamentary democracy. It is an egalitarian society in which respect for individual rights is accorded top priority. This respect for individual rights has been enhanced by the institution of the ombudsman. Sweden is top in the representation of women in Parliament (2000).

HEALTH/WELFARE

Health, welfare, pensions, social insurance, job training, and education are all covered by Sweden's cradle-to-grave system of benefits and services. Infant mortality in Sweden is exceptionally low.

ACHIEVEMENTS

Sweden was the first country to institute the office of ombudsman, which is designed to protect the individual from the government. Cultural activities of all types are subsidized by the state, a fact that has contributed to Sweden's already rich culture. To ensure editorial diversity, the state subsidizes many newspapers.

Switzerland (Swiss Confederation)

GEOGRAPHY

Area in Square Miles (Kilometers):
15,941 (41,288) (about twice the size of New Jersey)

Capital (Population): Bern (129,000)

Environmental Concerns: air and water pollution; acid rain; loss of biodiversity

Geographical Features: mostly mountains (Alps in the south, Jura in the northwest), with a central plateau of rolling hills, plains, and large lakes; land-locked

Climate: temperate, but varies with altitude

PEOPLE

Population

Total: 7,262,500

Annual Growth Rate: 0.3%

Rural/Urban Population Ratio: 39/61

Major Languages: German; French; Italian; Romansch; others

Ethnic Makeup: 65% German; 18% French; 10% Italian; 7% Romansch and others

Religions: 47% Roman Catholic; 40% Protestant; 5% others; 8% unaffiliated or not indicated

Health

Life Expectancy at Birth: 77 years (male); 83 years (female)

Infant Mortality Rate (Ratio): 4.5/1,000

Physicians Available (Ratio): 1/592

Education

Adult Literacy Rate: 100%

Compulsory (Ages): 7–16

COMMUNICATION

Telephones: 4,800,000 main lines

Daily Newspaper Circulation: 418 per 1,000 people

Televisions: 370 per 1,000 people

Internet Service Providers: 115 (1999)

TRANSPORTATION

Highways in Miles (Kilometers): 44,378 (71,118)

Railroads in Miles (Kilometers): 2,785 (4,492)

Usable Airfields: 67

Motor Vehicles in Use: 3,640,000

GOVERNMENT

Type: federal republic

Independence Date: August 1, 1291

Head of State/Government: President Moritz Leuenberger is both head of state and head of government

Political Parties: Radical Free Democratic Party; Social Democratic Party; Christian Democratic People's Party; Swiss People's Party; Swiss Liberal Party; Green Party; many others

Suffrage: universal at 18

MILITARY

Military Expenditures (% of GDP): 1.2%

Current Disputes: none

ECONOMY

Currency ($ U.S. Equivalent): 1.79 Swiss francs = $1

Per Capita Income/GDP: $27,100/$197 billion

GDP Growth Rate: 0.3%

Inflation Rate: 1%

Unemployment Rate: 1.7%

Labor Force: 3,800,000

Natural Resources: hydropower; timber; salt

Agriculture: fruits; grains; vegetables; meat; eggs

Industry: machinery; chemicals; watches; textiles; precision instruments; tourism

Exports: $98.5 billion (primary partners EU countries, United States, Japan)

Imports: $99 billion (primary partners EU, United States, Japan)

http://www.admin.ch/ch/index.en.html
http://www.odci.gov/cia/publications/factbook/geos/sz.html
http://www.yoodle.ch//en/defaultClassical_800x600.asp

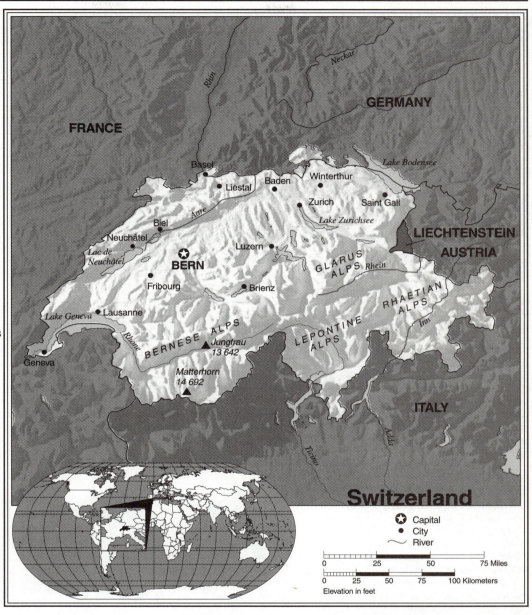

Switzerland

⭐ Capital
● City
〰 River

Elevation in feet

THE SWISS CONFEDERATION

In the words of the country's official Web site, "Switzerland is a nation shaped by the resolve of its citizens: it is not an ethnic, linguistic or religious entity."

In 1291, three tribal chieftains from the areas that later came to be known as the cantons of Schwyz (from which the German name for Switzerland, *Schweiz*, derives), Uri, and Unterwalden, met and signed what might be called a collective-defense treaty. By implication, this treaty amounted to a proclamation of independence from the Habsburg Empire, which hitherto had taken care of security in its own fashion. The town of Lucerne was invited two decades later to join the loose confederation. The centuries that followed did not bring peace to the new structure, or even constant goodwill among its inhabitants.

Although it is exceptionally beautiful, the country has very few natural resources—so few, in fact, that for several centuries a great many Swiss men found a livelihood by leaving the country and hiring themselves out as mercenaries. Europe always had plenty of wars, and it was not hard to become a soldier in a foreign army. The Swiss guards in the papal palaces in the Vatican constitute a relic of what used to be a widespread practice. Each canton guarded its independence fiercely and refused to surrender any of its sovereignty to the intercantonal authority.

This condition changed, however, when the Swiss Confederation suffered defeat at the hands of the French in 1515. The Confederation states then agreed to coordinate their external relations and to pursue a common objective of optimal independence from other European states. An effective guarantee of the Confederation's independence was achieved as early as 1521, when members signed a treaty with France. In this accord, which was to survive until the French Revolution, the Swiss states promised to provide soldiers and arms for the French king; in return, they were to receive the protection of the French Army as well as free access to French markets.

Whereas defeat in battle had produced greater unity in the area of foreign policy, the Reformation of the sixteenth century tore the Swiss Confederation apart. Switzerland, notably Geneva, became a haven for Protestant refugees, while at the same time its inhabitants largely remained Roman Catholic. After more than a decade of religious violence, pitting the Catholic rural cantons against their Protestant urban counterparts, the Peace of Kappel in 1531 forged a settlement that divided the country into Protestant and Catholic cantons. All citizens of a canton were to share the same faith. Cantons in which the denominations were equally represented were cut into halves.

In addition to identifying the country–city faultline as a religious divide, the Reformation also dramatically altered the ethnolinguistic composition of the Swiss Confederation. The immigration of thousands of French Protestants (Huguenots) fleeing persecution in Catholic France helped to transform Swiss society from a mono- to a multiethnic one. Moreover, the Huguenots, many of whom were accomplished merchants, bankers, and manufacturers, helped to transform the economic base of the Confederation, enabling it to pioneer the Industrial Revolution on the European continent.

The year 1648 provided an important milestone, in that the Peace of Westphalia decreed that the Swiss Confederation would henceforth be independent of the Holy Roman Empire. The next major transformation in Switzerland's political organization resulted from the French Revolution and its aftermath. During the Napoleonic Wars, the Confederation was temporarily absorbed into the French empire. However, the brief humiliation of French occupation was more than compensated for by the territorial gains and constitutional innovations enjoyed by the Swiss Confederation under Napoleon Bonaparte's sovereignty. The French emperor nearly doubled the territory of the confederation by adding six new cantons: one French-speaking, one Italian-speaking, three German-speaking, and one Romansch-speaking.

Napoleon unsuccessfully tried to mold the Confederation into a highly centralized republic and then decided to provide the Swiss with the foundations for a federal democratic constitution, with the Act of Mediation in 1803. This act was inspired by the U.S. Constitution. It combined federalism, popular sovereignty, the separation of powers, and a bill of rights with a central government in charge of foreign policy.

Napoleon was defeated in 1815, and the Congress of Vienna devoted little time to the position of Switzerland. Nevertheless, some of its decisions were of great significance. The Congress recognized Switzerland's military/political neutrality, which has since been a tradition. More important, it completed the Swiss Confederation's territorial expansion by adding two more French cantons to its jurisdiction. In 1848, on the eve of the autocratic retrenchment throughout most of Europe, Swiss advocates of a democratic national constitution finally gained control of sufficient cantons to transform the cantonal alliance into a unified nation-state with an official federal Constitution and institutions. Thus, Switzerland is the world's second-oldest federal union, after the United States.

(Photo courtesy of Cornelia Warmenhoven)

Switzerland has a long and colorful history. Often the country is remembered for its famous, some would say infamous, banking laws. The country, however, has many magnificent cities and towns. Pictured above is a bridge over the River Aare in the capital city of Bern.

Although significantly revised in 1874, in order to delineate and constrain the powers of the federal government vis-à-vis the cantons, Switzerland's democratic Constitution and institutions demonstrated considerable resilience in peacetime as well as during the two world wars in the twentieth century. Armed neutrality during the wars prevented not only Switzerland's occupation and destruction from without but also averted a potentially suicidal conflict among the country's French, German, and Italian ethnic groups. The period between the world wars presented the Swiss democracy with its most serious test, as extreme right- and left-wing political movements emerged. However, democratic institutions were able to weather the storm in Switzerland, which was not the case in neighboring Germany, Italy, and Austria.

Switzerland mobilized more than 400,000 soldiers and communicated its determination to defend its territory to the last man if invaded by Nazi Germany and Fascist Italy. With invasion deterred, it continued profitable trade with these Axis powers and steered a diplomatic course that would not antagonize either side (Axis or Allies) during World War II. In recent years, it has become clear that Switzerland itself benefited from the money, artworks, and other treasures that had been deposited in Switzerland for safekeeping—most of which have never been returned by the banks to the real owners, such as the Jews of Germany.

Following the war, Switzerland retained its neutral stance. Since the United Nations was founded by the victorious nations, Switzerland refused to join that organization. Indeed, in 1986, Swiss voters rejected a parliamentary proposal to join the United Nations, in the belief that membership would subvert the country's policy of neutrality. Switzerland, however, is hardly isolationist. For example, it has been a member of many UN specialized agencies, the European Free Trade Association, and the World Trade Organization.

A UNIQUE BRAND OF FEDERALISM

As this brief overview of Switzerland's political history indicates, the transformation of the loose cantonal alliance into a unified nation-state was a long, drawn-out process. The national Constitutions of 1848 and 1874 committed all cantons to the principles of democracy and conceded the new federal authority jurisdiction over defense, trade, and many legal questions. The cantons, nevertheless, made certain that all powers not expressly granted to the national government by the Constitutions remained in their own hands. Fur-

thermore, the cantons ensured that national institutions were organized in such a manner that any further encroachment on cantonal autonomy by the central government would be prevented.

Following World War II, Switzerland was unable to completely escape the global trend toward increased intervention by central governments in the management of social and economic life. However, the Swiss have been more successful than other Western democracies at keeping that trend in check. Today, as was the case a century ago, the loci of Swiss political loyalty and activity remain the canton and the commune (township). As a result, the Swiss national political institutions are extraordinarily weak for a modern democratic state.

Switzerland is composed of 20 cantons and six half-cantons, which, in turn, are made up of some 2,900 communes (townships). Both the cantons and their constituent communes enjoy wide independence from the central government, seated in Bern. Each canton and half-canton has its own constitution and is free to choose its own form of government and electoral system, as long as these are consistent with democratic principles and the federal Constitution. In 1978, the voters in the French-speaking part of canton Bern voted in favor of self-government. As a result, the canton of Jura was established in 1979.

Currently, all cantons are governed by a directly elected, unicameral legislature with a collegial executive. While most cantons have instituted similar electoral systems, which extend the vote to all citizens above the age of majority, one half-canton, Appenzell Outer Rhodes, enfranchised women only as recently as 1991. Women have been allowed to vote in national elections by a national law passed in 1972. The powers and responsibilities of the cantons are considerable. Each canton has its own taxing authority, law-enforcement facilities, and independent school system. For example, Zurich and other German-speaking cantons have decided to make English, not French, the second language of instruction in their primary schools. The education officials of French-speaking cantons see their move as a threat to Swiss national identity.

The communes' right to self-government is also guaranteed by the federal Constitution. However, they must submit to cantonal supervision in a number of areas. Like the cantons, each commune can choose its own form of government and electoral system, as long as they are compatible with the federal or cantonal constitutions. Communes are responsible for

administering utilities, roads, schools, and fire and police forces, among other things. Perhaps even more noteworthy is the exclusive power of communes to grant Swiss citizenship. In other federally structured countries, questions of nationality and citizenship are resolved by the national government.

FEDERAL INSTITUTIONS

The organization of the national government also provides the cantons with an effective check on the centralization of political power. The national Legislature is called the Federal Assembly. Its lower house, the National Council, is made up of 200 members directly elected to four-year terms in national elections, according to a system of proportional representation. The National Council's federal aspirations, however, are effectively contained by the co-equal upper house, the Council of States. Modeled on the U.S. Senate, the Council of States is comprised of 46 representatives—two from each canton and one from each half-canton. In 1999, 23 percent of the members of the National Council were women, while 19.6 percent of members of the Council of States were. (These figures put them ahead of Italy and France in the representation of women, but behind Germany and Austria.) All Swiss parliamentarians are part-time politicians who must earn their own living.

Cantons vary in size from about 36,000 to more than 1 million inhabitants. Since both houses must approve all bills before they become law, the Council of States can block legislation that it considers inconsistent with the interests of a majority of cantons. The conservative bias of the upper house is reinforced by the fact that the more rural, less densely populated cantons have a disproportionately large political voice in it. Yet another check on federal excess is the legal requirement that constitutional amendments be approved by a "double-majority"—a majority of voters in a national referendum and a majority of the cantons.

The executive branch of the federal government, the Federal Council, is a collegial body with seven councillors, who are elected by the Federal Assembly (both houses) for four-year terms. Since 1959, when four leading political parties entered into a grand coalition, the seven Council posts have been distributed according to a formula that reflects the relative strength of the four parties in Parliament, while ensuring that the country's French and Italian ethnic minorities are represented. In 1984, a woman was elected to the Federal Council for the first time; in 2000, three of the councillors were women. The president of the Confederation, first-among-equals, is elected by the Federal Assembly from among the councillors for a one-year nonrenewable term. His or her role is to chair council meetings and to perform ceremonial and representative roles as the Swiss head of state.

Decision making within the executive body, as in most other branches of Swiss federal, cantonal, and communal government, is a collective process characterized by compromise and consensus building. The Federal Council is not collectively or individually responsible to Parliament; that is, it cannot be disbanded or voted out of office in mid-term. However, its powers are quite limited. For example, the Council can neither dissolve Parliament nor veto parliamentary bills. The Federal Court of Justice, with its 39 judges, elected by the Federal Assembly for renewable six-year terms, can declare cantonal law to be unconstitutional, but not statutes passed by the Federal Assembly. Unlike the U.S. federal judiciary, the Swiss federal judiciary has no lower-district and appeals courts.

THE PARTY SYSTEM

The weakness of national political institutions is demonstrated also by the Swiss party system. While political parties are relatively effective at channeling and representing the public will at the cantonal and communal levels, they are loosely organized and lack discipline at the national level. Moreover, the major national parties, which serve as umbrella organizations for their cantonal affiliates, have essentially been captured by the country's major organized economic interest groups (labor, business, agriculture, etc.).

One explanation for the weakness of Swiss political parties is the fact that the multiple divisions within the society (religious, ethnic, class, and regional) cut across and mute one another and thus have tended to de-ideologize the electorate. Other factors working to undermine party politics include the consensual style of Swiss politics and the tradition of direct democracy via referendum at the national and cantonal levels, while, on exceptional occasions, there is also public decision making at the *Landesgemeinde* level. (The *Landesgemeinde* is like the New England town meeting, where people come together to decide directly on issues.)

Since 1959, the four largest Swiss parties, have collaborated to dominate both houses of Parliament and the Federal Council. Until 1995, the largest of these, the Radical Free Democratic Party (FDP), typically won about a quarter of the vote in parliamentary elections. The FDP is a

(UN photo by M. Vanappelghem)

Switzerland has often needed to import labor in order for Swiss manufacturers to expand. Companies that elect to remain in Switzerland rather than setting up operations in other countries must often use so-called guest workers, such as those pictured here, to stay in business.

status-quo party with close ties to industry, finance, and the media. Two other coalition parties, the Christian Democratic People's Party (CVS) and the Social Democratic Party (SPS), also have controlled large blocks of seats in Parliament. The former is slightly left of the FDP, drawing most of its support from Swiss Catholics. The latter, traditionally the party of the working class, is further to the left but has essentially abandoned the Socialist platform of its founders. During 1959–1999, the smallest member of the coalition was the Swiss People's Party (SVP). It has traditionally represented the most conservative elements of Swiss society, including small-business people, farmers, and artisans.

Since forming the grand coalition, the four member parties have converged along the political spectrum. Their combined share of the national vote has never fallen below 73 percent. They have, therefore, never experienced serious opposition in Parliament from a smattering of smaller parties—some moderate, some more extreme. While the overwhelming majority of the four governing parties has been good for consensus politics, their convergence has had a debilitating effect on national political life. As a result, voter turnout for national elections has fallen to

extremely low levels by European standards in recent years.

For the first time in several years, national voter turnout actually went up in the federal elections of 1999 (43.3 percent as compared to 42.2 percent in 1995). By Swiss standards, the party results represented an "earthquake." The SVP did even better than expected by gaining an additional 7.5 percent of the votes. This translated into 22.54 percent of the votes, making it the biggest electoral party and the second-largest parliamentary group, with 51 total seats in the National Council and the Council of States. The SVP's coalition partners, the SPS, the FDP, and the CVP, won 59, 61, 50 seats, respectively. Among the opposition parties, the Green Party (GPS) did the best in winning 10 seats in the National Council (but none in the Council of States). The remaining 15 National Council seats were split among seven minor parties.

The Zurich wing of the SVP, led by prominent "Switzerland-by-itself" nationalist Christoph Blocher, seized upon the results as the grounds for challenging the "magic formula" that has for the past four decades allotted the SVP only one seat in the Federal Council, as compared to its partners' two each. The SVP's campaign benefited from the media skills of Blocher in projecting himself as the underdog champion of the ordinary Swiss against the leftist establishment and foreigners. However, public-opinion polls indicated that a clear majority of Swiss favored retaining the "magic formula," and the party elites subsequently did so. In 2000, when the SVP's one (moderate) representative on the Federal Council resigned, the SVP candidates favored by Blocher were unable to obtain the necessary majority to be elected in the Federal Assembly. Thus, another SVP moderate was elected.

DIRECT DEMOCRACY

A distinctive reliance on direct democracy via popular initiatives and referenda at the national level has lent considerable legitimacy through the years to the Swiss political system. A popular initiative to change the federal Constitution requires 100,000 signatures to compel a referendum. Then, to pass, the referendum must obtain both the majority of the national votes and the majority of the votes in a majority of the cantons; historically, only a small minority of popularly initiated constitutional revisions have met these conditions.

With 50,000 signatures, citizens can challenge bills (and treaties) passed by the Federal Assembly in a referendum; similarly, new laws can be proposed by referendum. There are also referendums at the canton level. By threatening to mount a direct democratic campaign, interest groups can force governmental attention on their special concerns. In an average year, Swiss citizens vote about four times, on about 20 different issues. Turnout is usually around 40 percent, so majority decisions may reflect the views of little more than one fifth of the electorate. However, 69 percent of the voters turned out in a 1989 referendum on the abolition of the Swiss Army (easily defeated), and 79 percent turned out in a 1992 referendum on Swiss membership in the European Economic Area (narrowly defeated). Only around half of the proposals put directly to the voters win their approval. Also, direct democracy is not cheap; one source estimates that each national referendum costs the government $3 million. Observers have wondered aloud about whether direct democracy is up to handling complex issues, such as financial policies. On the other hand, years may pass between when sufficient signatures are collected and when the issue is voted, thus allowing ample time for the issue to be publicly discussed by the contending groups and individuals. Swiss direct democracy hardly involves precipitous decision making.

INTEREST GROUPS

The decentralization of the Swiss political system has been quite effective at reinforcing the bonds of community at the cantonal level as well as defusing and containing intercantonal disputes. Yet, in view of the anemic nature of Swiss parties and political institutions at the national level, one might assume that the Swiss find it difficult to act decisively on important national policy issues, other than those regarding foreign policy and defense of cantonal rights. In fact, the Swiss have proven that they are quite capable of formulating and instituting effective national economic and social policies.

To explain the paradox of strong federal governance without a strong state or strong political parties, one must pay attention to Switzerland's highly organized system of interest-group representation. Moving into the political vacuum at the federal level, a number of nationally organized interest groups have assumed many of the powers and responsibilities usually associated with national political parties and central government.

The most influential of Swiss interest groups are those that represent economically defined national constituencies, such as labor, business, agriculture, and finance. Their power derives from the fact that they have a monopoly of representation within their economic sector, encompass a broad section of Swiss society, are hierarchically organized and hence relatively disciplined, and are granted quasi-governmental status by the federal Constitution. It stipulates that they must be consulted by the government and the major political parties on all legislation concerning their interests.

In practice, however, the influence of these private interest groups on policy making and policy implementation is much greater than the Constitution would suggest. Their penetration into the national political parties has assured the major interest groups de facto representation in Parliament. Furthermore, the constraints placed on the central government by the Constitution and the small size of the federal bureaucracy have allowed the private interest groups to assume many of the functions of national government.

For the most part, organized labor, business, and finance regulate themselves. They have also collaborated with one another to develop a private social-welfare system on the national level. This system has included, among other things, unemployment insurance, health insurance, and private-pension schemes for the Swiss workforce.

ECONOMIC FOUNDATIONS OF THE POLITICAL CONSENSUS

The result of Swiss political and industrial stability has been an economic prosperity envied for many years throughout much of Western Europe. The fact that Switzerland was able to stay out of World War II granted the country a head start. Blessed with peaceful industrial relations, a skilled workforce, and abundant capital, the Swiss economy grew at a rapid rate for the first two decades after the war. Inflation was low and unemployment virtually nonexistent.

During the 1970s and early 1980s, however, Switzerland's main economic indicators turned sour, along with those of much of the rest of industrialized Europe. After the second oil shock in 1979, economic growth stagnated and inflation rose to nearly 7 percent. By the mid-1980s, however, things had begun to turn around again. Economic growth resumed, albeit at a lower rate, and inflation began to fall. Switzerland's economic growth rates began to compare favorably to those of most of its Western European neighbors.

The pillars of the Swiss economy are a large international banking and financial-services sector and a highly competitive and diversified industrial sector. Favorable banking laws with strict secrecy requirements, a stable currency, and a minimum of government regulation have allowed Switzerland to develop into one of the world's leading centers of international banking and insurance. In regard to the renowned secrecy, one must note that the Swiss Supreme Court ruled in 1971 that Swiss banks must show U.S. tax officials their records of U.S. citizens suspected of tax fraud. Secrecy was further compromised when the drug trade started to become rampant throughout Europe. In 1998, Parliament passed anti–money-laundering laws that extend new controls from banks to all "para-banking" organizations that handle money, even lawyers and trustees. Staffing problems have thus far hampered the implementation of the laws. On the other hand, in recent years, magistrates in Geneva have shown great zeal (not matched in other cantons) in tracking down money from crooked sources, even without foreign prodding.

Swiss industry has prospered since World War II, thanks to well-developed domestic capital markets, a relatively docile yet highly skilled labor force, and virtually unlimited access to the prosperous markets of Europe and North America. Precision engineering and timing devices have long been Swiss strengths; other important export industries include chemicals, pharmaceuticals, and heavy engineering.

Due to a limited domestic market, trade has played a particularly important role in Swiss industrial growth. The government has been aggressive at negotiating bilateral trade treaties and other agreements designed to open markets to Swiss exports. Switzerland was a founding member of the European Free Trade Association in the late 1950s, and, though not a member of the European Union, it has cultivated extensive commercial relations with EU member countries. In 1999, 63 percent of Swiss exports went to EU countries, while 80 percent of Swiss imports came from EU states.

The most serious constraint on Swiss industrial development has been the country's perennially limited supply of labor. In order to expand, Swiss manufacturers have either had to import foreign labor or locate more and more of their production facilities abroad. Since the late 1950s, the number of foreign "guest workers" in Switzerland (mostly from Southern Europe) has expanded rapidly. At the same time, Swiss industry has been in a good position to pursue multinational strategies of direct foreign investment abroad.

Switzerland has made the transition from an industrial to a postindustrial society. The rural areas have lost population; urbanization has proceeded steadily. Only a little more than half a century ago, in 1941, the agricultural sector accounted for 21 percent of those in employment, but by 1970, that proportion had sunk to 8 percent. It is now under 3 percent. The number of persons employed in industry and services rose markedly, although some 200,000 jobs were lost in the recession years, predominantly in industry. The proportion of individuals employed in services started to exceed 50 percent (today that percentage is more than 66 percent), and Switzerland thus became a postindustrial society. Currently only about 31 percent of the Swiss workforce are employed in industry; most people are engaged in services, including banking, finance, and marketing. In 2000, the Swiss economy grew at a healthy rate of 3 percent. Consumer price increases (1.6 percent over 1999) were lower than in the United States, Britian, and the Euro Zone. The early 2001 unemployment statistics indicated that only 1.7 percent of Swiss workers were unemployed, the lowest level in Europe.

IMMIGRATION
The Swiss government has taken steps to reduce the inflow of foreign workers. Some of these measures were caused by public apprehension concerning the phenomenon of *Überfremdung*—a German

term that may be translated as "overalienization." Switzerland is a small country, with a population of only about 7.26 million. In 1945, foreigners made up only 5 percent of the resident population; by 2000, this proportion had climbed to 19.3 percent, among the highest in Western Europe. Taking into account that Switzerland has not just one homogeneous culture but, rather, several different cultures, it was feared that large waves of foreigners would seriously dilute all that was typically Swiss.

A series of referenda were prepared to curb the influx of aliens and to permit the forcible removal of those whose work contracts had expired. Although the most far-reaching referenda were ultimately rejected, many Swiss have remained more than a little sensitive on this point. (The stormy protest movement, it should be noted, was most virulent in Switzerland's German-speaking areas.) The Swiss government feared that the country might eventually be stigmatized as racist by the outside world, and endeavored to regulate foreign immigration to prevent such referenda. During the economic recession of 1974–1976, tens of thousands of foreign workers had to leave for their homelands. In spite of these massive departures, apprehension continued, and in 1981 another referendum was held, this time against a law that was to improve the legal status of the foreign worker. That too was defeated.

In September 2000, Switzerland drew international media attention when its voters went to the polls to vote on a proposal, initiated by a Radical Party (FDP) maverick, to cap the country's foreign population at 18 percent. (In recent months, many Swiss linked increases in crime to the influx of 140,000 Albanian refugees from Kosovo.) With the current percentage at 19.3, this proposal would have forced some foreign residents to leave. Sixty-four percent of the voters rejected the referendum proposal. The Swiss political and economic leadership had strongly opposed the proposal. One fourth of the workforce are foreigners, and even more will be needed in the future to compensate for the low birth rates of native Swiss. Furthermore, foreigners now contribute twice as much into the national pension fund as they withdraw. The outlook is for new laws that restrict future immigration by professional qualification and that encourage the integration of immigrants already in Switzerland. Fifty-nine percent of foreign residents in the country are from EU countries, and the implication of draft proposals is that tightening restrictions will affect mainly non–EU immigrants.

INTERNATIONAL RELATIONS
Since World War II, Switzerland has continued to adhere strictly to the precepts of armed neutrality. All male Swiss citizens must perform a year of compulsory military service. Thereafter, each soldier becomes part of the country's large citizens' militia and participates in periodic refresher training exercises. Little Switzerland, with 360,000 soldiers, has proportionally one of the largest armies in the world. Despite its militia tradition, Switzerland's military possesses state-of-the-art weaponry, such as Leopard 2 tanks and F/A-18 fighter jets. Critics point out that in terms of cost per soldier, the Swiss Army is one of the most expensive in the world. While other Western European countries in the post–Cold War era have generally cut their military expenses by one third, Switzerland has cut back only one tenth. Supporters argue that armed conflicts (some not far away from Switzerland) are also part of the post–Cold War world and that Swiss independence necessitates defense commitments. Furthermore, military budget cuts would jeopardize thousands of civilian jobs. In a November 2000 referendum, 62 percent of voters rejected a proposal to cut the military budget by one third over 10 years.

In accordance with absolute political neutrality, the Swiss are committed to maintaining diplomatic and economic relations with all countries of the world, regardless of the type of political regime they have or the foreign policies they conduct. Thus, though clearly of a Western economic and political orientation, Switzerland has refused to participate in any regional or international organization that might require it to take sides in international disputes. In 1996, despite its traditional neutrality, Switzerland allowed NATO troops to pass through its territory on the way to Bosnia for peacekeeping deployment. Switzerland also joined NATO's "Partnership for Peace" framework for cooperation.

Switzerland's traditional neutralism has not led to the country's disengagement from the international system. Although it has never been a member of the United Nations, it has maintained "permanent observer" status and has been active in a number of UN specialized agencies, many of which have their headquarters in Switzerland. In its "Foreign Policy Report 2000," the Federal Council made clear that its goal is to become a full member of the United Nations by 2003, in order to strengthen the country's international role. Switzerland has become a donor nation (but not a member) of the International Monetary Fund and the World

Switzerland emerges as an "oath association" of three cantons A.D. 1291	A treaty with France guarantees Swiss independence 1521	The Peace of Kappel divides Switzerland into Roman Catholic and Protestant cantons 1531	The Peace of Westphalia; Switzerland becomes officially independent from the Holy Roman Empire 1648	Switzerland is absorbed by the French Empire; Napoleon decides to enlarge the Swiss territory 1799–1804	A democratic national Constitution, endeavoring to counter centralization, gives major power to the cantons 1874	Switzerland manages to remain neutral during two world wars 1914–1918; 1939–1945	Women are accorded the vote in national elections 1972	East–West arms-control talks are based in Switzerland 1980s	Scandal explodes over Swiss banks' ties to Nazi Germany 1990s

2000s

Switzerland deals with its role in the Holocaust

Bilateral sectorial agreements are to be implemented to draw Switzerland and the EU closer together

Bank. The headquarters of the World Trade Organization and European Free Trade Association are in Geneva, which has hosted numerous international conferences, such as East–West arms-control talks. Important international nongovernmental organizations, such as the World Council of Churches and the International Committee of the Red Cross, are also based there. Swiss diplomats frequently offer their "good offices" to nations that are estranged from each other or lack diplomatic representation of their own in certain parts of the world.

Switzerland has sometimes been a source of irritation to other European countries. In the mid-1980s, for example, a large pharmaceutical company spilled toxic waste into the Rhine River. The pollution killed thousands of fish, and West Germany and the Netherlands, which were grievously affected, embarked upon lengthy lawsuits, which ultimately were settled out of court. Also, banking practices do not always incur approval, particularly if the customers happen to be unscrupulous dictators who have enriched themselves through kleptocracy.

In the mid-1990s, it was revealed that billions of dollars in assets had never been returned by Swiss banks to Jewish Holocaust survivors and their families. As the scandal blossomed, other aspects of Switzerland's relationship with Nazi Germany were scrutinized on the world stage, and the country's once pristine reputation was further tarnished. Switzerland had always denied that any of the approximately 25,000 Jewish refugees who were permitted to enter the country during the war, along with other non-Swiss Jews, some of whom had lived in Switzerland for years, were subjected to forced labor. But the mounting body of evidence, supported by recently released official documents and backed by victims' statements, makes it clear that a network of more than 100 work camps was established by an official

decree on March 12, 1940. Those who were interned do not equate the Swiss labor camps with the Nazi concentration and death camps, but they do say that Jews were held, against their will, in harsh conditions.

In 1997, Swiss companies and the Swiss Central Bank settled claims by setting up a $183 million fund to benefit needy victims of the Holocaust. In 2000, a U.S. federal judge gave final approval to a $1.25 billion settlement of legal claims of owners of dormant bank accounts, slave laborers, and refugees denied entry, to be paid by Swiss banks and companies. The country's moral accounting continues. A late 2000 report implicated Swiss officials, who expelled or denied refuge, in the deaths of thousands of Romani (Gypsies) at the hands of the Nazis.

THE SWISS AND EUROPE
In the early 1990s, the European Economic Area emerged as a halfway point for the small affluent democracies of the EFTA seeking entry into the European Union. While Sweden, Switzerland, and Austria moved from the EEA to the EU in 1995, Swiss voters had stopped their government's pursuit of membership by narrowly rejecting (50.3 percent) the EEA in a 1992 referendum. The Federal Council then shifted to a strategy of seeking sector-by-sector agreements with the European Union for closer cooperation. In 1999, after four years of negotiations, seven bilateral sectorial agreements were signed, covering civil aviation, overland transport, free movement of persons, research, public procurement, agriculture, and elimination of technical barriers to trade. Both houses of the Parliament overwhelmingly approved them, but sufficient signatures were collected by opponents to set the stage for the May 2000 referendum on the agreements. Sixty-seven percent of the voters—more than expected— approved the agreements more closely

linking the country's economy to the European Union. The "free movement of persons" agreement requires the approval of the European Parliament and the 15 EU member states; only then will the legally interlocked set of seven agreements come into force (anticipated after mid-2001). Before this can occur, Swiss voters are scheduled to make another trip to the polls to vote on a pro–EU initiative group's "Yes to Europe" proposal, which would force the government to begin negotiations for full membership. The Federal Council opposed the proposal (which was soundly defeated by voters in early 2001) because it has favored waiting until the next parliamentary session (2003–2007), by which time the public may recognize the benefits reaped from the sectorial agreements.

Europhiles feel that the strength of the positive vote on these agreements indicates that the corner has been turned in relations with the European Union. Euroskeptics, on the other hand, see the agreements as providing economic advantages without any loss of national sovereignty. Thus, the Swiss appear ambivalent on "Europe," with many French-speaking, urban, and younger Swiss welcoming further European integration; and many German-speaking, rural, and older Swiss thinking that the optimal deal has already been struck with the EU.

DEVELOPMENT

Switzerland has made a rapid transition from an agricultural society to a postindustrial society. Its glowing worldwide reputation has been tarnished by revelations of its bank ties and other connections to Nazi Germany.

FREEDOM

In spite of the cultural differences within Swiss society, there is little overt discrimination. The country has a very favorable human-rights rating.

HEALTH/WELFARE

Switzerland has high life expectancy and literacy rates, and its welfare system is sound. The WHO 2000 survey rated Switzerland's overall health-system attainment as second in the world only to that of Japan.

ACHIEVEMENTS

Switzerland plays a disproportionately prominent role in international affairs, often assisting in mediating efforts. It hosts numerous international organizations, conferences, meetings, and summit talks.

United Kingdom
(United Kingdom of Great Britain and Northern Ireland)

GEOGRAPHY

Area in Square Miles (Kilometers): 94,251 (244,111) (about the size of Oregon)

Capital (Population): London (7,074,000)

Environmental Concerns: air and water pollution

Geographical Features: mostly rugged hills and low mountains; level to rolling plains in the east and southeast

Climate: temperate

PEOPLE

Population

Total: 59,512,000

Annual Growth Rate: 0.25%

Rural/Urban Population Ratio: 9/89

Major Languages: English; Welsh; Scottish form of Gaelic

Ethnic Makeup: 81.5% English; 9.6% Scottish; 2.4% Irish; 1.9% Welsh; 4.6% other

Religions: 76% Anglican; 14% Roman Catholic; 5.5% Presbyterian; 2.5% Methodist; 2% other

Health

Life Expectancy at Birth: 75 years (male); 80 years (female)

Infant Mortality Rate (Ratio): 5.6/1,000

Physicians Available (Ratio): 1/629

Education

Adult Literacy Rate: 100%

Compulsory (Ages): 5–16

COMMUNICATION

Telephones: 32,000,000 main lines

Daily Newspaper Circulation: 383 per 1,000 people

Televisions: 612 per 1,000 people

Internet Service Providers: 364 (1999)

TRANSPORTATION

Highways in Miles (Kilometers): 230,394 (371,603)

Railroads in Miles (Kilometers): 10,464 (16,878)

Usable Airfields: 498

Motor Vehicles in Use: 29,000,000

GOVERNMENT

Type: constitutional monarchy

Independence Date: January 1, 1801 (the United Kingdom established)

Head of State/Government: Queen Elizabeth II; Prime Minister Tony Blair

Political Parties: Labour Party; Conservative and Unionist Party; Liberal Democrats; Welsh National Party; Scottish National Party; Ulster Unionist Party; Democratic Unionist Party; Social Democratic and Labour Party; Sinn Fein; Alliance Party; Green Party

Suffrage: universal at 18

MILITARY

Military Expenditures (% of GDP): 2.7%

Current Disputes: Northern Ireland issue; dispute with Spain over Gibraltar; other territorial disputes with Argentina, Mauritius, and Antarctica; Rockall continental shelf dispute

ECONOMY

Currency ($ U.S. Equivalent): 0.70 pound = $1

Per Capita Income/GDP: $21,800/$1.29 trillion

GDP Growth Rate: 2.6%

Inflation Rate: 2.3%

Unemployment Rate: 5.1%

Labor Force: 29,200,000

Natural Resources: coal; petroleum; natural gas; tin; limestone; iron ore; salt; clay; chalk; gypsum; lead; silica; arable land

Agriculture: wheat; barley; potatoes; vegetables; livestock; poultry; fish

Industry: production machinery; aircraft; petroleum; metals; food processing; paper and paper products; textiles; chemicals; clothing; motor vehicles; tourism; others

Exports: $271 billion (primary partners EU, United States)

Imports: $305.9 billion (primary partners EU, United States)

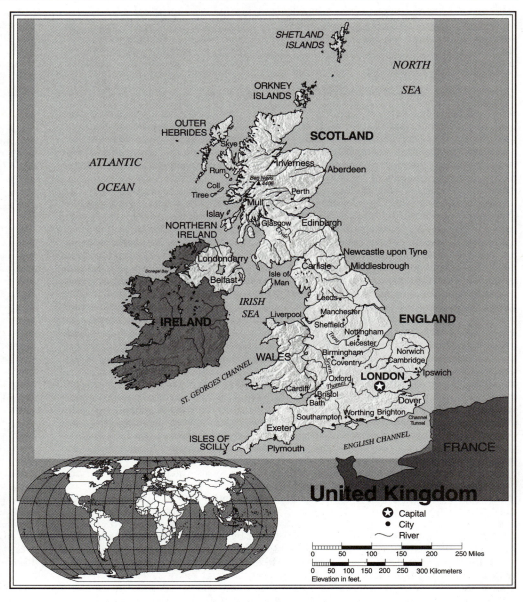

THE UNITED KINGDOM

Although this country is variously called "England" (which in fact refers only to the larger part of the main island) or "Britain" and "Great Britain" (the combined area of England, Scotland, and Wales), today its official designation is "United Kingdom of Great Britain and Northern Ireland" (which includes that part of Ireland that was retained by the British). Nearby are the Isle of Man and the Channel Islands, which enjoy a separate territorial status, being governed by lieutenant-governors on behalf of the British sovereign. The Channel Islands are closer to France, and their inhabitants speak a dialect that is closer to French than to English. The islands are the only area officially under British control that was occupied by the Germans during World War II.

HISTORY

Throughout its rich and eventful history, Britain has enjoyed the strategic advantage of being an island. The natural barrier that the seas around it afford constitutes a protection against invasions. However, the barrier has also generated a degree of insularity that has often led to blatant ethnocentric attitudes. A classic example of this is the headline in the British press reporting a dense fog all over the English Channel, noting "Continent Isolated."

It is interesting that the natural barrier has proved much more effective in the last two centuries than was the case earlier. It was inadequate in Roman times, when the Roman general Agricola experienced little difficulty in crossing the waters between Gallia (as France was then called) and Britain, thus adding England to a vast empire in A.D. 43. When, some centuries later, the Roman Empire collapsed and its armies withdrew from the island, northern Germanic tribes, known as the Jutes, Angles, and Saxons, embarked upon large-scale invasions.

Then it was the turn of the Vikings, who not only periodically raided the coastal areas but also established settlements. Indeed, for some time a large part of England served as an extension of Viking bases in Scandinavia. In the ninth century, a tribute known as *Danegeld* had to be paid to these usurpers.

Finally, in 1066, William the Conqueror led the Norman Invasion. He too experienced no difficulty in crossing the English Channel, and he defeated King Harold in Hastings. However, this turned out to be the last successful invasion of Britain. Centuries later, both France's Napoleon Bonaparte and Germany's Adolf Hitler were frustrated in their attempts to subdue England.

The early invasions (by the Romans, Danes, and Normans) proved beneficial to England, with the blending of cultures proving enormously enriching. Shortly after the Norman Invasion, there were two cultures: that of the Anglo-Saxons (in a way the original inhabitants) and that of the new overlords, the Normans. The latter had by that time developed certain societal refinements, such as an aristocracy, and they tended to look down upon the rugged Anglo-Saxons.

After a few centuries, however, the distinction between conquerors and conquered faded and a new nation was born, which, in the Hundred Years' War, became quite powerful. This war, which lasted from 1337 until 1453 (thus longer than its name suggests), resulted from the complicated relationship between England and France, whose royal dynasties were closely related. The English claim to the French throne sparked the hostilities. The war consisted of fervent attempts, notably by King Henry V, to make English kings rule the kingdoms on both sides of the English Channel. Ironically, the war had the opposite effect, establishing two rival kingdoms that were in conflict with each other for prolonged periods. These conditions were to last until the eighteenth century, by which time both countries had built large empires. The Hundred Years' War itself also provided a measure of irony: Although England came out triumphant, it lost sizable territories across the Channel. However, England would retain numerous holdings on the European continent until the seventeenth century.

Britain experienced religious persecution from time to time as well. Although one might have expected that the establishment of its own state church would result in religious independence, a climate of tolerance was absent. But the overall quality of life in England was superior to that of continental Europe. There can be little doubt that the English royalty greatly contributed to these superior conditions. In particular, the powerful Tudor Dynasty (1485–1603) introduced a distinct English civilization. During that period, England became successful in trade; it became prosperous as a result, and, immediately following the vanguard in explorations and discoveries (Portugal and Spain), it embarked upon expansion. Whereas continental Europe remained bogged down in religious warfare (caused by the Reformation and its aftermath), England was relatively peaceful, especially after the Spanish Armada had been defeated. Indeed, the Elizabethan Age (1558–1603) was a period in which a very rich culture, most notably in literature, flourished.

The seventeenth century witnessed political difficulties. The monarchy was set aside, and Britain experienced republican rule for a brief time. After the Restoration by Charles II in 1660 and the subsequent ouster of his brother, James II, in 1688, resulting from the so-called Glorious Revolution, a foreign king, William III of Orange, was invited to sit on the British throne. This episode weakened the power of the monarchy and, conversely, strengthened the sovereignty of the British Parliament, which became even stronger in the

(Photo courtesy Lisa Clyde)

Stonehenge is at the center of one of the richest prehistoric landscapes in the world. It is generally considered to have been a place of worship. Over the millennia, it has gone through three distinct stages of evolution, the first taking place around 3000 B.C., the second in roughly 2000 B.C., and the third in approximately 1500 B.C. Ongoing archaeological investigations continue to seek its meaning.

NORTHERN IRELAND: AN INTEGRAL PART OF THE UNITED KINGDOM OF GREAT BRITAIN AND NORTHERN IRELAND

"Home rule is Rome rule." "A Protestant Parliament for a Protestant people." "Ulster will fight; Ulster will be right." Each of these phrases portrays the intensity of feelings of the Protestant community in Northern Ireland regarding the creation of an Irish state, with its government in Dublin, in the period 1870–1920. These same slogans have been heard in modern times, in response to both the terrorist tactics of the Irish Republican Army (IRA) and the decidedly more peaceful meetings between the prime ministers of the Irish Republic and Britain.

The history of this rather small region (less than 17 percent of the island from which it has been carved) has not differed markedly from that of the remainder of the island—that is, the Republic of Ireland, or Eire. Since there was no boundary or other type of demarcation, the vicissitudes of the two parts were very similar, at least until an accident of history caused Northern Ireland to remain an extension of Britain.

In the distant past, this British foothold was—like the entire island, other British isles, and various areas on the continent—inhabited by Celts, whose presence in Western Europe has generated a great deal of speculation. Their religion, which is identified as Druidism, seems to have included human sacrifice. However, in A.D. 432, Patrick, who subsequently became the patron saint of Ireland, arrived and embarked upon the conversion of Celts to Christianity, a labor of love that was completed by his successors.

Danish invasions took place a few centuries later. While these raids produced few permanent settlers (as they had in England, where Danelaw became the system of government), the eastern part of the island continued to be colonized from south Wales, an Anglo–Norman stronghold.

It being closer to the European continent, England was much more subject to population waves than was Ireland, both before and after the year 1066 (the time of the Norman Invasion). Furthermore, the differences between the two regions became more pronounced after the Reformation made itself felt throughout Europe. In England, the Reformation met with a degree of success even before King Henry VIII created the Church of England, whose Protestant creed became the official state religion. However, Ireland's relative isolation caused it to remain unaffected by the Reformation; it remained almost universally Roman Catholic. By that time, the island had come under English domination, but the English overlordship was fairly tolerant with regard to differences in religion.

But early in the seventeenth century, a large "plantation," consisting of nine counties in the north jointly called Ulster, was opened to Presbyterian Scots and other Protestant groups in northern England. Most of these people had been eking out a living in the barren parts of northern England and northeastern Ireland, so it appeared to grant them new opportunities. However, as time went on, the Ulster plantation gradually became an enclave and its society a distinct minority. In religious respect, the Ulster settlers were surrounded by Roman Catholics. This led to the emergence of a "fragment mentality"—a stridency that derives from being vastly outnumbered.

As might be expected, relations between the Protestants and the Roman Catholics were already poor during the time when the greater part of Western Europe was embroiled in religious warfare. But since the Protestants had been imported and were in effect part of the occupation, they remained under the protection of the British.

As if religious segregation were not enough of an irritant, class distinctions emerged, with the Protestants being better off on the occupational and social scales. They made no attempts to assimilate or to integrate; and when, in the nineteenth century, relations between the British and the Irish steadily deteriorated, the "Unionists" (a label referring to their union with Britain) remained staunchly pro-British. The Anglo–Irish War (1919–1921) tested the loyalty of the Ulster plantation. At no point did the Unionists involve themselves with the revolutionary action that the Irish had initiated. Nevertheless, they started to experience considerable apprehension lest Britain should abandon them. To guard against that eventuality, they established their own Constitution and made elaborate plans for a government of their own, a government that would be protected by loyal military forces.

BRITAIN GUARANTEES THE UNION

The pressures exerted by the mobilized Unionists translated into the Government of Ireland Act, which explicitly excluded the northeastern part of the island from the new political entity, the Irish Free State. Naturally, now that the self-determination of the Irish was at hand, borders needed to be drawn. Of Ulster's nine counties, three were heavily populated by Roman Catholics; these three were allocated to the new state. Their elimination caused the remaining six counties to tilt very heavily toward Protestantism. Thus, the enclave that was left to the Unionists may have had a smaller area, but it was at least cohesive in religion.

From that point on, Britain affirmed its guarantees to Northern Ireland with every step that the Irish Free State (subsequently, Eire) took to dissociate itself further from Britain and the Commonwealth. When in 1939 Ireland declared that it would remain neutral in World War II, the British immediately mobilized Northern Ireland. British military strategists considered it highly fortunate that part of the island to the west of Britain was still in British hands: It shortened the risky route of the ships carrying supplies from the United States to the United Kingdom.

When Ireland decided to leave the British Commonwealth after the war, the British Parliament passed the Ireland Act of 1949, which guaranteed that Northern Ireland was to remain part of the United Kingdom unless its own Parliament voted otherwise. That Parliament was dissolved in 1972, and the successive assemblies that followed proved so ineffective that they had to be suspended. Since it would be difficult to rely on a parliamentary vote in these circumstances, the British government decreed that the status of Northern Ireland could be changed only through a popular referendum. Referenda being fairly alien to British politics, it was an unusual suggestion. But it might well work, since it seemed obvious that the Protestant majority would not vote itself out of its union with the United Kingdom.

EXERCISES IN STALEMATE

For a considerable period, the United Kingdom practiced "salutary neglect" in Northern Ireland—it allowed Northern Irish institutions to administer the area. In due time, however, it became clear that the Unionist government harbored inherent biases against the Catholics in Northern Ireland and that tensions and clashes between the Protestants and the Roman Catholics were more often than not provoked by the former. Also, measures that the Northern Irish government took were often directed against the Roman Catholic minority. Britain thus decided to dissolve the Parliament in Belfast (in which the Roman Catholics had disproportionately little representation) and to terminate salutary neglect. Subsequently, Northern Ireland was directly ruled. Local government was largely left alone.

Law and order became precarious, as the security forces, largely identified with the Protestant government, were unable to maintain civil order. Therefore, the United Kingdom stationed several thousand British troops in Northern Ireland beginning in 1969, the year in which massive sectarian violence erupted. A large part of this violence was aimed at Catholic civil-rights activists.

At first the Catholic community welcomed the British troops, naïvely assuming them to be neutral, or at least more objective, in their approach than the Royal Ulster Constabulary (the only armed police force in the United Kingdom) and the auxiliary part-time Special Force (a vigilante organization established by the Protestants). But it did not take long for the Catholic community to conclude that the British troops had been dispatched to restore and maintain the law and order of the Union. While essentially protecting the status quo of the Northern Irish territory, these troops were, as a matter of course, more supportive of British and Northern Irish interests. The introduction of British Army units amounted to an escalation, which was matched by the re-activation of the Irish Republic Army, a paramilitary organization that had been semidormant for nearly 15 years.

The IRA initially merely protected Catholic areas from attacks by armed Protestants before British troops could respond. However, it soon split into two factions: A Socialist faction, basically ideological and almost nonviolent, and a Provisional faction, dedicated to the use of total force to "rid the island of the British." But since the British Army was numerically and technically superior, "total force" implied terrorism. Other militant groups subsequently came into being on both sides, often extending terrorist activities into Eire and England.

PEACE INITIATIVES

Warlike conditions ("The Troubles") have prevailed in Northern Ireland since the late 1960s. Some initiatives seem to bear fruit, such as the so-called Peace Lines, which crisscross working-class Belfast, separating the Protestants from the Catholics and creating a sectarian map that has no relation to the city as it appears in official atlases.

The New Ireland Forum was established in 1983. Participation was open to all democratic parties that rejected violence and that had members elected or appointed to the Irish Parliament or the resurrected Northern Ireland Assembly. Its first session was held in Dublin in 1983. The Forum published various reports, with one central operational conclusion: "The validity of both the Nationalist and Unionist identities in Ireland and the democratic rights of every citizen of this island must be accepted; both of these identities must have equally satisfactory, secure and durable, political, administrative and symbolic expression and protection."

At the highest level, an Anglo–Irish Agreement was established after long and painstaking negotiations between the governments of the United Kingdom and Eire. Here a new element was added: an Intergovernmental Conference, meant to enable the Irish government to express views and put forward proposals on various aspects of Northern Ireland affairs. The Conference intended to make determined efforts to resolve any differences between the two governments. It also hoped to improve cooperation in ending terrorism. Yet the validity of these and other peace initiatives was still in question. The fighting in Northern Ireland continued, and the IRA remained very active. There were repeated bomb attacks in London. In March 1991, terrorists succeeded in firing three mortar rounds at the very heart of the British government, 10 Downing Street.

In 1993, the IRA called a cease-fire in order to begin peace talks concerning the rule of Northern Ireland; Sinn Féin, the pro–IRA Northern Irish political party, was supposed to be included in these talks. February 1995 witnessed the issue of a "Framework Document," which proposed a basis for negotiations. On February 9, 1996, the IRA, out of frustration because it was not included in the peace talks, called an end to the cease-fire by bombing the Canary Wharf in London. In response, both the Nationalists and the Loyalists in Northern Ireland marched onto the streets to show their support for a resumption of the peace process.

At the end of February, the governments agreed that all-party talks would begin in June and be open to all parties disavowing violence. In May 1996, elections were held to determine participation in the talks. Sinn Fein won nearly 16 percent of the vote, but the party was turned away from the negotiations when they began on June 10 because of the IRA's continued campaign of violence.

Through early 1997, the negotiations made little progress. The May 1997 election of Tony Blair and the Labour Party government, however, re-energized the process and led to increasing pressure on the IRA to restore the cease-fire. After some negotiation, the IRA restored its cease-fire in July 1997, and Sinn Fein was admitted to the talks process in September 1997.

In a final marathon push in April 1998, which included the personal intervention of President Clinton, all parties, on April 10, signed an agreement. The "Good Friday Agreement" (April 10 was Good Friday) was put to a vote, and strong majorities in Northern Ireland and Ireland approved it in simultaneous referenda on May 22. The agreement provided for an elected 108-member Northern Ireland Assembly, to be overseen by a 12-minister Executive Committee in which Unionists and Nationalists would share responsibility for governing. The agreement would institutionalize cross-border cooperation with Ireland and would create mechanisms to guarantee the rights of all.

Finally, it seemed that there would be a true, sustained peace. Two major players, David Trimble (leader of the Ulster Unionists) and John Hume (the Catholic Nationalist leader), were even awarded the 1998 Nobel Peace Prize for their part in the peace process. (This was the second Nobel Peace Prize offered in the Northern Ireland question. The first, in 1976, was awarded to two Northern Irish women—Mairead Corrigan and Betty Williams—who led the Northern Ireland peace movement.)

Bitter disillusionment set in beginning in August 1998, however, when the self-styled "Real IRA," a breakaway group opposed to the Good Friday Agreement, set off a blast in Omagh, Northern Ireland, in a crowded market, killing 28 and injuring 220. It was the worst single terrorist incident in Northern Ireland's history.

For a time it seemed that the terrible event would actually strengthen the peace process, as all of Ireland—North and South, Nationalist and Unionist, Irish Republican and Ulster Loyalist—was appalled. But additional scattered terrorist incidents further undermined the peace process. By March 1999, the blockage was acknowledged when the deadline for devolution of power to a new power-sharing administration in Northern Ireland, originally set for March 10, was formally postponed.

In November 1999, the Ulster Unionists agreed to participate alongside the Sinn Fein in a regional government, with the condition that the IRA begin giving up its weapons by the end of January 2000. On December 2, 1999, the British Parliament devolved power to the new regional cabinet, and Ireland renounced its claim to Northern Ireland. Mainstream Protestant Trimble, of the Ulster Unionists, became the first minister, with moderate Catholic Seamus Mallon of the Social Democratic Labour Party as the deputy first minister; their parties each received four cabinet seats. The Sinn Fein, often called the political arm of the IRA, received two. The reactionary Democratic Unionist Party of Ian Paisley, which opposed the peace accord, also received two seats; however, his party's ministers boycotted cabinet meetings. After the IRA failed to meet the deadline regarding its weapons, the British government suspended the regional government of Catholics and Protestants in February 2000.

The IRA's compromise whereby it would put weapons out of use and allow periodic inspection by international observers of its caches allowed the British to reinstate the power-sharing government in May 2000. Although the major Protestant and Catholic armed groups maintained their cease-fire, low-level violence nevertheless continued. In July, attempts by security forces to keep Protestant Loyalist marchers from their traditional parade routes through Catholic Nationalist strongholds provoked civil disobedience, which escalated into violence. The following months saw "turf battles" between supporters of rival Loyalist militias, the Ulster Volunteer Force and the Ulster Defense Association. There were also bloody feuds within the Nationalist camp between supporters of the IRA and its splinters, the Real IRA and the Continuity IRA. For example, the Belfast head of the Real IRA was assassinated in October 2000. Trimble's hold on power grew more precarious in the face of criticisms from within his own party not only over the lack of progress on IRA arms decommissioning, but also over police reforms that would make the Royal Ulster Constabulary, historically a Protestant force, more acceptable and open to Catholics. The IRA linked its disarmament with acceleration of British-troop withdrawals.

The outlook for the success of this particular power-sharing experiment is uncertain. In March 2001, the threat of terrorist attacks on the main island was made clear by the Real IRA's bomb explosion outside the BBC's headquarters in West London. However, many would agree with Trimble's long-term optimism, expressed in a February 2001 interview with *Süddeutsche Zeitung,* that the corner has been turned politically and economically and that though there may be relapses, there will be no going back to the bad old days in Northern Ireland.

---

eighteenth century. A notable setback at the end of that century was the secession of the American colonies.

However, in 1815, England defeated Napoleon, and the entire nineteenth century may be called "the century of Britain." During this period, the ascendance of the British Empire was truly spectacular. Even before the scramble among various European countries for territories in Africa, the extensive British possessions were referred to as an empire "where the sun never sets." Britain ruled the waves, commerce and trade prospered, and the Pax Britannica was a global concept. Although the French language may have retained its supremacy as a diplomatic language, England was indisputably the most powerful nation in the world from the Congress of Vienna in 1815 until World War I. That era has been called "a century of peace." It may not have been peaceful all over Europe, but at least there was no conflict engulfing numerous nations in the period from 1815 to 1914. Apparently the Vienna Congress had succeeded in establishing a balance of power, and Britain had become the main balancer. World War I (1914–1918) is often termed "the Great War," and once it ended, another English-speaking nation emerged as a major world power: the United States of America.

The period between the two world wars witnessed a steady diminution of British power, and the liquidation of the British Empire after World War II further weakened the United Kingdom.

THE TWO MAINSTREAMS OF DEMOCRACY
The political cultures of the United States and the United Kingdom have a great deal in common. In spite of the American Revolutionary War, it is possible to speak of a sizable heritage, in that values, visions, perspectives, and other cultural facets of the colonial power were inherited by the successor government. After all, there were no substantial linguistic or ethnic differences between the United States and Great Britain. However, whereas the United States became the first country to possess a modern constitution, thereby putting itself in the vanguard of democratic ideology, its British counterpart is not only considerably older but in large part unwritten. The British Constitution cannot be traced back to a historic assembly, such as the Constitutional Convention in Philadelphia. Instead, it consists of customs, traditions, and conventions that have evolved over the centuries. But it is not entirely unwritten either: Important laws such as the Magna Carta, the Habeas Corpus Act, various reform bills, and other

major pieces of legislation have in the course of time gained constitutional status.

If this amounts to a difference in format, there are also differences in substance between the two Constitutions. Thus, in the United Kingdom, the judiciary lacks judicial review power, which has been a principle of constitutional validity in the United States since the early nineteenth century. In Britain, parliamentary sovereignty prevails, to the point that the judiciary cannot declare legislation unconstitutional; in other words, an act of Parliament can be overturned only by a subsequent act of Parliament.

The principal difference between the presidential and the parliamentary systems into which the two mainstreams of democracy have eventually materialized is that the presidential system has an explicit separation of powers. It contrasts with the fusion of the executive and legislative branches of government, which has been the hallmark of the British system of governance.

In the United Kingdom, the executive branch of government is seated in the House of Commons. In fact, the British prime minister and his or her cabinet ministers *are* members of Parliament and are elected as such. At the same time, the highest judicial organ is part of the House of Lords. The Law Lords, who constitute

a committee in Britain's upper chamber, exercise functions similar to those of the U.S. Supreme Court.

An additional contrast between the presidential and parliamentary systems involves the tenure system. In the United States, all executive and legislative officeholders have a fixed tenure (e.g., the president, four years; a senator, six years; a representative, two years). In the British parliamentary system, there is no fixed tenure. It is obviously absent in the House of Lords, since that chamber historically has depended on hereditary succession, and more recently also on lifetime appointments. In the House of Commons, the rule prevails (a rule that as a matter of course also applies to the executive) that an election should take place within five years after the last election. But the government can call an election at any time it deems fit. The year 1974, for example, was marked by two national elections. Subsequently, there were no elections for five years; and it was then, in 1979, that the leader of the Conservative Party, Margaret Thatcher, called a vote of no-confidence, which carried. This forced an election that ousted the Labour government from power and made Thatcher prime minister.

The governments of the United States and the United Kingdom also differ as far

as the executive structure is concerned. Great Britain has a dual executive, whereas the United States has a single chief executive, the president, who performs both head of state and head of government roles. In the United Kingdom, the post of the head of state is taken by the monarch, who for practical purposes has no political powers and who serves the country as a symbol of national unity. The monarch is, of course, not elected, but succeeds to the throne on the basis of heredity. He or she must be "above" politics and in fact may not make any political utterances in public. The Speech from the Throne, comparable to the U.S. State of the Union address, is prepared by the prime minister and his or her cabinet. The monarch, in effect, acts only as a mouthpiece of the government.

The prime minister is the most powerful person in the British system. Some political analysts, when comparing the U.S. president and the British prime minister, believe the prime minister to be more powerful in his or her context. Indeed, some people have suggested substituting the term "prime ministerial" government for the customary phrase "parliamentary" or "cabinet" government.

The British Parliament offers one more anomaly: It is bicameral only in a strictly formal sense. The House of Lords has lost much of its power and influence. It has, in effect, become a debating club and revising chamber. The House of Commons is where the action is; it is the hub of British politics. All of the House of Lords can do is delay non-money bills.

HURDLES AND HIGHLIGHTS

British trends and tendencies have nearly always been marked by approaching a desired end in gradual stages. A strong resistance to change—no doubt enhanced by British insularity, which withstood outside influences—has always predominated.

British society is often qualified as deferential, since class distinctions are fairly large and conspicuous. It is interesting that class *distinctions* prevail as opposed to class *conflict*. Karl Marx, who lived in London for many years, frequently despaired of the possibility that the British proletariat would ever allow itself to be mobilized into a revolution.

However that may be, Britain has, in spite of its penchant for gradualism, often been in the vanguard of material innovations. One of these concerns the Industrial Revolution, beginning in the mid-eighteenth century. It may be said that the following century and a half constituted the watershed between the traditional, predominantly rural life and the fast pace of

(Photo courtesy Lisa Clyde)

About 80 miles southwest of London is the town of Salisbury. A dominant presence in the history of the town and in its skyline is the Cathedral of St. Mary, construction of which began in 1220 and ended in 1258. Its spire is 404 feet tall, the highest church spire in England.

an industrialized, technological society. Britain has never been self-sufficient in agriculture. This shortage in agricultural production was offset by an unparalleled commerce and trade in early times and, subsequently, by the sale of new industrial products.

Nearly two centuries after the start of the Industrial Revolution, the United Kingdom again had to adapt to bewildering changes, this time brought about by decolonization. During World War II, but before the United States became a participant in that conflict, U.S. president Franklin D. Roosevelt and British prime minister Winston Churchill jointly issued the Atlantic Charter. This document was a series of ambiguous statements concerning self-determination, which, according to Churchill, referred to the victims of recent Nazi aggression: There could be no doubt that the war was also fought to help them to regain self-determination. However, Roosevelt clarified that it referred to colonies as well—in fact, to all non-self-governing territories. The positions of the two leaders were antithetical, and Churchill once declared that he had not become His Majesty's prime minister to preside over the liquidation of his empire. But that liquidation came inexorably after the war and, contrary to British development, it did not proceed gradually. Today, Great Britain has a few small colonies, including Bermuda, Gibraltar, the Falkland Islands, and some less important territories. (Hong Kong reverted to China in 1997.)

THE COMMONWEALTH

Despite the avalanche of change in imperial matters, the British people have been able to cope and compromise. In part, this flexibility may have been aided by the "substitute" that British statesmen provided for the Empire: the Commonwealth of Nations. The Commonwealth is a large and global organization, formally headed by the British monarch. But he or she has little or no power in the domestic context, and there is no political supremacy on the part of the United Kingdom either. The Commonwealth thus concentrates on the commercial and trading advantages that accrue to its members. Membership is entirely voluntary and usually is offered as soon as a colony gains its independence. Very few countries have refused to join. One such exception is Burma (currently called Myanmar), which in 1948 feared that its nonalignment policy would be compromised by joining. And South Africa was thrown out as a result of apartheid and other racist policies; it was allowed to return to the fold only when Nelson Mandela came to power. Fiji, a small group of islands in the Pacific,

was expelled in 1987 because the Fijians themselves (who ironically constituted a minority) usurped power after staging a coup.

Before World War II, the colonies of settlement (such as Canada, Australia, and New Zealand) had attained a larger degree of autonomy within the British Empire/Commonwealth. But they were so attached to Britain that they refused to surrender the "God, King, and country" ideal that had characterized their relationship in the past. Naturally, such sentiments of loyalty were absent in the colonies that had suffered exploitation.

But these colonies, when the question of whether they wanted to join the British Commonwealth was raised on the eve of their independence, almost universally agreed to continue the link with Great Britain. However, they added that they nevertheless wanted to become republics. Since the British monarch is the head of the Commonwealth, this was an unusual request, but Britain showed itself to be remarkably flexible in not resisting such an anomalous structure. In part, that flexibility may also have been promoted by the absence of political affinity between the United Kingdom and some of its former colonies: Britain's traditional political model (the so-called Westminster model) has never taken root in those countries that became independent after World War II. It is true that there are some superficial resemblances in parliamentary procedure, but many ex-colonies have never been able to commit themselves fully to democratic ideals.

The Commonwealth meets on an annual basis, rotating the venue among the capitals of members. A permanent secretariat has been established in London (which is logical, since the British monarch is, after all, its ex-officio head). Commonwealth matters assume an important place in British policy making. They are usually dealt with at the cabinet level and are considered closely related to foreign affairs.

THE ECONOMY

One hurdle that Great Britain has always faced, since it started to mature into a modern nation-state, is the extreme vulnerability of its economy, resulting from its lack of self-sufficiency in food production. Enemies have naturally attempted to capitalize on this weakness, trying to starve the British by cutting off their supplies.

Napoleon's so-called Continental System, which outlawed all trade between Britain and the European mainland, represented such an attempt. And after the fall of France, in 1940, Adolf Hitler not only tried to bomb England into submission—he also cut the

supply lines on which the United Kingdom depended. A great many supply ships were sunk on their way to Britain by Germany's U-boats (submarines).

Modern wars, therefore, have been extraordinarily hard on the United Kingdom, even if it managed to win them. The two world wars were devastating to the country; both struggles consumed so much of Britain's strength that it finally ceased to be a major world power.

Ironically, the powers vanquished in World War II, Japan and Germany, managed to rebuild their economies in a relatively short period of time, whereas the European Recovery Plan, better known as the Marshall Plan, launched by the United States, proved an inadequate remedy for the exhausted British economy. Admittedly, the United Kingdom was very heavily bombed during the war and its capital equipment was, in many cases, damaged beyond repair. But Germany and Japan suffered massive destruction, too. The question thus remains: Why couldn't Britain recover?

Some analysts have suggested that the loss of empire should rank high among the factors that contributed to Britain's decline. This argument, however, is not very convincing, since, in a few cases, former colonial powers (such as the Netherlands and Portugal) were able to improve their economies after the ties with colonies had been severed.

Other analysts conclude that the extensive welfare program that Britain embarked on shortly after World War II was its undoing. Was this not a luxury that it could ill afford? But again, this theory is weak: Welfare systems had become a general trend among industrialized democracies in Western Europe. At that time, the Scandinavian countries, Ireland, and other nations also viewed welfarism as a top priority, but none of them was as adversely affected as the United Kingdom.

The rapid deterioration of the British economy became a matter of international record when the British government requested in the late 1940s that the United States take over certain responsibilities in the eastern Mediterranean region. Washington was prepared to do so (a measure that was to lead to the Truman Doctrine). Nevertheless, the British economy remained relatively weak throughout the 1950s and 1960s.

THE EUROPEAN UNION

Before it finally joined the European Union in 1973, the United Kingdom's application for admission had been rejected twice, by French president Charles de Gaulle. (The term "European Union" became legal in 1993 with the ratification of the Maastricht Treaty. However, in order to cut down on confusion, it is used here also in reference to the EU's precursor, the European Economic Community, nicknamed the "Common Market" by the British and Americans, and later self-described as the European Community.) Most analysts believed that de Gaulle's negative attitude vis-à-vis the United Kingdom stemmed from the strong bias that he had developed against the Anglo–American coalition during the war years. However, the French president officially based his rejection of the British application on the backwardness of Britain's industrial capacity. According to de Gaulle, the British factories that had been bombed during the war had not been properly repaired or replaced. Thus, in his mind, the United Kingdom, if admitted, would drag the European Union to lower levels of competitiveness and output.

A few years after de Gaulle left the scene, the British admission was renegotiated. This time, the application met with success. The United Kingdom was admitted in 1973, along with Ireland and Denmark. For the British, this was a historic moment—it was, in fact, the first time that the country in a sense became an integral part of "Europe."

Winston Churchill had many years earlier explained British affiliations in terms of circles. The first circle, the one closest to the United Kingdom, constituted its imperial legacy, the Commonwealth, at that time dominated by former settlement colonies such as Canada, Australia, New Zealand, and South Africa, whose affinity to "King and country" was unquestionable. The second circle had reference to the strong bonds with the United States. These ties had been strengthened as a result of a common culture as well as a common front in wartime. Their alliance related to heritage and shared values rather than to explicit treaties. The third, and outer, circle referred to Europe, of which the United Kingdom had always been a part, in a geographical sense. The country had always interfered in European politics through balances of power, but it had never participated in a joint endeavor. Many Britons realized that, whatever the bonds with the United States would be, the Commonwealth had lost out against the European Union. A belated referendum (held in 1975, two years after the British admission became official) indicated that a clear majority of Britons favored the step.

However, the integration of the British economy into the EU did not stem the negative trends. Inflation and unemployment were extraordinarily high in comparison to those of its fellow members.

THE THATCHER ERA

After World War II, the two major British parties—the Conservative Party and the Labour Party—alternated in power without offering significant changes in their economic policies. True, there was a degree of discontinuity, in that a Conservative administration might denationalize, or privatize, selected large industries that had been nationalized by Labour governments. But the welfare system was considered sacrosanct. No Conservative administration would tamper with it, for fear of incurring public indignation.

In 1979, the Conservative Party assumed office, and the Thatcher era started. Margaret Thatcher was the first British prime minister since the 1940s to introduce major changes in the economy. Her government made cuts in the costly National Health Assistance System and in other welfare projects, in an attempt to come to grips with the wages and prices spiral.

To that end, Thatcher had to tackle the trade unions, whose influence in economic matters had often been very decisive, yet at times dysfunctional. The methods that Prime Minister Thatcher used in the realm of trade-union reform greatly reduced union influence in the economy. She also proved innovative when it came to coining names for controversial policies that aimed to boost the economy, such as "People's Capitalism" and the "enterprise revolution" (an economic shake-up that was to generate an "enterprise culture"). A relatively strict monetary policy and freedom from exchange-rate controls were introduced. Finally, Thatcher aimed for lower taxation rates. There was thus a definite departure from the past, which was marked by what is known as the Social Democratic (Keynesian) consensus. One of her chief goals was to make the middle class large and confident again. The term "Thatcher revolution" is often used in this context. In some ways, Thatcher's revolutionary policies ran parallel to Reaganomics, another set of controversial policies that was initiated on the other side of the Atlantic during the same period. Yet she avoided the massive deficit spending that characterized the Reagan and Bush years.

In her attempts to get the economy on its feet again, the British prime minister was aided by three factors. First, during the early part of her administration, North Sea oil exploitation started to pay off. Not only did the United Kingdom become self-sufficient in oil (a fact that naturally translated into savings), but there was enough surplus to sell to other countries, thus generating revenues. Second, as in many other countries in the West, the public mood experienced a distinct shift to the right; socialism and welfarism had oversold themselves. Finally, the Conservative Party had little to fear from the opposition. The British Labour Party, traditionally factionalized, had become more fragmented than ever, as a result of ultra-leftist penetrations.

Thatcher's economic policies met with some success. Inflation and unemployment rates slowed down and appeared to stabilize, and exports went up, in spite of the strengthening of the British currency. Also, the exploitation of oil fields proved a windfall that was only partly offset by the worldwide oil glut. In 1988, however, the economy became a victim of its recovery, since its strong growth had led to renewed inflation and record trade deficits (resulting from consumer demand).

The prime minister also fought hard in the European Union, believing her country to be shortchanged. She insisted that the United Kingdom should get as much out of the European Union as it put in; she occasionally threatened to withdraw Great Britain's membership. These threats were conceivably part of a hard bargain by the "Iron Lady," who would have realized that the British withdrawal would harm Britain more than the Union. However, the membership issue was still very much alive in the United Kingdom even a decade after the referendum. Indeed, the Labour Party had indicated that it favored withdrawal from the Union.

Margaret Thatcher became prime minister in 1979 as the result of a successful no-confidence motion. Elections would normally have been due in 1984. However, the decisive victory of the British forces in the Falkland Islands War added greatly to her popularity. (In 1982, Argentina had suddenly taken this group of British islands off its coast; Thatcher's immediate response was to send an expeditionary force, which reconquered the islands in a matter of weeks.) Thatcher decided to advance the parliamentary elections by a full year, and in June 1983, she increased her party's majority in the House of Commons. The next elections witnessed the same procedure—they were held not in 1988 but a full year earlier, the prime minister's reasoning being that the economy had favorably responded to a series of government measures.

Again Thatcher did very well. But it soon became apparent that she had reached the apogee of her popularity. Her rating in public-opinion polls started to decline, first gradually, and then somewhat precipitously. Some observers believe that the "community charge," popularly known as the poll tax, ultimately proved her undoing. This type of regressive taxation enhanced inequality in an already highly stratified society. If the poll tax was a matter of substance, Thatcher's style too had become increasingly irritating; a large part of the public had become tired of the self-congratulatory and sanctimonious tone that Thatcher adopted when discussing her causes. Her anti-Europe views had caused her foreign minister to resign. She reiterated her belief that the European Union insidiously promoted socialism. She wanted no part of the direction in which the Union was heading. If monetary unification would result in a single European currency, Great Britain would lose much of its sovereignty, if only because it would no longer be completely free to put its own budget together. Still, amazingly, in spite of all her criticism of the monetary measures within the "ever closer union," Thatcher, on October 4, 1990—that is, shortly before her loss of power—did allow Britain to enter the EU's Exchange Rate Mechanism. But this U-turn had come too late.

When the end came, it was not the opposition that forced Thatcher to step down, but her own party. The only face-saving gesture that the Conservative Party was prepared to make in the fateful days of November 1990 was that her successor would not be someone whose policies would contrast with hers. John Major seemed to fit that description nicely.

Margaret Thatcher will occupy a unique place in the pantheon of British prime ministers. Without irony, one must note that, her Euro-phobia notwithstanding, her prime ministership witnessed the completion of the basic work on the "Chunnel," the tunnel under the English Channel that connects Great Britain with continental Europe.

POLITICAL PARTIES

A party system is usually regarded as a two-party system if the two largest parties secure 75 to 80 percent of the total vote. The British party system consequently has always been regarded as a two-party system, since its two largest parties (first the Conservatives and the Liberals; and after the turn of the century, the Conservatives and Labour) traditionally received the qualified majority vote. However, in 1981, the new Social Democratic Party emerged as an offshoot of the Labour Party. It was the first new major party in Great Britain in 80 years. Although its manifesto differed considerably from that of the Liberal Party, an agreement was made that it would run on the same ticket. The centrist coalition, intended only as an electoral device, was named the Alliance. The Alliance worked to the advantage of these two small parties, which jointly received more than a quarter of the total vote in 1983.

The British Labour Party has very close ties with the various trade unions, in particular the Trade Union Congress (TUC), the British equivalent of the American AFL-CIO. During the Industrial Revolution, the trade unions were the only organizations to counter capitalistic excesses. But the unions acted from the outside and lacked direct influence in Parliament and its proceedings. After a time, labor believed that it should also be represented by a political party that participated in elections and competed for seats in Parliament. Thus the British Labour Party was born after 1900, and it made rapid strides in parliamentary politics, increasing its voting totals at the expense of the Liberal Party.

In 1988, the majorities of the Social Democrats and the Liberals, responding to mutually waning prospects, voted to merge their centrist parties as the Social and Liberal Democrats (SLD). Dissidents attempted to resurrect separate parties, but these soon faded from the scene. After a poor showing in the 1989 European elections, the SLD renamed itself the Liberal Democrats and began to consolidate itself under the leadership of Paddy Ashdown. In 1992, the Liberal Democrats won 17.8 percent of the votes, behind the Conservatives' 41.8 percent and the Labour's 34.4 percent, but ahead of combined votes for other parties, 6 percent. The Liberal Democrats have tended to be stronger in local elections, in some places controlling the councils. Nationalist parties have strongholds in Wales (Plaid Cymru), Scotland (Scottish National Party), and Northern Ireland (Sinn Fein and Social

(British Information Service)

A photograph of the Chamber of the House of Commons in session, and the first to feature John Major's government. Shown to the right are opposition members of Parliament, to the left the government, and in the center, the speaker (in this photo Betty Boothroyd, the first woman speaker).

Democratic and Labour Party). Among the minor parties, the Green Party, which has won a number of local council seats since its origins in 1973 (and two seats in the European Parliament in 1999), is the most noteworthy.

THE ELECTORAL SYSTEM
The British electoral system, although attempting to reflect democracy, has through the centuries included certain anomalies. Universal suffrage has come only gradually. Some reform acts in the nineteenth century greatly extended the electorate. Sir Walter Bagehot, the famous author and commentator on the English Constitution, called this process "the most silent of revolutions." Women over age 30 were allowed to vote in 1918, and in 1928, women over age 21 won suffrage. (However, it should be noted that the current representation of women—18.4 percent—lags behind that in most other Western European parliaments.

In addition, the United Kingdom became known for its plural vote, meaning that one person could have more than one vote. University graduates, for example, could have an additional ballot at their alma mater. Businesspeople could locally vote on behalf of their businesses. Only as late as 1948 was the principle of "one man, one vote" fully adopted.

A discussion of electoral anomalies would not be complete without mention of the so-called rotten boroughs, which existed in the past. No census was taken for the purpose of elections in earlier years, and consequently the boundaries of electoral districts were never redrawn. As a result of mobility and other factors, it could happen that a member of Parliament represented a very small number of persons. Interestingly, in this modern age, the British have become very generous with respect to voting rights and even granted them to citizens of Commonwealth nations who happen to reside in Britain (Indians, Nigerians, Malaysians, Ghanaians, and others). This right is also accorded to citizens of the Republic of Ireland who live in the United Kingdom. In 1984, Ireland reciprocated, granting the right to vote in Irish elections to British citizens residing in Ireland.

One of the most controversial issues surrounding British elections is the method according to which votes translate into parliamentary seats. The British system, based on the "first-past-the-post" concept, grants victory to the candidate who secures a plurality of votes (more votes than any other candidate, but not necessarily a majority). This electoral system has made possible incredible discrepancies and distortions. A party may achieve the largest number of votes and still end up with fewer seats than its rivals. (This happened, for instance, in 1951, when the Labour Party secured 235 seats for its 13,948,605 votes, whereas the Conservative Party obtained 321 seats for its 13,717,538 votes.) While either of the bigger parties may on occasion benefit from the first-past-the-post rule, it inevitably operates to the disadvantage of the smaller parties. In 1983, the Alliance received more than a quarter of the total votes, but it was accorded only 23 seats.

In the 1990s, as part of its reform agenda, Tony Blair's "New Labour" government instituted proportional-representation election laws for the new regional assemblies in Scotland and Wales, the British seats in the European Parliament, and the new London Assembly. Thus the seats won would be more closely in line with the percentages of the votes won by the parties. The Commission on electoral reform for the House of Commons recommended a new, hybrid two-vote system whereby 80 to 85 percent of members of Parliament (MPs) would be elected as constituency representatives reflecting the rank-order preferences of voters, with weaker candidates eliminated and their votes transferred until someone receives an absolute majority. The remaining MPs would be elected from party lists whose seats would be allocated so as to ensure that the overall outcome of the election was proportional representation of the parties. Conservative leader William Hague denounced the proposal as a "complicated and confused" irrelevance. Liberal Democratic leader Paddy Ashdown supported the proposal, whose most likely effect would be to increase his party's parliamentary seats. Prime Minister Blair was noncommittal; in any case, the promised referendum on the issue would be delayed until after the next election.

THE POST-THATCHER YEARS
In 1988, having been invited to speak at the European College in Bruges, Thatcher delivered a scathing attack on both the European Union and its Commission president Jacques Delors, a French Socialist. In this speech, she showed herself very critical of a federated Europe; she made reference to what she called the "hated F-word." The prime minister played on the nationalist sentiments of her audience at home. The Bruges speech was widely publicized and increased her popularity among Euro-skeptics in the Conservative Party (some of whom founded a society that called itself the Bruges Group, an association that apparently aims at enshrining the anti–EU ideas and thoughts of the revered former prime minister). But in the end, Thatcher's hostility vis-à-vis the EU caught up with her.

Her downfall came in November 1990. The Conservative Party's main fear was that Thatcher would be a liability rather than an asset in the upcoming election. An internal ballot seemed the best solution to determine who should face that election as prime minister. After several inconclusive attempts, a solution emerged that was to reconcile the party, since the person selected was also favored by the outgoing prime minister: John Major. The obvious question would be whether or not he was the "Son of Thatcher," as some suggested.

However, it did not take long to find out that Major was his own man. It is true that, in general, he endeavored to follow the broad outlines of Thatcherism, most of which amounted to the "neo-liberal" formulations of the right wing of the Conservative Party. Major's crucible came in December 1991, when the European Council (as the summit meetings of EU's prime ministers and foreign ministers are called) met in Maastricht, the Netherlands, to discuss the Treaty of Political and Economic Union.

Since Britain was known to object to some of the principal points in the treaty, a couple of pre-Council meetings had been convened to give Britain the chance to explain its position. The EU found itself in an awkward position. On the one hand, unanimity was required. On the other, the Union's leadership did not want the draft treaty to founder on the legacy of Thatcher. In the end, it was decided to accord Britain the right of "opting out," at least temporarily. The device simply implied that the United Kingdom did not need to sign those parts of the draft that it objected to. There was thus no need for Britain to dissociate itself from the EU altogether. Major handled the complexities of the situation well, displaying expert diplomatic skills.

Domestically, too, he did not fare badly, although the British economy remained weak. In 1992, elections were held. Although the incumbents were at a grave disadvantage as a result of the poor economy, Major's astute handling of problems and conflicts helped to prolong the Conservative leadership of Britain in spite of the recession.

Then there was the coal-mine situation. Margaret Thatcher had scored a victory over the miners and their militant union in the mid-1980s. The majority of the British mines were old, if not obsolete, and most of their operations, particularly deep mining, had become prohibitively expensive. In an attempt to cut govern-

The Norman Invasion; Battle of Hastings A.D. 1066	The Magna Carta; noblemen wring concessions from King John 1215	The Hundred Years' War 1337–1453	The Spanish Armada is defeated; supremacy over the known world shifts from Spain to England 1588	Civil war; establishment of a republic 1642–1649	The Bill of Rights 1689	Loss of the American colonies 1776	Extensions of the franchise 1832, 1867	The Anglo–Boer War in South Africa 1899–1902	World War I 1914–1918	The first minority Labour government 1924

ment expense, Major decided in October 1992 to close almost all of the mines. The proposal caused an uproar, since a great many people work in the mines, and it seemed doubtful that other work could be found for them in a time of high unemployment. As a result, numerous rallies and protest meetings were organized to denounce the measure; the miners handed "Coal not dole" stickers to passersby.

Major's popularity subsequently plummeted to the lowest level ever for a British prime minister, with only 16 percent of the British public expressing confidence in him. Major quickly put together a revised plan, which in effect would allow a majority of the mines to remain open. This plan was passed by the House of Commons.

LABOUR TAKES THE REINS

However, Major was unable to stave off the rising popularity of the Labour Party and the party's leader, Tony Blair. As the May 1, 1997, elections date approached, it became apparent that Major had trouble asserting his authority in the Conservative Party. After the votes were tallied, the Conservatives, with 30.7 percent, had experienced their worst performance since 1906. Eighteen years of Conservative control was ended as Labour won a historic victory, with 43.3 percent of the vote, which translated into a huge 179-seat majority. Blair was confirmed as the new prime minister. At age 43, he was the youngest to hold that post since 1812.

The new prime minister's center-left government took the reins of power during a complicated economic and political era. He stated that the bulk of his time has been taken up with dealing with the Northern Ireland problem. But he has also had to deal with persistent economic concerns, including unemployment and underemployment, and pressures to trim the social-welfare budget while retaining services and the "safety net"; difficult social issues, such as increasing single motherhood and rising crime and truancy in poorer urban neighborhoods; the political challenges stemming from the steady rise of nationalism in Scotland and Wales, both of which voted in September 1997 referenda for increased political autonomy; and in 1998 elections were held for their new regional assemblies.

Devolution of power to Scottish and Welsh assemblies, a power-sharing experiment in Northern Ireland, election of a London mayor and assembly, less secretive government (more freedom of information), and electoral reform were key components of Labour's ambitious program of constitutional reform. Blair sought a "democratic renewal" of British democracy, embracing themes long monopolized by the Liberal Democrats and Charter 88 (a constitutional reform–advocacy group). From his view, a modernized Britain requires a more participatory citizenship to counter the erosion of popular consent in recent decades. For Labour, the House of Lords, a relic of premodern, undemocratic Britain, was a prime target. Blair's initial plan was to remove all hereditary peers (lords); however, he compromised so that 92 could remain during a transitional period. The second stage, a reconstituted upper chamber, which might, for example, include regional representatives, is still pending. Blair's government also took a step toward a written constitution by incorporating the European Convention on Human Rights into British law. (Parliamentary sovereignty is maintained, however, because the House of Commons can set aside a judge's decision). As of October 1, 2000, British subjects have a (quasi-) Bill of Rights for the first time in history.

Critics say that Blair has adopted most of Thatcherism's substance while utilizing style and rhetoric to present his domestic program as new. The concept that Blair utilized beginning in 1997 to portray his approach was the "Third Way," a path between the "old" left of centralized bureaucracy, collectivism, and social welfare, and the "neo-liberal" right of *laissez-faire* capitalism, privatism, and social *in*security. In part because of overlaps with the policies of U.S. President Bill Clinton (a self-described "New Democrat"), whose economy was booming, European center-left (and even center-right) politicians paid a lot of attention to Blair's initiatives. For example, Blair emphasized moving people from welfare to work through investments in retraining. While not standing in the way of markets, Blair talked about morality, family, and community. He and his home minister (who in Britain controls the police) took strong law-and-order stands. Critics

argued that there was no big idea at the core of the Third Way and that it was all public relations. Yet in the late 1990s, the Blair government acted boldly on a number of fronts, from the Bank of England to Northern Ireland. In contrast to French leaders, Blair embraced the globalizing economy. As political scientist James Cronin pointed out, the factor that explained Blair and his colleagues' performance was their determination to "not be seen as 'old Labour,'" (whose obsolete ideology kept the party out of power for 18 long years).

The New Labour government enjoyed a relatively long honeymoon. Blair's energy and charm stood him well in the early years, and his chancellor of the Exchequer, Gordon Brown, maintained a firm hand on the budget: New Labour was not to come across as a tax-and-spend leftist party. But also, there was no serious competition from the Conservatives, who were having troubles finding their footing in opposition. Post-1997 Conservative leader William Hague failed to project an image of competence as an alternative prime minister. More recently, however, the media began to label Blair a "control freak" wanting everything his way in the party and government. This was most conspicuously evident in his tactics regarding the first direct mayoral elections in London. Blair moved heaven and earth to sideline the candidacy of leftist Labourite Ken Livingstone (the infamous "Red Ken," who, in the 1980s, headed the Greater London Council until Thatcher abolished it). Ultimately Livingstone, running as an independent, defeated both the official Labour and Conservative candidates.

Nevertheless, Blair remained far ahead of Hague in popularity and Labour ahead of the Conservatives in voter preferences except for a period in the autumn of 2000. Huge protests about soaring fuel prices, which began with direct actions by French truckers, spread to Britain truckers, who began to do un-British things like blockading oil refineries. Although gasoline became scare, the motoring public expressed general support for their civil disobedience. Blair's government scrambled to retain its authority in the crisis. Although taking an initial hard line, ultimately it made tax concessions to the truckers. "Prudent" budgeting in previous years allowed Gordon Brown to have plenty of

The Battle of Britain; Nazi Germany attempts to bomb England into submission
1940

Election of the first majority Labour government
1945

British India, the cornerstone of the British Empire, becomes independent (India and Pakistan)
1947

The United Kingdom becomes a member of the European Union
1973

Margaret Thatcher is elected as the first female prime minister
1979

Britain defeats Argentine forces in the Falkland Islands War
1982

Labour's Tony Blair becomes prime minister; both Scotland and Wales win greater autonomy
1990s

Resolving the Northern Ireland problem remains the leading political issue in Britain

2000s

Britons support truckers in fuel-price protests

An epidemic of hoof-and-mouth disease devastates British agriculture and tourism

extra revenues to spend, with the next election looming.

It is hard to know whether the agricultural crises of first BSE ("mad cow disease") and then hoof-and-mouth disease plus a series of railway disasters will be on the minds of voters when they next vote on parliamentary candidates. The economy has remained sound: in December 2000, unemployment stood at 5.3 percent (compared to 8.7 percent in the Euro Zone), consumer prices were up 2.7 percent over the previous year (2.1 percent in the Euro Zone), and Britain's gross domestic product was growing at the annual rate of 2.4 percent (3.3 percent in the Euro Zone).

In the last years of the Major government, there was a wave of scandals. There have been few to trouble Blair, with the notable exception of the financial improprieties that forced Blair's friend and adviser Peter Mandelson twice (!) from the cabinet. Electoral behavior in Western democracies has become more volatile in recent years, making any predictions precarious. However, most pundits would be surprised if Blair and Brown's New Labour did not easily win the next general election.

BRITAIN AND EUROPE'S FUTURE

Prime Minister Blair has favored "constructive engagement" with the European Union, in contrast to the more defiant (and insular) orientation of his Conservative predecessors, John Major and Margaret Thatcher. Under William Hague's leadership, the Conservatives have become overwhelmingly Euro-skeptical. However, a Conservative-led Britain would be unlikely to pull out of the EU, due to the resulting economic disruptions; at the same time, it would be likely to resist firmly any and all moves toward "an ever closer union." At the Nice summit in December 2000, when he refused to relinquish the national veto in tax policies, Blair let it be known that there were limits on how much economic integration his constructively engaged government would accept.

In the June 1999 European Parliament elections, only 24 percent of British voters bothered to vote; from those who did, big gains were scored by the Conservatives. (In the May 1997 House of Commons elections, 71.5 percent of British voters had turned out.) British European Parliament turnout was easily the lowest among the EU national electorates. Polls in late 1999 indicated that the British trail only the Swedes in having a negative assessment of how much EU membership has benefited their country.

The vast majority of British citizens surveyed since their government exercised its treaty right to opt out of the Economic and Monetary Union (which got under way on January 1, 1999) oppose abandoning the British currency, the pound. The weakness of the Euro in its first two years has reinforced their reservations about turning monetary policy over to the European Central Bank in Frankfurt. Yet some polls also indicated that many Britons are resigned to their country's eventual membership in the EMU, which as of 2001 included all but three of the EU member states. Blair's view has been that Britain would be better placed to influence developments from inside than outside, but that Britain should join the EMU only when economic conditions are ripe. Gordon Brown, the powerful chancellor of the Exchequer, has been even more cautious about the issue, which the government has pledged to put to a referendum at some point in the future.

In contrast to his Conservative predecessors, Blair has supported the move to give substance to the second pillar of the European Union—the common foreign and security policy. Britain has pledged to contribute more than 12,000 troops to the European rapid-reaction force scheduled for possible deployment in global "hot spots" after 2003. The British government has envisaged the force as a supplement to NATO and in no way the foundation stone for an autonomous European army. The new Bush administration's plan to develop a nuclear missile–defense shield, which would rely in part on British installations, has put the British in an awkward position between the United States and EU countries that are apprehensive about a new arms race.

The British government appeared basically satisfied with the Nice summit's outcome in December 2000. The EU member states had agreed to enough institutional change to allow the entry of new members from Central/Eastern and Southeastern Europe over the next decade. However, the draft treaty's awkward compromises have set the stage for yet another EU intergovernmental conference in 2003–2004. In anticipation of a growing constitutional debate, Prime Minister Blair has advanced the proposal of adding a second chamber, consisting of national MPs, to the European Parliament in order to counter the continuous push of the European Commission and the (current unicameral) European Parliament toward "ever closer union." Thus, even under New Labour, the British vision of Europe's future is hardly federal.

DEVELOPMENT

Britain was the first country to enter the Industrial Revolution. Britain's industrial output is still very large, although much of its capital equipment has become outdated.

FREEDOM

Britain was the first European country to create democratic conditions within its borders. Today the United Kingdom rates high in human rights.

HEALTH/WELFARE

The Labour government, coming into office immediately after World War II, introduced the welfare system. The cornerstone of this system is the National Health Service.

ACHIEVEMENTS

The exploitation of North Sea oil has been a success. Britain is self-supporting in the realm of energy and is able to export and sell some of the surplus to neighbors. In 1994, the Channel Tunnel opened between Britain and France.

Annotated Table of Contents for Articles

Regional Articles

Topic Guide to Articles

TOPIC AREA	TREATED IN	TOPIC AREA	TREATED IN
Coexistence	1. European Integration: Past, Present and Future 2. Europe 2007 3. EU Enlargement	European Union (EU)	1. European Integration: Past, Present and Future 2. Europe 2007 3. EU Enlargement 4. Our Constitution for Europe 5. Seeing Green 8. Danes Say 'No' to Euro 14. Survey of Portugal 15. Survey of Spain
Culture	6. Poverty in Eastern Europe 7. Europe: Is There a Fourth Way? 9. New Golden Age 15. Survey of Spain		
Current Leaders	3. EU Enlargement 7. Europe: Is There a Fourth Way? 8. Danes Say 'No' to Euro	Foreign Investment	1. European Integration: Past, Present and Future 2. Europe 2007 3. EU Enlargement 11. Domestic Issues Dominate German Agenda
Development	1. European Integration: Past, Present and Future 2. Europe 2007 5. Seeing Green 6. Poverty in Eastern Europe 9. New Golden Age 10. German Unification 14. Survey of Portugal	Foreign Policy	2. Europe 2007 3. EU Enlargement 7. Europe: Is There a Fourth Way? 17. Britain, Europe and America: Keeping Friends
Economy	1. European Integration: Past, Present and Future 2. Europe 2007 3. EU Enlargement 4. Our Constitution for Europe 6. Poverty in Eastern Europe 9. New Golden Age 13. Hot and Sticky in Ireland 14. Survey of Portugal 15. Survey of Spain	Foreign Relations	1. European Integration: Past, Present and Future 5. Seeing Green 7. Europe: Is There a Fourth Way? 17. Britain, Europe and America: Keeping Friends
		Germany	10. German Unification 11. Domestic Issues Dominate German Agenda
Environment	5. Seeing Green	Health and Welfare	5. Seeing Green 6. Poverty in Eastern Europe 9. New Golden Age 10. German Unification 12. Germany's New Identity
Equal Rights	12. Germany's New Identity 16. 209 Years Later, the English Get American-Style Bill of Rights		
Ethnic Roots	6. Poverty in Eastern Europe	History	1. European Integration: Past, Present and Future 2. Europe 2007 15. Survey of Spain
Euro	1. European Integration: Past, Present and Future 2. Europe 2007 4. Our Constitution for Europe 8. Danes Say 'No' to Euro	Human Rights	4. Our Constitution for Europe 12. Germany's New Identity 16. 209 Years Later, the English Get American-Style Bill of Rights

TOPIC AREA	TREATED IN	TOPIC AREA	TREATED IN
Immigration	9. New Golden Age 12. Germany's New Identity	**Politics**	1. European Integration: Past, Present and Future 2. Europe 2007 3. EU Enlargement 4. Our Constitution for Europe 5. Seeing Green 7. Europe: Is There a Fourth Way? 8. Danes Say 'No' to Euro 13. Hot and Sticky in Ireland
Industrial Development	5. Seeing Green		
Integration, European	1. European Integration: Past, Present and Future 7. Europe: Is There a Fourth Way? 8. Danes Say 'No' to Euro 12. Germany's New Identity		
		Race Issues, European	12. Germany's New Identity
International Relations	1. European Integration: Past, Present and Future 2. Europe 2007 3. EU Enlargement 7. Europe: Is There a Fourth Way? 8. Danes Say 'No' to Euro 17. Britain, Europe and America: Keeping Friends	**Social Unrest**	15. Survey of Spain
		Social Welfare	6. Poverty in Eastern Europe 9. New Golden Age 11. Domestic Issues Dominate German Agenda
Language	4. Our Constitution for Europe	**Standard of Living**	6. Poverty in Eastern Europe 12. Germany's New Identity 13. Hot and Sticky in Ireland 14. Survey of Portugal
Monetary Issues	1. European Integration: Past, Present and Future 8. Danes Say 'No' to Euro 13. Hot and Sticky in Ireland		
		Unemployment	11. Domestic Issues Dominate German Agenda
Nationalism	1. European Integration: Past, Present and Future 2. Europe 2007		

Great Decisions, 2001

European integration: past, present and future

How will the next steps in European integration affect the U.S.–EU relationship?

by Amie Kreppel

THE EUROPEAN UNION (EU) currently consists of 15 member states ranging from Greece to Sweden and Ireland to Italy. It covers over 1.2 million square miles, with some 375 million citizens. Its gross domestic product (GDP) is the second-largest in the world (only marginally less than the U.S.). It is currently the largest market in the world and the largest trading partner of the U.S. In addition, the member states of the EU are among America's closest political and military allies. Clearly, the political, economic and military futures of Europe and America are closely linked just as their histories have been closely connected in the past.

Because of its size and influence, the future development of the EU will have an impact on not just America, but the world. The EU is in the process of deepening its ties to Africa and Latin America, expanding toward the newly democratizing countries of Eastern Europe, completing the project of monetary union, and initiating an independent security and defense policy. Already the 15 member states of the EU speak with one voice within the World Trade Organization (WTO) and the G–8 (seven most industrialized nations plus Russia) summits. As the process of European integration continues, it is likely that the international role of the EU as a single entity will increase still further.

Understanding the development and history of the European Union is fundamental to comprehending the current issues and dilemmas that face the member states today and the international role of the EU in the future. But it is also important to understand the process of supranational integration in the EU because in many ways it serves as a benchmark and a model of what is possible in supranational institution-building. In the new era of "globalization," regional integration projects are becoming increasingly common. From the North American Free Trade Agreement (NAFTA) to the Association of Southeast Asian Nations (ASEAN) and the common market of South America (MERCOSUR), regional trade associations are emerging across the globe. None of the current trade organizations or customs unions approaches the level of internal EU integration, even in the economic sphere, and their future development is difficult to predict. This makes it imperative to understand the process of EU integration over time and its very humble beginnings to try to comprehend supranational transformations occurring across the globe.

Looking at the EU today it is hard to believe that just 55 years ago the countries of the Continent and Britain were climbing out of the ashes of the second catastrophic war in as many generations.

In 1945 no one could have guessed that in just five short years, Germany, Italy, France, and the Benelux countries (Belgium, the Netherlands, Luxembourg), working together, would begin a process of integration that would lead to the current EU. Over the last 50 years the project of European integration has undergone a massive transformation. From an original plan that was meant only to link German and French coal and steel production to a supranational organization including 15 member states with political competencies ranging from economic and monetary policy to foreign and security policy to environmental and social welfare policy. The rapid and extensive development of the EU is unparalleled. At no time in history have so many countries willingly given up so much of their national sovereignty to a supranational authority.

EU's evolution

Throughout history there have been numerous attempts to unify the European Continent under a single ruler through force. However, from the ancient Romans to Charlemagne to Napoleon the goal has proved either elusive or only temporary. Yet, in the immediate aftermath of World War II (yet another example o the cost of attempting to unify Europe by conquest), the idea of creat-

had allowed it to pursue the territorial conquests that led to war.

However, the U.S. and Britain, who still occupied the Western zones, worked against France in support of German re-industrialization. The U.S. in particular wanted Germany to rebuild its industry because it was believed that the economic recovery and continued independence from the encroachment of Soviet influence on all of Europe depended in part on an economically powerful Germany. In the end, the U.S. and Britain were successful and France had to find some way to deal with the potential of a revitalized German industrial sector.

It was Jean Monnet who eventually proposed the solution. A long-time proponent of European integration and in charge of economic planning in France at the time, Monnet was also well respected in the U.S. and Britain. During the war he had been quite effective in helping to organize the Allied supply line. In a private proposal to Robert Schuman, then foreign minister of France, Monnet proposed the creation of a common coal and steel community between Germany and France and any other European nations that wanted to join. The proposal was quietly presented to the German chancellor, Konrad Adenauer, and visiting U.S. Secretary of State Dean Acheson. After garnering their support and that of the rest of the French cabinet, the proposal was announced publicly on May 9, 1950. Thus, just five years after the end of World War II, two of the primary combatants plus Italy and the Benelux countries were proposing to unify their coal and steel production under a supranational authority. Because at the time coal and steel were not only the fundamental building blocks of industry but also of war, the proposal was essentially guaranteeing that there would be no more war between Germany and France.

The new ECSE (officially launched in 1952) included four institutions: the High Authority, the Council of Ministers, the Common Assembly and the Court of Justice. The most important was the supranational High Authority, with Monnet himself as its first president, in charge of creating a true common market in coal and steel between the six member states. This included

ing some kind of pan-European entity once again flourished. Support for some sort of European unification was found among international leaders such as Britain's Winston Churchill, intellectuals like Count Coudenhove-Kalergi, and entrepreneurs like Jean Monnet of France, who would eventually become the "father of Europe."

Within just a few years of the end of World War II, a series of attempts had been made to unify Europe. These included the Organization for European Economic Cooperation in 1947 (the predecessor to the Organization for Economic Cooperation and Development, OECD) and the Congress of Europe held in 1947 (which led to the creation of the Council of Europe in 1949). Although neither of these was able to effectively create a "united Europe," they did serve to keep the flame of European

integration alive. Adding fuel to the fire was the successful creation of a customs union among the Benelux countries in 1948.

In the end it was a much less grandiose plan that proved to be the first step on the path to true European integration. The European Coal and Steel Community (ECSC) was born out of an attempt to resolve a long-standing dispute between Germany and France over the highly industrial Ruhr and Saar regions on the border between the two countries. Aside from the border disputes between Germany and France over this region, France had also hoped to limit the redevelopment of the German coal and steel industries to protect its own producers (who were still in the process of rebuilding after the devastation of the war). In addition, France as well as the rest of Europe, was well aware that it was Germany's industrial strength that

supervising prices, wages, transport, investment and competition. National representatives appointed to the High Authority were to be economic and sector experts, not politicians. They were also required to work toward the good of all member states, not just their own.

The Council of Ministers was more nationally oriented and consisted of representatives from the governments of the six member states. The Court of Justice was created as an outlet for those who felt they had been wronged by a decision of the High Authority (either an individual, a company or a member state). The Common Assembly was a kind of parliamentary body with the task of supervising the High Authority, although it had very little real power and was fundamentally a chamber of debate.

The institutional framework created by the ECSC remains the basic structure of the European Union of today, although there have been numerous changes in the relative balance of power between the institutions and some name changes. The Common Assembly became the European Parliament and the High Authority became the Commission after 1957.

Euratom and EEC

The next successful step forward on the path toward European unification was the joint creation of Euratom and the European Economic Community (EEC) in 1957. Euratom was an attempt to repeat for atomic energy the success of the Coal and Steel Community and because it related to a specific sector of the economy it was, in fact quite similar to the ECSC. The EEC on the other hand was a significant department from the past. The EEC had as its goal the creation of a single market between the member states for all goods and services. This was a much larger task than either the ECSC or Euratom and proved to be both more difficult and eventually more successful than either.

The joint creation of Euratom and the EEC represented yet another compromise between Germany and France. By the middle of the 1950s Germany was already experiencing its "economic miracle" and was hungry for the larger market for its goods that a general EEC would create. France, on the other hand,

was doing less well economically and feared the effect that an influx of inexpensive German goods (particularly agricultural products) would have on its own producers. At the same time France was the only country with real nuclear power potential on the Continent and the creation of Euratom would push the other member states to contribute toward the development and expansion of France's nuclear energy capacity.

In the end both communities were created and eventually (in 1965) merged with the previous Coal and Steel Community. However, there were some differences between the institutional structure of the new Communities and the ECSC. Most importantly, the balance of power between the High Authority and the Council of Ministers was reversed so that the Council of Ministers was more influential and had the final say on all legislative decisions. As a symbol of this change, the High Authority's name was changed to simply "the Commission." The Court and the Assembly remained largely unchanged.

Between 1957 and 1969 the EEC proved highly successful. The member states experienced unprecedented periods of sustained economic growth and development. So much so tat Britain, which had originally turned down membership in both the ECSC and the EEC, asked to join in 1961 and again in 1967 (their application was vetoed both times by French President Charles de Gaulle). The lowering of intra-member-state trade barriers went forward ahead of schedule, as did the creation of common external tariffs. Despite the optimism of the late 1960s and early 1970s, the burgeoning EEC was headed for a very difficult period, due largely to changes in the global environment. The end of the Bretton Woods system (which guaranteed the convertibility of the dollar to gold) in 1971 led to a period of global-exchange-rate instability. In the European Community, where further market integration depended on stable currency exchange rates, this proved devastating. Adding to the difficulties of this period were the oil crisis of the mid-1970s and a heightening of cold-war tensions. By the late 1970s, Europe seemed destined to stagnate. In spite of the negative attitudes of the 1970s, the process of integration was moving forward, albeit not

with the same energy as it had in the 1960s. Britain, Ireland and Denmark joined in 1973 and by 1979 negotiations were well under way for the newly democratized Greece, Portugal and Spain to join. Also the member states agreed to hold biannual "summits" of the national leaders to discuss political and foreign policy issues not covered by the treaties, with the goal of presenting a united European position in the global political arena. Finally 1979 also witnessed the launching of the European Monetary System (EMS), with an exchange-rate mechanism (ERM) that finally aided the member states in controlling currency fluctuations. The EMS worked so well that it was the springboard for the eventual move to monetary union in the 1990s and the creation of the euro.

By the early 1980s, the prognosis for European integration was improving. In 1984 the European Parliament passed a Draft Treaty on European Union that, while failing to be endorsed by the member states, reopened discussions about future integration and institutional reform. The Parliament's proposal caught the interest of France's President Francois Mitterrand. With his support and the hard work and determination of then president of the Commission Jacques Delors, a new proposal was put forward for the completion of the single market. The member states agreed to convene an Intergovernmental Conference (IGC) to discuss the matter further and work toward treaty reform.

This IGC, held in 1985, led to the creation of the Single European Act (SEA) which revitalized the European Community and reformed its institutions. The primary achievements of the SEA were the introduction of a clear program for the creation of the single market, internal decisionmaking reform (including the introduction of qualified majority decisionmaking in the Council and an increase in the role of the European Parliament) and the formalization of the earlier biannual summit meetings between the heads of state of the member states.

Revitalization

The SEA revitalized the Community and marked the beginning of what has become the most stunning period of rapid integration and expansion in the history of the European integration project. The SEA was fully implemented in 1987 and by 1991 it was clear that further reforms would be necessary to keep the momentum of integration going to complete the single market. To create an effective single market, the countries of the European Community would also have to create a single currency and a true monetary union. Another IGC was convened; this time the goals were more expansive and the resulting Maastricht Treaty (1992) created a new entity: the European Union.

The new EU included the original EEC (including Euratom and the ECSC) but it also formally incorporated elements that did not fall under the jurisdiction of the original treaties, such as foreign and security policy and issues related to immigration, crime, etc. These were incorporated through the creation of a three-pillar structure.

In the first pillar are the original EEC activities; the second pillar includes a formalized European Council to deal with common foreign and security policy (CFSP); and the third pillar deals with justice and home affairs (JHA). The first pillar remains highly supranational with frequent recourse to qualified majority voting in the Council and an active and increasingly powerful Parliament. The second and third pillars, however, are primarily intergovernmental. Unlike the first pillar, most decisions in the second and third pillars are made by unanimous consent. This requires that the member states give up far less national sovereignty because all member states must agree before any action can be undertaken.

Economic agreement

Perhaps the most important aspect of the Maastricht Treaty was that it created a European Monetary Union (EMU) with a single currency (the euro) and a European Central Bank (ECB). The member states agreed to a strict set of economic performance criteria that would be required before any country could join the new "Euroland," including limiting inflation, budget deficits and low long-term interest rates. A schedule for implementation was agreed to with the permanent locking of exchange rates to begin between member states meeting the criteria by January 1, 1999, at the latest.

Eleven member states (out of 15) joined at the beginning of 1999. Three countries (Denmark, Sweden and Britain) decided to wait, while one country (Greece) wished to join but failed to meet the criteria initially. Greece has since met the criteria and became a member of Euroland on January 1, 2001. Denmark, on the other hand, narrowly rejected membership once again by national referendum in October 2000.

The rapid process of European integration did not slow down after the Maastricht Treaty. In 1996 another IGC was held to address further institutional reform, especially in light of probable expansion of the EU to the newly democratized countries of Eastern Europe that had formally applied for membership.

Just as in the late 1970s Europe had been eager to assist the new democracies of the Mediterranean, in the early 1990s the EU wanted to help the countries of Eastern Europe consolidate their democracies and economic regeneration through EU membership. However, enlargement in the east has proved to be more difficult than the Mediterranean enlargement. In part this is because the number of applicant countries is much larger (as many as 13 versus only 3) and in part because EU integration is much further along, making it much more difficult for applicant countries to meet the minimum requirements for accession. All countries that wish to join the EU must adopt the bundle of EU laws and regulations that have already been passed, known as the *acquis communautaire,* or simply the *acquis.*

Eastward expansion is further complicated by the lack of effective external border controls in many of the applicant countries. Europol estimates that as many as 500,000 illegal immigrants enter the EU each year, most from the poorer states of the former Soviet Union (but also from Bangladesh, Iraq and even China). Additionally, 350,000 potential immigrants apply for political asylum in the EU each year. As a result of growing concerns over a possible labor shortage in the EU in the future due to low birthrates, the Commission has recently called for an end to the "fortress Europe" policy of zero immigration. This does not, however, allay member-state fears about the potential for uncontrolled and overwhelming illegal immigration flowing through the weak border controls of the applicant countries in the east.

The 1996 IGC, which led to the 1997 Amsterdam Treaty, failed to achieve many of the reforms necessary before eastern enlargement can move forward. As a result another IGC began in February 2000 with the goal of significant reform of the institutional structure and procedures of the EU. Since these issues directly affect the balance of power between member states, they are highly contentious and many still remain to be resolved.

The issue of institutional reform can no longer be ignored, however, if enlargement is to occur successfully. The EU of today not only covers a much greater array of policy areas then the original EEC, it also includes more than twice as many member states and may soon include more than four times as many. The institutions and decisionmaking structures that were created to allow six member states to move toward integration of their coal and steel markets are proving increasingly incapable of functioning with the current 15 member states and the broader policy competencies of the EU today. They would fail completely if the EU grows to include 20, 25 or more member states.

Institutional dilemmas of enlargement

When the EEC was created in 1957 it included the same original six members states as its predecessor the ECSC: Germany, France, Italy, Belgium, the Netherlands and Luxembourg. The institutions of the EEC were in part designed to function effectively with this limited membership. The Parliament had only 142 members, distributed proportionally among the member states, based on population. The Commission had just nine members. The Council of Ministers was smaller, with just one member from each member state.

In 1973, when Britain, Ireland and Denmark joined the EEC, the number of members in the Parliament increased to 198, the Commission to 13 and the Council of Ministers to 9. The introduction of direct election for members of Parliament in 1979 increased that body's membership still further to 410, while the other institutions remained unchanged. By 1986 the EEC (renamed simply the European Community) included 12 member states, twice the original number. The Parliament increased its membership to 526, the Commission to 17 and the Council of Ministers to 12.

With such a large and diverse membership, unanimous decisionmaking became exceedingly difficult. The SEA addressed that by introducing limited use of weighted qualified majority voting in the Council, but it was still hard to reach agreement. In the Commission the goal of collegial decisionmaking stretched nearly to the breaking point, and often decisions had to be made by vote. While the Parliament was better able to function effectively, it faced a space shortage. All of the institutions struggled under the increasingly heavy burden of working in all of the Community's languages. The amount of time, money and energy required to provide simultaneous translations of debates, meetings and official documents skyrocketed as the number of languages increased from four in 1957 to nine in 1986.

These problems were reemphasized when Austria, Finland and Sweden joined the EU in 1995. By that time, the Maastricht Treaty had already been ratified, so all three countries had to massively restructure their national policies before they could complete the accession process. In addition, their membership brought the total number of member states to 15, with 11 different languages. The Parliament increased to 626 members, the Commission to 20 and the Council of Ministers to 15.

The three waves of enlargements have all been significantly different yet all posed very real problems for the countries joining and the Community as a whole. Britain's membership was contentious, having been vetoed twice by de Gaulle before finally succeeding. Britain brought with it not only a "special relationship" with the U.S. that was initially mistrusted by the other member states, but also a political culture that was significantly different from the other members and far less supportive of the supranational ideal. The Mediterranean enlargement was difficult not only because of economic inequalities, but also because all three countries had comparatively weak internal democratic structures and produced agricultural products in direct competition with France and Italy. The most recent enlargement has brought into the EU fold countries with a clear social democratic bent and extremely high social-policy costs that have been difficult to reconcile with the other member states.

East European wave

As difficult as these past enlargements have been, institutionally, politically and economically, they are nothing compared to the problems raised by the current East European wave of enlargement. All of the current member states share much of their cultural history and West European traditions. Although the Mediterranean countries were certainly economically backward compared to the other member states, they were far more developed than most of the current applicant countries. In total there are 13 applicant countries that were formally acknowledged by the Commission in March 1998. These are Bulgaria, Cyprus, the Czech Republic, Estonia, Hungary, Latvia, Lithuania, Malta, Poland, Romania, the Slovak Republic, Slovenia and Turkey. Of these, five are considered to be closest to meeting the necessary conditions for membership: the Czech Republic, Estonia, Hungary, Poland and Slovenia. Even these more advanced countries fall far short of even the least economically developed current member state.

Aside from the obvious economic difficulties that the next wave of enlargements will pose, there are significant political dilemmas as well. Almost all of these countries have only recently been freed from the burden of Soviet domination. Most continue to have comparatively weak party systems and unstable governments. The ability of these political leaders to function effectively within the institutions of the EU must be seriously questioned. Not only will they bring with them very different political cultures, but they will most likely also have very different political and economic interests from the current member states. In an institutional environment that still requires unanimity for many decisions, the result could be stagnation.

If all of the applicant countries were to successfully complete the requirements for application and gain acceptance, the EU would balloon to 28 members. Although in the short term that number is likely to be closer to 20, under the current rules this would cause the Council and Commission to become largely dysfunctional. The Council of Ministers with 20 members ranging from Estonia to Germany and Slovenia to Britain would probably find it next to impossible to reach unanimous decisions. The Commission would swell to 26 members, seriously threatening its continuation as a collegial body.

Broad institutional reform is clearly essential. Two possible solutions have been discussed. The first is the "Europe à la carte" design that would allow member states to select what aspects of union membership in which they want to participate. This would, it is argued, allow some countries to move forward with deeper integration without being held back by other member states that are less advanced or more reluctant to integrate. It would also free applicant member states from the full burden of the *acquis*, since they would not have to agree to join all aspects of the EU.

EU MEMBERS	
COUNTRY	**YEAR JOINED**
Belgium	1951
France	1951
Germany	1951
Italy	1951
Luxembourg	1951
Netherlands	1951
Denmark	1973
Ireland	1973
Britain	1973
Greece	1981
Portugal	1986
Spain	1986
Austria	1995
Finland	1995

Many believe, however, that a Europe à la carte would spell the end of the European integration project since the impetus for member states to work together to achieve mutually acceptable programs for further integration would be lost. European unity would slowly dissolve as some groups of member states distanced themselves through further integration while others rejected further integration or even reversed some of the steps already taken.

As a result a second model of future integration has been proposed that would allow member states more latitude in determining *when* they would move forward with the integration project, but not allowing them an "opt-out." This model is often referred to as a "tiered Europe," in which a "core" group of member states would continue increasing the supranational aspects of the EU right away, while allowing the other member states additional time to acclimate their peoples and/or their economies to the necessary changes.

The key difference between the two models is that a tiered Europe would not allow permanent "opt-outs"; instead it would simply allow some member states additional time to catch-up to the core group. This model will have to be used to some extent with the next wave of enlargement if the applicant countries are to join on schedule. Most of the East European countries have no hope of being able to adopt the full *acquis* or achieving the euro criteria in the near future. These new member states will have to be given time to slowly adapt their national policies to the realities of EU membership.

The institutional dilemmas faced by the EU in light of past and future enlargements make the success of the current IGC so imperative. Without institutional reform, enlargement toward the east will effectively cripple the decision-making institutions of the EU, potentially bringing future integration to a standstill. Given the already enormous economic and political significance of the EU globally, this could prove to have significant effects far beyond the EU's borders.

AMIE KREPPEL *is an assistant professor of political science at the University of Florida. Her research focuses on political institutions such as parliaments and parties in Europe and the U.S., with particular focus on the European Union.*

Article 2

The Washington Quarterly, Autumn 2000

Europe 2007:
From Nation-States to Member States

Simon Serfaty

Think back. The time is June 1957. A few European heads of state and government are meeting in Rome to launch a small common market as a down payment for an "ever closer" community that many of them question and none truly understand. What has brought them together is the evidence of their predecessors' failures, none as tragic as the two world wars they have just waged and from which they have been partially rescued by their new senior partner across the Atlantic. The peace they now hope to achieve will be sought a piece at a time. Their goal is not to dissolve the nation-states they represent but to save them from themselves as well as from each other.

Think ahead. Fifty years have passed, and the time is June 2007. The small community of six has grown into a larger union of many. Along the way, its agenda became ever more complex: deepen in order to widen, but also widen in order to deepen, and reform in order to do both. More than once, the process seemed hopelessly stalled. Yet, the fiction imagined earlier as "Europe" has turned into a reality: with an identity distinguishable from that of the nation-states that constitute it, with regional institutions distinct from the national institutions that gave them birth, and with a collective discipline often enforced in opposition to its members' preferences. Now, in 2007, the nation-states of Europe have been recycled into member states. Their former condition is a memory and a conviction more than a fact—the lingering memory of the national sovereignty they used to enjoy within their own boundaries and the illusory conviction that it has not been overtaken within the new common space they have agreed to make their own.

By the very nature of history, moments with a predictable impact on the future cannot be identified until the moment is gone. The future must become past before it can be told. By its very nature too, a defining moment remains open-ended even after it has been uncovered. Hindsight determines when it began and even how it closed. Yet, because so much has already happened since 1957, much of what will be celebrated in 2007 can already be anticipated. It is as if history, at last, could be denied its imagination as it brings Europe into an endgame whose outcome can be ascertained before its time has been lived:

- reform of EU governance negotiated through three Intergovernmental Conferences (IGCs)—in July 1997 in Amsterdam, in December 2000 in Nice, and, most likely, in 2004;
- a completed euro-zone, started in 11 countries in 1999, perfected at 12 or more in 2002, and completed at 15 by 2005, with all EU members and many other non-EU states wanting easy access to the fully operational single market that began in 1992;
- confirmation of the EU commitment to enlargement to the East, beginning in 2005, but also possible accession talks with such recalcitrant states as Norway and even Switzerland;
- fulfillment, in 2003–2007, of the post-Kosovo "headline goals" as a down payment for a European Security Policy that would set the stage, at last, for a Common Foreign Policy and, ultimately, a Common Defense Policy; and
- by 2007, too, new forms of U.S.-EU and EU-NATO relations, including new accords between the United States and the EU, as well as a reformed Atlantic Council that coordinates the complementary roles played by the two institutions.

In 2000, no country in Europe, and few of its political leaders, would dare acknowledge the scope and even urgency of the decisions they contemplate and the transformation they face. Historians will marvel. This, they will write, was a time neither for grand designs conceived from the top down, as after World War II, nor for revolutions enacted from the bottom up, as after World War I. This, historians will explain, was a time when the leaders' will to be, as opposed to their predecessors' will to do, was matched by their followers' urge to join and enjoy the system rather than to leave or destroy it. Yet, historians will conclude, so it had been for the process of European integration during and after the Cold War, when every step in a specific area caused unexpected progress in another area—from a supranational coal and steel community to a common defense, from the aborted common army to a modest common market, from an expanding common market to a challenging monetary system to a single market to a single currency and back to a common defense.

Completing Europe

No defining moment can escape the unpredictable event, the unexpected decision, or even the unforeseeable act of God that shaped all subsequent developments. Who could have expected, at the start of each of the past decades, the enormity of what followed—the military conflicts, the economic crises, the political upheavals, the technological revolutions, and much more? Still, several interrelated variables are most likely to shape Europe in the coming years.

- the economic cycle and the pace of growth;
- the quality of political leadership and public support;
- the relative effectiveness of Europe's institutions in addressing their members' common agendas for stability within Europe and on its periphery; and
- the steadfastness of America's engagement in, and commitment to, Europe.

Economic growth is essential. The economic cycle is the starting point to think about Europe's future. A severe economic crisis in 2001–2002 could cause serious political instabilities and sharp societal disruptions that would delay or even reverse earlier decisions by and about the EU. Conversely, a sustained period of growth would help: the construction of Europe is especially effective when it can rely on affluence to produce even more affluence. The 2.5 percent annual rate of economic growth projected by the European Commission's Agenda 2000 for the six-year period until 2006 is a minimum requirement; more, as anticipated for 2000–2002, would be better, but less could be harmful, especially in Germany and Italy. More growth will help each state respond to its citizens' needs—including, for example, education for the young, jobs for the adults, and pensions for the old. But growth will also help the EU institutions fulfill its members' commitments, as it would create additional funds for the provision of subsidies to struggling sectors or underdeveloped regions, as well as for enlargement.

> # Much of what will be celebrated in 2007 can already be anticipated.

Although there is every reason to be bullish on Europe for the coming years, downside risks abound. Whether the recession in Japan will bottom out beyond 2000 will affect this country's neighbors in Asia, but it will also have consequences on the United States and Europe, including Germany and France, which have invested heavily in China and Japan, respectively. Further risks can be found in the emerging markets in Latin America (with immediate consequences for Spain), as well as out of an economic collapse in Russia. Most significant, a hard fall of exceedingly high levels of U.S. equity and share prices in 2000–2002 would affect consumer spending and business investment in the United States and feed through the European economies.

Finally, up to the elimination of national currencies in 2002 and beyond, as the euro goes, so will Europe. Started in the early 1990s against a background of weakening growth, high unemployment, and much public skepticism, the euro emerged, late in the decade, as Europe's most ambitious project for the new century. Its achievements immediately prior to and after January 1999 should not be minimized. The euro was an unprecedented act of collective will: the heads of state and government gambled on Europe's future, and they won, despite odds that seemed insurmountable, especially in the southern countries. Inflation was conquered, budget deficits reduced, and long-term interest rate differentials narrowed. The euro also transformed policymaking in Europe by reducing the range of national sovereignty, diluting the authority of the European Commission, and even challenging the prerogatives of the European Council and the sacrosanct right of representation for all member states in the European Central Bank (ECB). It stood as a rampart against monetary instabilities in Asia, Latin America, and Eastern Europe, against military conflicts in Southeastern Europe, and even against political turbulence in key European countries and within the EU. In 1999–2000, its weakness relative to the U.S. dollar was temporary and helped explore the fluctuating bands between the two currencies, below and above parity.

The role of the euro as a global currency must be clarified.

To be sure much remains to be done. Although Europe's fiscal and monetary policies changed dramatically in anticipation of the euro, these policies will need to be maintained at least until 2002—which would prove politically difficult without a robust economic cycle. National currencies must be formally withdrawn. The euro zone must be extended to all EU states, including the United Kingdom. An effective common economic policy must be devised and enforced. The ECB's full independence of national governments and related pressures must be asserted. Reliable exchange rate arrangements with other global currencies, including the dollar, must be developed, and relations with other centers of monetary authority, including the U.S. Federal Reserve Bank, must be coordinated. The role of the euro as a global currency must be

clarified. As these goals are achieved, new questions about Europe's ability to move on with the euro will emerge. Yet, these are not merely questions about Europe's ability to have a single currency; these are questions about Europe's ability to have a future, as well as America's interest in a flourishing future for Europe. The EU states and their governments have invested too much in the euro to let it go, and the United States has invested too much in Europe to let it falter.

Political centrism is necessary. In coming years, long-term institutional goals of the EU will test the ability of its members' weak centrist coalition majorities to postpone the instant gratification sought by their respective constituencies. For example, the budgetary discipline imposed through the Commission or the ECB may become politically too costly if unemployment remains too high. Moreover, EU governments can call on the same logic of austerity imposed upon them by the EU to demand from the EU a budgetary discipline that the Commission could no longer withstand. For example, after 2006 the ceiling placed on its budget (1.27 percent of its members' gross domestic product) may have to be revised upward to meet the costs of enlargement. Other perennial issues that will demand action even before enlargement include reforming further the Common Agricultural Policy (CAP), abandoning protected national champions to the forces of the free market, and even including in the EU budget an appropriate level of defense spending to permit progress in the security area.

On the way to 2007, such questions—who pays, who gains—could spur a public debate that would erode the political orthodoxy that progressively reconciled Left and Right into a centrist Third Way throughout the EU. "Europe" is an idea that citizens agree to embrace, but it is not yet an ideal for which they are prepared to suffer and, therefore, not a reality on which member governments may be ready to stand for a long time against the will of their constituencies. For much of the 1990s, it is Europe's agenda, defined from the top down, that drove the political calendar of the EU states. In coming years, however, the risk will be that doing

more for Europe might be viewed as higher than the risk of doing less. "I was turned out," complained Margaret Thatcher in June 1995, "because I said 'No' to Europe, 'No, no, no.' " For Tony Blair and others, political risks may ultimately emerge out of an exaggerated " 'Yes' to Europe, 'Yes, yes, yes.' "

The economic cycle is the starting point to think about Europe's future.

Twice in 1999, national governments acted on extraordinarily sensitive issues without credible support from their national constituencies but with a legitimacy borrowed from the institutional label placed on their actions. So it was with the euro, and so it was, too, with the war in Kosovo. In both cases, the centrist consensus held because the policies worked. Risks were high, however. In France, Italy, and Spain, the governing majority would not have survived a failure to meet the Maastricht criteria of economic convergence for inclusion in the core group of states that launched the euro in January 1999. In Italy, Germany, and Greece, the government might not have survived a NATO decision to escalate the air war into a potentially nightmarish land war. Conversely, however, the policies worked because they received the broad centrist stamp of nearly nonpartisan approval often led by center-left heads of government that dared challenge large parts of their constituencies. In the end, the quality of Europe's leadership will determine how well and how quickly the EU can proceed with the difficult agenda set for the years ahead. That such high expectations would be placed on a post-Cold War generation that has often appeared to lack the visionary drive of the post-World War II generation add much irony to the urgency of the moment.

Institutional effectiveness is imperative. The salvation rather than the erosion of the nation-state was the goal of European integration. Indeed, the states of Europe repeatedly sought to reassert their authority and influence over the institution to which they gave birth. Thus, much of what the EU becomes by the year 2007 will depend on its capacity to address effectively the common concerns and priorities that its members are not able or willing to do separately. Current trends in jobs and job creation, income distribution, shrinking populations, depleted pension funds, immigration flows, and education, for example, may not be manageable by any single member. These trends are not, and do not have to be, the same for all members. Nor are they, nor do they need to be, evenly shared. But these trends and the questions they raise are sufficiently common to EU states to suggest needs and priorities that the EU must address on behalf of all its members:

- Inequalities within the EU and EU states, and between EU members and nonmembers, cannot be allowed to persist, let alone widen. Otherwise, there might come a point when the societal and political consequences of these inequalities could no longer be borne by the EU states relative to each other and by the rest of Europe relative to the EU. Especially as enlargement to the east gets under way, the EU will have to engage its neighbors creatively and generously.

- The EU must help its members reconcile those aspects of their economy that are world class and those that drag them toward the bottom of the new global economy. For example, Europe must be ready to service the 60 million people who are expected to have or seek Internet access by the end of 2002, with an electronic commerce estimated at $223 billion.

- European firms must be able to compete globally for new markets, and the EU must help ease the privatization that is necessary before further consolidation within Europe and even across the Atlantic. Yet, such needs should remain sensitive to the uniqueness of an economic space that became common before its political integration could be effectively completed.

- Shrinking and aging populations everywhere in the EU area may erode the domestic cohesion of each EU state and distort the solidarity of the Union. The EU will have to find ways to ensure social peace by assuming the state's traditional role in such areas as the provision of welfare and health care and the management of pensions.

- As boundaries are blurred, the EU should also assume greater responsibilities in balancing Europe's structural dependence on many more immigrants (with Europe's need estimated at 135 million by 2025) and the Europeans' desire to close the borders on any new foreign arrivals.

How well the EU addresses or anticipates the needs of its members will shape the public reaction to EU initiatives. In other words, ask not what the nation-states of Europe can do (and actually do) for the EU, but what the EU does (and can do) for its member states. After World War II, the idea of Europe relied on a public mood that was generally permissive because it seemed to grant political redemption for the state, economic affluence for its citizens, and security for the nation. After the Cold War, however, the idea of Europe must rely on governments that give the EU credit for what it does rather than blame it for what they cannot do.

Security is fundamental. Western Europe entered the new century safer than at any time in the twentieth century and, arguably, any time in its history. The European space is irreversibly civil, meaning that war among the EU states is no longer thinkable. Yet, first Bosnia, and then Kosovo, served as reminders that there is also much unfinished security business in twenty-first century Europe, including not only unresolved ethnic and territorial conflicts that predate the Cold War, but also Russia and a new security agenda of risks imported from adjacent areas outside the European continent.

With regard to Russia, other bursts of military brutality at home, *à la* Chechnya, or outbursts of geopolitical revisionism in the former Soviet space, are matters of primary concern. Developments in the direction of democracy, a market economy, and a community of interests with the West are also important. Yet, Russia is too big, too strong, and too nuclear to be managed by the Europeans alone, but it is also too close, too engaged, and too European to be left to Americans alone. There, as in so many other areas, the states of Europe and the United States, as well as the EU and NATO, play roles that must be made, and will have to remain, both compatible and complementary. At least in the period ahead, the United States will continue to provide the security guarantee of choice through NATO as the primary security institution. The main EU role in Russia will be to contribute to its assimilation into the open, affluent, and democratic space to which Brussels holds the key. For both the EU and NATO, engaging Russia stops far short of a membership that will remain highly unlikely (and even not desirable) for at least one full decade beyond 2007—coinciding, as it so happens, with the 100th anniversary of the Soviet revolution.

Conditions are more complicated (but not more dangerous) on Europe's periphery, along a wide arc of Islamic states extending from the Maghreb region through the Middle East into the Persian Gulf and to Turkey and its neighbors in Cental Asia. These complications do not result from contradictory goals, competitive interests, or adversarial policies among EU countries, as well as between them and the United States within NATO. On the whole, the EU countries maintain comparable goals and compatible interests in the region: deterring the spread of terrorism and containing the rise of radical Islamic governments; avoiding the interruption of oil supplies and the manipulation of oil prices; sustaining the peace process between Israel and the Arab states; preventing the proliferation of weapons of mass destruction; and often, but not always, protecting and enhancing human rights. Even when these goals and interests are not shared evenly, differences among EU states are usually lesser than differences with the United States, especially when they entail the use of military force (mostly, U.S. military force).

Beyond the traditional security agenda, new security risks could disrupt trends in some EU states and, by implication, their position within the EU. Bursts of transnational terrorism, locally grown or imported from non-EU states south of the Mediterranean, could unsettle current centrist conditions in the EU. Elsewhere, renewed conflict in the Balkans or civil wars in North Africa could raise pressures for unilateral interventions that would not seek "legitimization" from the EU, with or despite other NATO countries. After centuries of wars, however, Europe has lost its taste for military conflicts and violence, including politically motivated terrorist acts. Many such crises, therefore, could invite appeasement from EU states, causing discord within the EU and with the United States. That these tensions could erupt in the context of ongoing institutional debates over enlargement to the East is an additional complicating factor.

U.S. engagement is indispensable. After World War II, Europe's fate depended on the two superpowers. Both the United States and Russia took a time-out from their own history to reinvent Europe—one power obviously more creatively and effectively than the other. Besides what the superpowers did or failed to do, the fate of the nation-states of Europe also rested on their ability to overcome their divisions. They, too, needed to take a time-out from their own history if they were to be saved from each other, as well as from themselves. The idea of Europe was hardly born then. But however ancient the idea, it gained life only after the need for unity could no longer be ignored, even as the urge for revenge lingered for many more years.

Now, despite fading Cold War memories, every EU decision has a U.S. dimension, and every U.S. decision can have an EU dimension. That such would be the case is one of the most enduring legacies of the Cold War, not because it means the resurrection of the United States as a European power but because it confirms its status as a power in Europe. Arguably the only Great Power left on the continent, the United States is a nonmember member state of the EU. Thus, the indispensability of U.S. engagement is hardly a matter of false sentimentality. Rather, it is a question of genuine interests that are not matched, *in toto,* in any other region of the world outside the Western Hemisphere. In 1999, the war in Kosovo confirmed what had already been demonstrated

during the war in Bosnia and the negotiations in Dayton in late 1995, namely, the centrality of U.S. leadership and power in a continent where U.S. interests have become too significant to be left to Europeans. This need not be a matter for domestic debate, as shown by the ease with which Congress and the U.S. public endorsed the decisions to enlarge NATO, deploy peacekeeping forces in Bosnia indefinitely, and wage war in Kosovo and participate in its postwar reconstruction. That it would seem to be a matter for debate in Europe betrays a troubling European ambivalence about the United States that will have to be addressed and resolved during the coming years.

The U.S. commitment to a strong and united Europe, however, is not unconditional. The commitment to a united Europe presupposes a transatlantic community built on compatible social values, democratic practice, liberal economic structures, and free-trade policies. The commitment to a strong Europe also anticipates allies that would readily assume a larger share of the common defense burden in and beyond the NATO area. Although common sense would suggest that 50 years of transatlantic solidarity and more than 40 years of European unity have taught both sides of the Atlantic that there are no alternatives to staying the course, some ambivalence continues to blur each side's vision of the other. European allies that follow the U.S. leadership selectively and on the cheap, a fragmented U.S. Congress that emphasizes domestic issues and does not fear divorce with Europe, and a public that does not care much about interests that are not explained well are possible catalysts for transatlantic fragmentation. Such potential for drift would be especially serious if Europe's new strength on its way to 2007 were to cause an assertiveness that would induce the EU and its members to challenge rather than share the global leadership of their senior partner. A significant item on the transatlantic agenda is the development of a restructured relationship between the EU and the United States that would reflect the U.S. recognition of Europe as a copartner and Europe's acknowledgment of a privileged status for the United States within the EU.

Farewell to History

Admittedly, it is possible to imagine circumstances that would stall Europe and its relations with the United States to the point of derailment by the year 2007. Both the European and the transatlantic agendas are ambitious, and their overlapping calendars raise serious risks of overload. Reforming the EU institutions in 2000 and beyond, managing the transition to economic and monetary union in 2002 and beyond, launching a European Security Policy in 2003 and beyond, completing the EU reforms in 2004 and beyond, and enlarging to the East in 2005 and beyond will not be easy. That such a European agenda will have to be addressed from the top down without causing or exacerbating public disappointment, and even anger, from the bottom up adds to the complexity of the coming years. Simultaneously, the road to new U.S.-EU and EU-NATO relations is fraught with many obstacles. Early skirmishes prior to a millennium round of trade negotiations in 2000 and beyond, debating the future of nuclear weapons in 2001 and beyond, defining euro-dollar relations in 2002 and beyond, and ensuring the integration of the "headline goals" and the NATO mandate in 2003 and beyond are all serious matters.

A reversal of Europe's postwar quest for an "ever closer union" could unfold as follows: After 2000, the risks of renewed conflict in the Balkans (starting, say, in Montenegro) and worsening conditions within Russia and in Ukraine would make some key EU members (especially the UK) more fearful of European defense initiatives that might weaken NATO without providing for alternative protection. Simultaneously, new instabilities at the periphery of Europe could also leave the main EU states too distrustful of their respective national goals to permit progress in the area of a common foreign policy. Renewed violence over the unilateral declaration of a Palestinian state, conflicts of political succession in key Arab states (especially Egypt, but also Saudi Arabia and Syria), a drift toward civil war in Iran, renewed conflicts with or within Iraq, instabilities in the Caspian region, and even some antisecular turbulence in Turkey are only a few of the issues that could escalate quickly into a significant crisis for both Europe and the United States, but also for U.S.-European relations. Because of Europe's rising dependence on oil imports, renewed volatility in the greater Middle East would be especially critical for the EU economies under pre-crisis conditions of high oil prices. Farther away, Europe's indifference to other crises involving significant U.S. interests—for example, a war between India and Pakistan over Kashmir or a crisis with a desperate North Korean regime, both with sharp nuclear overtones—would cause even more exasperation in the United States and desperation in Europe.

'Europe' is not yet an ideal for which citizens are prepared to suffer.

In 2001–2002, too, a hard landing in Wall Street could ignite the kind of deep economic crisis (recession and inflation) that disrupted transatlantic and intra-European relations in the 1970s. The euro's failure to shield the EU states from the U.S. economy and to deliver its goals of rapid growth and full employment would make it easier for non-euro states (including the UK) to stay away. A sense of euro-failure would also encourage public debate for withdrawal or, at least, a postponement of the final launch of the single currency past 2002, especially in the context of national elections in Italy and the UK (in 2001) and in France and Germany (in 2002). With most EU members prepared to take a time-out from the aging Maastricht agenda, ratification of the IGC reforms negotiated in 2000 would be stalled in 2001, causing crippling bilateral tensions within the EU, especially between France and Germany. With geopolitical turmoil at the EU periphery, monetary turbulence within the EU area, and open disagreement between the two key EU states over institutional reform, a postponement of EU enlargement, too, would seem inevitable—thus extending the political crisis to applicant countries

in the East. In such a context, the widespread growth of public opposition to Europe would bring the EU to the forefront of the political debates for the next cycle of national elections in 2005–2007, when populist alliances would aim their dissent at "Europe" and the pain imposed in its name.

Haunted by the reported sights of Europe's historic ghosts, a new U.S. administration, halfway through the one-or two-term mandate started in January 2001, might reappraise its commitments to both Europe and its Union. Emphasis would be placed on the unilateral protection of U.S. interests at a lesser cost—meaning less NATO (and certainly no further enlargement) and less EU (and certainly fewer trade concessions), but also less America in Europe (including Kosovo and Russia) and even elsewhere. Denied the United States' leadership, NATO would fade, causing new instabilities in Europe, across the Mediterranean, and beyond. Questions about NATO would come together with more questions about the EU, prompting the United States to return to special bilateral relationships for a more effective management of its interests in Europe.

The United States is a nonmember member state of the EU.

None of these developments, and more, can be excluded. They might not force a collapse of the EU, but they could cripple its institutions to the point of confining the European construction to an unfinished and unsatisfying single market—open in some areas and protected in others. Even a modern Cassandra, however, would view these developments as eminently doubtful. They certainly would mark a brutal departure from the trends observed over the past 40 years. What is doubtful, too, is that such developments would be left to themselves and escalate and spread as readily as they might have during the Cold War, after World War II, or be-

tween the two world wars. In Europe, as well as between Europe and the United States, there is now a self-control that is an especially enduring legacy of the twentieth century. The total war that erupted in 1914 (and again in 1939), the mythical disengagement from Europe that became the U.S. option of choice in 1919 (and was briefly considered in 1945), the economic depression and tariff wars that erupted after 1929, the totalitarian insanity that spread in and beyond Europe after 1933, and the ideological bid for hegemony that was initiated in 1945 are no longer likely. Even the collapse of European integration or a U.S. disengagement from Europe escape definition. In the end, imagining Europe in 2007 without the EU, or the United States without Europe, is a challenge.

Thus, the second half of the twentieth century has left Europe with few alternative futures. By 2007, Europe will not be finished, but it will have continued to move closer, and even close, to completion. On that part of the continent where the EU will have gained a reality nearly tantamount to that of a state, the emotional burdens of history have been put to rest. This is as much of an end game as history can play. Although June 2007 may come too quickly for the EU agenda to be completed, specific dates matter less than the general trends that underline them. Admittedly, the process of turning the nation-states of Europe into the member states of the EU will remain erratic, with moments of national despondency and institutional stagnation. In the end, however, what matters most about the construction of Europe is the *relance* that follows delays and setbacks. For the past 40 years, no member ever pulled out, and the time is gone when any could, as France and the United Kingdom occasionally assumed to be the case. During the coming years, the very few countries that declined an invitation to join (like Norway) or refused to seek it (like Switzerland) are likely to review their decision, even as many other applicants await their entry or the opening of access negotiations.

Assuming a robust economic cycle, effective national and EU political leadership, relative stability in Eastern Europe and at the southern periphery, and continued cooperation with the

United States, the EU will display, by the year 2007, many characteristics often gained at the expense of its members:

- a more muscular Commission, with reduced influence of the smaller members relative to their larger partners, but also of the latter relative to the Commission president; and a more assertive Parliament, relative to the Commission but also in relation to national legislatures;
- a changed weighting of votes in the Council of Ministers devised to extend qualified majority voting while protecting the more populous members against blocking minorities in a wider EU;
- a deeper range of policy responsibilities that would give the EU further authority for priorities (from the job and industrial policy to food safety and visa policy) that individual members cannot address on their own for lack of political will or economic resources;
- a fully operational single market with a completed euro-zone grouping most if not all current EU members as well as various non-EU and even non-European members that might enter the zone for reasons of their own;
- new steps, beginning with the 2003 Headline Goals, toward a European Security Policy, a Common Foreign Policy, and, ultimately, a European Security and Defense Policy that would give Europe more autonomy of action within the context of an enlarged NATO; and
- between 20 and 25 members, including at least half of the current applicants from the East, plus, possibly, smaller states like Cyprus and Malta and even, arguably, Norway and Switzerland.

What about the other Europe, however, where the virtues of integration have not been recognized yet? There, possible futures do abound, and which future comes to pass may affect the pace of what comes to be elsewhere. Late in the 1990s, the ghosts sighted in the Balkans were those of an earlier time when nations were looking for a state and states were looking for their borders. This old-fashioned anarchy still prevails in some

parts of Europe, but resulting conflicts, as in Bosnia and Kosovo, have seemed both out of place and out of time—there, in Europe, and now, at the close of the twentieth century. Unlike previous European wars that often started in the Balkans, these conflicts could, therefore, end with relative ease. In short, historic legacies in Southeastern Europe cannot be neglected, but their earlier defining characteristics—namely, a potential for the spontaneous escalation of a local limited conflict into a general and total war—no longer apply.

If not in the Balkans, might these legacies of strife and violence be more significant farther in the East, especially Russia? There, the case for a return of history is more worrisome. Entering the twenty-first century, the problem of Russia is Europe's problem—the hole in the doughnut of the increasingly thicker dough that shapes Europe's integration. What will be of Russia—moderate and well integrated, alienated and combative, broken and collapsed, or in a worse combination, well integrated and combative, or collapsed and alienated? No one knows, and no one can even imagine it yet. For Russia, 2007 is just one date in the near future. It will take many more years, indeed several decades, for this country to come to terms with its place in Europe, whatever that place may be, and its role in the world, whatever that role may become. In the meantime, neglecting its needs, offending its pride, ignoring its interests, or challenging its power can only be done at everyone's peril, in Europe and across the Atlantic.

Staying on Course

There is a need for thought and prudence, then, but no great need for imagination. A more united Europe is good for the states of Europe and it is good for the United States too: whether now or by 2007, there is no better plausible alternative. Any other conclusion would fundamentally challenge what has been the most consistent goal of U.S. policies toward Europe during the past 50 years, as well as Europe's most consistent objective for more than 40 years.

As the debate over the boundaries of permissible differences between the ascending influence of Europe and the

peerless power of the United States unfolds, the most compelling vision statement is to complete the vision that guided both sides of the Atlantic after World War II and throughout the Cold War. Since the North Atlantic Treaty was signed in 1949, and the Rome Treaties in 1957, U.S. policies in Europe and the integration of Europe have served U.S. and European interests well. More Europe, but also more America in Europe will continue to serve these interests well too. The central lesson of the twentieth century is that America's problems in Europe result from Europe's failures: a war that cannot be ended, a revolution that cannot be controlled, or, closer to us, a currency that would not be stabilized and sustained.

Entering a new century, America's main fear about Europe should be a Europe that is weak and divided, and our main hope should be a Europe that becomes stronger and more united. After the old Europe had caused far too much pain for too long, the birth of a new Europe has been relatively quick and extraordinarily civil. Midway through the century, a "good" European like Stefan Zweig agonized over the many futures he had been forced to outlive: the Europe of the *fin du siècle*, extraordinarily vibrant and prosperous, but also prepared to spread its dynamism and prosperity to others on the continent and beyond; postwar Europe, willing to bear the burdens of the slaughters for which it had been responsible during the previous few years; and interwar Europe, overwhelmed by the tyrants, torn by its passions, betrayed by its intellectuals, ready to fall once again in the abyss that it had dug for itself. "How many lives must I live," moaned Zweig after an existence that he found too hopeless to endure. The past 50 years have also had their share of killing, but now at least Zweig would no longer feel hopeless. At last, Europe's long and brutal struggle with its history may be ending.

Simon Serfaty is professor of U.S. foreign policy at Old Dominion University and director of the Europe Program at CSIS. His most recent book is *Memories of Europe: Farewell to Yesteryear* (CSIS, 1999).

Article 3   *Europe,* June 2000

EU Enlargement

Continent could be unified by the end of the decade

By Bruce Barnard

The Europe Union's historic enlargement to the former communist nations on its eastern flank is picking up speed and paving the way for the unification of the continent by the end of the current decade.

After the successful launch of the single currency, the euro, enlargement is the major challenge facing the European Union. Moreover, EU leaders raised the stakes at their Helsinki summit last December by agreeing to allow five additional applicants to start accession negotiations alongside the five frontrunners that began talks more than a year ago.

The talks with the "first wave" applicants—Poland, Hungary, Czech Republic, Estonia, and Slovenia as well as Cyprus—are moving onto extremely sensitive issues, such as farm subsidies and the free movement of labor. These prickly issues are testing the tempers of the negotiators, slowing progress, and prompting speculation that the EU's enthusiasm for enlargement is waning.

The negotiators knew the going would get tough when the controversial farm and labor dossiers were put on the table because of the wide gap in their respective positions. Brussels doesn't want the applicants, some of whom like Poland have extremely large farm sectors, to have full access to its farm subsidy program, the Common Agricultural Policy (CAP), because it would bankrupt the EU budget. Agriculture already accounts for more than half of EU's $90 billion annual spending. The applicants,

EU Members
Possible First Wave of New Members*
Possible Second Wave of New Members**
*Cyprus not shown
**Malta not shown

however, are pushing for full benefits from the CAP.

The two sides are also deeply divided over labor rights. The EU, fearing a flood of cheap labor from the East, wants a long transition period before freedom of movement of labor while the applicants say their citizens should be allowed to work anywhere in the EU once they are members.

The standoff soured relations between the EU and some applicant countries, whose officials were already annoyed by European Commission President Romano Prodi's remarks that Brussels must take a tough line in the negotiations. Poland's president, Aleksander Kwasniewski, warned enlargement could be threatened by a growing "virus of selfishness." Jerzy Buzek, Poland's prime minister, cautioned that delay in enlargement could cause disillusionment in the applicant countries. Support for EU membership in Poland has fallen to around 50 percent from a peak of 80 percent. Hungary has accused the EU of employing delaying tactics.

The remarks are aimed at domestic audiences on both sides and do not accurately reflect the state of the negotiations. However, the Austrian government's inclusion of the far-right, xenophobic Freedom Party, which has expressed concern about enlargement, has stirred unease in the applicant countries.

Even without these outside pressures, the negotiations are facing enormous difficulties as they confront an EU rulebook that runs to more than 30,000 pages, spanning everything from environmental standards and antitrust policy to farming and monetary union. The previous enlargement, involving Austria, Sweden, and Finland, which were all richer than the average EU member country when they joined, involved three years of sometimes difficult and acrimonious negotiations, while the accession of Spain and Portugal, both poorer than the EU average, lasted seven years before culminating with their memberships in 1985. The entry of even poorer East European countries, which have large farming sectors and are in the throes of painful transitions to market economies, were expected from the outset to be much more complicated.

However, even as the negotiations with the frontrunners stumble (the talks with the second wave applicants are only just getting underway), there have been developments on both sides. The EU will hold a constitutional conference this year to reform its treaties and working practices, particularly voting systems, so that policymaking will not grind to a halt when membership expands from fifteen countries at present to twenty and then twenty-five and more.

The EU will hold a constitutional conference this year to reform its treaties and working practices so that policymaking will not grind to a halt when membership expands.

The negotiations have been complicated by the uneven rate of economic reform among the applicants, with Poland and the Czech Republic falling behind Hungary, Estonia, and Slovenia.

The Czech Republic, in particular, is skidding. The Commission said government reforms were "unsatisfactory" in its latest evaluation of the country's suitability for EU membership. Liberalization of the telecommunication industry has been postponed for three years; most utilities remain state-owned; and antitrust and other business legislation remains on the drawing board.

The EU is also expanding into the Mediterranean. Cyprus is negotiating with the first wave, and Malta began talks alongside the second tier of East European and Baltic applicants—Bulgaria, Romania, Slovakia, Latvia, and Lithuania.

The applicant countries also are changing rapidly as they strive to complete the transition to a market economy with sweeping privatization and deregulation programs and refocusing trade to the West, especially the EU, which now accounts for more than half of their imports and exports. Change is everywhere: Poland floated its currency, the zloty, in April. Bulgaria is selling 51 percent of state-owned Bulgaria Telecommunications to a Dutch-Greek group. Germany's Deutsche Telekom paid $565 million for a controlling stake in Radiomobil, the Czech mobile phone operator. MOL, the Hungarian oil and gas company, is paying $262 million for Slovnaft, a Slovakian oil company, and US Steel is investing $700 million to transform VSZ, the Slovakian high-quality steel maker it recently bought for $160 million, into an export launch pad into the European Union.

American, European, and Asian investors are bullish about the region. Foreign direct investment in Poland hit a record $8 billion last year, boosting the total to $39 billion, or 40 percent of flows into the former Soviet Bloc. Estonia and Hungary also are favorites for outside investors, while the Czech Republic, the former hot spot, is losing its attraction.

Despite this progress, the EU's earlier talk of the first new members joining in 2002 or 2003 has given way to a vaguer horizon, with 2005 regarded as the most realistic date. The EU is also grappling with a difficult problem: what to do if smaller countries are ready to join before Poland, the largest and most strategically important applicant country. Poland's progress has slowed, and Slovenia and Hungary are favorites to qualify first for membership. However, letting them in first could trigger a backlash in Poland, undermining the EU's enlargement strategy. Should the EU allow Poland in whether it is ready or not or should it delay letting other countries in until Poland is ready?

The smart money is betting that enlargement might not take place until 2006, allowing the first countries to join the EU on its fiftieth birthday.

Article 4   *The Economist,* October 28, 2000

OUR CONSTITUTION FOR EUROPE

America's founding fathers met in Philadelphia to draft their constitution.
More humbly, and to provoke debate rather than, necessarily, to please the
constitutional lawyers, ours have met in St James's Street

A WELL-DRAFTED constitution for the European Union should, if nothing else, impose clarity on confusion. The Union does have constitutional arrangements, but they are woven into a series of treaties so long and complicated that nobody can understand and remember all of them. Something much shorter and simpler will be needed if the nature and purpose of the Union is ever to be appreciated by even a fraction of its citizens. Even more important, the current treaties set an objective—"ever closer union"—which alarms many Europeans.

From an ideological viewpoint, the idea of a new, clearer constitution may appeal most readily to supporters of a centralising Union. They will see it as a means for installing more sovereignty at the Union level, so making the Union less reliant on member states for its legitimacy and its powers.

But the idea of a constitution appeals also to some defenders of the nation-state, such as President Jacques Chirac of France. They see it as a way of limiting explicitly the powers of the Union and reasserting those of national governments.

That leaves, as the natural opponents of a constitution, Eurosceptics who think the Union should not exist at all, or that it should be as weak as possible, and who thus oppose something that might help it function better. But since the European Union does exist, and has been increasing its powers over the years, the case in principle for a constitution defining those powers is strong.

The tricky question is, of course, what a constitution should say. Here is our suggestion. It relies for its provisions and its language mainly on the treaties of the Union; and on a proposed "Basic Treaty" for the Union, drafted this year by the European University Institute in Florence at the request of the European Commission. It also borrows from the American constitution.

Nonetheless, it seeks to define the basic aims and capacities of the Union not as they are today, but as we think they should be. It is designed to co-exist with the Treaty of Rome and with the other treaties of the Union. However, these would require substantial amendments. Our constitution proposes fundamental reforms to the institutions of the Union. It changes, and is intended to change, the balance of power among them.

It recognises the European Council, in which heads of government meet, as the highest policy-making body of the Union. It subordinates the European Commission to the Council, as a civil service. It fixes new voting rules for the Council of Ministers, and obliges it to disclose how ministers voted there. It shrinks drastically the size of the European Parliament, in the hope that this will produce a more effective assembly whose members are recognised as more substantial figures in their home countries. It provides for a new chamber of representatives of national parliaments, the Council of Nations, and charges this body with the task of constitutional oversight. This helps to entrench the principle that the Union can be held properly accountable to its citizens only if the established democratic institutions of member states play a greater role.

Our constitution recognises that governments can both give powers to the Union, and take them away. It recognises the right of a country to leave the Union. These sovereign rights are incompatible with a constitutional obligation to pursue "ever closer union"; so that phrase is not retained.

This document thus acknowledges the intergovernmental character of the Union, as well as its supranational aspect. But it does not fix limits to the integration that governments may wish to pursue. It presumes that governments will go on making treaties to modify the powers of the Union in particular fields and to fix the modalities by which these powers are exercised. But it demands that any new treaties be agreed unanimously among member states; or, if they are not agreed unanimously, then that they are binding only on their signatories.

Furthermore, it seeks to discourage further integration of certain kinds. Notably, it calls on the Union to uphold a "principle of subsidiarity" which favours the devolution of power and would be policed by the Council of Nations. Also, it requires the Union to continue relying on national governments for its financial resources rather than being able to raise taxes on its own behalf. The constitution itself can be amended, but only if governments agree unanimously to do it and if citizens back the idea of referendums in every country.

It should also be noted that, despite the role of the new Council of Nations in constitutional oversight, the enacting of a European Union constitution will tend to increase the influence of the Court of Justice. Governments will need to choose their judges wisely. Read on—and respond.

A *Constitution* for the European Union

We among the states of Europe, seeking to encourage peaceful, open and constructive relations between our peoples, and seeking to advance our common interests in the world, ordain and establish this Constitution for our European Union. This constitution shall prevail over other European and national law, including treaties of the Union, should conflict arise.

ARTICLE 1
Founding principles

The Union is established by the Treaty on European Union signed in Maastricht on February 7th 1992, and founded on the European Communities.

The Union shall uphold the principles of liberty, democracy and the rule of law.

The Union and its Member States shall respect the fundamental rights of citizens, including, but not limited to, those rights guaranteed by the European Convention for the Protection of Human Rights and Fundamental Freedoms signed in Rome on November 4th 1950, and rights common among Member States. (1)

All powers, other than those clearly delegated to the Union by this constitution and by the treaties of the Union, are reserved to the Member States.

The Union and the Member States shall uphold the principle of subsidiarity. (2)

ARTICLE 2
Languages

English, French and German shall have equal standing as the sole official languages of the Union institutions. (3)

ARTICLE 3
Citizenship

The Union shall have legal personality. Every person holding the nationality of a Member State shall be a citizen of the Union. Citizenship of the Union shall complement and not replace national citizenship.

Any citizen of the Union having the right to move and reside freely within his own Member State shall have the right to move and reside freely within the territory of all Member States.

A citizen of the Union residing in a Member State of which he is not a national shall have the right to vote in his country of residence only.(4)

ARTICLE 4
Institutions

Heads of government, one from each Member State, shall meet at least once every six months, as the European Council. (5) This European Council shall be the high policy-making body of the Union. It shall give instructions and guidance to the Council of Ministers and to the European Commission, which shall be published after each meeting along with voting records. (6)

The Union shall be served by the following common institutions: a Parliament, a Council of Nations, a Council of Ministers, a Commission, a Central Bank, a Court of Justice and a Court of Auditors. These institutions shall possess those powers, and only those powers, granted to them through treaties ratified by all Member States.

ARTICLE 5
Parliament

Parliament shall consist of representatives of the peoples of the Member States. Representatives shall be elected by direct universal suffrage for terms of five years. Their number shall not exceed 100. Seats shall be allocated among Member States in reasonable proportion to population. Parliament shall fix its own rules of procedure. (7)

Parliament shall debate the policies and the legislation of the Union. It may strike down legislation and it may propose amendments to legislation, where this is authorised by the treaties of the Union. It may request the Commission to propose legislation to the Council of Ministers. It may bring actions before the Court of Justice.

Parliament, including committees of the Parliament, has a general right to question in public hearings any member of the Commission, or any proposed member of the Commission.

Parliament may, acting by a two-thirds majority of its members, dismiss any member of the Commission. (8)

No judge shall be appointed to the Court of Justice without approval from the Parliament.

Save as otherwise provided here and in the treaties of the Union, Parliament shall act by a simple majority of the votes cast.

ARTICLE 6
The Council of Nations

The Council of Nations shall consist of representatives drawn from the parliaments of Member States, according to procedures devised by the respective parliaments. Seats shall be allocated with reference to population, save that every Member State shall have at least two representatives and the number of representatives shall not exceed three times the number of Member States.

The Council of Nations shall act as a constitutional council. It shall have power to overrule the Court of Justice. It may strike down legislation. The Council shall act by a simple majority of the votes cast.(9)

ARTICLE 7
The Council of Ministers

The Council of Ministers shall be the legislature of the Union. It shall consist of one representative of each Member State. Each representative shall have the rank of government minister, and shall be authorised by his government to make commitments on its behalf.

The Council of Ministers shall consider, and, when it so decides, enact, laws and resolutions furthering the aims of the Union as set down in this Constitution and in such other treaties as Member States may from time to time enact, provided always that any such treaties have been ratified by all Member States.

The Council shall act by unanimity where the treaties of the Union require it to do so. At all other times it shall act by a double-majority system: to carry, a vote shall be supported by a majority of Member States, containing a majority of the Union's population. (10)

An agenda shall be published before each Council meeting. A voting record shall be published immediately after it. (11)

In all other respects, the Council shall fix its own rules of procedure.

ARTICLE 8
The Commission

The Commission shall be the secretariat of the Union. (12) It shall consist of a president and 12 commissioners (13), having authority over a civil service. It may propose and draft legislation for the Union, at the direction of the European Council or at the request of the Parliament. It shall have the right to bring cases before the Court of Justice. It shall have the general task of ensuring that the laws of the Union are respected.

The European Council shall appoint the President of the Commission, acting by a simple majority. The President shall be appointed for a term of five years, which may be renewed. Member States shall propose candidates for commissioners' posts from among their own nationals. The President of the Commission shall choose his commissioners from among those candidates, also for terms of five years, and shall decide their responsibilities.(14)

Commissioners shall act in the general interest of the Union. They shall neither seek nor take instruction from any Member State nor from any private interest.

The Commission shall act by a simple majority of its members.

ARTICLE 9
The Court of Justice

The judicial power of the Union shall be vested in the Court of Justice, and in such inferior courts as Member States may ordain and establish through treaties. The Court shall be the supreme court of the Union in matters of Union law only, save that it may be overruled by the Council of Nations on matters which the Council of Nations considers to be constitutional in nature. The court shall have appellate jurisdiction over inferior courts, including those of Member States, in matters of Union law only.

Each Member State shall appoint one judge to the Court of Justice, save that no appointment shall be made without the approval of the Parliament. A judge in office may be dismissed only by a vote to that effect by both the Parliament and the Council of Ministers. The retirement age for judges shall be 70.

The judges shall elect a President of the Court from among their number, and shall fix their own rules of procedure.

The Commission, the Parliament and the governments of Member States have the right to bring actions before the Court. The Court may choose to hear actions brought by private and legal persons.

No judge shall seek or take instruction from any Member State or from any private interest.

ARTICLE 10
The Court of Auditors

The Court of Auditors shall examine the revenue and expenditure accounts of the Union and its institutions. At least once each year it shall provide the Parliament and the Council of Ministers with a statement of assurance as to the reliability of the accounts, and the legality and regularity of the underlying transactions. This statement shall be made public.

Each Member State shall nominate one member to the Court. Each member of the Court shall act in the general interest of the Union. None shall seek or take instruction from any Member State or from any private interest.

ARTICLE 11
The Central Bank

The Central Bank shall be governed solely by an executive board consisting of a President, a Vice-President, and five other members. (15) Each shall be appointed by the European Council, by simple majority vote, save that heads of governments representing countries outside the Monetary Union shall not participate in voting on these appointments. Each executive board member shall be appointed to an eight-year term, which shall not be renewable.

The Central Bank shall define and implement the monetary policy of the Monetary Union, this Monetary Union consisting of all, and only of, Member States that have adopted the euro as their sole legal tender.

The primary aim of the monetary policy of the Central Bank shall be the maintenance of price stability within the Monetary Union.

The Central Bank shall hold and manage the official foreign reserves of those Member States within the Monetary Union. It shall have the exclusive right to authorise the issuing of banknotes and coins within the Monetary Union.

Members of the executive board shall neither seek nor take instructions from any government nor any private interest.

ARTICLE 12
Taxation

The Union shall levy no taxes.(16)

ARTICLE 13
Commerce

The Union, and Member States, shall strive to remove all obstacles to the free movement of goods, and services, and capital within the Union. Save that governments may disallow the free movement of specified goods and services where there is a clear and significant risk to public health, or public order, or national security. No national law regulating the taxation of income or profit shall be construed as an obstacle to the free movement of goods, or services, or capital. (17)

The Union shall fix common rules on competition to assist the proper functioning of free markets.

The Union shall fix a common regime for trade between Member States and other countries.

ARTICLE 14
Monetary Union

Membership of the Monetary Union is open to all Member States, save that Member States within the Monetary Union may impose reasonable, objective and non-discriminatory entry criteria on Member States wishing to join the Monetary Union.

ARTICLE 15
Justice and home affairs

The Union shall fix a common policy for the entry of foreign nationals on to the territory of Member States. The Union shall fix a common policy for the granting of asylum by Member States.

A person charged with a criminal offence in a Member State carrying a sentence of imprisonment shall be given up for extradition, on the demand of a high court, by any other Member State in which he may be residing or in which he may have taken refuge, promptly, or on completion of any prison sentence he may be serving, or about to serve, when the extradition request is made. (18)

ARTICLE 16
Foreign and defence policy

The Council of Ministers shall appoint a High Representative, authorised to speak for the Union in matters of foreign and defence policy on which the Council of Ministers has agreed a common position. The High Representative shall have the right to seek decisions from the Council in matters of foreign and defence policy. (19)

Member States shall seek to agree common positions when acting in international organisations, save on questions of national representation.(20)

ARTICLE 17
Other policy areas

Member States may, through treaties, grant powers to the Union in other policy areas, and take back powers granted

previously, so long as such treaties are ratified by all Member States. (21)

ARTICLE 18
Treaties made among groups of member states

Member States may make treaties among themselves to which some but not all Member States are signatory. Parties to any such treaty may choose to make the treaty justiciable before the Court of Justice, provided that:

i) Nothing in the policy content of the treaty contradicts anything in the main policy content of this Constitution or any existing treaty of the Union;

ii) The parties include at least half the Member States of the Union at the time of signature, and no other party at any time;

iii) Any other Member State may accede to the treaty at any time, subject only to reasonable, objective and non-discriminatory criteria.

ARTICLE 19
Accession

Member States, acting by a three-quarters majority of states, may agree to admit to membership of the Union other countries that are able and willing to meet the obligations of membership. (22)

ARTICLE 20
Suspension and secession

The Council of Ministers may suspend the voting rights of a Member State, if that Member State departs from the basic values or violates basic rules of the Union. (23) In such cases the Council of Ministers must act by a three-quarters majority of states, exclusive of the Member State that is the subject of the vote.

A Member State may leave the Union at any time. (24)

ARTICLE 21
Amendment

This constitution may be amended only by all Member States acting unanimously, and after a referendum in each and every Member State on the proposed amendment or amendments.

1. There is thus no need for an additional charter of fundamental rights, currently under discussion.
2. This enshrines in the constitution a principle that governments have often proclaimed, but which has rarely been used in practice as a way to judge or justify new initiatives.
3. This shifts the main burden of other translation to member states that want it.
4. There may be arguments for assigning the voting right to the country of nationality, rather than the country of residence. But taxation is based mainly on place of residence. The formula chosen here preserves the link between taxation and representation. A citizen may, on the other hand, run for office in any country that will allow him to do so.
5. One country, one representative. France will have to decide whether to send its president or its prime minister.
6. This describes, and so institutionalises, current practice. It denies the European Commission the status of "high policy-making body" that federalists would wish to assign it.
7. Including its choice of seat, so ending the monthly commute between Brussels and Strasbourg.
8. A new power. Until now Parliament has been able to dismiss the commission en bloc only. This new power for Parliament checks the greater power given to the commission president in choosing his commissioners.
9. We favour separating constitutional oversight from the other duties of the Court of Justice, and charging a chamber of parliamentarians with this responsibility. The reason is that the court would have a stronger tendency to extend the reach of European law than would such a chamber. This is the tendency which our constitution aims to resist.
10. A new formula. The current system is one of "qualified majority voting", whereby a proposal must command at least 62 out of 87 possible votes in the European Council. Member states agreed in 1996 that the voting system needed reform before the Union could add many more new members. This formula links voting power more directly to population.
11. A new requirement, which obliges governments to reveal to the public how their ministers have voted on Union business.
12. A new description, recognising a shift in political power away from the commission and towards the European Council.
13. A new formula. At present the commission has 20 members, including the president, two from each bigger member state and one from each smaller member state. Governments agree that reform is needed here before another enlargement of the Union. "Capping" the commission at a fixed size, regardless of the number of member states, is one option.
14. A new mechanism, designed to encourage member states to "compete" for commission places and so to offer better candidates.
15. There will be no national bank governors. With enlargement, the inclusion of national bank governors on the governing council of the central bank will become unwieldy and (since big countries will

have the same representation as small ones) inequitable.

16. A short article, but an important one. It underlines the standing of the nation state within the Union. Any transfer of power to tax should require not merely unanimity among governments, as for treaty changes, but a constitutional amendment, which demands also the direct endorsement of citizens through referendums.

17. This puts flesh on the bones of subsidiarity in an area of policy where the issue is especially likely to be fudged.

18. The minimum of guaranteed co-operation that would be needed as an alternative to harmonising national systems of criminal justice in a Union of open borders.

19. A minimal mechanism for ensuring that member states can be obliged to consider common positions and common actions in respect of world events. In practice, the Union is evolving mechanisms for institutionalising common foreign and common defence policies which are likely to be the subject of future treaty provisions.

20. Thus national governments with, for example, permanent seats on the UN Security Council, cannot be obliged to support the abolition of those seats.

21. An "ever closer Union" is not, therefore, a constitutional obligation. Powers can also be returned to member states, if all member states agree.

22. Action by majority vote is proposed here so that accessions cannot be blocked by a local squabble, or by one country's threat of a tactical veto; and also on the grounds that the addition of any one new country to the Union is unlikely to be a matter of vital national interest to any one country already within the Union. This clause also removes Parliament's right to block an accession through a simple vote. But Parliament could always challenge an accession before the Court of Justice, on the grounds that the candidate country was not "able and willing to meet the obligations of membership".

23. A simplified version of current rules, and one that leaves the Court of Justice to decide, if asked, what is a "basic value" or a "basic rule" at the time in question.

24. A new provision, perhaps surprisingly.

Article 5

Europe, February 2001

Seeing Green
Forecasting Europe's Environmental Future

By Ariane Sains

It is spring in Warsaw and a gray veil hangs above the city. Below, the murky Vistula River flows through the city, carrying debris and factory wastewater. Downtown, cars spew fumes from unleaded, low-octane gasoline. With the clanging of the trams and honking of outraged motorists trying to get around the city's gridlocked streets, it's hardly a silent spring on the ears, but it comes close environmentally.

In Stockholm, it's also spring, and fishermen in their hip boots are out in the Strömmen, the Baltic tributary that flows through the city, hoping to catch salmon for dinner. Although air pollution is there, the sky looks clear, and there is no gasoline smell. Flowering trees are blooming in Stockholm's myriad parks. People with cloth satchels full of empty glass and plastic bottles are conscientiously carrying them to the recycling bins located throughout the city.

As it heads into the next century, Europe is a continent environmentally divided. On one end of the environmental spectrum stand the Nordic countries, Germany, and the Netherlands with their stringent regulations and strong interest in the environment, followed by other EU countries with somewhat lesser environmental records. At the other end of the spectrum are the countries of the former Soviet Union and Eastern Bloc, who face serious environmental problems ranging from nuclear waste contamination to toxic waste and the burning of brown coal in power plants that spew greenhouse gas emissions into the air. As much as they may want to, the people of these countries don't have the money for cleanup. At the same time, countries such as Poland, Estonia, and the Czech Republic are heading for European Union membership, and that makes their environmental problems the EU's problem.

West Versus East

When the Iron Curtain fell in 1991, it revealed environmental problems in the former Soviet Union and the East Bloc virtually incomprehensible to Westerners. On the Kola Peninsula, at Murmansk, nuclear submarines are leaking radioactivity into the North Atlantic. The waste scrap from nickel factories lies in heaps, slowly polluting both soil and water. In middle Europe, the infamous Black Triangle where the coal-mining industries of Germany, Poland, and the Czech Republic meet is a horrible example of industrial progress with environmental neglect. The air, soil, and

Heavy car pollution threatens air quality in many large European cities.

Cherepovets in Russia to clean up water and reduce the dust that is clogging the town's air.

Separately, the Finns started a bilateral program with St. Petersburg to improve drinking water and water management, in an effort to stop waste. Work is also underway to improve the handling of sewage and industrial waste. The programs are open-ended. The Finns, who have long had one foot in both parts of Europe, made environmental questions in Eastern Europe and the former Soviet Union one of the key issues of their EU presidency, which began July 1, 1999. Their idea was to make enough of a start during their six-month reign that work can continue regardless of who has the presidency. In 2000, the EU created an action plan for the program dubbed the Northern Dimension. The cleanup will take years. It's estimated that in northwest Russia alone cleanup costs for nuclear waste, poisonous chemicals, and toxic metal scrap will top $1 billion. At least part of that money has to come from the West.

The Finns see environmental issues as going far beyond the environment. In the case of the former Soviet Union, many environmental questions go to the heart of security policy, says Finnish President Tarja Halonen.

Another critical question for the EU is the state of nuclear reactors in potential member countries such as Lithuania and Bulgaria as well as in Russia. EU and European Commission officials pressed Lithuania to shut down its two Chernobyl-type reactors in return for the right to begin EU membership negotiations, and Lithuania has agreed to shut one of them in 2005. However, that takes money, and since the reactors produce about 80 percent of Lithuania's electricity, they have to be replaced with something. The most likely solution is to fire up existing power plants. But they are old, inefficient, and burn heavy oil. Instead of improving the environment, the Lithuanians argue shutting down their nuclear power will only make greenhouse gas emissions worse. On top of that, the average cost for electricity would double, something neither Lithuanian industry nor householders can afford.

The question of cost goes to the heart of the environmental schism between

water in the region are so polluted that virtually nothing can grow there.

In an effort help solve some of the problems, Western European environmental ministers worked with their Eastern counterparts to create an action program based on Agenda 21, the EU's program for environmental improvement into the new millennium. They also began working with the Eastern European ministers to set environmental priorities and help create environmental regulations, which most of the countries lacked. Even when they were there, they didn't conform to EU regulations. Under the TACIS program, local governments in Western and Eastern European countries and the former Soviet Union are paired in projects designed to improve the environment in the East. In one project, local councils from Mendip and Somerset in the United Kingdom are working with their counterparts in Svetlogorsk, Belarus, to develop a sustainability strategy, which includes helping to educate people in Svetlogorsk about the environment and why they should care about preserving it. In another project, the French communities of Grande-Synthe and Dunkerque are working with

Eastern and Western Europe. Beyond the need for cleanup money is the difference between what people in Western countries with high standards of living are willing—and, more importantly, able—to pay for environmental improvement compared with people in Eastern Europe and the former Soviet Union. It's not that people in those countries aren't environmentally conscious. Many are. The Green movement is gaining strength and Greens already sit in the governments in Slovakia and Georgia. But others say that paying higher taxes to fund recycling plants or research into renewable energy is a luxury they can't afford. Keeping Chernobyl-type reactors, with their cheap electricity, running is the path of least resistance. Dumping waste and metal scrap into the ocean is also cheap, and the way things have always been done. As they move toward EU membership, however, it will be up to the EU to convince the countries of Eastern Europe and the Soviet Union that they not only can clean up the environment, they must.

More than 300 European cities and towns are part of the Sustainable Cities Program, in which the Europeans are paired with municipalities in developing countries. In contrast to other, earlier programs in those countries that focused on specific areas such as air pollution or waste management, Sustainable Cities aims at just what its title infers: creating an overall program that takes all environmental aspects into account in a coherent policy. Essentially, it's an outgrowth of the way EU policy for member states is moving.

History and EU Policy Today

Western Europeans have slowly come to the realization that environmental protection is no luxury. As far back as 1973, the EU had a tough environmental program. In 1992, the European Union crafted Agenda 21, a framework program intended to improve the environment well into the new millennium. The program covers sustainability, clean air, clean water, and industrial environmental improvement.

In 1993, the European Union adopted its fifth Environmental Action Program, intended to carry out the specifics of the Agenda 21 framework program for sus-

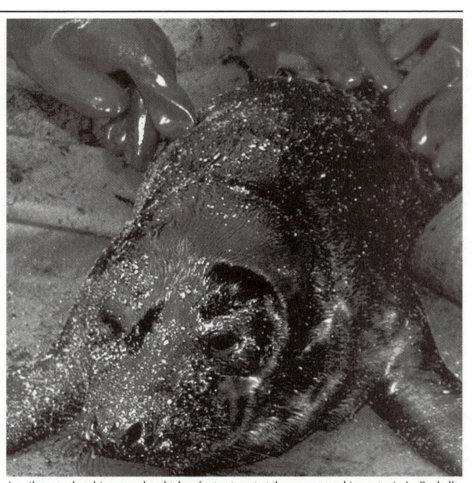

An oil-covered seal is rescued and taken for treatment at the oceanographic center in La Rochelle, France following the sinking of the Maltese oil tanker *Erika,* which sank in the Bay of Biscay on December 12, 1999.

tainable development. Under five main groups—Industry, Energy, Transport, Agriculture, and Tourism—the program outlined a series of ambitious goals and policy instruments for trying to meet them. Key among those policies are economic methods. In other words, making environmentally friendly choices and sustainability attractive financially while making the opposite more expensive (usually through taxes) and, therefore, less attractive. Three years after the program began, a review indicated that progress was being made in reducing the release of chemicals that deplete the ozone layer and in reducing emissions of sulfur dioxide. That was the good news. The bad news was that little had happened in the transportation sector, one of the areas with the heaviest impact on the environment in Europe.

The new Treaty for Europe, crafted in Amsterdam in 1997, has sustainable

development as one of its main goals. One of the most important provisions is that member countries can introduce new regulations that are more stringent than EU standards. The treaty also gives more power to the European Parliament to legislate environmental policy. In 2001, a sixth environmental program is finally expected to be ready.

For member countries, incorporating EU regulations into national law is a complex, time-consuming, and expensive process. Moreover, despite efforts to forge a common EU environmental policy, member states clash regularly with the Union. Many of the disagreements have to do with protectionist trade policy cloaked in environmental garb. Denmark, for instance, has been waging a years-long battle to continue a domestic ban on metal cans, claiming they cannot be properly recycled or disposed. But while they want to keep cans out of Denmark, the Danes

An operator at Kozloduy nuclear power plant in Bulgaria checks the terminals in the control center of a reactor. The state of nuclear reactors in potential member countries poses a critical question for the EU.

have no problem with exporting their Tuborg and Carlsberg beer in the dreaded metal. That, said the European Commission in 1998, violates both EU environmental policy and the principle of the open European market. In response, the Danes said they had no intention of lifting the ban or changing Danish law. The Commission has applied to the European Court for Justice for a decision against Denmark.

The Danes are not alone. Greece has been cited for allowing sea turtles to be hunted, in violation of EU policy on endangered species, and Finland for hunting endangered birds.

The Greens

In Western Europe, the Greens began gaining visibility in the 1970s. In many cases, the movement grew out of the student protests of the late 1960s. For some years after that, Greens in most European countries felt it was their duty to work outside the establishment. But as the movement developed, even the more radical Greens began to realize that without being in the political inner cir-

cles, there was a limit to what they could accomplish.

They began to organize politically for election to local, regional, and national legislative bodies. Their success is perhaps most notable in Germany, where Green leader Joschka Fischer was appointed foreign minister following the 1998 election and Jürgen Trittin was made minister of environment and nuclear safety. While Fischer is from the more mainstream wing of the German Green Party, Trittin is from what's known as the fundamentalist wing. He has clashed sharply with Germany's heavyweight nuclear industry and with other cabinet members over decommissioning Germany's nuclear power (see sidebar).

The Greens' surge has not been limited to Germany. Its members are also part of the governments in France, Italy, and Finland. In Sweden, the Greens are part of an alliance formed by the minority Social Democratic government, which chose to enlist Green and Left support to form a parliamentary majority. In Finland, the Greens joined the so-called rainbow coalition of Prime Minister Paavo Lipponen in 1995, with

party leader Pekka Haavisto winning the dual role of environment/ development minister. Haavisto hammered out an agreement with the rest of the cabinet that put the building of new nuclear plants in Finland on hold for four years. The Greens also succeeded in raising taxes on less environmentally friendly forms of energy, notably fossil fuel. Furthermore, Greens hold local or regional offices in many more European countries, an important step on the road to national office.

Besides influencing policy in their home countries, Greens have gone to the European Parliament. One of the most notable is France's Daniel Cohn-Bendit, who led the 1968 French student protests that reverberated throughout Europe.

Taxation

Using taxes as a way to steer environmental policy is part of the EU's strategy and one that is being adopted in many member countries as well. The Netherlands, for instance, has a wide range of taxes covering energy, recycling, and fossil fuels, part of its second national environmental policy plan, which took effect at the beginning of 1996. On the energy side, the point of the taxes is to encourage individual consumers of electricity to conserve. At the same time, tax revenue is used to offset income tax payments in what the Ministry of Housing, Spatial Planning, and Environment says is an effort to "shift the tax burden away from labor and capital based-income and toward use of the environment." But the Dutch policy also takes practical business considerations into account. The horticulture sector, which not only produces Holland's ubiquitous tulips but also vegetables exported year-round through.out Europe, lives on very slim profit margins and uses a large amount of energy compared to the number of employees in the average company. That means energy taxes have a greater impact on those companies than on others. While still required to pay the electricity taxes, the horticulturists receive a break on natural gas, which is vital to their businesses in order to keep their greenhouses running. The system is a practical compromise between good environmental policy and industrial need. In addition, horticulturists have agreed to improve energy efficiency by 65 percent between 1980û2010. Other countries, and various political parties, are also trying to use taxation as a policy tool.

After beginning with the so-called "command and control" approach—in other words, dictating requirements to member states—the EU's policy is evolving into more of a carrot-and-stick approach. EU officials say that is necessary to meet new environmental goals and forge a truly common policy. They say too that common policy will help eliminate the temptation for member states to act in their own interests at the expense of the union and will also make for smoother operation of the borderless internal market. Joint regulations governing transportation, for instance, and the kind of fuels vehicles can use would eliminate the problems created by some countries having less stringent pollution control standards for vehicles or allowing lesser grades of fuel.

Another reason for moving away from command and control is that it tends to fragment policy. Regulations are generally aimed at specific problems or areas, such as air or water pollution, but don't work in a broader context. Individual regulations often don't take into account their effect on other areas. By broadening the way policy is made, EU officials hope they can create a more coherent environmental approach.

An essential step here is more industrial involvement. Former Environment Commissioner Ritt Bjerregaard maintains that "telling industry what to do, and in which way it must be done, does not encourage a proactive approach." Bjerregaard claims that with relatively small, short-term investments industry can make major environmental improvements and win positive public relations. The idea of industry voluntarily changing its practices for environmental betterment, says Bjerregaard, "is no utopic dream."

Green Sells

At least in some EU countries it appears she's right. Clear-cutting in Swedish, German, and Finnish forests didn't stop because the forest-products companies in those countries were altruistic. It stopped because it was legally banned. But, in the process, the companies discovered that green sells. Western European consumers, in fact, are often willing to pay more for products they consider to be environmentally friendly, everything from unbleached paper to recycled glass.

In Sweden, even electricity has become green. When the electricity market was deregulated in 1996, utilities suddenly realized that they were competing to sell kilowatts that were the same as every other utility's kilowatts. They also realized that Swedish consumers were highly environmentally conscious. In cooperation with the Swedish Society for the Conservation of Nature, a private, well-respected environmental organization, a program was developed to give an environmental seal of approval to types of electricity generation that don't harm the environment. Utilities are also trying to convince consumers to pay more for environmentally friendly electricity, out of social conscience. With electricity market deregulation starting on the Continent, the approach will likely spread there. France's giant state-owned utility Electricit de France is already studying the Swedish model.

However, consumers not only want to be told that products are green, they want proof. And that applies to commercial customers as well as household consumers. When Germans, for instance, began asking Swedish forest-products companies how their goods were produced, the Swedish companies analyzed every step of their production processes. That meant looking at everything from how trees were cut to what type of electricity was being used to produce finished goods to what kind of packaging materials were used. The process was long and expensive but, say company executives, worth it. Market share actually increased in Germany because both industrial and household consumers liked the idea of buying green products.

And nothing happens in a vacuum. What the forest-products companies did impinged on the utilities. They also analyzed everything about their production processes, to show consumers that they were environmentally responsible from beginning to end.

Swedes and Germans aren't the only ones sensitive to export markets and the best way to increase sales. Finland exports more than 80 percent of its forest products, with 70 percent going to the European Union. There, notes Hannu Valtanen, director of natural resources and forest policy for the Finnish Forest Industries Federation, "environmental and ecological aspects have become important."

Clean Up After Yourself

Buy a new toner cartridge for a laser printer in more than ten Western European countries and it's likely to come with a self-addressed label for postage-paid return to the manufacturer. A number of countries require that producers take responsibility for recycling everything from those cartridges to shavers, cars, and electronic equipment. Plans are to significantly increase all forms of recycling in the twenty-first century.

But recycling often also means consumers end up paying more. In Sweden, producers are required to take back and properly dispose of electronic and electrical equipment that consumers previously have thrown out. To finance that, producers will charge take-back fees, which could total nationally as much as $118 million per year. For consumers that translates into higher prices of anywhere from about $3.50 more for a cel-

Not Easy Being Green

Germany's Green Party makes a tricky transition from the opposition to part of the governing coalition.

The Chernobyl nuclear power plant is located in the Ukraine, more than 1,000 miles from Berlin. Yet, this facility is intimately connected with the history of Germany's Green Party. Because it was an accident at Chernobyl in 1986—the worst of its kind in history—that prompted German voters to begin taking Green policies seriously. By the time Chernobyl was finally shut down last December, the Greens were in control of three government ministries and the country had committed itself to phasing out nuclear power.

The Green Party's roots run deep into the soil of the 1960s protest movement. The arms race was heating up, and cold war rivalries were being played out on German territory. Warsaw Pact and NATO troops faced off along Germany's internal border. Mass protests were held in West Germany, opposing the government's nuclear policy and demanding disarmament. When Ronald Reagan visited Bonn in 1982, he was met by nearly half a million angry demonstrators. The following year would see the first Green representatives elected to the German parliament.

Parallel to their antiwar efforts, the Greens continued to push their agenda on environmental protection and sustainable development. Their "No Nukes" battle cry translated into an energy policy favoring renewable energy sources and rejecting nuclear power. Environmental issues would gain added significance after the end of the cold war. However, even before that, Green activists in Germany had made opposition to nuclear energy a top priority. In 1985, they began occupying nuclear power plant construction sites. It was against this backdrop that the news came of a nuclear accident at Chernobyl.

It didn't take long for Chernobyl's radioactive (and political) fallout to reach Germany. Having first played down the danger to public health, Helmut Kohl's conservative government issued belated warnings for people to stay indoors and not eat foods at risk of contamination. Public confidence was shaken. Environmental awareness rose, and the Greens credibility grew.

In the 1998 general election, the Greens won just more than 6.5 percent of the vote and formed a coalition government with the Social Democrats. Both parties agreed on the goal of phasing out nuclear power. Moreover, after months of negotiations a deal was struck requiring the power industry to shut down its nineteen nuclear plants gradually over a period of more than thirty years. The compromise arrangement—negotiated in part by Germany's Green Environment Minister Jürgen Trittin—disappointed party traditionalists who wanted the reactors taken off-line much faster. It was a Pyrrhic victory for the Greens.

The responsibilities of government have also forced the Greens to compromise their traditional pacifist stance. In order to stay in the government coalition, the party had to endorse the participation of German troops in the Kosovo conflict. This issue caused upheaval within the party. The measure was supported by the country's foreign minister, Joschka Fischer—himself a Green—who won the backing of the party's "pragmatic" wing. Nevertheless, traditionalists were outraged, accusing the leadership of abandoning their principles. Power has come at a price.

To some degree, the Greens are victims of their own success. Since entering government to phase out nuclear power, introduce an ecological tax scheme, win legal recognition for same-sex marriages, and push through limited dual citizenship. Ironically, by making progress on these issues the party has robbed itself of several important rallying points.

Now, looking ahead to the next election in 2002, the Greens are struggling to redefine their agenda. The party is expected to campaign for a liberal immigration law, greater reliance on alternative energy, and improved public transportation. Their declared goal is to defend their status as Germany's third-strongest political force and continue governing with the Social Democrats.

After two years in government, the one-time protest party has become an integral part of the political establishment. If a Chernobyl-like accident were to happen in Germany today, the Greens would be forced to share responsibility. The realities of power politics occasionally place the Greens on the wrong side of the environmental fence.

—*Terry Martin*

lular phone to as much as $30 for a television or personal computer. Some 150,000 tons of equipment is expected to be collected annually.

There is also concern that if fees are set too high, retailers may be tempted to pocket the money instead of using it to finance recycling because profit margins on their goods are low. If that happens, it will be a classic example of the difficulty throughout the EU in creating a common, sustainable environmental policy.

What Next?

As part of Agenda 2000, the EU's sweeping policy strategy for the new millennium, the Transport, Energy, and Agricultural councils need to revamp their regulations so that they are geared toward sustainability. In addition, there is Agenda 21, which deals specifically with promoting sustainable development well into the next century.

In the new century, the buzzword for Europe is climate control. The EU countries have agreed to reduce greenhouse gas emissions by about 8 percent from 1990 levels between 2008 and 2012, and carbon dioxide emissions 15 percent by 2010. While EU officials say the target will be met, they also warn that further reductions won't be possible until a blanket carbon dioxide tax is adopted. Agenda 21 specifies carbon taxes as a crucial way to help further reduce emissions, but such a tariff is one of the most bitterly debated issues among EU members. The failure of countries to agree during the climate conference in the Hague in November 2000 on a system for reducing emissions doesn't help the situation. Furthermore, while the EU may have a common reduction goal, member countries have different individual requirements. Some have set more ambitious national goals. Denmark, for instance, wants to reduce carbon dioxide emissions 20 percent from 1988 levels by 2005. To do that, the Danes have begun a sweeping program to replace old coal plants with new, clean technology, but it won't be completed until well into the next century. Other countries, such as Sweden, however, have dispensations to increase levels somewhat.

Energy and the kinds of fuels used to generate electricity are an essential part of Europe's emissions problem. Increasingly, Europe is considering natural gas as an alternative. An EU study envisions a natural gas network that would incorporate Russia, the Baltic countries, and Northern Europe to bring gas to the Continent. Deregulation of the European gas market means high gas prices are beginning to fall, making it economically feasible to build more gas-fired electricity plants. But there is sharp division about just how environmentally friendly gas is. Often, opposition to gas comes from countries, such as France, that rely extensively on nuclear power. The French argue that using gas will, in fact, increase their greenhouse gas emissions, and their official line is that there are no negative consequences from nuclear power.

If It Rolls, It Probably Pollutes

Another major part of Europe's emissions problem is leaded gasoline, which is still widely used in Western Europe and is even more common in Eastern Europe. The Nordic countries are far ahead here, since they began phasing out leaded gas in the early 1990s. But take a deep breath in central London or Paris and it's a different story. The high cost of gasoline is one reason the changeover is taking so long in many parts of Europe. Leaded gasoline is expensive enough; unleaded costs even more, which means motorists have no incentive to use unleaded. The Swedes solved this problem by subsidizing unleaded gasoline for a number of years. Other sorts of alternative fuels are also subsidized, including ethanol and rapeseed oil. However, while that has helped reduce emissions from cars, truck transport in the Nordic countries makes for heavy carbon dioxide pollution. Throughout Europe, truckers pay special taxes under an EU program designed to help pay for pollution control and research into alternative fuels, but they still get to drive. Moreover, subsidies will only work for so long. Eventually, alternative fuels will have to pay their own way.

The same is true of alternatives such as electric cars. Along the lines of its unleaded gas subsidy, Sweden is proposing tax breaks to help offset the higher cost of electric cars, especially when their use is measured against how much regular cars can be driven. France has also tried to promote electric cars, especially in traffic-clogged Paris, where the fumes from idling cars can be overwhelming. But consumers resist, largely because electric cars aren't convenient.

Fewer Species

The European Environment Agency says that measures aimed at safeguarding Europe's biodiversity—keeping the range of plants and animals we now have—have fallen short. Financial greed—short-term economic gain is a more polite term—instead of concern for the environment in the long term is behind the problem. Again, the problem is most severe in developing countries or the countries of the former Soviet Union and Eastern Europe, which tend toward short-term environmental thinking.

To try to mitigate the problem, the EU is setting up protected areas throughout Europe, under a program called Natura 2000. The system is supposed to be completed by 2004. Again, however, member states lag behind EU vision. Some have been slow to designate protected areas, which will hurt the overall program as well as delaying much needed protection for fragile ecosystems. Beyond that, in poorer, more rural areas, short-term farming interests take precedence over long-term environmental protection. That is likely to be an even worse problem as the EU begins to assimilate Eastern European and former Soviet countries that have largely rural economies.

EU policymakers claim that reduction of farm-price subsidies helps the environment because it reduces incentives for overproduction and means less use of chemical fertilizers. But even so, about a quarter of Europe is considered rural, and farming still plays a key role in affecting the environment.

Into The Woods

Coupled with biodiversity is forestland management. Because so much of

Europe is covered by forests and because the forest industry is a central part of the economies of so many member states and candidate countries, including Sweden, Denmark, Finland, Germany, Italy, Latvia, and Lithuania, protecting the forests is a key environmental issue for the European Union. Some 315.8 million acres of land in the EU is forest, up more than 10 percent from 1960. Forest policy goes hand-in-hand with agricultural policy and since it touches on everything from air, water, and soil pollution to jobs, greenhouse gas emissions and protection of endangered species.

The Finns have spent several years developing a system for certifying their forestland, and they want the EU to adopt it. The Finns also want the Union to step up forestry research and air pollution control related to the industry.

It Takes More Than Legislation

Whether it's energy or leaded gas, recycling or clean water, European officials are increasingly realizing that policy isn't enough. The citizens in every European country, as well as industry, have to be convinced that improving the environment makes sense. With that in mind, when Europe's environment ministers convened in Arhus, Denmark, in June 1998, they invited non-governmental organizations and citizens' environmental groups to join them, as well as representatives of the European Federation of Green Parties.

The EU has concluded voluntary agreements for improving cars' fuel efficiency, which would in turn help cut carbon dioxide emissions, with the European, Japanese, and Korean automobile manufacturers associations. And officials believe that good environmental policy is also good economic policy, since it will help create jobs and stimulate the right kind of tourism.

But major problems remain in Western Europe and the environmental dilemma of Eastern Europe and the former Soviet Union is monumental. Their problems may well turn out to be the EU's greatest environmental challenge in the twenty-first century.

Article 6　　　　　　　　　　　　　　　　　　*The Economist,* September 23, 2000

POVERTY IN EASTERN EUROPE

The land that time forgot

Sorting out the impoverished rural parts of Eastern Europe is one of the stiffest challenges facing the European Union. All the odder that it seems not to care

DOLNA, MOLDOVA

A FATHER and his two small sons sit beside an emerald pond, netting frogs amid the rising dragonflies and falling swallows. The father trades the frogs for sugar or cooking oil. "It's one way of staying alive," he says. Beyond the pond, a valley unfolds: Pushkin's woods in the heights (somewhere in those trees, in 1821, the exiled poet fell in love with Zemfira, a Gypsy princess) then cow pasture, vineyards, orchards, and at last the blue-washed houses and rutted dirt lanes of Dolna, a village of 2,000 souls, lost in central Moldova.

Maria, arms crossed against her ample chest, stands at the end of a dung-spattered lane where Dolna peters out into the orchards. A toddler son and a ten-year-old daughter squat at her feet, displaying the tell-tale listlessness of malnutrition. A softly bearded elder son stands off to one side wearing a curious, vacant expression. "He's epileptic," Maria explains. She works as a day-labourer in the fields, when she can, for the equivalent of 80 cents a day. Sometimes she works just for food. She receives occasional handouts of pasta from an American government food pro-

gramme, but no other state support. Her son's medicine costs half her monthly earnings; often she saves by giving him herbs instead. What would make her life better? "A piece of butter for the children once in a while," she says.

Just one story, among millions like it, in Eastern Europe's forgotten villages. It is difficult, criss-crossing the region, to credit that the 21st century has arrived at all. Rather, the millennium seems to have marked a return to the 19th century. Out of necessity, villages have reverted to survivalism. Pathetically, the symbol of post-communism there is the hoe, the cheapest farming tool available. People have replaced tractors in the fields, rows of them bent double in the sun.

Under communism, poverty was covered up. Official figures were artificially enhanced and are deeply unreliable. But even allowing for that, Eastern Europe's villages are in fast and visible decline. This state of affairs—the slide of tens of millions of Europeans into the third world—may end up costing the European Union, in the long run, far more than any war.

Unplugged from the state

In these villages, as in the towns, life-expectancy has been cut by five years or more since communist times. Illiteracy, in the poorest villages, has tripled. Again, allowance should be made for falsified communist figures; but signs of deterioration are evident. Malnutrition is rife. Villagers eat less food and of lower quality: the OECD estimates that the consumption of meat in Eastern Europe has halved since 1990. Since then, too, the incidence of tuberculosis and hepatitis has doubled in the region. Poverty is on the rise. Half of all Moldovans, according to a recent World Bank study, now earn less than $220 a year, down from $2,000 in 1992. "Moldova," the study notes, "has possibly the highest endowment of human capital for a country at its level of income."

Not for much longer. Human capital—the skills of tractor mechanics, nurses, winery managers, indeed, of almost everyone with an education—is dissipating at an alarming rate. Economic collapse and government neglect have seen a rapid erosion in the basic social services villages provided, albeit poorly, under communism. The notion of the village as a primary care unit offering basic schooling and health services, limited employment and rudimentary community life is being lost.

The situation east of the so-called Belgian curtain—the new Brussels-imposed line that skirts the former Soviet border, excepting the slightly more prosperous Baltics—is especially dire. Judged by cash and in-kind income, almost all of rural Moldova, Belarus and Ukraine has seen a collapse in living standards since 1991. These economies were not so much unshackled as unplugged from the Soviet Union, and many economists wonder whether Moldova and Belarus are viable nation states at all.

Elsewhere, the picture is not uniform. In Slovakia and Hungary, the poorest villages are often welfare-dependent Gypsy communities that lack leaders and farming skills. In Poland, apart from the poor south-east, villages are better-off than in 1990. Communities along the Bug river in eastern Poland, a region which most Poles consider to be in terrible shape, look prosperous compared with those on the other side of the Belgian curtain. On the Bug, the land is worked by machinery; most homes have a reliable, if expensive, supply of gas and electricity; villagers shop in well-stocked stores and, most important, the state is visible and strong.

Large and inefficient agricultural sectors—in human terms, impoverished villages—are one of the main obstacles to admission into the European Union for Poland, Romania, Bulgaria and, possibly, Slovakia. Some 24% of Poles and 36% of Romanians still work in agriculture. In Moldova, that figure is closer to 50%. Across the region, agricultural inefficiency is the rule. Ukraine has some of the best farmland in the world—freshly ploughed, the soil is jet-black—but exports pitifully little produce. Poland has a $600m trade deficit in agricultural products, Romania a deficit of $311m. Raw material is often scarce: only 10% of Moldova's cattle stock is now on large farms. The rest, owned one or two to a family by villagers, are too scrawny and diseased to interest agroprocessors. Polish supermarkets prefer to fly in watermelons from Morocco rather than ship them from neighbouring Ukraine.

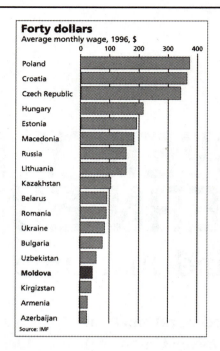

Forty dollars
Average monthly wage, 1996, $

	0	100	200	300	400
Poland
Croatia
Czech Republic
Hungary
Estonia
Macedonia
Russia
Lithuania
Kazakhstan
Belarus
Romania
Ukraine
Bulgaria
Uzbekistan
Moldova
Kirgizstan
Armenia
Azerbaijan

Source: IMF

Strip farming, flower seeds

Across the region, formerly collectivised land has been reduced to a kind of medieval strip-farming, with slices of field just large enough to support a family but not to compete in any market. Even for farmers with the skills and capital to succeed, the economics are crushing. Polish farmers saw a 5% decline in gross revenue last year, while the cost of their inputs—mostly fuel and machinery—rose by over 10%.

Closer to the bottom is Anatol Ojog, a farmer in Slobozia Dusca, a village of 3,500 on the banks of the Dniester river in eastern Moldova. Before the Russian financial crisis of 1998, Mr Ojog sold his tomatoes for almost a dollar a kilo; now he gets ten cents. Over the same period, he reckons, the price of his dollarised inputs—plastic, electricity, gas to heat his greenhouses—has risen by 600%. To survive, Mr Ojog has stepped

up productivity and has joined a co-operative intent on exporting flower seeds to the EU. If Mr Ojog, a trained agronomist and well travelled, cannot find a way of succeeding, few will.

Where Dolna is average, Slobozia is rich and educated. It managed to preserve its human capital and benefited from its closeness to Chisinau, the capital, an hour to the west. Before 1998, several farmers there had made enough money (mainly by growing fruit and vegetables under plastic) to buy cars and start building new houses.

In Soviet days, the village collective was an experimental farm for Moldova's agricultural university. It had the best livestock and technologies and the know-how to make them work. Fine fruits were grown here and sent, hand-wrapped in tissue, to the Soviet space programme. "Comrades", a cosmonaut wrote back, "your pears were most delicious. We ate them in orbit."

That was 20 years and a civilisation ago. Now the remains of Slobozia's collective farm poke up behind the village like a scuttled battleship. Its dairy, built with American equipment in heady perestroika days of the mid-1980s, has been ransacked. Everything has been stolen, down to the metal hinges on the door. Slowly, the bricks are being pilfered. The water pipes that once irrigated the fields have been dug up—from four feet underground—and sold for scrap. This means serious problems for Slobozia, since wells in the village are drying up. Only yards from the Dniester, the crops are now dependent on rainfall. Water is a problem across the region, even in basic ways. The communist-built public baths where people used to wash and relax have closed in most villages.

Communism offered women equal opportunities, to a point; but there was always a dark side to it, and that side

Better under communism
New developments in rural social services in Moldova

	1990	91	92	93	94	95	96	97	98
Dwellings, 1,000 m² of space	254	na	na	na	na	98	70	68	53
General-education schools, '000 places	12.3	6.0	2.7	1.2	0.6	0.2	0	0.3	0.1
Pre-school institutions	5.0	5.1	2.6	0.9	0.7	0.05	0.09	0.2	0
Clubs and palaces of culture, '000 places	1.2	5.4	3.0	1.1	0.9	0	0.7	0.4	0.4
Polyclinics, '000 visits per shift	1.4	na	na	na	na	0	0.07	0	0.1
Hospitals, number of beds	319	106	235	0	20	0	0	0	0

Sources: National statistics; World Bank

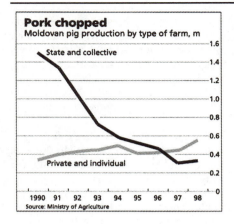

Pork chopped
Moldovan pig production by type of farm, m

State and collective

Private and individual

1990 91 92 93 94 95 96 97 98
Source: Ministry of Agriculture

has endured. Alcoholism deprives many women of support at home. Moonshine vodka also contributes to increasing levels of physical and sexual abuse: according to a women's rights group in Ukraine, 30% of women in Ukrainian villages have been raped. As for the elderly, many survived Stalinism and the second world war only to see their pensions become worthless. Mihail Arsene, a 74-year-old in Slobozia, lost both his legs in the war and receives a disability pension of about $10 a month. His wooden legs are broken; but the only factory that manufactured prostheses in Moldova closed five years ago.

Across the battered parts of Eastern Europe, children are the greatest hope and the biggest cause for concern. The education system is failing them. Never mind Internet access; many village schools have to close in the winter for lack of heat and electricity. Children used to be fed two or three meals a day by the schools; now they get nothing. Hungry children, teachers passionately point out, find it hard to concentrate. Even well-fed children must tend animals and pick crops at the expense of homework.

Teachers are little better off. Their salaries, particularly in Ukraine and Moldova, often go unpaid and, in any case, rarely exceed $50 a month. Supplies are often pitiful, with books and equipment dating from the Soviet period. "We need financial and moral support," says Constantin Nicula, the headmaster of Dolna's village school. Yet his school is one of the lucky ones. A World Bank programme has paid for a new heating system, another scheme gives the school free coal. Teachers cheer up their classrooms with fresh

flowers. "But even if we do a good job," says Mr Nicula, "these children have nowhere to go." In communist times, 90% of children from Dolna went on to some kind of vocational or further education. Now only 5% go on to further education, and these are the richest rather than the brightest children.

Medical clinics are in even worse condition than the schools. At first glance, Dolna's clinic is reassuring. The doctor, Vera Mutu, wears a clean white uniform; everything around her is neat and tidy. It soon becomes apparent, however, that she has little to offer the sick. Even thermometers and bandages are in short supply. The Dolna clinic received $90 from the government last year, not including Mrs Mutu's monthly salary of $15 (another theoretical figure—she has not been paid since August last year). To survive at all, village doctors take small payments from patients. The death rate in the village has increased alarmingly over the past ten years.

No one, save the critically ill, can expect medicines without first paying for them. This was common under communism too; but UNICEF reckons Moldova now has only 20% of the insulin it needs for diabetes patients. There is talk of setting up local health co-operatives, administered and paid for by villagers themselves. But time is running out. Destitute village doctors are leaving their rural practices in large numbers. Within five years, some analysts say, there will be no rural clinics worth saving in the poorest areas.

The atrophy of state institutions presents similar problems. Where there was once some authority, albeit communist, there is now just a vacuum. The state, when it reaches the villages at all, is too often hopelessly underfunded or corrupt, or both. Sometimes the situation is almost laughable. Oleg Bulat, a stout police captain from near Slobozia, is trained to investigate rural car crashes. He used to attend a hundred prangs a year, sometimes cutting people from the wreckage. Nowadays, he says, he can never get to a crash because there is no petrol for his car. Policemen everywhere rely on small bribes to survive, undermining the credibility of law enforcement for decades to come.

Why, compassion aside, should Europe care? Perhaps because Eastern Europe's villages have become a wellspring of illegal immigration into Western Europe. Around 600,000 out of 4.3 m Moldovans are now estimated to be working abroad, mostly illegally. Some of this movement is temporary; people work abroad for a while, then return. But between Belarus and Bulgaria it is hard to meet a young person who is not desperate to start a new life in the West. Villages should not, and cannot, hold back their bright young people; but they need to retain at least some of them to survive. As it is, Dolna and Slobozia's best are working as hotel maids in Italy, prostitutes in Turkey and construction workers in Portugal.

They go through the Belgian curtain into Western Europe with the same persistence as Mexicans who smuggle themselves into the United States, their desperation intensified by the high interest rates on the money they have borrowed to get there. In Eastern Europe's small towns, Western Union offices have sprung up to disburse funds from relatives working overseas. This movement of people presents the EU with a conundrum: a continued flow of undocumented workers threatens European stability, but shutting off the flow has terrible consequences. The $50 Maria gets from her brother in Germany once or twice a year, for example, helps to keep her children alive.

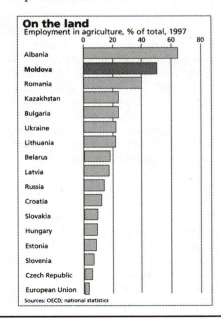

On the land
Employment in agriculture, % of total, 1997

0 20 40 60 80

Albania
Moldova
Romania
Kazakhstan
Bulgaria
Ukraine
Lithuania
Belarus
Latvia
Russia
Croatia
Slovakia
Hungary
Estonia
Slovenia
Czech Republic
European Union
Sources: OECD; national statistics

In considering how to help, EU officials have to unravel how Eastern Europe's villages have got this way. In truth, they are dogged not just by communism but by longer history. In contrast to Western Europe, where villages have an ancient tradition of inherited land, many villages in Eastern Europe were released from serfdom only with the land reforms of 1863 and were swallowed by communism a few decades later. The murder and expulsion of the brightest and wealthiest landowners, especially during the Stalinist period, is still acutely felt in former Soviet lands. "Our best died in Siberia," says a melancholy elder, when asked what has held his village back.

In some places, it is possible to retrieve a sense of local pride and enterprise. In western Ukraine, the spectral legacy of Austro-Hungarian rule, with its stronger sense of civic participation and private ownership, can still be felt. Community life in Cherche, a village 80 miles south of Lvov, is strong. The village hall there is being renovated. In place of Lenin on the walls are portraits of notable Ukrainian nationalists. The village mayor, Ivan Antoniv, a retired physics professor, says the situation is hard but not critical. He sees hope in the transformation of the abandoned collective farm into a state-of-the-art pig farm, underwritten by Dutch and British venture capital. Colonel Myron Pankiv, the farm's director, used to command a nuclear-missile silo in Kazakhstan. Now he is in charge of feeding pigs by computer.

Even in places like Dolna, a new sense of ownership and responsibility is slowly taking root. The worst of the poverty is offset by the fact that villagers own their own homes, grow most of what they need and receive assistance, at critical times, from relatives. Technical-service stations are being developed where small farmers can buy simple equipment against future earnings.

Back to the kulaks

Yet much more needs to be done. First, land reform. Formerly collectivised land has, or is about to be, parcelled out to villagers across Eastern Europe, except in Belarus. But because the land parcels are so small—rarely more than a few hectares—this privatisation has created

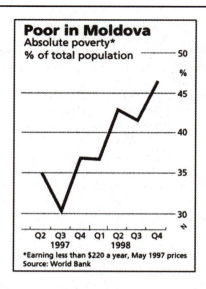

Poor in Moldova
Absolute poverty*
% of total population

*Earning less than $220 a year, May 1997 prices
Source: World Bank

a new kind of dependency. Where a state does not provide welfare payments, land becomes what the OECD dryly calls "food security". The emphasis now is on consolidating land into medium-sized farms capable of surviving in the market place. Villagers are encouraged to lease their strips of land to enterprising farmers, a sort of neo-kulak class, in return for a share of the harvest. "An entrepreneur finds markets," says Patricia Orlowitz, a Chisinau-based specialist in land reform.

She is right; but Eastern Europe still has to compete against EU producers. To help level the playing field, the EU has created a fund for rural infrastructure and agro-processing initiatives called SAPARD (Special Accession Programme for Agriculture and Rural Development). Some $1 billion of the $3 billion Poland plans to spend on rural development up to 2006 will come from this scheme. Most of it will be used to bring the country's agro-processors up to speed; at present, only 6% of Poland's dairies and 1% of its meat-processors meet EU standards.

But SAPARD is not capable of engineering change where it is most needed: in the villages in countries that will not be joining the EU in the near future. What can be done for them? First, decision-makers must spend time in the countryside, examining the destitution for themselves. Western judgments on Eastern Europe are too often made from the perspective of a comfortable hotel in, say, Prague, without taking into account the actual conditions in rural ar-

eas where over a third of the population lives. East European governments need to try harder, too. Almost half of Romanians live in the countryside, and more than half its farmers live below the poverty line; yet the country only came up with a rural-development policy in 1998.

Educating villagers to exploit agricultural niche markets, as Mr Ojog is doing with his flower seeds, could yet bring results. Organic farming has potential, though production on a scale that would interest western supermarkets would require well-administered co-operatives. Tourism is a potential earner, especially in the unspoilt Carpathians. The hedgerows and wildlife lost and mourned in Western Europe still flourish in the east; wolves still lope in Pushkin's woods. With imagination, villages could benefit from "leapfrog technologies" such as mobile phones and computers, while enhancing rather than obliterating the best of their folk traditions. But they cannot do any of this without considerable outside help.

That help would have to include technical expertise and enough money to kick-start rural savings-and-loan associations, rebuild rural roads and sewers, bring in new technology and underwrite the schools. Villages would also have to play their part by rebuilding community structures, assuming responsibility for the competitiveness of their local agro-processing plants, tackling corruption and supporting co-operative efforts based not, as before, on collectivisation from above, but on enterprise from below.

Perspective, however, should not be lost. Fiscal indicators mean nothing to a shepherd who measures his wealth in sunny days. The needs of villages are primary ones. In Dolna, villagers have a humble wish-list: cheaper medicines, more meat to eat, more jobs, a chance for young people to attend college or get training after school, the return of the little village "culture palace" with its occasional films and plays, and, perhaps, the reopening of the rustic house where Pushkin stayed. In communist times, when clubs and schools were instructed to visit, it attracted up to 20 busloads of tourists a day; now it is boarded up. "I used to earn a little selling lemonade to the tourists," says one old lady. "Will they come again soon?" Here's hoping.

Article 7

The Nation, November 6, 2000

Europe: Is There a Fourth Way?

IF WESTERN EUROPE IS TO BE TRULY INDEPENDENT IT MUST DEFEND ITS WELFARE STATE

DANIEL SINGER

Wonder batteries, claims a famous French advertisement, only wear out if they are being used. The opposite is true of democracy. It is now withering away spectacularly in our "advanced" countries, where it has become a money-dominated ritual, thanks to which every four or five years we may abdicate our sovereignty and pick among the candidates of the establishment.

Democracy can only gain ground when people take matters into their own hands—when, at all levels, from the bottom to the very top, they collectively try to gain mastery over their work, their lives and their fate. So why, then, did I feel a certain unease when Europe's protesting truck drivers were using blockades to put pressure on their governments to lower fuel costs? Because to be really genuine, people's power must have a purpose and aim at a more just society.

In the case of the truck drivers, the protest was a reflection of general discontent and the absence of a coherent policy of officialdom. One could imagine a left-wing administration with an ecologically sound energy policy defending the fuel tax as a conservation measure. But that would have meant a government capable of attacking both OPEC and the oil companies, one developing public transport, having a general fiscal policy that penalizes profit and reduces taxes affecting mass consumption, giving working people the feeling that their interests are being defended. The bulk of the population does not feel this. They are told that the West is getting more and more prosperous, yet they perceive that this only applies to a thin layer at the top. A sharp rise in the price of gasoline hit them in their pocketbooks.

In fact, what this crisis showed is that the so-called left-wing governments of Western Europe have no project, no vision, no progressive alternative. We are seeing the final funeral of that nine days' wonder, the fairy tale of the "third way." In Western Europe the period from 1945 to 1975 was one of unprecedented growth (about 5 percent a year of gross national product), and its people probably did better than elsewhere in terms of collective social benefits. This "social democratic" interlude, while not as attractive as it is now being painted in retrospect, did provide advantages worth defending. But, naturally, the miracle of "capitalism with a human face" did not last. After twenty years of defeats of labor around the globe, the United States—with some lessons from Japan—emerged with another potential model, one based on the unquestioned and undiluted dictatorship of capital. It is this model that for several years now has been peddled to Western Europe.

Many voters saw the third way as some sort of combination of America's new ruthless dynamism with the welfare state and social democracy. The leftist label attached to this US model was supposed to convince Europeans, keen on retaining the social gains won in the years of prosperity after the war, that the welfare state would not be dismantled too brutally. The purpose of the third way was to dismantle it, but to carry out the process without promoting a radical response.

Now the leftist governments in Europe may well have fulfilled this task. Last year eleven out of the fifteen governments on the Continent were still run by leftist parties. Since then, Austria has swung to the right, and the odds are that Italy will follow suit in next year's general election. A triumphant Jörg Haider in Austria, and the formerly open fascist Gianfranco Fini and jingoistic regionalist Umberto Bossi in Italy, were a lot to swallow. And the extremist Vlaams Blok for a third of the vote in the municipal election in Antwerp, Belgium's second-largest city. But when in quiet, civilized Denmark the xenophobic People's Party of Pia Kjaersgaard gets 7.4 percent of the vote and now claims double that share in opinion polls, something is rotten in the kingdom of Europe. Fortunately, the most dangerous, Jean-Marie Le Pen's National Front in France, lost momentum through a split. Still, if we don't do our duty, there are plenty of candidates to exploit the growing discontent. But the dividing line between Europe's left and right has become so blurred that it requires an expert's eye to draw the political distinction between the conservative José Maria Aznar in Spain and the progressive Tony Blair in Britain or Gerhard Schröder in Germany.

The quickening pace of events in the past few years, however, while a common currency was being set up by eleven members of the European Union, has confirmed the views of the doubters, namely, that a Europe that does not have a different social project, a model of its own, has no chance of standing up to the United States and of really defending its interests. I am not referring here to the fall of the euro, which in twenty months has lost over a quarter of its value. This is merely a symptom. What is more significant is the way in which economic concentration has been proceeding during this period. This process has been speeded up by the disappearance of national frontiers and the use of a joint currency under the EU.

Mergers and acquisitions are multiplying in all sectors. They are no longer limited to the big eating the small. Now giants are swallowing giants, and, increasingly, they are doing it across frontiers. But they are not necessarily, or even predominantly, doing it within the borders of the European Union. When a big Swiss or German bank, having absorbed its neighbors, looks for a target, it is most probably some specialist investment bank, quite likely to be Ameri-

can—for instance. Deutsche Bank has taken over Bankers Trust in New York. With the dollar riding high, Europeans have been pouring their money into the United States, while the Americans are picking plums in Europe. (Incidentally, rumor has it that the rather reluctant US decision to bolster the euro in September was imposed on US Treasury Secretary Larry Summers by US mutual funds worried by the drop in value of their holdings.)

The British are still suspected of being a US Trojan horse bent on turning the European Union into a vast free-trade area.

Last year the United States was the main target of direct foreign investment ($276 billion), and the reason is crucial. Why should a European banker or manufacturer disdain a US partner if he thinks the American company is the most profitable? What means has a European government to impose its controls over the free flow of money, the pattern of investment—let alone some form of democratic planning—if at the same time it encourages and praises "globalization," which is to say the spread of the US model? Naturally, there will be conflicts between European and American interests, notably over agriculture. There will be deals and confrontations. But a Europe that does not consciously build a different kind of society, that relies on the US Treasury to back the euro, cannot be taken seriously as a potential equal partner.

There are, to be sure, other domains, like diplomacy and military collaboration, in which Europe may try to assert itself. But here too, especially after Kosovo, any claims of growing inde-

pendence are illusory. True, the European Union now has a man in charge of common policy, but he is Javier Solana: In his youth a protester against US weapons in Europe, Solana was Secretary General of NATO in his last job. German Foreign Minister Joschka Fischer, another ex-revolutionary protester turned pillar of the establishment, claims openly that for him the long-term aim remains a "European federation." But the EU for the moment is traveling in the opposite direction, and its very existence may be at stake. Its size, shape and institutional structure are all uncertain. Started with six members, now with fifteen and expecting to more than double that number with the entry of Eastern European nations and others over the next few years, the EU doesn't even know whether it has a future. If it really aims at a federation, it should build a closely integrated core and a loose periphery. In other words, strengthen the institutions that govern it before admitting new members.

Otherwise, the entry of new members into an organization with weak institutions could kill the very idea of a federal structure. The British are still suspected of being a US Trojan horse bent on turning the whole construction into a vast free-trade area that will be extended beyond the Atlantic Ocean. At the December summit meeting of the ruling European Council in Nice, the fifteen must, in principle, agree on their institutional project so as to be able to proceed with the enlargement of the EU. It would be surprising if the deliberations at Nice, which will mark the end of the six-month French presidency of the EU, closed the year with any significant results. A provisional compromise is much more likely.

But if Western Europe is to try to be truly independent and stand up to the United States, it cannot do it on capitalist lines. It will have to start all over again by defending its welfare state. Because, despite the spread of the working poor, of precariousness and of uncertainty, despite the growing attitude against public pensions, Europe's welfare state is still more attractive than the American, and you can start the struggle only when people have the feeling that they are fighting for their interests. A European New Left would thus have to begin by coming to the rescue of the welfare state, but not fraudulently, as under the third way. It must do so by showing how this welfare state must be broadened, made universal and rendered really democratic. It must show quickly—almost at once—that you cannot carry out such a progressive strategy and at the same time allow the savage extension of the US model of globalization. That model, by its very inspiration—profit above all—renders any such developments impossible.

Europeans will have to start the battle at once, because it is only in the renewed struggle for higher wages and less precarious working conditions that the labor movement can understand that its demands clash with the very foundations of existing society. The same is true for feminists and radical ecologists. It is through their own experience that they must learn that their aspirations and dreams cannot be fulfilled within the established order. Only then will a genuine New Left be able to seek a "fourth way," one that does not conceal the established disorder but tries to move beyond the confines of capitalist society.

Controls over the movement of capital; the use of state power not to boost big corporations but to combat them; capitalism; socialism; democratic planning from below—all these have become unused or dirty words. So it would be naïve to expect such a policy to be invented, let alone applied, overnight. It is difficult to determine which tendency

will prevail, because we are living in a strange period affected by two contradictory trends. On the one hand, the American model is spreading seemingly inexorably, with little effective resistance, and certainly none by European governments. On the other hand, its ideological domination has been shaken and weakened. More and more people, particularly in the younger generation, are now convinced that globalization brings about not prosperity but polarization, social uncertainty, inequality, the strains and stresses of uncertainty about the future.

There is something deeper, a growing feeling that the society it promises has no attraction. The extraordinary success in France of José Bové and his revolt against McDonald's and the *malbouffe*—junk food—that it symbolizes is only one small example. With mad-cow disease, with the rushed abuse of genetic modification in the interest of big chemical corporations and agribusiness, there is increasing awareness that a system driven forward by the accumulation of profit cannot take time to study the use to which we should put our extraordinary progress in science and technology; that it is unable to assure the necessary precautions; that, condemned to permanent but uncontrolled growth, it may soon threaten the very future of our planet.

And yet, despite this awareness, especially among the young, the search for an alternative is still very timid. Shall

we then wallow in gloom and doom? In some sections of the British former New Left there is now a temptation to retreat into ivory towers and rely on the old Marxist argument that "capitalism contains within itself the seeds of its own destruction." The greatest illusion of all, however, would be to assume that capitalism will accomplish this task of self-destruction on its own, without the help of a vast movement from below, which in its struggle would also be forging the vision of a different society. In 1995 the French "winter of discontent" reminded us that one could resist, at least for a time, even without a clear alternative.

Historically, the lesson of Seattle—if we don't freeze it as a fetish and reduce it to repetitive demonstrations—is even more important. With the Americans taking the lead in the struggle against the world seen as merchandise, we are reminded that globalization is not the only form internationalism can take. What is at stake, it is now clear, is neither the imposition of the US model on Europe nor the defense of Europe's welfare state. It is our common struggle—from below and on a worldwide scale—against a capitalist system both triumphant and in deep crisis. Amid the present confusion, we may actually be watching the early phase of a new historical period.

Daniel Singer is The Nation's *Europe correspondent.*

Article 8   *Europe,* November 2000

DANES SAY NO TO EURO

Voters elect not to adopt single currency for now

By Leif Beck Fallesen

The Danes gave the thumbs down to the euro in an attempt to slow the political integration of the European Union. But the Danish 'no' may have the exact opposite effect, instead speeding the deepening of European integration desired by Germany and France, notably expressed earlier this year in Berlin by German Foreign Minister Joschka Fischer and French President Jacques Chirac.

The 'no' appears as if it could entrench the move toward a two-speed, or rather a multi-speed European Union, which is anathema to the position held by the Danish government and others outside Euroland, including the Central and East European countries applying for membership.

Swedish Prime Minister Goran Persson tried to play down the Danish referendum's impact on Sweden, but the government-appointed economist responsible for the official report on Sweden, Lars Calmfors, said the influence of the Danish vote would be profound. A delay of Swedish entry is a very likely result. The 'no' vow was a person" defeat for Minister Prime Minister Poul Nyrup Rasmussen.

A majority of the Swedes oppose euro participation, although the prime minister has indicated that he may call a referendum before 2004. In the United Kingdom, the euro is even less popular, with polls running three to one against the euro, and the tabloid the *Sun* eulogizing the Viking Danes in a rerun of the Danish 'no' to the Maastricht Treaty in 1992.

In the United Kingdom, a delay of a decision on the euro is also likely. Despite the insistence of a government spokesman that the Danish outcome would not change the British government's position, that a referendum may

The 'no' vote was a personal defeat for Danish Prime Minister Poul Nyrup Rasmussen.

be held after the next general election. Chancellor of the Exchequer Gordon Brown said that this depended on the assessment of the five economic tests on euro membership early in the next Parliament. Sweden also has set economic tests for joining, but like in the UK, the essential determinant is likely to be political, not economic.

The fact that economics are less important than politics in the present phase of European integration is the real lesson of the Danish 'no'. Sweden and the United Kingdom will ignore it at their peril if they proceed with their referendums. But polls in many EU countries, certainly in Germany, indicate that the euro would not survive electoral scrutiny, and the Danish 'no' should perhaps inspire EU leaders to worry more about popular support for European integration henceforth. The low rate of partici-

pation in the elections for the European Parliament is a similar warning sign even to the core EU countries that don't intend to hold euro referendums.

The pollsters had prepared the European Union and the Danes for a 'no' at the September 28 vote. However, they had not forecast that the 'no' side would win by more than 6 percent—53.1 percent versus 46.9 percent for the 'yes' camp. When Poul Nyrup Rasmussen, Denmark's Social Democratic prime minister, announced the referendum in March, he had hoped that the issues would be predominantly economic, like the referendum that endorsed Danish membership of the then European Communities in 1972 and the referendum on the Single European Act in 1986.

This year, however, the campaign almost immediately turned into a dispute on the future of political integration in

Pia Kjaersgaard, leader of the right-wing Danish People's Party, which opposes joining the euro, celebrates with her supporters after the 'no' vote.

the European Union. The 'yes' side was never able to dispel convincingly the 'no' camp's claim that the euro is part of a master plan to make the EU a federal super state.

When the Danes voted in 1972, 1986, and 1993, the Danish economy was weak, and the economic arguments in favor of closer union with its European neighbors were potent. This year, the strength of the Danish economy and the image of the euro as a weak currency blunted the economic arguments for joining the currency.

The 'no' side also successfully introduced harmonization of EU fiscal and social policies as a threat to the Danish welfare state. The 'yes' campaign, however, was marred by tactical blunders and a lack of enthusiasm, and a joint statement by the 'yes' parties that pensions were not in danger and an attempt to procure support from other EU countries with a similar statement made little impression on the voters.

National sovereignty was a major issue in this year's referendum, and this is the principal explanation of the dif-

ference between this and earlier referendums.

Pia Kjaersgaard, leader of the right-wing Danish People- Party, which opposes joining the euro, celebrates with her supporters after the 'no' vote. Almost a fifth of liberal and conservative voters cast ballots against the euro-more than twice the number that voted 'no' to the Maastricht Treaty in 1992 and almost certainly the single-most important demographic determining the outcome. Otherwise, little has changed in Danish voting patterns on EU issues since 1972.

The archetypal 'no' voter is still a low-paid, female public sector employee with minimal education. The archetypal yes' voter is an educated male earning an above average income. The typical 'no' voter is older than the typical yes' voter, but gender is the best predictor. During the 1990s, young Danes have tended to vote pro-European, but in the euro referendum, it appears that a majority of young voters sided with the anti-euro forces.

The 'no' victory was a personal defeat for Prime Minister Poul Nyrup Ras-

mussen, a strong pro-European in a historically deeply divided Social Democratic Party. Nevertheless, the vote will have no immediate parliamentary repercussions.

A poll conducted on referendum day showed a marginal decline in support for the prime minister's party and a marginal increase in support for the Liberal Party, which joined the government in campaigning for the euro. The same polls showed the five parties that sided with the 'no' camp (which together control less than a quarter of the seats in the Danish parliament) saw only minimal gains from the outcome. These numbers seem to confirm the historical evidence that Danish voters separate domestic party preferences from their European preferences.

The 'no' side has demanded that Danish EU policies be redefined in a more euro-skeptic stance. However, apart from not joining the euro, this will not happen in any formal sense. There is no change in the parliamentary majority supporting the Danish position at the Nice Summit, which will debate and perhaps decide on the institutional changes necessary for the EU to add new member nations.

Currently, no transfer of sovereignty to the European Union is envisaged, allowing the Danish government to ratify a new treaty. EU leaders will almost certainly tread warily to avoid a situation where the Danish voters will be in a position to reject a treaty, like they did in 1992. However, if the Nice Treaty is ratified, there will be more treaties or constitutional amendments. The Danish government unofficially hopes that a new referendum will allow Denmark to join the euro in four to six years' time, but before that can happen, there must be a fundamental positive shift of Danish attitudes toward political integration. That is a daunting prospect, and success is by no means guaranteed. By that time, a pioneer group of EU countries may have moved so far ahead that the Danes probably would not dare try to catch up in one fell swoop.

Leif Beck Fallesen, EUROPE's *Copenhagen correspondent, is publisher of the Borsen newspaper.*

Europe, February 2001

A New Golden Age

Prosperous Dutch Face Tough Challenges

By Roel Janssen

In matters of sex, drugs, and death, the Netherlands has a reputation of relative permissiveness. This reputation was illustrated late last year with the parliamentary approval of legalized euthanasia. To many Dutch observers it came as a surprise that this landmark decision drew much international attention. Not only did the Vatican disapprove publicly but other European governments voiced their opposition to the policy as well. In an editorial in its European edition, the *Wall Street Journal,* paraphrasing James Bond, called it a "license to kill."

Until recently, Dutch law regarded euthanasia as a crime, but if applied under strict conditions on patients who find themselves in a situation of "unbearable and hopeless suffering," doctors applying lethal assistance were not prosecuted. These rules had been broadened further to allow for cases of psychological suffering in addition to physical pain. Now, both of these scenarios have been covered under the new euthanasia law, making the Netherlands the first country where assisted death under certain conditions is a legal act.

The legalization of euthanasia follows last year's legalization of prostitution and an open tolerance for soft drugs, such as marijuana and hashish. With soft drugs, however, there remains a snag: While the public sale of small quantities is allowed, the abundant 'coffee shops' that sell the drugs are theoretically not allowed to buy large quantities of soft drugs. This legal inconsistency helps keep the soft drug trade a highly lucrative criminal business and simultaneously has turned the Netherlands into one of the most advanced producers and international traders of marijuana.

Aside from the issues of drugs and death, the New Year started with a pleasant surprise for the Dutch. In January, their incomes were substantially up due in large part to the largest tax reduction ever applied with a single stroke. Income taxes were slashed, raising personal spending power. These across-the-board cuts stem from the gains made from the government and parliament's adherence to prudent fiscal policies and clearly signals the economic policy shift the Dutch have undergone in the last few years.

To be sure, the Netherlands still boasts a high marginal tax rate—declining from 60 to 52 percent—and both energy and sales taxes were increased. But with a budget surplus in 2001, the first surplus in thirty years, it looks probable that the Netherlands is heading for further lowering of taxes in the years ahead. It signals a far cry from the so-called "Dutch disease" that ravaged the economy in the seventies and eighties with bloated government spending, lavish social security outlays, and fast rising unemployment.

Wealth has shifted from the public to the private sector, both on the corporate and personal level. During the past decade, total public spending has been reduced relative to the growth of the economy. Nowadays, it counts for less than half of the national GDP.

While the Netherlands remains the European leader in tax reduction, other EU countries—such as Germany, France, Belgium, and Italy—are taking their own steps toward lowering taxes. There is broad political consensus that high taxes and social security burdens hamper economic growth. The process of fiscal consolidation and discipline

(front, from left to right) German Chancellor Gerhard Schröder, British Prime Minister Tony Blair, and Dutch Prime Minister Wim Kok attend a signing ceremony at the EU summit in Nice in December.

imposed by the EU's program to implement a single currency has created the headroom to reduce taxes. The Dutch tax rebate of 2001 amounts to about 1 percent of the country's GDP.

> The seventeenth century is described as a golden age in the Netherlands, a time when merchants brought unprecedented wealth to the country and Rembrandt van Rijn (1606–1669) painted his precious canvases, including *The Stone Bridge.*

Boosted by record-level real estate prices (especially in the major cities), the broadening of equity ownership, and the high levels of the stock market, capi-

tal remains abundant. There is even ample talk about the Netherlands enjoying the beginning of a new golden age in the twenty-first century. The golden age of the seventeenth century saw merchants bring unprecedented wealth to the country, Rembrandt paint his precious canvases, and the world's first financial market blossom in Amsterdam.

Indeed, the Netherlands' economy has been growing briskly now for almost a decade. In the past ten years, the net growth of Dutch jobs amounted to 20 percent. Business and consumer confidence continue to be strong. The European Commission recently praised the "impressive macroeconomic performances" of the Dutch economic policy management.

Wim Kok, the leader of the Dutch government since 1994—who had served previously as the finance minister and leader of the country's largest labor union—was hailed in 2000 as one of the founding fathers of the "third way" by President Bill Clinton. Like other social democratic government leaders in Europe, Kok is a proponent of moderate economic reforms, seeking to strike a balance between socialism and unfettered capitalism. In 1994, his first government composed of social democrats (who are associated with the color red) and market liberals (known for their trademark blue) was quickly labeled the "purple coalition" by the Dutch press. Although the present government,

which is the second purple coalition, looks less vibrant than the first, there is no clear alternative in sight. General elections are not due until 2002.

On moral issues, the government has reconfirmed the liberal stance of the Netherlands. None of this has shaken Dutch society. It has, however, increased drug tourism and crime, particularly offenses related to drug and human trafficking. Last year, fifty-eight illegal Chinese immigrants were found dead in a Dutch truck attempting to smuggle them to the British port of Dover. But it is fair to say that all European countries are struggling with drugs, illegal immigration, and human trafficking.

The national mood is strong in a broad sense. Being a country that considers itself to be the biggest of the smaller states in Europe (ranking fifth in economy and sixth in population of the EU countries), the Netherlands is extremely sensitive to its international reputation. Thus, it came almost as a national victory when Ruud Lubbers, the former prime minister, was recently appointed as the new high commissioner of the United Nations refugee organization. Furthermore, the announcement that Dutch troops will form the core of a UN peacekeeping force in the war-torn border zone between Ethiopia and Eritrea was seen as recognition of the Netherlands' historically strong international outlook. That mission, it is hoped, will also wipe out the traumatic memory of the inglorious role the Dutch army played at the Srebrenica enclave during the Bosnian war when thousands of Muslim men were abducted (and subsequently killed) by the Serbs under the eyes of the Dutch soldiers. Furthermore, at the European Summit in Nice, in early December, the Netherlands was granted a slightly heavier voting weight than neighboring Belgium.

Last summer, finishing eighth in the medal ranking of the Sydney Olympic Games, right after the world's major athletic powers and surpassing countries like Cuba and the United Kingdom, was a further cause for national pride. Gold medal-winning swimmers Pieter van den Hoogenband and Inge de Bruijn and bicyclist Leontien van Moorsel gained popularity normally confined to television stars and helped to assuage the Dutch national soccer team's humiliat-

Olympic heroes such as Leontien Zijiaard, who won the gold medal in track cycling, gave the Dutch plenty to cheer about at this summer's Sydney Games.

ing defeat in the European Championship earlier in the summer.

The Dutch psyche aside, several major structural problems loom in the near future. Education and health care, which are both publicly managed services in the Netherlands, are in crisis due to bureaucratic management, lack of funds, and lack of personnel. The privatization of public utility monopolies—electricity, natural gas, cable television networks, the power grid, the railways, and local public transportation—has stagnated. State monopolies have been reconfirmed in some sectors, while in others private monopolies have emerged and market competition has failed to materialize. Likewise, reforms and partial privatization of the social security system have

not brought about the desired improvements in labor participation. Even with the current tight labor market, more than a million people (out of a total population of 16 million) of working age are receiving some form of social security.

Another issue that looms: chronic road congestion has become an intrinsic part of daily life. Regularly, the total distance of Dutch morning traffic backups exceeds 220 miles, which is longer than the actual distance between the country's north and south borders. It has become all the more acute because people increasingly are moving to the suburbs, more work is being done part-time, and more women have moved into the labor market. Despite proposed measures like paying tolls to drive during the rush hours, there is no agreement on how to reduce traffic in the long term.

A third issue is the continued influx of foreign asylum seekers. More than 70,000 are now awaiting entry in asylum centers; many more cases are pending appeal at the Dutch high court. Of course, this is an issue with which all European countries are coping, and it is complicated further by the EU's open internal borders. However, in the Netherlands the severity is aggravated because of the strong economy. Pressed by labor shortages employers are urging the government to give entry to specific groups of foreign workers, such as Indian computer technicians and Philippine nurses. At the same time, illegal immigrants are much sought after to work in the lower strata of the labor market.

Meanwhile, corporate wealth continues to rise. Early last year, the Netherlands plunged into the new economy of Internet startups. Not all of them were successful, though. Most infamous was the case of Internet provider World Online. Its founder, Nina Brink, was caught covering up secret deals to sell her shares in the company prior to its initial public offering. The ensuing scandal and the resulting collapse of World Online's

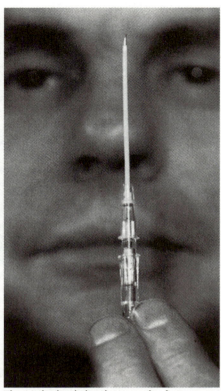

The Netherlands has become the first country where euthanasia is legal under certain conditions.

share price shook the confidence of many Dutch investors in the dot-com sector. However, the affair also had the effect of shifting investors' attention to the traditional heavyweights of the Dutch economy, such as the financial giants ING Barings and ABN Amro, the oil business of Royal Dutch, the electronics maker Philips, the processed food concern Unilever, and the publishing conglomerate Elsevier. These, and many others, continue to expand vigorously throughout the international markets. Just like in the previous golden age of Dutch merchant expansion.

Roel Janssen is EUROPE*'s Netherlands' correspondent.*

Article 10

The Economist, September 30, 2000

GERMAN UNIFICATION

Togetherness: a balance sheet

Ten years ago, Germany embarked on one of the most ambitious projects in its history: the absorption of the east. How far has it succeeded?

BERLIN

WHEN, on October 3rd 1990, the 62m Germans in the west were formally united with the 16m in the east, the two parts could hardly have been more different. The west was one of the richest, most highly industrialised and technologically advanced nations in the world; the east was near-bankrupt, economically and psychologically shattered after nearly 60 years under two successive totalitarian regimes. Yet West Germany's chancellor, Helmut Kohl, assured his compatriots that the east would be transformed within a matter of years into a "flourishing landscape" with a standard of living comparable to the west's—and relatively cheaply, too.

Ten years and the net transfer of DM1.2 trillion ($540 billion) later—more than double this year's entire federal budget—many on both sides of the former barbed-wire and walled divide are wondering what happened to that "flourishing landscape". After a four-year spurt of rapid transformation, with growth rates well above those in the west, the east's "catching-up" appears to have stopped. The building industry, which fuelled the east's earlier boom, is in deep crisis. Economic growth over the past three years has slowed to a sluggish 1.2–2% a year. The east's GDP per head is still only two-thirds of west Germany's. Unemployment, stuck at over

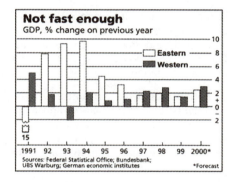

Not fast enough
GDP, % change on previous year

Sources: Federal Statistical Office; Bundesbank;
UBS Warburg; German economic institutes   *Forecast

17% for the past three years, is twice western levels. Average gross wages are 25% lower. And another DM300 billion–400 billion in public money is still needed to bring the east's infrastructure into line with the west's.

Small wonder, perhaps, that some top economists, such as Hans-Werner Sinn, the head of Ifo, a Munich-based think-tank, describe German unification as "an economic failure". But it is all a matter of perspective. If you consider where eastern Germany has come from, rather than where it is meant to be heading, a much brighter picture emerges. After the collapse of communism, East Germany found itself in political and economic ruins. So, of course, did many other Soviet-block countries. But, unlike those countries, East Germany was forced to adapt overnight to a completely new system not of its own making nor necessarily suited to its needs. Despite what politicians claimed and easterners had

hoped, unification was not a merger but a takeover. Indeed, after the initial euphoria had worn off, many easterners began to wonder whether it did not more resemble a defeat. Everything was thrown out, the good with the bad. All had to be rebuilt from scratch on the West German model: political structures, education system, laws, health and welfare services, police, currency, industrial fabric, institutions, managerial and political elites.

Rushing towards freedom

Unification was not the fruit of mature reflection by political leaders. It came in a rush, propelled by an unstoppable wave of popular longing from the east for the dreamed-of freedoms and material benefits of the west. Few politicians desired, or dared, to point out the risks. Yet for many easterners, the shift from a centrally-planned to a market economy came as a tremendous shock. As the antiquated and indebted factories sputtered to a halt and the inefficient collective farms were broken up, three-quarters of East Germany's 10m workers found themselves out of work.

Matters were made worse by the decision to set the exchange rate for the east German currency at one to one against the mighty west German D-mark. Politically, it was probably justifiable: economically, a disaster. Less than a year earlier, at the time of the fall

of the Berlin wall, the East German mark had been exchanging at an official rate of nine to one, and at up to 20 to one on the black market. The crazy new parity sent wages and prices spiralling, making east German products even less competitive. East Germany's traditional export markets in the former Soviet block collapsed, while its domestic customers, armed with their new D-marks, rushed to buy western products. At the same time, some 2m easterners moved west in search of better jobs. The labour force shrank by a third, and over 1m remained jobless.

That is where east Germany has come from. Yet by that yardstick, enormous strides have been made since unification. Wages, pensions, and GDP per person have almost doubled. Net household income is now 90% of that in west Germany. Productivity has leapt from 41% to 67% of western levels. Fixed capital per job has risen from 20% of the west's to 75%. Unit labour costs are just 12% higher. Manufacturing, which lost 70% of its jobs after unification, is now growing at a double-digit rate, and manufacturing exports jumped by a third in the first half of this year. They now account for 21% of output, almost double the level in the mid-1990s.

Housing conditions have vastly improved. The *Plattenbau,* the grim high-rise inner-city cement blocks of communist times, are being abandoned for neat little villas on the outskirts of towns. More than 750,000 new homes

have been built since 1993, most of them with every modern convenience (see table). Some 40% of easterners now own their own home, the same proportion as in the west. There is some way to go: hundreds of thousands of abandoned buildings still litter the landscape, many of the streets are still cobbled and full of potholes, and weeds grow up through deserted railway lines. But the fresh-painted, orange-roofed villas are taking over, and town centres gleam with shopping malls, fashionable cafés and beautifully restored old buildings.

On the outskirts, spanking new factories, high-tech research centres and exhibition complexes have sprung up, outshining much of what the west has to offer. More than three-quarters of the east's industrial plant has been installed since 1990. New motorways weave their way across the countryside; not a single kilometre was built during 40 years of communist rule. The railways have been electrified and modernised, and east Germany now boasts one of the most modern telephone systems in the world.

Pollution, one of the horrors of the communist era, has all but disappeared. The east's stinking rivers are once again teeming with fish and plant life. The choking smog that used to hang over every town, engendered by the use of brown coal in houses and factories and by dirty two-stroke-engine cars, has gone. (The old "Trabi" has become so rare that it is now a collector's item.) Dust particles in the air have been reduced by 99%, sulphuric acid by 88%. Thanks to an injection of DM50 billion, east Germany, formerly the world's third-most-wasteful consumer of energy, now has the most modern, efficient and clean brown-coal-fuelled power stations in the world. Billions more have been spent in bringing the east's lamentable

sewage-treatment and water-purifying systems up to western standards.

A sense of loss

What has been achieved has been staggering. But unification has produced losers as well as winners. Although 60% of easterners admit that their lives have improved over the past ten years, 16% claim they have got worse. They miss the certainties of centralised organisation, the solidarity between neighbours in the face of shared deprivations, the lack of envy in a classless society where all earned more or less the same wage, even the sense of personal security in a supervised police state. Women miss the free, universal creche and kindergarten places that made it so much easier to combine a career with family life. Although only a small minority (6%) would like to see the return of the German Democratic Republic, two-thirds complain that they do not yet feel "totally at home" in a united Germany.

A similar, and growing, proportion of easterners complain that they feel like second-class citizens. They resent being paid lower wages than their western colleagues for the same work, often with longer hours. A bus driver, university lecturer or hospital doctor in east Berlin, for example, earns 13% less than his counterpart just down the road in the west of the city, but is expected to work one and a half hours more a week. In the private sector wages on average are 20% lower and, in some industries, 40%. The fact that easterners are often earning triple what they were ten years ago is no consolation. Despite the huge transfers of public money, six in ten complain that the government is "not doing enough" to bring their living standards into line with those in the west.

The unknown other

Adding insult to injury, most of the top jobs in business, banking, public administration, the universities, law and the media are in the hands of westerners, brought in after unification when the east's elites were thrown out. Ten years later, they are still there. Even in the political domain, two of the five east German states are led by westerners—Kurt

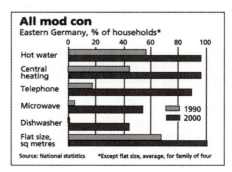

All mod con
Eastern Germany, % of households*

Source: National statistics   *Except flat size, average, for family of four

Biedenkopf in Saxony and Bernhard Vogel in Thuringia—and around half of all eastern state ministers are westerners. The same is not true of easterners in west Germany. In the federal government, only two out of 17 ministers are from the east, and it was only this year that a mainstream political party, the Christian Democrats, chose an easterner, Angela Merkel, as their leader.

This dismissive attitude can be warranted; after 40 years in an economically backward place, east Germans sometimes simply lack the skills for the top jobs. More often, it stems from ignorance. An estimated 80% of west Germans have never set foot in the former Democratic Republic. They still tend to look down their noses at their eastern cousins, regarding them as uncultured, inefficient and ungrateful. The easterners' relatively low level of productivity is put down to laziness rather than lack of investment. Why should they, the hard-working westerners, continue to bail them out?

The German media, almost exclusively in western hands, tends to present the east's communist past as a period of unrelieved misery, where everyone was either a spy or a victim, and where children were brought up in a godless society with no ethical values and no tolerance for others, providing a breeding-ground for the neo-Nazi excesses that are emerging today. But Axel Schmidt-Gödelitz, a former West German diplomat in East Berlin, points out that most people in East Germany were simply apolitical and made the best of what they had. "They had good lives," he says. "Most did not feel they were living in a prison."

Outwardly, there is no longer much difference between the Germans of the east and the west. They dress the same, watch the same television, read the same newspapers, learn the same things in school, eat pretty much the same food. But many can still detect a difference in style and attitude. "I can tell east from west within five minutes of entering a room," says Mr Schmidt-Gödelitz, who now organises weekend conferences between easterners and westerners in an attempt to remove mutual prejudices. "Easterners tend to be warmer, simpler, more reticent, more honest about themselves, less willing to put on a façade."

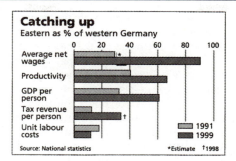

Catching up
Eastern as % of western Germany

Source: National statistics   *Estimate   †1998

(Bars for Average net wages, Productivity, GDP per person, Tax revenue per person, Unit labour costs, shown for 1991 and 1999.)

Westerners are more self-assured, assertive, better able to sell themselves in a way that many easterners take for arrogance." Wolfgang Thierse, the first eastern president of the Bundestag, the parliamentary lower house, complains that his more direct, unsophisticated approach is often mistaken for naivety and a lack of political cunning. Angela Merkel suffers from the same misconception.

Far from finding east German workers any less diligent than their colleagues in the west, west German and foreign investors often judge them as more reliable, harder-working and more willing to learn new skills. That should be no surprise: over the past ten years, two-thirds have had to change their job at least once to something quite different from what they did before.

Dynamism or violence?

Among many of the younger generation, in particular, there is now a new zip, curiosity and optimism. They are eager to embrace the new technologies and ready to take risks. Tens of thousands of new companies are being set up every year, most by young entrepreneurs from the east itself. Many fail. But some 500,000 firms created over the past ten years have survived, bringing jobs for 3m people. Many are in the forefront of their field—in data-processing, office equipment, biotechnology, optical precision instruments, electrical and medical engineering. Around towns such as Jena in Thuringia and Dresden in Saxony, impressive mini- "Silicon Valleys" have been set up.

Not all the young, however, have been so willing to take up the challenges of the new world. Some, deprived of the social structures and ideological certainties of the communist regime, simply feel lost. Their parents, often themselves jobless, depressed and disillusioned, can provide no help. Their teachers are no longer so close to them as in the communist era. Their communist-organised sports clubs, leisure activities and cheap collective holidays have gone.

Some, a tiny minority, have sought refuge—and an identity—in far-right and neo-Nazi groups. Such groups are not exclusive to east Germany. But easterners are responsible for nearly half the racist violence in Germany, although they represent less than one-fifth of the population. Foreigners, who account for less than 2% of the east's inhabitants compared with 10% in the west, provide a convenient scapegoat for the easterners' fears and feelings of inferiority. A recent survey of 16,000 14-year-olds in Rostock, a northern port in Mecklenburg-West Pomerania, in the former east, showed 40% blaming foreigners for unemployment and 18% considering violence an acceptable solution.

East Germany may not yet be quite the "economic powerhouse" proclaimed by Rolf Schwanitz, the (eastern) minister for the five eastern states. But it is on the move. The apparent stalling of the catching-up process was linked to the contraction of the overheated building industry, which is expected to continue to shrink for another couple of years. But manufacturing industry, which surged ahead by 12% in the first six months of this year, is now the main motor of the east's economy, and has overtaken construction in terms of the numbers employed. The east German economy is expected to grow by 2.5–2.7% this year, still slower than the west's forecast 3% or more, but almost double last year's 1.5%.

Rüdiger Pohl, a western professor who heads an economic think-tank in the city of Halle, in Saxony-Anhalt, reckons that the east is now well on its way to establishing a competitive market economy. "It is a massive success story," he concedes. "The east had good reason to be proud." But it has not yet drawn level with the west. Indeed, some wonder whether it ever will. Mr Biedenkopf, the premier of Saxony, has estimated that even with economic growth of 4% a year in the east and 2% in the west, it would take another 30 years for eastern GDP per person to equal that in the west.

But perhaps equality is the wrong target. There have always been differences between regions. No one expects the GDP in agricultural Schleswig-Holstein to equal that of industrial Baden-Württemberg, though both states are western. Putting the east on a self-sustaining footing is all that should be aimed at in the short term. And how long will that take? Manfred Stolpe, the premier of Brandenburg, reckons just six or seven years. The government estimates another ten. Others, like Mr Pohl, say a whole generation may be needed.

And how much longer will the west be willing to go on footing the bill? Last year, a net total of DM144 billion (after the deduction of the tax revenue collected from the east) of public money was transferred to the east, the equivalent of nearly 5% of the west's GDP. Westerners are made all the more aware of the burden by having to pay an earmarked 5.5% "solidarity" surcharge on their tax bills—also paid by easterners—to help pay for the reconstruction in the east. In fact, two-thirds of that goes on things like motorways, which the government also finances in the west. But most westerners are unaware of that. Those who venture into the east are often horribly envious of some of the east's superb modern installations, paid for with "their" money. The *Frankfurter Allgemeine Zeitung,* one of Germany's weightiest dailies, recently carried a front-page editorial arguing that it was "neither necessary, nor desirable" to continue to give special treatment to the east. It should learn to stand on its own two feet.

That debate will certainly continue for years to come. In the meantime, a growing number of eastern leaders feel that it is time to proclaim their achievements a little more loudly to the world. "Perhaps we don't speak so much about our successes because we're still only halfway there," says Reinhard Höppner, the dynamic premier of Saxony-Anhalt. "We've got to persuade people to continue to support us. But we've also got to teach the easterners to have more pride in what they have achieved . . . With the enlargement of the EU, we will soon no longer be on the margins of Europe, but at its centre. I am very optimistic."

Article 11

Europe, September 2000

Domestic Issues Dominate German Agenda

Schröder Government focuses on economic growth

By Terry Martin

What a difference a decade makes. It's as if the grand hand of history was busily checking off items on its "to do" list in Germany: tear down the Berlin Wall, reunite East and West Germany, elect a new government, move the capital, scrap nuclear energy. And other monumental tasks lie ahead. Having overcome its cold war division, Germany is now preparing to take on the challenges of EU enlargement and globalization. First, however, the government must attend to some pressing domestic concerns.

One of the biggest problems remains unemployment. The country's jobless rate has been hovering around 10 percent for years. It was the single biggest issue in the 1998 general election and played a significant role in the demise of Helmut Kohl's government, which had been in office for sixteen years. The Social Democrat Party (SPD) attacked

German Chancellor Gerhard Schröder (right) and Finance Minister Hans Eichel celebrate after winning a vote in the Bundesrat on July 14 to cut taxes by $24.25 billion.

Kohl's Christian Democratic Union (CDU) for failing to invigorate the labor market, and the strategy worked. Now voters are expecting Chancellor Gerhard Schröder to make good on his promise to create more jobs.

Fortunately, the German economy is picking up pace. The weak euro is driving an export boom that is helping to boost economic activity across the board. First quarter economic growth rose to 3.3 percent and is expected to continue rising. Consequently, unemployment is falling. In the spring, the jobless rate dropped below 10 percent for the first time in four-and-a-half years. These figures may not match the stellar statistics of the US economy, but in Germany, they are cause for celebration.

Relief on the economic front has given Schröder's government the latitude to concentrate on other issues such as tax reform. The sober fiscal package drawn up by Finance Minister Hans Eichel marked a radical and welcome departure from the Keynesian views of his predecessor Oskar Lafontaine, who resigned after just five months in office. At Eichel's urging, taxes on personal income and business revenue are being lowered. And much to the surprise of big business, the capital gains tax on corporate shareholdings is being abolished. That move alone should stimulate a wave of economic activity as companies unload their extensive cross-shareholdings and pour capital into new ventures.

But one should not get the impression that the government's approach is all carrot and no stick. Heavy use of non-renewable forms of energy, for instance, is being discouraged with a special ecology tax. It is exactly the sort of measure voters were expecting from the SPD's environmentally minded coalition partners. But the Greens' timing left something to be desired. The ecological tax was introduced just as crude oil prices began to skyrocket. Businesses and consumers (especially the car lobby) have blamed rising fuel bills largely on the government, costing the Greens and the SPD precious political points.

Considering that the Greens had never before served in a federal government, they've proved remarkably efficient in pushing their political agenda. Right at the start they succeeded in snapping up three cabinet posts, gaining control of the health, environment, and foreign affairs portfolios. Despite an early series of blunders largely attributable to inexperience, they've managed to put their stamp on vital sections of government policy. By far their biggest success in office has been persuading the energy industry to pull the plug on nuclear power. Though the agreement allows the power companies broad flexibility in shutting down their plants (it could take two decades before the last of the country's nineteen reactors goes off line), the deal amounts to a political watershed for the Greens. Their opposition to nuclear power was a founding party principle and key rallying point.

Some believe the Greens would not have survived this long as a coalition partner were it not for the opposition getting embroiled in a party financing scandal. The CDU has been rocked to its core by revelations that it secretly collected millions of dollars in cash during the Kohl era and used the money to finance its political campaigns. By accumulating slush funds while in power,

Jürgen Willner posts a letter of application at an employment center in Bremen, Germany. Germany's jobless rate is currently at around 10 percent.

the Christian Democrats have opened themselves up to accusations of influence peddling. Helmut Kohl himself has admitted accepting millions on his party's behalf while he was chancellor but still refuses to name the donors. Although Kohl insists he never traded political favors for cash, the affair has seriously tarnished his reputation.

Needless to say, the party financing scandal proved to be political poison for the CDU and a godsend for the "Red-Green" coalition. CDU chairman Wolfgang Schäuble was forced to resign, and the CDU's popularity plummeted. Chancellor Schröder's governmentùwhich had been languishing in opinion pollsùwas given a new lease on life. Now, however, the Christian Democrats, under the leadership of Angela Merkel, are slowly clawing back support. Recent polls show the CDU again gaining the upper hand on the SPD.

Chancellor Schröder's government is hoping the changes it has introduced will begin bearing fruit before the next general election in 2002. By that time business leaders and wage earners alike should be able to make an informed

judgement about the merits of tax reform. Meanwhile, the government is practicing fiscal austerity, curbing spending, and chipping away at public debt. With the deficit, taxes, and unemployment falling, Germany is poised to make the most of Europe's predicted economic upswing.

A healthy economy in two years time would force the opposition in Germany to search elsewhere for a viable campaign issue. The CDU might be inclined to concentrate on topics involving foreignersùa strategy they've pursued successfully over the past two years in state elections. In Hessen, the CDU campaigned against dual nationality and gained control of the regional parliament. In North-Rhine Westphalia, the party focused on the theme of immigration. If politicians feel they can exploit voters' anxieties about immigrant labor from Poland and the Czech Republic, EU enlargement could end up a major campaign issue in Germany's next election.

Terry Martin is a EUROPE contributor based in Berlin.

Article 12

In These Times, June 26, 2000

Germany's New Identity
For immigrants, there is power in a union

By David Bacon

FRANKFURT, GERMANY

Twenty-six years ago, as a young theology student, Manuel Campos fled Portugal one step ahead of the secret police. Just before the fascist dictator Marcelo Caetano fell in 1974, Campos discovered his name on a list of people about to be arrested. A priest got him out of the country, and Campos suddenly found himself in Germany, a young man with no prospects, few skills and a head full of radical ideas.

He arrived at the end of a long wave of immigration, promoted by big companies that advertised for contract workers throughout southern Europe. Asylum seekers like Campos were part of the mix, welcomed at a time when Germany's labor supply was low, and the need for educated workers was high. He wound up in an auto plant. "I saw the assembly lines filled with immigrants like myself," he remembers. "When I came here there was nothing for us. We had either fled our countries, like me, or we were looking for a way to send enough money home so that our families would survive. Lots of us were here for both reasons."

Campos didn't forget the experience. Today he heads a unique department in the big German industrial union, IG Metall, where he organizes immigrant workers. He moves with frenetic energy—his fingers race through piles of paper as he talks a mile a minute, pulling out charts and numbers to back up his point: Immigrants have had a big impact on the German workplace.

About 7 million of Germany's 80 million people are immigrants, who make up about 2 million of its 34 million workers. IG Metall is the largest single union in the world, with 2.8 million members, but its membership has been declining as industry leaves high-wage Germany. Still, the number of immigrant members has remained relatively constant—about 275,000. The numbers indicate the important role union membership plays in immigrant life here.

Campos tends to make speeches when he describes the work of his department, but he has some justification for crowing. With his help, IG Metall has fought with Germany's largest corporations to develop unique agreements combining affirmative action with protection against discrimination and harassment.

In Germany, like the United States, federal law forbids discrimination. But Campos says the law doesn't really protect immigrant workers, who are still referred to officially as foreigners. "It's almost impossible for immigrants to file complaints and get them enforced," Campos charges. "The court system is very conservative, and many judges are racists. . . . As a result, it's much better for workers when we negotiate these agreements, which are then enforced by the works council in the workplace."

German unions don't have locals that correspond to individual workplaces. Instead, elected works councils, which can include union members and non-members alike, negotiate over most conditions in each factory and enforce these agreements. Only wages and the hours

of work (which in the West includes a 35-hour work week) are negotiated nationally, between unions (like IG Metall) and industrial employer associations. But IG Metall has an organization in each city and region. In most of them, workers have elected commissions that promote the welfare of immigrant workers. There are more than 1,000 such elected commission members throughout Germany.

IG Metall has pressured four major German corporations, including Volkswagen, to adopt standalone agreements outlawing discrimination or harassment based on immigrant status, along with other forms of racial and sexual mistreatment. The anti-discrimination agreements bar "mobbing"—racist harassment designed to make people quit. The agreements also require the company to provide training for unskilled workers at the bottom of the work force, and then to hire them into new positions. Tests for the new jobs, which discriminated against immigrants, have been changed. "Hundreds of thousands of immigrant workers are trapped at the bottom, in jobs which are likely to disappear," Campos says. "So they need training desperately."

The first wave of "foreign workers," who arrived in the early '60s, had it the hardest. They were called guestworkers and worked under contract. Mahmut Aktas' father came from Turkey in 1963. "He was all alone," Aktas says. "The people didn't accept him. He didn't have his family with him, and he didn't speak the language well."

"I'm a German citizen now, and they still see me as a foreigner," says Turkish immigrant Mahmut Aktas, a chief steward at DaimlerChrysler. "They don't like it if I have a more-skilled and better-paid job than many native Germans have."

Seventeen years after his father took that first labor contract, Aktas finally arrived himself, as a teen-ager. Aktas, who now works at the huge DaimlerChrysler plant here, says discrimination is common. "I started out as a skilled worker, but to my foreman, I was just a foreigner," Aktas says. "Officially, the company says discrimination doesn't exist, that we can go into any job in the factory. But if you want to get into a really skilled position, they discourage you. I'm a German citizen now, and they still see me as a foreigner. They don't like it if I have a more-skilled and better-paid job than many native Germans have."

Aktas says IG Metall makes a real effort to stop discrimination, noting that several officers of the union now are foreign workers themselves. "If a worker is prevented from getting a better job," he says. "He or she can talk to a steward, and the odds are good that the steward will be Turkish or Kurdish."

Aktas is one of them—a chief steward at DaimlerChrysler. Yet of the 36,000 workers in his plant, only 2,500 are Turkish or Kurdish, and he describes his experience as very contradictory. On the one hand, discrimination is something he confronts every day. "For example, if I or one of my Turkish co-workers want

to become a spokesperson for our group in the factory," he says, "many of our German co-workers just would not vote for us."

Yet he has been elected steward three times, and believes he fights for the interests of all his fellow workers, of whatever nationality, sometimes in the face of considerable company hostility. "I've taken on their cause," he says. "My foremen and supervisors don't like this one bit, and they make it clear. They don't like the union, and they particularly don't like the fact that I'm a Turkish shop steward."

In fact, immigrants make up a much larger percentage of the union's stewards than their percentage in the general membership. "We have the worst jobs and the lowest pay. The jobs that are disappearing the fastest belong to us. But we remain union members in much greater numbers than others because we look at the union as our political homeland."

This political support stretches beyond the workplace. In the past decade, since the reunification of Germany, unemployment has soared, particularly in the former East Germany. There, groups of young neo-Nazis say immigrants have taken their jobs, and have even burned the hostels where many immigrants live. "The danger to immigrants in Germany is constant—you feel it in the streets every day," Aktas says. "You have to watch your back at every moment. I can't say the union is 100 percent behind Turkish workers, but it's a lot better than anywhere else in German society."

Every year the union mounts a campaign to coincide with the U.N. International Day of Action Against Racism. It organizes demonstrations in coalition with the German Intercultural Council and the Committee of Turks in Germany. Campos was one of many IG Metall members who went to Berlin to demonstrate against a recent neo-Nazi march. "I don't think Germany is about to become fascist again," he says. "But these groups, although they're small, are very aggressive and protected by the police. If there's no visible opposition, it's a signal that hating immigrants is acceptable."

Not all German labor is equally committed to defending immigrants. Other German unions still see immigrants as a threat. To more conservative unions, like those in construction, immigrants are viewed as low-wage competition at a time of high unemployment. Some say immigrants are deliberately hired by employers to drive down wages.

They point to Berlin, where the federal government is building a new city center. German construction unions have been locked out of what has become the country's largest construction project. Instead, a network of subcontractors has hired an immigrant work force primarily from Eastern Europe, where unemployment is higher and wages much lower. These workers are almost all undocumented. "On our side, a worker can earn as much in an hour as the same person can earn in a day over there," says IG Metall cultural affairs director Kurt Schmitz.

A wage wall, like the iron fence between Tijuana and San Diego, has replaced the old brick wall that used to divide the East from the West in Berlin. German sociologist Boy Leutje, who has made comparative studies of labor in the United States and Germany, says that unions in both countries confront the same choice: "Are they going to fight a losing battle to keep immigrants out of their trade and their country, or are they going to see them as potential union members and try to organize them?"

With unions like IG Metall in the lead, Germany is only now beginning to see its immigrant population as permanent residents, rather than foreign workers who will someday go home. But the concept of immigration, of Germany as a country of immigrants, doesn't exist yet, Campos says. Until the law was finally changed this year, the child of an immigrant, born in Germany, was still ineligible for citizenship. "Instead of an immigration law, we have a law which calls us foreigners. We need the law to see us in a new way, to set out the conditions under which we can immigrate, and to spell out the rights the government guarantees us."

"My children were born here," Aktas adds. "For us, our future is here in Germany."

Article 13   *The Economist*, July 29, 2000

Hot and sticky in Ireland

Ireland has a shaky government and a booming economy. Can either last?

DUBLIN

NOW that Ireland's legislators have taken off for their summer holiday, the government of Bertie Ahern can heave a sigh of relief after a string of scandalous embarrassments that would have toppled many an administration elsewhere in Europe. Just before the three-month recess, it survived a no-confidence vote, thanks to a tiny junior coalition partner, the Progressive Democrats, staying sheepishly loyal.

But its tribulations are not entirely over. The Irish people are still being treated to the tawdry if sometimes sadly comical spectacle of Charles Haughey, a former Taoiseach (roughly pronounced "tea-shock", as Irish prime ministers are called), being humiliated before an investigative tribunal that is making Mr Ahern's conservative-nationalist Fianna Fail party look distinctly sleazy.

A once irrepressibly chirpy bounder with a deftly populist touch who ran Ireland in three spells between 1979 and 1992, Mr Haughey has already been castigated for misleading another of several tribunals looking, among other matters, into his finances and into land development in Dublin. And he has been ridiculed by the entire nation for his vanity and high-living shenanigans; in 1991, it turned out, he spent nearly $22,000 on shirts, hand-made in Paris, courtesy of the public purse. Now 74 and blighted by cancer, he has been allowed by the party-payments tribunal judge, Michael Moriarty, to limit his attendance to only two hours a day. He looks a forlorn figure; he still faces criminal proceedings, and could yet go to prison.

But the bigger question is whether any of the dirt dished on Mr Haughey will stick to Mr Ahern, who was the party's chief whip for much of the time that Mr Haughey was soliciting party contributions from Irish businessmen, many of whom—it transpires—salted their money away, untaxed, in a bank in the Cayman Islands. Mr Ahern, now 48, signed a string of blank cheques on Mr Haughey's and the party's behalf, including some for those shirts.

So far, however, Mr Ahern, who seems eminently to merit his sobriquet of "the Teflon Taoiseach", is soiled but not so badly that he looks like being tipped out of office. The impression is that he was sloppy but not dishonest about accounting for contributions to party coffers. Most of the events under scrutiny took place a decade ago. But the phrase "I can't recall" has fallen embarrassingly often from Mr Ahern's lips.

Unlike Mr Haughey, however, the current prime minister is not a high liver. Mr Ahern is modest sort; another nickname is "the Anorak Man", for his amiably drab taste in fashion. He hails from the old Fianna Fail school of hard-nosed, nationalist-minded, back-scratching political managers. Nobody accuses him of being a visionary, but few insiders think he is in politics for material gain. So he will probably survive the ordeal. In the autumn, he will oversee a generous budget, then perhaps go to the country in the spring.

With the economy booming, Fianna Fail (pronounced "Feena Foil") would very probably win the most seats, but that would by no means guarantee it another stint in office. Thanks to Ireland's complicated system of proportional representation and transferable votes, governments are very likely to be coalitions. Pollsters say that the popularity of the two big parties, Fianna Fail and the main opposition, Fine Gael ("Finna Gale"), which is fairly similar in economics (if a bit more market-minded) but less nationalistic, especially over Northern Ireland, has slumped; the fortunes of the small parties, along with mavericks and independents, has soared. So new combinations could pop up, with or without Fianna Fail still in office.

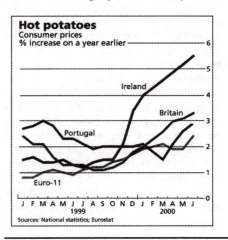

Hot potatoes
Consumer prices
% increase on a year earlier

Ireland
Britain
Portugal
Euro-11

J F M A M J J A S O N D J F M A M J
1999   2000

Sources: National statistics; Eurostat

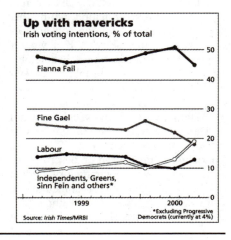

Up with mavericks
Irish voting intentions, % of total

Fianna Fail
Fine Gael
Labour
Independents, Greens,
Sinn Fein and others*

1999   2000

Source: *Irish Times*/MRBI   *Excluding Progressive
Democrats (currently at 4%)

But if politics has been warming up, the economy has been even hotter. Ireland has been growing at breakneck speed—faster than any other country in the EU for each of the past three years: last year, by nearly 10%; the year before, by nearly 9%. The latest guess for this year is 8%. Depending on how you measure it, Irish wealth per person may now be greater than Britain's (see article).

Unemployment, which was 15% in the early 1990s, is under 5%. Labour is in short supply, even though people, reversing an old trend of history, are pouring in from all over the EU, from Central Europe and back from the United States to get jobs. Foreign investors love the high level of Irish skills, especially in computing, and the country's low corporation tax, due to go even lower, to 12.5%, by 2003. Nearly a third of American investment in the EU is said to be going to Ireland, much of it into high-tech companies. The "emerald tiger" is veritably roaring.

The only real worry is inflation. At latest count, it was running at an annualised rate of 5.5%, and could, concedes Mr Ahern, pass 6% this summer. It is already double the EU average, and easily the highest rate in the Union. What to do?

The lever of jerking up interest rates has gone, since Ireland joined the single-currency euro-zone. At a mere 1% of the EU's GDP, the European central bankers in Frankfurt do not lose much sleep over Ireland's heat-up. Mr Ahern's government is loth to cut spending. The country sorely needs to revamp its creaking infrastructure to cope with its new wealth; last year, the Irish bought a quarter more new cars than the year before.

The government is equally reluctant to raise taxes; most of the inflation is coming from food, drink, housing and services. As a dampener, it is trying to fix the price of alcohol, Guinness included. More plausibly, it is trying to pass a law that would loosen the housing market by speeding up planning and building procedures; house prices have been zooming by 20% a year for the past few years. In some Dublin suburbs, they have quintupled in a decade.

In short, the government, which sounded dozily complacent about inflation a few months ago, is nervous—but, in essence, still hoping that the problem

Richer than the Brits?

DUBLIN

ARE the Irish really richer, these days, than the British? A couple of decades ago, even to have asked the question would have seemed ludicrous. The answer, today, depends on your yardstick. If you go by gross national product (GNP), the answer is no—but the gap is closing. If you go by gross domestic product (GDP) per person, the answer is yes, just. And if you go by GDP adjusted for purchasing power (to account for cost-of-living differences), the answer is a still plainer yes. Looking at figures for last year, Eurostat, the EU's statistics-cruncher, puts an Irishman's GDP, unadjusted for purchasing power, at 23,410 Euro ($24,970) a year versus a Briton's 22,670 Euro. Your average EU citizen's is 21,130 Euro, to an American's 31,820 Euro.

The reason the GDP answer makes an Irishman look richer than a Briton is that it does not cater for income earned abroad, nor does it deduct income paid to foreign creditors. Those tweaks produce GNP. Britain has a lot of foreign earnings; conversely, a lot of foreign-owned companies in Ireland declare their earnings there—but send a big chunk of them abroad. Hence Britons, judged by GNP, are probably still about 14% ahead.

However you look at them, the figures dumbfound most Ulster Protestants, for whom it was long an article of faith that people in "the republic" got more on the dole in Britain than they did by staying at home to "dig potatoes". In 1987, Irish incomes, measured by GDP per head, were less than two-thirds of those in Britain. But one thing, these days, is undoubted. People in the south are a lot richer than those in the northern, British, bit of the island. At last glance, Northern Irish GDP per head was a mere 76% of the United Kingdom's. That gap is far wider than can be explained by the bonus which the Republic of Ireland gets from the EU's budget hand-outs, which have lifted GDP, at various annual estimates since Ireland joined the Union in 1973, by between 4% and 7%.

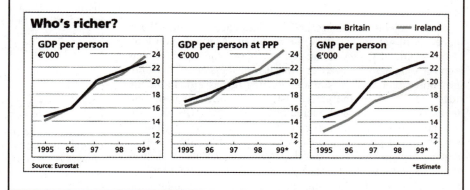

Who's richer? ▬ Britain ▬ Ireland

GDP per person €'000 | GDP per person at PPP €'000 | GNP per person €'000

1995 96 97 98 99*

Source: Eurostat   *Estimate

will gradually fade away, as competition at home and from across the EU bites and wage pressures ease off. Maybe that will happen; no one, in fact, is sure. More reassuringly, inflation apart, all the other basic economic figures have rarely, if ever, been better.

Ireland's oddity is the mismatch between its astoundingly bright and modern economy, which has left its old

rural roots far behind, and the old, village-hall tackiness of its politics (witness Mr Haughey), which harks back to the days when glad hands and scratched backs was how all business, political and otherwise, was done. But if the tribunals now digging up the dirt do their job of naming and shaming, that part of Irish life might well start to change faster too.

Article 14   *The Economist*, December 2, 2000

A Survey of Portugal
Half-way there

In only 15 years, Portugal has halved the big gap in living standards between itself and the rest of Europe. Finishing the job, says Patrick Lane, could take longer

IF YOU fly to Lisbon, you get a glimpse, even before you land, of how much Portugal has changed in a very few years. Stretching for about 17km (11 miles) across the river Tagus is the Vasco da Gama bridge, Europe's longest, opened in 1998 and named after the explorer who 500 years earlier became the first European to sail to India.

On the west bank of the river, you might also catch sight of Nations Park, the site of the 1998 World Fair (Expo). The Expo was a grand party which attracted 10m people, about as many as live in the whole of Portugal. It also rescued a corner of Lisbon from dereliction. From the terrace outside his fifth-floor office, Antonio Mega Ferreira, the president of Nations Park, points out what was there before work started: an oil refinery, a barracks, an abattoir and a rubbish tip. The Expo's 150 pavilions have gone, but they have left behind a legacy of a shopping mall, a futuristic railway and metro station, flats and offices, in one of which people are already preparing Portugal's next big party, the 2004 European football championship.

Such projects exude freshness, confidence and pride, and indeed Portugal has plenty to be proud of. Any way you look at it—politically, economically, socially—the country has come a long way in a short time. From the late 1920s until the almost bloodless revolution of 1974 it was a dictatorship, ruled until 1968 by the paternalist, arch-conservative Antonio de Oliveira Salazar and thereafter by Marcelo Caetano. To emerge from that, and from a post-revolutionary decade of economic and political chaos, as a stable democracy is a huge achievement in itself. To develop economically the way Portugal has done in the past 15 years is close to miraculous.

When Portugal joined the then European Community on new year's day 1986, its GDP per head, in purchasing-power-parity terms, was a mere 53% of the EU average. Closing the gap has been the stated aim of Portuguese governments ever since, and progress has been swift: GDP per head is now 75% of the European mean. No other hopeful entrant to the European club—not the big next-door neighbour, Spain; not even Ireland, the Celtic tiger of the 1990s—has made up so much ground in its first few years of membership (see chart 1).

Yet the gap has merely narrowed, not disappeared. "In New York or London there are pockets of poverty," says Antonio Barreto, director of the Institute of Social Sciences at the Uni-

versity of Lisbon. "You have to say that Portugal is poor with pockets of wealth." Within Portugal, the speed of development has created stark juxtapositions of old and new, rich and poor. Just 100 yards from the spanking new stock-exchange building on the outskirts of Lisbon is a collection of makeshift dwellings with corrugated-iron roofs and walls. In the steep streets around Oporto cathedral nestle tumbledown houses festooned with satellite-TV dishes. There is an obvious difference in living standards between the old, who live on pensions as low as 25,000 escudos a month, and the affluent, mobile-phone-toting young. Whereas GDP per head in Lisbon is about 90% of the EU average, in the neighbouring Alentejo it is only 60%, and in the Azores 50%.

And everywhere, there are signs that the journey is far from complete. The road from Lisbon to Sintra is lined with one half-built apartment block after another. Here and there, a reasonable road turns into a test track for a tank. And, says a grinning Lisboan, "If you go to Oporto, don't mention the metro." Portugal's second city has been waiting several years for its urban rail system to be completed. Other projects in next year's European City of Culture, such as the House of Music on the Boavista roundabout, are also far from finished.

Under construction

Having narrowed the economic gap between themselves and the rest of Europe so quickly, the Portuguese might be expected to eliminate it in similarly short order. But progress seems to have slowed down. In part, this is the backlash of a consumer boom in the late 1990s. But Portugal may also be pausing after a long spending spree. When economic liberalisation began in the mid-1980s, says Luis Filipe Reis, a director of Sonae.com, a telecoms company, "It was as if people had at last been let into a cake shop. They ate a lot of cake." Now indigestion is setting in.

The economic evidence confirms the impression of a slowdown. Between 1987 and 1991, says Abel Mateus, an economics professor at the New University of Lisbon, Portugal narrowed the gap with the EU in GDP per head by 10.7 percentage points. In the

Exchange rates
November 20th 2000

$1=	Esc 235
€1=	Esc 201
£1=	Esc 335

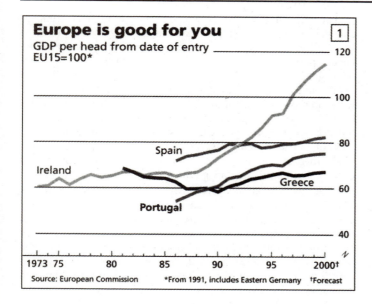

Europe is good for you

GDP per head from date of entry
EU15=100*

Ireland

Spain

Portugal

Greece

120

100

80

60

40

1973 75   80   85   90   95   2000†

Source: European Commission   *From 1991, includes Eastern Germany   †Forecast

If Portugal wants its GDP per head to grow faster than in the EU as a whole, it must do one of two things. The first is to increase the proportion of its population at work by more than the European average (or to ensure that it falls by less). This is not out of the question, because Portugal's population is ageing less quickly than that of the rest of the EU. But the country already has full employment, as well as a higher female participation rate in the labour force than most European countries, so such effects are likely to be small.

The second, likelier route is to maintain a higher rate of productivity growth than the rest of Europe. There are plenty of reasons to think this might be possible. Portuguese productivity levels are low, so there is scope to catch up. Each generation of Portuguese workers is better educated than the last. The prime minister, Antonio Guterres, whose Socialist Party holds 115 seats in the 230-seat parliament, has made productivity his mantra, and is trying hard to improve his country's performance. But lately output per head has been growing more slowly.

Whether the required increase in productivity will be achieved is a genuinely open question. A lot will depend on Portuguese business, but Mr Guterres could certainly help matters along. At present, the Portuguese government spends more than half of the country's GDP, a bigger proportion than in most EU countries, and much more than would be expected of a country where income per head is relatively low. The public sector's share has been rising sharply, thanks to a mixture of deliberate policy choices and lamentable spending control. In short, the government itself is soaking up a bigger share of the country's resources than it should. If it has an excuse, it is that change in Portugal has been too rapid to sit back and think. The next section will show just how relentless the change has been.

following four years it shaved off only another six points, and in the four years after that, to 1999, only a further 3.4. With that sort of recent record, and given the European Commission's growth forecasts for 1999–2003 (an average of 3% for the whole EU, a little less for Portugal), convergence could take 70 years, says Mr Mateus.

Most economists expect the economy this year to grow by less than the EU average, for only the third time since Portugal became a member. Some think that the below-par performance will continue in 2001. Is this merely a temporary lull in Portugal's progress, or has the rate of catch-up slowed for good? The answer, this survey will argue, will depend on the country's rate of productivity growth.

Pause for breath

Both Portuguese society and the economy have changed at bewildering speed

WHEN Portugal joined the EU in 1986, the motorway from Lisbon to Oporto, Portugal's second city, was still unfinished. These days, if you put your foot down (and most do), you can easily cover the 300km (190 miles) in two-and-a-half hours. Or you can take the nearly-new road east from the capital, speed past the stripped cork oaks of the Alentejo and be in Spain in under two hours. If you travel on Portugal's new motorways at the speed limit of 120km (75 miles) an hour, someone will sweep past you as if you were standing still. No wonder that Portugal has one of the highest road-death rates in Europe. But it is not only on the roads that the Portuguese are in a hurry. In fact, it sometimes seems as though they have spent much of the 20th century in a blur of ever-accelerating change.

The political and social transformation speeded up dramatically after the revolution of 1974, but the economic makeover

came much later. Jose Freire Antunes, a journalist and historian, says that the revolution was followed by "a lost decade" of state control of the economy, as well as political chaos. Jorge Braga de Macedo, a former finance minister, puts the time lost at more like 15 years, because it was not until 1989 that reprivatisation and other economic reforms really got going.

Then again, the roots of Portugal's social and economic transformation can be traced to well before the revolution. The country began to open up to trade in 1959, when Portugal became a founder member of the European Free-Trade Area. Salazar, who once declared, "I want this country poor but independent," must have had misgivings. "Salazar knew that if you opened up to free trade, it was like with a woman: either she's pregnant or she's not pregnant," says the University of Lisbon's Mr Barreto. And in the 1960s colonial wars and large-

The second example is women, whose lot has arguably changed more in Portugal than anywhere else in Western Europe. When Salazar was in charge, women needed their husbands' permission to take a job, to go into higher education or to get a passport. They did not get the vote until 1976. In 1960, only about 20% of women of working age were in paid work, the lowest rate for any of the countries that are currently in the EU. Now the proportion is 63%, towards the top end of the EU range. Forty years ago, women made up only about 25% of university entrants, and those who enrolled were less likely to graduate than men. Now women account for around 60% of both students and graduates.

Changes in women's private lives have also been dramatic. Thanks to better education, easily available contraception and creeping secularisation, fertility rates have dropped from the highest in Europe to among the lowest, much the same as they have done in Italy, Ireland and Spain. Infant mortality in Portugal has plummeted. In 1960, 80 out of every 1,000 babies died in their first year, the highest rate in Western Europe (and the same as in Britain in 1920). Now the figure is eight per 1,000—still above the EU average, but a vast improvement.

The third instance is migration. Historically, the Portuguese have been an emigrant race. Besides the 1m in Brazil, there are large Portuguese communities in France, Canada and South Africa, each of them more than half a million strong. In the last decade or so of the dictatorship, low incomes at home persuaded an average of 120,000 people a year to go abroad. The money these emigrants sent home propped up not only family budgets but also the Portuguese balance of payments.

Now the flow of people has gone into reverse, so that Portugal has become a net importer of labour. In 1974, only 32,000 foreigners resided legally in Portugal, most of them from other European countries. By 1997, their number had swelled to 175,000 (out of a population of about 10m), over half of them from Africa. This has broken up Portugal's racial homogeneity, though to a lesser extent than in Britain, Germany or France, and so far has been accompanied by much less racial tension than in those countries. But besides the Angolans and Cape Verdeans on the country's innumerable building sites, there are also increasing numbers of people from Central and Eastern Europe, often with valuable skills, who would rather take a menial job in Portugal than none at all at home. It is not unusual to hear stories of Romanian or Ukrainian doctors or nurses working as domestic servants. According to one recent estimate, 20,000 people from Ukraine alone are in Portugal illegally.

Virtuous circle

Social and political changes have fed into economic development, and vice versa. The rise in paid female employment has expanded the workforce and boosted the productive capacity of the economy; and the increase in women's incomes outside the home has made them more independent. The switch from emigration to immigration has largely been caused by economic growth. As the economy has developed, there has been less incentive for the Portuguese to go abroad to find work; and as Portugal's spare workers have been soaked up, the country has developed a chronic shortage of labour, both skilled

scale emigration helped push more women into paid work. "There were no men in Portugal," he explains.

You can argue about the dates, but not about the speed of change, which has been bewildering. After the 1974 coup, the far left was in charge for a year. Then came democracy, but not stability. Between 1975 and 1985, the average life of a Portuguese government was a mere ten months. The army continued to have a political role (although a waning one) until 1982, although a general, Antonio Ramalho Eanes, was president until 1986. Now not only is democracy firmly established, but stable government has become the norm. Since 1987, every government has served out its full four-year term.

And changes in society more broadly? Here are three examples, backed by figures from a comprehensive database compiled by Mr Barreto.[1] First, religion: Portugal is still a Catholic country, but less Catholic than it used to be. Pilgrims continue to flock in huge numbers to Fatima, the pope's favourite holy place, where Our Lady is said to have appeared to three shepherd children in 1917. Yet when people get married (in decreasing numbers), fewer of them want a church wedding: only about two out of three now, against nine out of ten in 1960. Divorces, which before the revolution added up to only a few hundred a year, now run to 14,000 a year, about a fifth of the number of weddings. Like young people everywhere else, modern Portuguese are less likely than their parents to marry before having children. In 1970, only about one in 14 babies was born outside marriage. Now it is almost one in five.

and unskilled. What with the half-built blocks of flats and motorways, the government's plans for further public works, and the need to build football stadiums for the 2004 European championships, there should be plenty of work for immigrant construction workers for years to come.

Economic growth has also closed off the labour market's other traditional safety valve: agriculture. Portugal has always had a low unemployment rate, because when times were hard its people either emigrated or returned to the land. Even at its worst in the 1990s, unemployment was only 7.5%, a figure that many European countries would be glad of. But the farm sector has shrivelled. In 1960, it accounted for 44% of total employment and 24% of GDP. Now those figures are down to 13% and 4% respectively—although that still leaves Portugal with proportionately the biggest farm sector in the EU.

The other big structural shift has been the decline in the state's importance as a producer. After the revolution, the government took over a quarter of the economy. Fifteen years later, the constitution had to be changed to allow privatisation

to begin; a further change in the law was needed to allow the state to reduce its stakes in nationalised companies below 50%. Since the 1990s the state has been running down its interests in banks, oil, petrol, gas, electricity, telecoms, even motorways, and it continues to offload more assets. In October, it sold another 20% of Electricidade de Portugal, the national power company. It is about to shed almost the last of its shares in Portugal Telecom (PT). In the new year, the delayed sale of its remaining 10% stake in Cimpor, a cement company, should at last go ahead. For all that, relations between the government and the formerly state-owned companies remain close, perhaps too close. The state keeps "golden shares" that allow it to veto mergers, even when it has only a minority stake. It is keeping a golden share in PT.

Amid all the economic statistics, one stands out: that Portugal's living standards since joining the European Community in 1986 have risen from about half the European average to three-quarters. In the ten years before that, Portugal's GDP per head had struggled to keep pace with the EU average. That is no coincidence.

Article 15   *The Economist,* November 25, 2000

A Survey of Spain
A country of many faces

Twenty-five years on from the end of dictatorship, there are many Spains, writes Stephen Hugh-Jones

THIRTY years ago, the writer of this survey went to work on a new European business monthly. It was to be published in four languages: English, French, German and—no, Italian. Why bother with Spain, a backward little economy fenced off from the European Community, run by a decrepit, tinpot dictator? To such inconsequence had General Francisco Franco, *caudillo de España* (por la gracia de Dios, indeed, or so he proclaimed) reduced a country that once was the greatest power in Europe and held the grandest overseas empire in the world.

The fault was not his alone: socially, politically and economically, Spain, torn by coups and civil wars, had been dropping behind the rest of Europe for 150 years. A semi-fascist dictatorship that survived the defeat of that system in 1945 was just its latest misfortune. But the result was undeniable: except to the United States, eager for military bases, to outsiders Spain did not matter.

And what a change there has been. Twenty-five years ago this week, Franco died. Into his place stepped the man the *caudillo* had chosen—one of his few wise decisions—as his heir, the surviving grandson of the king who had fled the country in 1931. The new King Juan Carlos soon, and unexpectedly, proved a pillar of democracy: first in the transition to it, and then, in 1981, using his status to stifle one more attempted coup. He is still there, a respected monarch who over 25 years has earned that respect. Spain's democracy is young, but today it looks firmly rooted and certainly operates

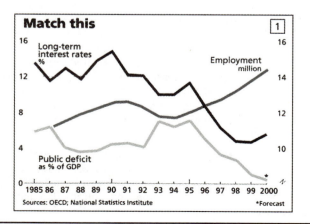

Match this   1

Long-term interest rates %

Employment million

Public deficit as % of GDP

1985 86 87 88 89 90 91 92 93 94 95 96 97 98 99 2000

Sources: OECD; National Statistics Institute   *Forecast

as well as any other. The economy has been transformed. In 1975, one-fifth of Spain's workers were still on the land; now it builds more cars than Britain, and has one of the global leaders in the mobile-phone business. It has privatised, and opened to the world, notably since it joined the European Union in 1986. Though unemployment is still high, since 1994 Spain's economy has been creating jobs faster than America's. Its public finances are in good shape. Its society, for better and for worse, has leapt 60 years in 25. Spain has reclaimed its place among the nations.

Yet what is this "Spain"? History alone has left several of them. It belongs to the Mediterranean, not just by geography but by eight centuries of Muslim conquest and Christian reconquest. It is once again an integral part of that Europe whose Holy Roman Emperor Charles V in the 1500s was a son of Spain—and the mightiest ruler in the continent after Charlemagne and until Napoleon. And it is part of, indeed one creator of, the Atlantic world.

Within its own borders lie many Spains. The land is large: about 1,000km (620 miles) from west coast to east, as far as from London to Prague; and that leaves out the Balearic Islands, and the Canaries, 1,200km away off West Africa. The country has temperate rainforests in the far north-west, deserts in the south-east. Its peoples are varied, their dialects—languages, some can claim—not just diverse to the ear but to

the eye as well: Galician sounds and reads much like Portuguese, Catalan like a mixture of Castilian (the official language) and medieval French; and then there is Basque, as mysterious to most Spaniards as if it came from Mars. Internet addicts coexist with peasant farmers, who in turn coexist with sectors of agriculture as advanced as any in Europe. Roman Catholicism reigns, but Muslim immigrants have begun to flow in.

Government, not least, typifies this diversity. Beneath (or rather, beside) a central government and parliament much like any others, lie (or stand) those of 17 regions, the comunidades autonomas, with wide powers. Real ones too, for the regions, like the German Länder or Canada's provinces, to use as they decide, not (like English counties) as the central government in Madrid tells them to. Whether they should have so much power, and be so ready and able, some of them, to thumb their noses at the centre, is disputed, and has been for centuries. But after the extreme centralism of Franco's regime, it was bound to happen, and it reflects the reality of centuries. Spain is one thing, each of its regions, for varying reasons, to varying degrees, another.

It is this change from an isolated Spain, artificially and (even if imperfectly) homogenised, to a Spain united but diverse at home and abroad that this survey will try to track.

Into the European market

An opened economy works wonders

IF TODAY'S Spain is deeply different from that of 1975, it owes a big debt, often unacknowledged, to the 14 years of government (1982–96) by the Socialist Party (PSOE) under Felipe Gonzalez. For a start, he showed that Spain could swing left without arousing scares about reds under the bed. He dismantled much of the state ownership and some of the bureaucratism inherited from Franco. He kept Spain in NATO, and then led it into the European Union. Yet today's Spain is, in turn, deeply different from that of Mr Gonzalez: it is a solid economic success.

For that, thank four things: EU membership and the opening of the economy; successive devaluations in the difficult early 1990s; the psychological shock of the 1993–94 crisis, when unemployment hit almost 25%; and now first-round membership of the euro zone. And one thing more: the election, in 1996, of the conservative People's Party (PP), under Jose Maria Aznar, a true believer, not just a pragmatic one, in shrinking the economic frontiers of the state. He did not win an absolute majority in parliament then, needing help from, notably, the Catalan "nationalists". But last March he won again, this time with an overall majority; and he deserved it.

The economic figures need no spin-doctor: real GDP growth of 4%, or very near it, for four years in a row; employment up from 11.7m in 1994 to 14.6m now, and unemployment down from 24% to less than 14%; long-term interest rates down from 15% in 1990 to less than 6%; not least, public spending, which hit 49.5% of GDP in 1993, now below 42%, and a general gov-

ernment deficit (financial operations apart) shrunk from 7% in 1995 to very close to zero.

Not all the figures are perfect. About one-third of all workers are now on short-term contracts, which spell flexibility to employers but insecurity to those they employ. The fall in unemployment is slowing, and at 13.7% (by the usually accepted measure, though a rival one puts it several points lower) it is still huge. Wages this year have not even kept up with prices, let alone added the share of growth that a workforce can reasonably expect. The current-account deficit would be a worry were Spain not tied into the euro. Inflation is higher than expected: 3.7% in September and rising, though a dip is due by year-end; with credit, even after a recent slowdown, still swelling at about 10% a year, and the oil-price rise on top, the Bank of Spain must be privately glad at the latest euro interest-rate rises. The government's inflation forecast, used as a basis for next year's budget, is 2%. The trade unions are not alone in thinking that moonshine. Yet all these are scratches and chips on the paintwork, not cracks in the engine-block. Spain is catching up with the rest of the EU, as it has been eager to, and in time the rewards of growth will surely, if slowly, trickle down.

How has it been done? The opening of the economy began long before Mr Gonzalez. The Opus Dei technocrats to whom Franco turned in the 1950s may have had old-fashioned ideas about the integration of religion and politics, but their economics was up-to-date. They could see that the dreams of autarky and all-wise government dear to men ranging from Mussolini to Nehru did not work. Spain's take-off began in the 1960s. The oil shock of 1973 was badly handled, however. "Nothing is worse than a weak regime plus administered prices," laments one businessman, and inflation was to plague Spain for years after Franco had ceased to do so.

Opened and jolted

The big opportunity came with entry to the EU. But the big jolt that made Spain ready to seize it was the 1993–94 crisis. Employers and trade unions alike saw they must change. Employers began to modernise in earnest, cutting staff, wages and prices to survive. Belatedly, unions realised that their own pressure for large wage rises and public spending, along with excessive job protection, had brought the roof in on themselves. The PSOE government legislated to make firing easier and less costly (and hiring therefore less risky), and the unions learned to live with it. Meanwhile, uneasy householders began to spend less (which lessened inflation), and to save more, and more productively: the 1990s saw the total of investment funds, life-assurance and pension funds (company ones included) soar from 5% of GDP to about 70%; stockmarket capitalisation, albeit concentrated in a handful of companies, rose likewise.

Then came a new government, determined to bring down the level of public spending into which its Socialist predecessors had been forced by recession and rising welfare costs. Next, by happy chance for Mr Aznar, came the euro, and, not by chance but by choice and effort, a Spain that was qualified for it. The prospect cut inflationary expectations. The reality sharply cut interest rates.

That was helpful for a state having to service public debt that, even now, totals 63% of GDP. It was even more helpful for companies eager to invest, as they knew they must, to match vigorous EU competitors. Spain's two-way trade in goods and services in 1991 totalled 37% of GDP; by last year, it had risen to 56%. That change has been arduous for business, but very salutary. And with memories of isolation not so far back, most Spaniards feel sure they were right to opt first for the EU and then (less surely: 51% for, 34% against, in a recent poll) for the euro.

Can the good times go on? This year's oil shock and aftershocks make short-term prediction doubtful. The government still hopes for 3.6% growth and 350,000 more jobs in 2001, but there are big clouds on the horizon. Recent figures show industrial output, housebuilding and car registrations all slowing. Lagging wages will cut demand. Mortgage rates are up nearly three percentage points since they bottomed out in 1999; no bad thing after two years in which house prices rose by more than a quarter, but this too must cut other spending. The long-term issues, though, are on the supply side: more privatisation, more competition, more mobile labour.

The state, after 20-odd years of sell-offs that began with the banks, has not much left to sell. Its holding company, SEPI, expects to raise $4.5 billion over the next four years, about one-third as much as in the past four. Its remaining 54% stake in Iberia, the national airline, has been heading for a sale for over a year, but has yet to be offered. The debt-laden state television service, RTVE, will soon be transferred to SEPI; a first step towards a sale, many suspect, although RTVE says

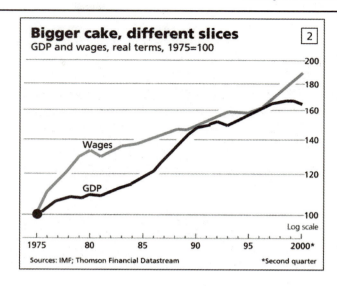

Bigger cake, different slices ②
GDP and wages, real terms, 1975=100

Wages

GDP

Log scale

1975   80   85   90   95   2000*

Sources: IMF; Thomson Financial Datastream   *Second quarter

not. The railways are even more indebted, and subsidised by about $1.2 billion a year; their infrastructure will remain state-owned, though by 2004 private trains may be using it. The state also still owns motorways (except for 1,700km that are privately run, under concession), some ports and airports; a Mediterranean ferry line; the historic parador hotels; a few other bits and pieces; and some shares left over from earlier privatisations.

Less noticed, but large, are the holdings of governments at lower levels. Regional ones supply water and run television stations; some run ports and airports. Many buses are run by regions or city halls. The savings banks, which make up about half of the banking sector, are mutually owned, but subject to public-sector boards. The result is not a swamp of universal inefficiency, crying out for sale and rehabilitation. Indeed, the city of Madrid had to take over failing private-sector bus lines. Some sales would surely help, however—yet look unlikely.

The real question, though, is competition. The PSOE, claiming to have led the way in privatisation, accuses Mr Aznar's men of failing to ensure real competition after it. That is hardly fair, and will be even less so if a liberalising package announced in June is vigorously enforced. It will stop the two big petrol-station operators, Repsol and Cepsa, opening new outlets for some years, and force Repsol to give up control of the one and only petrol haulier. Similar rules were to inhibit the two biggest electricity companies; and in 2003, four years before the date originally set, domestic consumers will be able to choose their supplier. Gas will be similarly freed, and Enagas, the distributor, is being hived off from Gas Natural, the de facto monopoly supplier. From this month, Telefonica, once a state monopoly, has been forced to open the "local loop" to rivals.

Fine, but consumers may wait to cheer until they see what happens with electricity. The two big boys (out of four that matter), Endesa and Iberdrola, do not want to fight each other, but hope to merge instead. The government has insisted that, if so, they must sell off a large chunk of their joint 80% of the market, making room for three newcomers to join the fray. The result would still be a giant, against five relative pygmies.

The better answer would have been a flat no; but the "national champion" notion, so dear in France, is alive even among Mr Aznar's team in market-opening, decentralising Spain.

The other supply-side issue is flexible labour. The government is arm-twisting employers and unions towards a pact on this. "We can't open the door to immigrant workers and yet do nothing about the labour market," says Rodrigo Rato, economy minister since 1996. It is indeed diverse, with wide regional variations in unemployment, and the women's rate twice the men's. Bosses and unions must agree by year-end, says the government, or it will legislate. Of that, more later: the issue of labour efficiency is more complex than it may seem.

Whatever the changes to come, though, the biggest has already happened. In three decades, a rather poor, rather closed country, its business and political enterprise alike tied down by an outdated philosophy, has become a middling rich one. Minds and markets have opened. This has generated money, and money is an astounding agent of change. If Spanish society today looks much like any other in Western Europe, that is not just because fascism went away. It is because money arrived.

National calm, regional turmoil

Political differences at the centre are not great. Those between centre and regions are

IN POLITICS, the most striking change is that today change strikes nobody. To the young, Franco's decades of power are ancient history. It is only 19 years since the last attempted coup, yet by now no one, young or old, can imagine another. Politics may be sharp—the bringing-down of Mr Gonzalez, eg—but the result is, well, just a different government. That is a change from days a lot further ago than Franco's. Much of Spain's political history since 1800 has been a tale of savage partisan strife, not least in the short-lived republic that he overthrew. The civil war, not 65 years past, saw countless atrocities; Franco's rebels, especially, butchered their political opponents with a zeal that might leave Chile's General Pinochet astounded at his own moderation. Today the Socialists and Mr Aznar's People's Party happily steal each other's clothes. The PP lives readily with the abortion law that it fought bitterly in the 1980s. Broadly, if not in detail, it accepts the health, pensions and welfare set-up that it inherited, just as Mr Gonzalez's men had earlier privatised, brought in short-term labour contracts and continued the subsidy of church schools.

Joaquin Almunia, a long-time lieutenant of Mr Gonzalez, was the unfortunate who unexpectedly had to lead the PSOE into last March's election. Thumped, as expected, it soon chose a new, young leader, Jose Luis Rodriguez Zapatero. He has already made a mark with the public. But the party has still to find policies that set it clearly apart from its rival. Like what? Well, suggests Mr Almunia, things like tougher rules for competition, less tough ones for illegal immigrants, a wiser choice of infrastructure priorities—better roads, say, but free ones, not the fee-charging *autopistas* dear to the PP. Over such differences are right and left to thunder into battle.

Spaniards may be grateful for the sheer ordinariness of it all. So they may for a far less ordinary step, one that suggests Spain has moved almost beyond the rest of Europe in political maturity. Barely re-elected, Mr Aznar let it be known that he would not aim for a third term. At only 47, he may justifiably have wider, European ambitions. Maybe he has drawn the lesson of the world's many leaders who have overstayed their welcome. The reason he himself gives is: "I don't want to think myself more important than I really am."

That is saying quite a lot. Mr Aznar was born into a country which, as he puts it, "had a liberal revolution, but never a national one", and which spent most of the 20th century paying

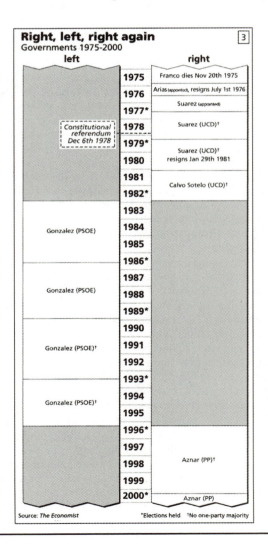

Right, left, right again [3]
Governments 1975-2000

left		right
	1975	Franco dies Nov 20th 1975
	1976	Arias (appointed), resigns July 1st 1976
	1977*	Suarez (appointed)
Constitutional referendum Dec 6th 1978	1978	Suarez (UCD)†
	1979*	Suarez (UCD)†
	1980	resigns Jan 29th 1981
	1981	
	1982*	Calvo Sotelo (UCD)†
Gonzalez (PSOE)	1983	
	1984	
	1985	
	1986*	
Gonzalez (PSOE)	1987	
	1988	
	1989*	
Gonzalez (PSOE)†	1990	
	1991	
	1992	
	1993*	
Gonzalez (PSOE)†	1994	
	1995	
	1996*	
	1997	
	1998	Aznar (PP)†
	1999	
	2000*	Aznar (PP)

Source: *The Economist*   *Elections held   †No one-party majority

for its consequent weakness. Now, this former tax inspector, still rather buttoned-up, for all his years in politics, can fairly hope to see the modernised Spain that he leads accepted as an equal among the five big countries of the EU. And—if he can avoid the hidden tripwires that await any leader—he will not be a lame duck on the way. He is content to see his party use its new absolute parliamentary majority to show who is in charge; and he, very firmly, is in charge of it.

Trouble with the regions

Tripwires there may be, and not only the normal economic ones, or the whiffs of scandal that helped to bring down Mr Gonzalez. Spain's diversity is expressed in 17 regional governments, plus provincial ones in some regions, and thousands of town halls. Over-expressed, say some critics; certainly visibly expressed, in massive regional parliament and government buildings, and payrolls to match. The centre, having transferred ever more functions to the regions, today has 45,000 fewer employees than in 1994; the regions have 235,000 more.

Yet that is only a symptom, say critics. Devolution of power was inevitable after the days of centralist dictatorship, when even regional languages were banned. The 1978 constitution duly allowed for a more representative structure, gradually built up in law and still under construction. In some cases, there were strong reasons. Basques and Catalans think themselves distinct nations, as, to a lesser extent, do Galicians. Navarra too had historical claims, and the Canary and Balearic Islands geographical ones. But why was the process extended to all, those critics ask, and where will it stop?

Devolution, in fact, is far from uniform. The Basque Country and Navarra collect their own taxes and pay the centre for services rendered. Other regions live on central handouts. The transfer of powers (over education, health and local roads, notably) varies in degree and is still incomplete. But the critics' question is fair. Some, including Mr Aznar, argue that devolution is now in its final stage and need go no further: the regions have all the powers they need. Indeed, regional parties may fade away, their *raison d'étre* gone. More probably they will demand that ever more functions be handed to them, say others: look at Jordi Pujol, the longtime Catalan-nationalist premier (to borrow the Canadian word) of Catalonia. And if one of them gets extra powers, others will want the same, and so on.

Maybe, but it does not have to be so. The Catalans and Basques really are distinct. Their nationalists' true aim is not just power as such, but recognition of identity: hence Mr Pujol's outcry over the otherwise trivial recent decision of the centre that car number-plates would not, in future, show where the car is from. What counts is how such issues are treated. Catalonia's schools, for instance, teach in Catalan. The federalist notions of Pasqual Maragall, leader of its local Socialists, would mean big constitutional changes. Are such things a menace to Spain? Treat them as such—as the Catalan PP, alone among the parties there, did until this autumn—and you may increase the risk. Treat them as perfectly compatible with Spanish nationalism, and compatible they can very well be.

Still, the regions take delicate handling, and charges of mistreatment by the centre are always good for votes. Often these end up with the Constitutional Court. But the big issues are political.

One, of course, is money. Regional governments handle about 30% of all public spending (and lower-level ones another 20%). Yet, Basques and Navarrese apart, nearly all their cash comes from the centre: on top of specific grants, 30% of the income tax collected in the region goes back to it. But the small print is endless, complaints abound—and the system has to be renegotiated between centre and regions every five years. Last time round, Socialist-run regions such as Andalusia rejected the result. Much of next year will be spent haggling over changes to the system, to apply from 2002.

What changes? There has been much talk of "co-responsibility", with each region (within limits) setting and collecting some taxes, notably those on alcohol, tobacco and petrol. The aim is that those who spend should share the political price of so doing. Good for the centre; good for the regions too, you might think, since this would add to their autonomy. Some see it that way, including Catalonia and Valencia, whose (PP) chief launched the idea; both might get lots more money that way. Others note that spending wins votes, but taxing does not. The centre too is now in two minds. At election time, Mr Aznar's team backed the idea. But now he does not need Catalan votes, and the relevant ministers have been backtracking. In any event, given Spain's large regional disparities, some equalising mechanism will also still have to be agreed on, with rich regions screaming that they are being ripped off, and poor ones that they are being starved.

Overall, the centre has to ensure fiscal sobriety. It has mechanisms to that end, and every quarter the Bank of Spain publishes each region's accounts, potential ammunition for its opposition parties. Even so, at least two regions—rich Catalonia and poor Andalusia—are not noted for their thrift.

Haggling over regional revenues is matched by rows over central spending. Andalusia has its high-speed trains, say others; when do we get ours? This sort of thing happens in any country, but the degree of devolution makes it more sensitive in Spain.

For a truly spectacular case of regional uproar, look at the national water plan unveiled in September, a huge scheme to transfer a cubic kilometre of water a year from (mainly) the Ebro, in the north, to the dry south-east. Even the proposed beneficiaries at once began arguing over their shares. But that was nothing to the rage in Aragon, where much of the water would come from. It has a desert of its own, and many depopulated villages, some without piped water at all. A huge protest march in the regional capital, Saragossa, last month was led by the city's PP mayor, defying his party, and the region's PSOE premier. This row can run for years, souring regional relations with the centre. It is doubtful that any of that cubic kilometre will ever run at all.

There is another, far uglier, regional shadow over Spain: terrorism from the Basque Country. Yet, for all the headlines about that, it is the ordinary, unreported, non-Basque 95% of Spain that gives a truer picture of its past, present and future.

The murderous minority

Most Basques reject terrorism

IF ONLY all Basques would listen. They are a nation, even if its myths date mainly from the late 19th century; and its ancient language is a thing apart. Franco squelched nation and language alike. The late 1960s brought ETA, the terrorist group that now disgraces both. Today, over 700 deaths later—a score of them in renewed violence this year—the Basque Country is Spain's Northern Ireland.

Or is it? No way, say ministers in Madrid. To Mr Aznar, ETA is a security problem, period: one that can be cured (not just bandaged) by effective policing, solid laws firmly applied, and public support for both; plus international support, and getting the region's non-violent nationalists who now run its government voted out. And if all fails? It won't: the defence minister, Federico Trillo, "cannot imagine a time when the army would be used there".

Certainly there are big differences. One is this determination: plainly, though they do not say it, Spanish ministers cannot imagine the repeated surrenders of British ones to terror. And some things are easier for Mr Aznar. ETA} has no hyphenated-American lobby, laden with dollars and ethnic hate, to fill its coffers. Its safe houses across the border in France (which also has its Basques) are not safe at all: the French reject ETA's aims, not just its methods. Above all, hundreds of thousands of ordinary Span-

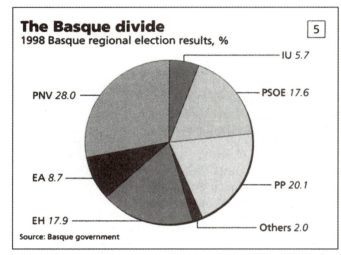

The Basque divide
1998 Basque regional election results, %

5

- IU 5.7
- PSOE 17.6
- PNV 28.0
- PP 20.1
- EA 8.7
- EH 17.9
- Others 2.0

Source: Basque government

iards, in the Basque Country and far from it, are ready to parade their rejection of ETA in the streets.

The biggest difference, though, is political. Only a third of Basque Country voters backed Spain's new constitution in 1978. But the Basque Nationalist Party (PNV), which abstained then, did back the later "statute of Guernica", the region's own basic law, and in sundry coalitions has ruled there ever since. Basques, unlike Ulster's Catholics, have suffered no discrimination; if anything, the reverse. They get good jobs. They man the police force. Their language is spreading; many schools teach only in Basque, some in both languages, under 15% in Spanish alone. This is Quebec, not Ulster—and with less ethnic antagonism than either.

So what do the nationalists want? The right to a country of their own. It is for this, and only this, that ETA kills: its other excuses are bogus. So too, however, is the Madrid version, shared by the national media, that ETA's thugs are not just vile murderers, which is true, but also fascists, Nazis and/or (it varies) Marxist-Leninists, and venomous ethnic haters with it. That is mostly false.

And the PNV? It got talking with the gunmen's political allies in 1998 and, in effect, made a deal. It signed with them (and Catalan and Galician nationalists) the now notorious "pact of Lizarra" proclaiming the right to self-determination; and the gunmen announced a ceasefire. In later re-

Spain and its regions

FRANCE
BASQUE COUNTRY
Santander
ASTURIAS
CANTA-BRIA
Bilbao
ANDORRA
GALICIA
NAVARRA
LA RIOJA
CASTILLA & LEON
CATALONIA
Oporto
Valladolid
Saragossa
Ebro
Barcelona
ARAGON
MADRID
Madrid
S P A I N
Valencia
Balearic Islands
PORTUGAL
EXTREMADURA
CASTILLA-LA MANCHA
VALENCIA
Lisbon
MURCIA
Seville
ANDALUSIA
Mediterranean Sea
Canary Islands 1,200 km
Malaga
GIBRALTAR (to Britain)
200 km

gional elections, the pro-ETA party, now known as EH, did well; and the PNV, previously allied with the region's Socialists, formed a coalition with it. But talks between the central government and ETA front men got nowhere. Last December ETA declared its ceasefire ended, and soon began killing again.

Now the heat is on the PNV. It has ended its links with EH, but it still believes in trying to talk—and gets fiercely abused for palling up with murder. That too is false, and may be risky. Mr Aznar wants early elections in the Basque Country, and would send his tough interior minister, Jaime Mayor Oreja, a Basque, to head the PP list. His aim is to get the PNV out of government. That is imaginable: the region's voters, once pro-nationalist by a margin of two to one, by last spring were evenly split, and recent polls show PNV support down. Surely, in a democracy, others should get a chance? Maybe, but when polled not 15% of the voters fancy a wholly non-nationalist government. And where would it lead? To fewer deaths, for a start, says Mr Aznar. Many Basques ask whether PNV supporters would not swing to radicalism, meaning more division and death, not less.

At the PNV's grandiose Bilbao headquarters, one who asks is Jose Urkullu, at 39 a coming man in the PNV; a possible leader, some say, when its durable head, Xabier Arzalluz, steps down by 2004. As a pact, Lizarra is dead, says Mr Urkullu. But its principles are not: Spain is a plurinational country, and needs a new model of the state based on what each nation decides. And what is that "Basque nation"? Anyone in the region's three provinces, he says.

Remarkably, EH says the same. In its tatty office nearby, you will get the full "we are a nation" story; a sharp account of the PNV's eternal problem of disliking the Spanish constitution, yet working within it; plus some tosh about "state violence and our response to it". What you will not get is ethnic hate. Yes, Basques must get self-determination (and EH still dreams, as the PNV in the real world does not, that this might include Basques in Navarra, even those in France). But just who are Basques? "All who live and work in the Basque country." Humph. And whether ETA would let them make their minds up in peace, let alone accept a No result, are other questions. But if this is fascism, it is an odd sort. No more fascist, let alone Marxist, are EH's economic ideas. The party talks of a participative economy, mixed, but with a strategic role for the public sector, and more co-ops (an old fact of Basque life). It does not love multinationals, "but they are there". This may be pie in the sky, but totalitarian pie it is not.

The ETA gunmen certainly use fascist methods, be it street thuggery, fire-bombing or murder. But it serves no good purpose to distort their beliefs—and far less to tar-brush the peaceable PNV. As Mr Aznar recently said, arguing—as distinct from shooting—for independence is legitimate. There must be a better way of arguing back than to lump together those who argue with those who shoot.

Article 16   *New York Times,* October 2, 2000

209 Years Later, the English Get American-Style Bill of Rights

By SARAH LYALL

England gets its first domestic bill of rights on Monday in a fiercely debated move that is likely to fundamentally reshape the centuries-old relationship between the citizens and the government.

The change comes with the scheduled enforcement of the Human Rights Act, a centerpiece of the Labor government's legislative program, which for the first time incorporates the European Convention on Human Rights into English law.

"It is a very significant departure," said Julian Knowles, an international human rights lawyer at Matrix, a law firm in London. He said England would now have a written bill of rights enshrined in a single document and containing the same sort of guarantees that Americans have had since 1791.

"We have always trusted the executive and judiciary to protect our rights, but it's been a matter of trust only," he said. "This document sets out in clear terms what these right are."

As far back as 1215, Magna Carta began to define the limits of the powers of Britain's rulers. But Britain does not have a written constitution, nor an American-style bill of rights. Instead, its citizens have always had what are known as negative rights—that is, they have been allowed to do anything they want, unless there is a law specifically forbidding it.

"We've never had positive rights," said Jenny Watson, deputy director of the Human Rights Act research unit at King's College London. "We've always relied on the silence of the law. If the law doesn't say that you can't do it, you can do it. But we started finding that rights we thought we had were cut back.

"What we are getting now for the first time ever are rights that are ours, that are written down, that are easy to understand, that you can teach in schools and that you can enforce in courts," she said.

But the human rights law, which the Labor Party championed in opposition, has provoked an enormous amount of controversy as the country prepares for it to take effect.

The judicial system is bracing itself for a flood of new legal challenges to laws governing areas like the news media, employment practices, the immigration service and the criminal justice system. And the government has been furiously preparing for the law's effect by training its judges and setting aside about $88 million to pay for the new cases.

Many conservatives are terrified that it will wreak havoc with many of England's treasured institutions, giving too much power to interest groups like gays, minorities, criminals and women at the expense of government authority.

Britain signed the European convention in 1953 but because of a lack of political enthusiasm has not made it a part of domestic law until now. The convention sets out a range of positive rights like the right to freedom of expression, the right to a fair trial and the right to privacy. In the past, Britons who felt that their rights had been violated by the government or by public authorities had to take their cases to the European Court of Human Rights in Strasbourg, an arduous process that could drag on for as long as five years.

Over the years, the Strasbourg court has ruled against Britain in a number of high-profile cases, forcing the government to rewrite British law time and time again. Among a host of other things, Britain has been forced to outlaw corporal punishment of children in schools, to limit its use of telephone wiretaps and to abandon its policy of banning gays from serving in the armed forces.

Human rights lawyers say there are a number of areas of English law that might be vulnerable to challenges under the new act. One law likely to be challenged on free-speech grounds is the Official Secrets Act, a sweeping measure that makes it a criminal offense to disclose any aspect of government work unless the disclosure has been officially authorized.

Similarly, there are also expected to be a challenges to various aspects of the country's stringent libel laws, and to laws setting out when it is appropriate for judges to issue restraining orders on newspapers.

In criminal justice, there is likely to be a challenge to the right of the Home Office, rather than the judiciary, to determine actual jail time for murderers sentenced to life in prison. And much of the Blair government's own legislation covering asylum seekers, suspected soccer hooligans, suspected terrorists and the like will also probably be challenged on human rights grounds.

"In almost all areas of our law, there are particular issues that people will want to challenge using the human rights act as a new tool," said Anne Owers, director of Justice, a British human rights group. "But nobody really knows how it's all going to fall out. It's going to be quite a shake-up."

Indeed, no one can say how many of the expected challenges are likely to succeed. Already, decisions in several courts have specifically referred to the human rights convention. In Birmingham, for example, a judge threw out a case against two men charged with speeding on the ground that requiring them to declare who was driving the car at the time violated their right against self-incrimination.

In Scotland, which incorporated the Human Rights Act into its own law earlier this year, some 60 cases of 600 have actually been successful. The new act extends the incorporation to English and Welsh law.

Meanwhile, conservative-minded newspapers have printed a stream of articles saying in effect that the act will upend everything from the ability of teachers to prohibit their pupils from having gay sex in school, on the ground that it would violate the students' right to freedom of expression, to the right of the government to ban polygamy, on the ground that it would violate some people's freedom of religion.

Opponents of the change are also concerned, they say, that the new law will force the government to cede too much of its power to the judiciary, which has traditionally been subordinate to the executive.

"Our Parliament will be powerless to stop unelected and unsackable judges making the law because the Human Rights Act is so loosely and vaguely worded," wrote Norman Tebbit, a former minister in the Thatcher government, in the Daily Mail. "They will give rights to some people—sex perverts, illegal immigrants, and criminals are top of the list—which will rob us of freedoms we have had for centuries."

But Jack Straw, the home secretary, has been at pains to argue otherwise, pointing out that the rights under the convention are all meant to balance one another, rather than be considered as absolute rights on their own.

"Nearly all the rights in the convention, except the 'absolutes' against torture and slavery, are qualified or limited in some way," he wrote in The Guardian. "They reflect a vision of a human rights culture in which individuals, while exercising their rights, still have an obligation to act responsibly to others and to the wider community."

Geoffrey Robertson, a human rights lawyer at Doughty Street, a law firm, said that for a people with a stirring civil liberties tradition, Britons have been surprisingly acquiescent as their own rights have been diminished by government encroachment.

"Britain is perhaps the country in the world that's produced the greatest rhetoric about liberty, from people like Milton and John Locke and Shakespeare, without giving people the rights that the poets and playwrights have articulated so forcefully," he said. "This will give people a new weapon to deploy against bureaucracy, and will help produce a better culture of liberty here."

Article 17   *The Economist*, February 10, 2001

BRITAIN, EUROPE AND AMERICA
Keeping friends

Tensions between the new American administration and the European Union over defence could put Britain in a pivotal position. Britain may not find that a comfortable place to be

BRUSSELS AND LONDON

EVEN before George Bush was in the White House, the British government was rushing into print to affirm that Britain's "special relationship" with the United States would not be affected by a change of president. In an article entitled "It could be the start of a beautiful friendship", Robin Cook, Britain's foreign secretary, argued that the United States would remain Britain's "closest ally and our biggest export market." This week Mr Cook was in Washington, meeting Colin Powell, America's new secretary of state, in an effort to get the new relationship off on the right footing. Before the end of the month Tony Blair will also travel to the United States to meet Mr Bush.

The British urgency betrays a certain nervousness. It is not just that New Labour felt ideologically more comfortable with Bill Clinton and the Democrats. It is also that developments in the triangular relationship between Britain, Europe and the United States threaten to undermine one of the basic tenets of British foreign policy—that there is no choice to be made between a close relationship with the United States and the intimate relationships entailed by Britain's membership of the European Union.

Mr Cook insists that it is an "anti-European myth" to suggest that such a choice might exist. He asserts that "any loss of influence in Europe would damage our economic relations with the US, and our strategic relations." In other words, Britain strengthens its voice in the United States only by playing a full part in the EU. But could that change?

The question has been raised in recent months by shifts in the political and strategic debate on both sides of the Atlantic. Up to now, part of the transatlantic bargain has been that Europe and America deal with one another as equals, closely intertwined but occasionally quarrelsome, in economics; whereas in defence questions the Europeans—albeit sometimes tardily and reluctantly—follow America's lead. This division of labour made it possible for Britain to be a more-or-less dutiful European over many economic questions (which sometimes meant taking the Union's side in disputes over aircraft subsidies or carbon emissions) while remaining loyal to America in the areas of defence policy that really count.

Certain elements in this careful balance are now changing. On the one hand, the European Union—long dismissed as an economic giant and a geopolitical dwarf—is sounding more serious about turning itself into a real force in world affairs, with a role to play in "security" as well as economics. In particular, the Union now wants to become more effective at "crisis management"—a term that can mean anything from sending a few aid workers to a war zone to the use of military force. The appointment of a senior politician, Javier Solana of Spain, to personify the Union's external policy is a token of the EU's desire to speak with a louder, though not necessarily anti-American, voice. So far, there is little to show for this effort, except in the Union's immediate neighbourhood; but the EU's declared intention to exercise a bit more

influence has already changed the transatlantic mood.

At the same time, some of the hardest strategic heads in the United States are suggesting that the Asia-Pacific region, and possibly the Gulf, rather than anywhere in Europe will pose the toughest challenges to American security in the years to come. This does not imply that America has any intention of disengaging from the defence of Europe. But it may mean that the terms on which America remains engaged will grow tougher, and there will be less patience with what seems, from an American viewpoint, like petty posturing or point-scoring by the Europeans. The resulting squabbles will be harder for Britain to finesse.

It is against this rather inauspicious background that the Union, late last year, produced the first tangible result of its efforts to become a serious player in the arena of international security. This was a formal agreement to create—albeit largely on paper—a 60,000-strong military force which could be assembled rapidly and sent to a distant trouble-spot for up to a year.

Tony Blair had invested a good deal of personal authority in this, believing at first that it could serve to nip transatlantic squabbles over burden-sharing in the bud while allowing the Union to acquire a small, but from America's viewpoint harmless, degree of independence. For Mr Blair, Euro-defence also offered a way to demonstrate Britain's central role in the Union at a time when its commitment is being questioned because of the country's reluctance to join the euro, the single

European currency. In the language of his friend Bill Clinton, this would have been a win-win-win outcome.

Yet the new American administration seems less confident than its predecessor that European muscle-flexing can be kept within harmless limits. Donald Rumsfeld, the new American defence secretary, told a security conference in Germany at the weekend that he was "a little worried" by European plans for greater military independence. "Actions that could reduce NATO's effectiveness by confusing duplication, or by perturbing the transatlantic link, would not be positive," he said. Unless these anxieties can be allayed, Mr Blair may find himself squeezed by competing pressures from his European and American allies.

So far at least, transatlantic squabbles over European-only defence operations have been more about impressions than reality. Even if Europe's hopes of acquiring the capacity to sustain a 60,000-strong force are fulfilled, there is no real challenge to American-led NATO as the main guarantor of European security. A more substantial issue is posed by the Bush administration's keenness to deploy missile defences to shield America's own territory against rogue attacks, to protect American bases overseas—and possibly also to shelter the territory of allies.

Several prominent European figures, particularly in France and Germany, have already made it clear that they regard the Bush administration's plan as ill-conceived and destabilising. Tony Blair will be put in a particularly tight spot if, as looks likely, the Bush administration asks for permission to upgrade the Fylingdales radar station in Yorkshire as part of an anti-missile defence system. His European allies will press him to refuse, or at least parry, the American demand. But if Mr Blair is asked point-blank by an American president to co-operate, could he really say no? So far, the British government is stalling—although Mr Blair is thought to incline privately to acquiescing in Mr Bush's request, if and when it is made.

These defence questions are the sharp edge of the most sensitive issue in Britain's foreign relations. Will the United Kingdom's membership of the European Union—with its self-proclaimed goal of "ever-closer union"—ultimately imperil its special relationship with the United States? It is a question that matters far more in London than in Washington. Yet Britain's future in Europe, and its consequences for its relationship with the United States, may also come to matter in Washington. Little doubt exists that there are voices within the European Union that expect (indeed, long for) a growing foreign-policy rivalry with the United States. If that emerged, the position of Britain would be both ambiguous and pivotal.

Anatomy of a friendship

The phrase "special relationship" is used far more in Britain than in the United States. When prominent Americans hear Britons use the term, they can be forgiven for looking slightly blank. The United States, with global interests and a polyglot population, has "special relationships" with many countries round the world. Arguably, its relationships with Israel, Japan, Mexico and Canada are as special as any with the United Kingdom.

It is also true that the relationship—forged during the first and second world wars—used to be a lot more "special". But despite Britain's diminishing global importance in the post-war years, it remains true that there is a special quality to its friendship with the United States in four particular areas. One, the most amorphous but perhaps the most tenacious, is historical and cultural. The other three are all defence-related, concerning intelligence, nuclear affairs and military matters.

This makes difficulties over NATO and anti-missile defences all the more awkward. The closeness of British intelligence ties with the United States continues to excite suspicion in the rest of the European Union. In recent months, for example, the European Parliament has been much exercised by the idea that the United States is committing large-scale commercial espionage through a satellite system code-named Echelon. The parliament has heard testimony that Echelon relies on satellite stations based in Britain. One French member of the European Parliament, Jean-Claude Martinez, fulminated that this proved that "Britain's real union is with America." Whatever the truth or otherwise of the Echelon allegations, Britain certainly shares intelligence with the United States, Australia, Canada and New Zealand that it does not share with its European allies, and has done since 1947. James Woolsey, a former director of the CIA, says that "although no one is a complete friend in the intelligence world, with Britain and America it is as close as it gets."

The closeness of the intelligence relationship is linked to Britain's nuclear relationship with the United States. Along with its permanent seat on the United Nations Security Council, Britain's possession of a small nuclear deterrent has been crucial to its claim to be more than just another middle-ranking power. Unlike France, which has insisted on developing its own nuclear *force de frappe,* Britain has been happy to buy all its Trident nuclear technology from the United States. Additionally, as two of the five "status quo" nuclear powers, Britain and the United States have had a shared interest in working together to block nuclear proliferation.

Intelligence and nuclear weapons lead naturally to the third element—close military and diplomatic co-operation. This was on display, of course, for much of the last century. The two countries were again close military allies during the Gulf war. Despite considerable differences over the handling of the conflict in Bosnia (America was covertly supplying weapons to the Bosnian army even as that army fought pitched battles with British and French peacekeepers around Sarajevo), Britain and America were once again in close harness by the time of the Kosovo campaign. Since the Gulf war of 1991, America and Britain have never ceased to co-operate closely in keeping a tight noose around Iraq—through bombing raids, sanctions or UN votes. While Britain's military contribution to this effort has been minor, its constant presence at America's side has been an important source of moral support and legitimacy.

The hinge of the Anglo-American military relationship in the post-war years has been NATO. That is why Mr Blair and his colleagues have been so keen to insist that the new EU military wing is intended to complement NATO, not to undermine it. Britain—which has offered to put more than 12,000 troops, an aircraft carrier and an amphibious brigade at the disposal of the nascent European force—insists that NATO will continue to be Europe's major security club and will retain its position as the main agency for military planning, even for European operations. The EU force will act only when NATO as a whole decides not to get involved. As such, argue the British, the European force will further an important goal of American foreign policy: that Europe take on a greater share of the burden of its own defence.

Not all Britain's European allies, however, share its limited view of the role of the new force. Romano Prodi, the head of the European Commission, has let slip that he regards it as a European army in embryo. And the French in particular clearly want the European force eventually to act as the military arm of an independent European foreign policy.

British Eurosceptics have rushed to raise the alarm that this "European army" will undermine NATO and drive a wedge between Europe and the United States. For many members of the Conservative Party, the issue is heaven-sent. They are suspicious of deeper European integration in all its forms, and have looked to the United States as an alternative to the European Union. Some have been trying to push the idea that Britain should join the North American Free-Trade Agreement. But, despite some interest from a couple of Republican senators, the idea that Britain should join NAFTA is not yet taken seriously in the United States.

Many Tories clearly see an opportunity in matching American anxiety about the future of NATO with British anxiety about the future of the EU. A just-published book called "Stars and Strife: The Coming Conflict between the USA and the European Union" by John Redwood, a right-winger who once ran for the post of leader of the Tory party, argues that Britain will face a crisis of loyalty sooner rather than later; and he has no doubt which friend it should plump for.

For a long time there was little public evidence of any such American worries about Europe's military initiatives. In a joint article with Robin Cook late last year, Madeleine Albright, then secretary of state, declared her support. But little more than a week later William Cohen, the then secretary of defence, remarked

in Brussels that NATO risked becoming "a relic," if the EU tried to develop its own military capabilities.

It is clear that some prominent Republicans are worried. Richard Perle, an adviser on defence to the Bush campaign, described the European defence initiative as "French manoeuvres aimed at sidelining the United States in Europe." Senator Gordon Smith, the Republican leader of the Europe subcommittee of the Senate Foreign Relations Committee, told a British audience last year that he suspected that the EU's real motive in building the force was to check American power, and urged the British to "never forget the vital British role as the linchpin in the Atlantic alliance."

Clearly Mr Smith's views do not represent the entirety of Republican opinion, let alone American opinion. For many years, support of closer European integration has been a fairly settled bipartisan policy in the United States. Both General Powell and Condoleezza Rice, George Bush's national security adviser, are keen on greater military burden-sharing. (And General Powell seemed blissfully happy as he met Mr Cook this week, declaring Britain and America "strong friends, staunch allies, forever into the future.") Robert Zoellick, America's new trade representative, has also been an eager supporter of closer European integration. These folk might well take a more relaxed view of the European defence initiative.

The uneasy European
Much will depend on how these issues are worked out in practice. If the more ambitious French-inspired visions of a European military force, with its own autonomous planning staff, gain ground, that will be a recipe for trouble. Britain might even be forced to rethink its commitment to the whole idea. An important subsidiary issue that will have to be resolved is the need to reconcile Turkey—a key member of NATO, but kept at arm's-length by the EU—to the new European defence arrangements. The Americans seem sympathetic to Turkish anxieties, while the Europeans will be extremely reluctant to give the Turks a veto over their military ambitions.

The whole debate also has to be seen in the wider context of Britain's uneasy relations with the EU. Much of Britain's

foreign-policy establishment believes that the choice between the EU and an "Anglosphere" of the United States and the Commonwealth countries was irrevocably made in the 1970s, when Britain joined the EU. So, despite Britain's palpable unease with the steps the Union has taken towards greater political integration, it would still require a fundamental change in British thinking for the United Kingdom to make a deliberate attempt to draw away from the European orbit, and closer to the United States. Only a crisis in relations with the European Union—perhaps an irreconcilable argument over the pace of European integration—might provoke such a shift. For this crisis to force a change in policy, it is likely that the fundamentally pro-European Labour Party led by Mr Blair would have to lose power to the Tories.

The decision would not be Britain's alone. Might it be conceivable that the United States would encourage such a move, perhaps by floating the NAFTA option? Again, such a development would require a big shift in American thinking. In particular, America would have to become much more wary of European political integration, and of the EU's foreign-policy and military pretensions in particular. That, in turn, would probably require three things: that the more ambitious French view of the European military force's future role had prevailed; that the Republicans would still be in office; and that the Eurosceptical wing of the Republican Party would be in the ascendancy.

It still seems highly unlikely that all these factors will come together and that the relatively stable pattern of the past 30 years will be disrupted. Most probably, Britain will continue to become more closely enmeshed in the European Union, with the continuing support of the United States, and consequently the significance of the special relationship will continue its gradual decline. But the idea that Britain has a fundamental choice to make between the United States and Europe is still too melodramatic at the moment. Instead, as a major review of Britain's defence strategy pointed out this week, Britain's most important role may well be to prevent misunderstandings between the two sides, while staying friends with both.

Credits

Page 186, Article 1. Excerpted from Great Decisions, 2001 edition, *European Integration: Past, Present and Future.* Reprinted with permission. Great Decisions is published annually by the Foreign Policy Association.

Page 191, Article 2. From the *Washington Quarterly,* Autumn 2000, pp. 15–29. © 2000 by the Center for Strategic and International Studies (CSIS) and the Massachusetts Institute of Technology. Reprinted by permission.

Page 198, Article 3. From *Europe* Magazine, June 2000, pp. 16–19. © 2000 by European Commission. Reprinted by permission.

Page 200, Article 4. From *The Economist,* October 28, 2000, pp. 17–18, 21–22. © 2000 by the Economist, Ltd. Distributed by The New York Times Special Features. Reprinted by permission.

Page 204, Article 5. From *Europe* Magazine, February 2001, pp. 16–24. © 2001 by European Commission. Reprinted by permission.

Page 211, Article 6. From *The Economist,* September 23, 2000, pp. 27–30, © 2000 by the Economist, Ltd. Distributed by the New York Times Special Features. Reprinted by permission.

Page 215, Article 7. From *The Nation,* November 6, 2000, pp. 20, 22–23. © 2000 by The Nation. Reprinted by permission.

Page 218, Article 8. From *Europe* Magazine, November 2000, pp. 16–18. © 2000 by European Commission. Reprinted by permission.

Page 220, Article 9. From *Europe* Magazine, February 2001, pp. 6–8. © 2001 by European Commission. Reprinted by permission.

Page 223, Article 10. From *The Economist,* September 30, 2000, pp. 25–26. © 2000 by the Economist, Ltd. Distributed by The New York Times Special Features. Reprinted by permission.

Page 226, Article 11. From *Europe* Magazine, September 2000, pp. 8–10. © 2000 by European Commission. Reprinted by permission.

Page 228, Article 12. From *In These Times,* June 26, 2000, pp. 18–19. © 2000 by In These Times; www.inthesetimes.com.

Page 230, Article 13. From *The Economist,* July 29, 2000, pp. 47–48. © 2000 by the Economist, Ltd. Distributed by The New York Times Special Features. Reprinted by permission.

Page 232, Article 14. From *The Economist,* December 2, 2000, Survey 2–16. © 2000 by the Economist, Ltd. Distributed by The New York Times Special Features. Reprinted by permission.

Page 235, Article 15. From *The Economist,* September 25, 2000, Survey 18. © 2000 by the Economist, Ltd. Distributed by The New York Times Special Features. Reprinted by permission.

Page 241, Article 16. From the *New York Times,* October 2, 2000. © 2000 by The New York Times Company. Reprinted by permission.

Page 243, Article 17. From *The Economist,* February 10, 2001, pp. 25–26, 29. © 2001 by the Economist, Ltd. Distributed by The New York Times Special Features. Reprinted by permission.

Sources for Statistical Reports

U.S. State Department, *Background Notes* (2000).

C.I.A. *World Factbook* (2000).

World Bank, *World Development Report* (2000/2001).

UN *Population and Vital Statistics Report* (January 2001).

World Statistics in Brief (2000).

The Statesman's Yearbook (2000).

Population Reference Bureau, *World Population Data Sheet* (2000).

The World Almanac (2001).

The Economist Intelligence Unit (1999).

Glossary of Terms and Abbreviations

Allies A coalition of countries during World War II, headed by the United States, Great Britain, the Soviet Union, and China.

Anschluss A German term denoting "union" or "joining," usually applied as a euphemism for the annexation of Austria by Nazi Germany in 1938.

Atlantic Charter An agreement concluded between Franklin D. Roosevelt and Winston Churchill "somewhere on the Atlantic" on August 15, 1941. Although the United States was officially neutral at that time, it provided assistance to a beleaguered Britain, which was fighting World War II on its own.

Axis A term generally used to denote the coalition of Germany and Italy during World War II that would, it was said, steamroll over Europe like an axis. Later the term came to include Japan when it, after Pearl Harbor, started to fight on the same side.

Basic Law The name for the West German Constitution when the Federal Republic of Germany arose in 1949. It was not called a constitution, since the West German leaders wanted to wait to promulgate a constitution until elections would have been held in a unified Germany. At that time, it was expected that such a unification would take place in the foreseeable future; the Cold War made the phase last until 1990. Contrary to all expectations, the unification did not provide a new beginning, and the reunited Germany still relied on the Basic Law of 1949.

Basques An ancient (pre-Indo-European) people who for a long time have lived in and alongside the Pyrenees. The French Basques number about 200,000 people. Their Spanish counterparts number approximately 2 million.

Benelux An acronym formed from *Bel*gium, the *Ne*therlands, and *Lux*embourg, having particular reference to the toll union among these countries, planned in London by their governments-in-exile in 1944 and materialized in 1948. Neither the term nor the concept itself was ever popular.

Bicameral Refers to a legislature that has two chambers.

BLEU A term that refers to the economic union between Belgium and Luxembourg, established in the 1920s, which had important financial implications, in that the currencies of these two countries became closely linked.

Canton A term used in Switzerland to refer to an entity that in the United States would be called a state. The Swiss confederation has 20 full and six half cantons. In many cases, the cantonal borders coincide with cultural frontiers.

Cohabitation A sharing of power between a president and a prime minister of different parties. The term originated in France.

Cold War A sharp deterioration of Soviet–U.S. relations following their alliance during World War II; caused by political and ideological rivalry. It lasted from 1947 to 1989.

Common Market *See* European Economic Community.

Commonwealth of Nations An association of the United Kingdom, its dependencies and associated states, and most of its former dependencies. As a successor of the British Empire, the Commonwealth has primarily trade and commercial implications. In a purely formal sense, the British sovereign heads the Commonwealth.

Communism An ideology whose followers seek to overthrow capitalism through revolution in order to establish an ideal classless society that eschews individual ownership of the means of production, including real property.

Congress of Vienna An international conference (1814–1815) that shortly after the Napoleonic Wars attempted to restore Europe to what it had been before those wars.

Consociational Democracy A political system in which the large cleavages (in ethnic or religious respect) that divide a society are purposely segregated by rigorous vertical organization. Only the top segments politically interact with each other.

Council of the European Union The main decision-making body of the European Union, made up of ministers from all member states.

Danelaw The law in force in the part of England held by the Danes before the Norman conquest; also, the area of England in which the Danelaw had been imposed.

Détente Relaxation of tensions between the Soviet Union and its allies and the Western countries. A decline in intensity of the Cold War.

Dirigisme One of the major bases of the French economy. It reflects an emphasis on mercantilism. The government "directs" the economy, i.e., is very involved in what products will be grown, what projects will be entertained, etc.

Enosis A Greek term meaning "induction." Refers to a policy to render Cyprus part of Greece.

Estado Novo Portuguese for "new state," a political system that Antonio Salazar designed. It amounted to fascism and was outlined in the Portuguese Constitution of 1933.

Euro The integrated currency of the European Union, launched January 1, 1999. It will replace participating countries' national currencies beginning in 2002.

Eurocommunism A movement emerging in Western Europe in the 1970s that announced national models of communism. The Italian, French, and Spanish Communist Parties made it clear that they would not follow the Soviet line if elected into power. They would also allow themselves to be voted out of power in democratic elections.

European Atomic Energy Community (Euratom) Founded by a Treaty of Rome at the same time (1957) that another such treaty established the European Economic Community. Euratom sought to develop the nuclear-energy resources of the six charter states, but as no solution was provided for the nuclear-waste problem, the body became progressively more obscure.

European Coal and Steel Community (ECSC) The first of the three Communities. It was established by the Treaty of Paris (1951) and was initially intended to pool French and German coal and steel resources. However, between its announcement and its establishment, Italy and the Benelux countries joined in the venture. Its aim, to create unified products and labor markets, proved such a success that the two other Communities (the EEC and Euratom) were established in March 1957.

European Community A name referring to the economic integration of Europe. *See* European Economic Community.

European Economic Community (EEC) Known as the Common Market; established by the Treaty of Rome in March 1957 to promote the integration of Western European economies through the removal of trade barriers. Initially, there were three distinct Communities; but, after the Merger Treaty of the mid-1960s, the three Communities were viewed as one organization. The qualification "Economic" gradually was dropped, to "European Community." The Maastricht Treaty (1992) changed the name officially to "European Union."

European Free Trade Association (EFTA) A trading bloc established in Stockholm in 1960, largely at the instigation of the United Kingdom. Since it operates on premises entirely different from those of the original European Economic Community, the membership of these two organizations has been mutually exclusive. As the EEC expanded, the EFTA correspondingly dwindled. The two blocs have agreed on a merger of their respective areas, which was defined as the European Economic Area (EEA).

European Monetary Union (EMU) European economic integration formalized by the Maastricht Treaty of 1992.

European Recovery Plan *See* Marshall Plan.

European Union The term most recently applied to the process of economic and political integration in Europe. Used broadly in this volume for ease of understanding. *See also* European Economic Community.

Fascism An ultra-right ideology that glorifies the state at the expense of the individual and opposes democratic and socialist movements. It became dominant after the March on Rome in 1922. Although initially not racist, fascism started to evidence a strong racist orientation as a result of the Italo–German alliance.

Fifth Republic Refers to France since 1958 when the Fifth Constitution was introduced and Charles de Gaulle came to power. The French give ordinal numbers to their constitutions and their republics. In reality, the number of constitutions that have been proposed and occasionally adopted has been much larger.

Franco–German Non-Aggression Pact and Friendship Treaty To end the intermittent hostility between Germany and France, this 1963 treaty prescribed semiannual consultations between the chief executives of France and Germany and listed a range of issues for further collaboration.

Free French Consisted largely of French civilians and military who managed to escape from France shortly before or during the German occupation, in the period 1940 through 1944. Headed by Charles de Gaulle, it assisted in the Normandy invasion (June 1944) and the liberation of Paris (August 1944). Although it cooperated closely with the French resistance, that collaboration was marked by rivalry when the war was about to end.

Führer Both Benito Mussolini and Adolf Hitler revealed a bias against the term *prime minister* or *chancellor*. They came up with the alternatives of *duce* (Italian for "leader") and *führer* (German for "leader") once they gained power.

Gastarbeiter A German term meaning "guest worker."

General Agreement on Tariffs and Trade (GATT) An important UN agency, established in 1958, that seeks to promote a global economy through reducing tariffs and eliminating other barriers to trade.

Gross Domestic Product (GDP) The total value of all goods and services in a country in a given year.

Holocaust The name later given to the period (1933–1945) of ruthless persecution and extermination of 6 million European Jews by Nazi Germany. Nearly 5 million other people were also murdered.

International Atomic Energy Agency (IAEA) A UN agency, established in 1957, that aimed at promoting the peaceful use of atomic energy. It lapsed into oblivion as a result of insurmountable difficulties with regard to the disposal of nuclear waste.

International Bank for Reconstruction and Development (IBRD) Commonly known as the World Bank, it constitutes a UN agency, established in 1945, that endeavors to promote economic development through guaranteed loans and technical assistance.

International Monetary Fund (IMF) A UN agency that endeavors to promote international cooperation and development. It issues money (either as gifts or as so-called soft loans) to less developed countries that have designed multiyear development projects.

Lager A German term meaning "camp." It is often used with reference to Austrian politics, where it has come to stand for a subculture.

Laissez-Faire Originally, *laissez-faire, laissez aller*, meaning "leave them alone." A concept of nonintervention developed by eighteenth-century French physiocrats in reaction against mercantilism. The concept became better known as a result of the classical economic writings of Adam Smith.

Lateran Treaties of 1929 Often referred to as the Concordat, these treaties were concluded between the Holy See and Italy and confirmed the sovereignty of Vatican City. They also recognized Roman Catholicism as the state religion of Italy. The late 1970s saw a revision of the clauses dealing with Roman Catholicism as the state religion and with the requirement of religious instruction at public schools.

League of Nations A peacekeeping organization established after World War I, considerably smaller than its successor, the United Nations. Although no longer a political force when World War II broke out, it was officially dissolved in 1946.

Low Countries   Belgium, Luxembourg, and the Netherlands.

Magna Carta   A charter, issued by King John in 1215, which curbed absolutism in England. Considered the most famous document in British constitutional history, it clearly reveals the viability of opposition to the arbitrary use of power. It is always cited as an example of the "written part" of the British constitution.

Marshall Plan   Officially known as the European Recovery Program (ERP). Nicknamed the Marshall Plan after George Marshall, the secretary of state in the Truman administration who designed and in 1947 announced the plan. It consisted of a massive transfer of goods and money from the United States to a war-devastated Europe. Almost all Western European countries were beneficiaries. The ERP dispensed more than $12 billion in loans and goods from 1948 to 1952 to be used for the industrial recovery of Europe.

Nazi   The official name of the party that assumed power once Adolf Hitler had been made chancellor was the National Socialist German Workers' Party. Its members soon came to be called Nazis.

North Atlantic Treaty Organization (NATO)   A collective-defense organization, founded in 1949 by the United States, Canada, and numerous Western European nations. Although its Central/Eastern European counterpart, the Warsaw Pact, was officially dissolved in 1990, NATO continues to exist and expand its membership.

Organization for Economic Cooperation and Development (OECD)   An international organization established in 1961 to supersede the OEEC, promoting economic growth, aiding developing nations, and working to expand world trade.

Organization for European Economic Cooperation (OEEC)   When announcing the European Recovery Program, the United States indicated that it did not want to deal with Western European countries on an individual basis. It wanted to deliver all the aid, goods, and moneys to one central point in Western Europe from whence the internal distribution could take place. The OEEC, headquartered in Paris, became responsible for the receipt and distribution of all Marshall Plan aid.

Organization of Petroleum Exporting Countries (OPEC)   An organization established in 1960 to set oil prices and coordinate the global oil policies of its members. It reached its apogee of power in the early 1970s, when it started to use oil distribution as a political weapon. After the mid-1970s, it was riddled with rifts and divisions, and it ceased to be a political force in world politics.

Ostpolitik   A term that originated with Chancellor Otto von Bismarck, who did not want the new Germany (1871) to have overseas colonies. Instead he wanted the country to expand overland in an eastward direction. The drive toward the East is a theme that may be compared to the American Manifest Destiny. Ultimately, Bismarck succumbed to the pressures for a "place in the sun," as colonialism was called in those days. As chancellor of West Germany, Willy Brandt recycled the term. It now came to mean the reestablishment of diplomatic relations between West Germany and all the countries to its east.

Paysantisme   One of the two major strands upon which the French economy is based; the term points to the close relationship of the French with the soil, a relationship that has continued through times of industrialization.

Postindustrial Era   This is the phase that many Western countries have attained in which manufacturing has to some extent been replaced by service industries.

Prime Ministerial System   A species of the parliamentary system in which the prime minister is no longer regarded as the *primus inter pares* (first among equals) but is considered far superior in power to the other members of the cabinet. In addition to the United Kingdom, Germany with its Chancellor's Democracy may serve as an example.

Proportional Democracy   A system whereby public appointments, from the cabinet level down, are distributed the various political parties in proportion to their representation in the legislature.

Proportional Representation   An electoral system that allows each party to have the proportion of seats in the legislature that it achieved in votes vis-à-vis the total vote. Put simply: a party that gets 10 percent of the total vote is entitled to 10 percent of the available seats in the legislature.

Reconquista   Spanish for "reconquest," the term has come to refer to the reconquest of territory held by the Moors in Spain. The Reconquista ended with the fall of Granada in 1492, a year that provides a watershed as the Age of Discovery then attained new heights.

Reformation   Although the term may appear very general, it usually applies to religion, notably the religious revolution that engulfed large parts of Europe in the early sixteenth century. Martin Luther became one of the foremost leaders of Protestantism, which resulted from his reformation efforts.

Renaissance   Literally "rebirth," this term refers to the rich period of Western European civilization marking the transition from the Dark Ages to modern times. The Renaissance shifted the emphasis from God and religion to man and individuality. It also emphasized worldly experience and produced brilliant accomplishments in scholarship, literature, science, and the arts.

Risorgimento   An Italian term, meaning "resurgence," that refers particularly to the liberation and unification of Italy in the nineteenth century. It was by no means a popular movement but, rather, the aim and ambition of aristocratic circles in an Italy that was still greatly fragmented.

Scandinavia   A term used to describe the Nordic countries: Sweden, Denmark, and Norway.

Social Democratic Party of Germany (SDP)   The only political party emerging in West Germany in 1949 that had pre–World War II roots. Its program was considerably diluted at the 1959 Bad Godesberg Convention. This was done to expand its constituency. The SPD, like many other socialist parties, suffered losses in the 1990s as a result of a negative coattail effect. While it was communism that came to be publicly discredited, socialism often shared its fate.

Socialism   A political and economic theory that aims at collective or government ownership and management of the means of production and distribution of goods.

Sottogoverno   An Italian term meaning "subgovernment" and referring to nonofficial sources of power. In Italy, politics often faces gridlocks that can be resolved only by sottogoverno, an outside force bent on compromise and reconciliation.

Thatcherism   A style of rigid right-wing governance, exemplified by Margaret Thatcher during the long period of her prime ministership (1979–1990). She exhibited a strong bias against compromise. The unions were soon muzzled. In domestic policy, she became known for her monetarist policies. In foreign policy, Thatcher repeatedly revealed suspicions that the EU bristled with socialistic endeavors.

Third Reich   The official name of Germany during the period of Adolf Hitler's dictatorship (1933–1945).

Trade Union Congress (TUC)   A British umbrella organization comprising most, if not all, trade unions in the United Kingdom. There is a connection between the TUC and the British Labour Party, in that the membership fees for the latter are reduced if one is a member of the former.

Truman Doctrine   A foreign-policy doctrine conceived and issued by President Harry S. Truman that made it clear that the United States would come to the aid of legitimate governments troubled by insurrections that were helped from the outside. The doctrine appeared to point to the multitude of liberation fronts that emerged all over the world and were often aided by the Soviet Union. Initially it applied to Greece and Turkey.

Unicameral   Refers to a legislature that has only one chamber.

United Nations (UN)   The term was initially used to denote the Allies. At various war conferences, plans were made to create a successor organization to the League of Nations. Hardly had World War II ended in Europe than a large conference was held at San Francisco to establish the United Nations as an organization. The United Nations endeavors to resolve conflict and maintain peace and security as well as to achieve international cooperation in international economic, social, cultural, and humanitarian problems.

United Nations Conference on Trade and Development (UNCTAD)   Convened as a special trade conference in 1964 (122 nations attended). That same year, the UN General Assembly granted it permanent status. It applies pressure on the advanced industrial states to lower their trade barriers so as to expand trade in primary commodities.

Value-Added Tax (VAT)   A tax imposed on all goods and services at every stage of their production, based on the increase in value of that good or service. The VAT started out in England but is now rapidly spreading through the European Community.

Vichy France   When Germany defeated France in 1940, it did not want to occupy the entire area of France. It created the *Etat Français,* a satellite that closely collaborated with the neighboring Nazis. Officially it was the successor of the Third Republic and was therefore responsible for the colonies as well. Vichy was dissolved when the Allied troops approached its area.

Vote of Confidence   Refers particularly to British politics. When a government feels that, in spite of reverses, it still has the support of the majority of the House of Commons, it can demand a vote of confidence, which will indicate that it still has the backing of the majority. Votes of confidence are always undertaken by the government.

Vote of No-Confidence   This refers to the traditional instrument of the opposition to oust the government. It is undertaken only when the opposition feels that the time has come to bring down the government on an important piece of legislation.

Weimar Republic   The German federal system that emerged after Germany as the successor of the Wilhelmine Empire. When Germany lost World War I, it needed a new constitution. The Weimar Constitution proved highly democratic, so much so that it rendered the position of government very precarious. The 1920s, moreover, were a time of extremist agitation, unemployment, depression, and severe inflation. The Weimar Republic was succeeded by the Third Reich.

Zollverein   A customs union established in nineteenth-century Germany to eliminate tariffs among the various states, principalities, and other political units. The Zollverein preceded the emergence of a unified Germany in 1871.

Bibliography

NATIONAL HISTORIES AND ANALYSES

Andorra

Barry Taylor, *Andorra* (Oxford; Santa Barbara: Clio Press, 1993).

Austria

Karen Barkey and Mark Von Hagen, *After Empire: Multiethnic Societies and Nation-Building: The Soviet Union and Russian, Ottoman and Habsburg Empires* (Boulder: Westview Press, 1997).

Gunter Bishof and Anton Pelinka, eds., *Austria in the New Europe* (New Brunswick: Transaction Publishers, 1993).

John Fitzmaurice, *Austrian Politics and Society Today* (London: Macmillan, 1991).

Alan Levy, *The Wiesenthal File* (London: Constable, 1993).

Dagmar C. G. Lorenz and Gabriele Weinberger, eds., *Insiders and Outsiders: Jewish and Gentile Culture in Germany and Austria* (Detroit: Wayne State University Press, 1994).

Bruce F. Pauley, *From Prejudice to Persecution: A History of Austrian Anti-Semitism* (Chapel Hill: University of North Carolina Press, 1992).

Anton Pelinka, *Austria* (Boulder: Westview Press, 1997).

Melanie A. Sully, *The Haider Phenomenon* (New York: Columbia University Press, 1997).

Belgium

John Fitzmaurice, *The Politics of Belgium: Crisis Compromise in a Plural Society* (London: Hurst & Co., 1996).

Lisbet Hooghe, *A Leap in the Dark: Nationalist Conflict and Federal Reform in Belgium* (Ithaca: Cornell University Press, 1991).

Ann Owen, "Belgium: Slimmer SHAPE," *Lancet* (North American ed.), 340: 1342–3, November 28, 1992.

M. A. G. van Meerhaeghe, ed., *Belgium and EC Membership Evaluated* (London: Pinter Publishers; New York: St. Martin's Press, 1992).

Cyprus

Vassos Argyrou, *Tradition and Modernity in the Mediterranean* (New York: Cambridge University Press, 1996).

Tozun Bahcheli, *Greek-Turkish Relations Since 1955* (Boulder: Westview Press, 1990).

Dan Hofstadter, *Goldberg's Angel: An Adventure in the Antiquities Trade* (New York: Farrar, Straus and Giroux, 1994).

Zaim M. Nedjatigil, *The Cyprus Question and the Turkish Position in International Law* (Oxford; New York: Oxford University Press, 1989).

John Reddaway, *Burdened With Cyprus: The British Connection* (London: Weidenfeld & Nicholson, 1986).

Rodney Wilson, *Cyprus and the International Economy* (New York: St. Martin's Press, 1992).

Denmark

Jens Henrik Haarh, *Looking to Europe: The EC Policies of the British Labour Party and the Danish Social Democrats* (Aarhus: Aarhus University Press, 1993).

Kenneth E. Miller, *Denmark: A Troubled Welfare State* (Boulder: Westview, 1991).

Gunnar Viby Mogensen, *Danes and Their Politicians: A Summary of the Findings of a Research Project on Political Credibility in Denmark* (Aarhus: Aarhus University Press, 1993).

Xan Smiley, "A Survey of the Nordic Countries: Happy Family?" *The Economist* (January 23, 1999).

Finland

Christian Bordes-Marcilloux, *Three Assessments of Finland's Economic Crisis and Economic Policy* (Helsinki: Bank of Finland, 1993).

Max Jakobson, *Finland in the New Europe* (Westport: Praeger, 1998).

Jyrki Kakonen, ed., *Politics and Sustainable Growth in the Arctic* (Aldershot; Brookfield: Dartmouth, 1993).

Risto E. J. Penttila, *Finland's Search for Security Through Defence, 1944–1989* (New York: St. Martin's Press, 1991).

Xan Smiley, "A Survey of the Nordic Countries: Happy Country?" *The Economist* (January 23, 1999).

H. M. Tillotson, *Finland at Peace and War, 1918–1993* (Wilby, Norwich: Michael Russell, 1993).

France

John Ardagh, *France in the 1980s* (New York: Penguin, 1987).

Richard Bernstein, *Fragile Glory: A Portrait of France and the French* (New York: Knopf, 1990).

Frank Costigliola, *France and the United States: The Cold Alliance Since World War II* (New York: Twayne Publishers; Toronto: Maxwell Macmillan, Canada; New York: Maxwell Macmillan International, 1992).

Julius Weis Friend, *Seven Years in France: François Mitterrand and the Unintended Revolution* (Boulder: Westview Press, 1989).

Steven Philip Kramer, *Does France Still Count? The French Role in the New Europe* (The Washington Papers, 1964; Westport: Praeger, 1994).

Peter Morris, *French Politics Today* (Manchester: Manchester University Press; New York: Distributed by St. Martin's Press, 1994).

Sophie Pedder, "A Survey of France: The Grand Illusion," *The Economist* (June 5, 1999).

William Safran, *The French Polity,* 5th ed. (New York: Longman, 1998).

Jean-Claude Scheid and Peter Walton, *France* (London; New York: Routledge in association with the Institute of Chartered Accountants in England and Wales, 1992).

Paul Stillwell, ed., *Assault on Normandy: First-Person Accounts from the Sea Services* (Annapolis: Naval Institute Press, 1994).

Ronald Tiersky, *France in the New Europe: Changing Yet Steadfast* (Belmont: Wadsworth, 1994).

Philip M. Williams, *Crisis and Compromise: Politics in the Fourth Republic* (New York: Anchor Books, 1966).

Germany

Karl Dietrich Bracher, *The German Dictatorship* (New York: Praeger, 1970).

Robert Burns, ed., *German Cultural Studies: An Introduction* (Oxford: Oxford University Press, 1995).

Marilyn Shevin Coetzee, *The German Army League: Popular Nationalism in Wilhelmine Germany* (New York: Oxford University Press, 1990).

David P. Conradt, *The German Polity,* 7th ed. (New York: Longman, 2001).

David P. Conradt et al., eds., *Power Shift in Germany: The 1998 Election and the End of the Kohl Era* (New York: Berghahn Books, 2000).

Robin Cross, *Fallen Eagle: The Last Days of the Third Reich* (New York: Wiley, 1996).

Marc Fisher, *After the Wall: Germany, the Germans, and the Burdens of History* (New York: Simon & Schuster, 1995).

E. Gene Frankland and Donald Schoonmaker, *Between Protest and Power: The Green Party in Germany* (Boulder: Westview Press, 1992).

Mary Fulbrook, *The Divided Nation: A History of Germany, 1918–1990* (standard title: *Fontana History of Germany, 1918–1990*) (New York: Oxford University Press, 1992).

Herbert Giersch, *The Fading Miracle: Four Decades of Market Economy in Germany* (Cambridge; New York: Cambridge University Press, 1992).

Mary N. Hampton and Christian Søe, eds., *Between Bonn and Berlin: German Politics Adrift?* (Lanham: Rowman & Littlefield, 1999).

David M. Keithly, *The Collapse of East German Communism: The Year the Wall Came Down* (Westport: Praeger, 1992).

Emil J. Kirchner and James Sperling, *The Federal Republic and NATO: 40 Years After* (New York: St. Martin's Press, 1992).

Rand C. Lewis, *The Neo-Nazi and German Unification* (Westport: Greenwood Publishing, 1996).

Peter H. Merkl, ed., *The Federal Republic of Germany at Fifty: The End of a Century of Turmoil* (London: Macmillan, 1999).

Stephen Padgett, ed., *Adenauer to Kohl: The Development of the German Chancellorship* (Washington, D.C.: Georgetown University Press, 1994).

Ernest D. Plock, *East German–West German Relations and the Fall of the GDR* (Boulder: Westview Press, 1993).

Gordon Smith et al., eds., *Developments in German Politics 2* (Durham: Duke University Press, 1996).

W. R. Smyser, *The Economy of United Germany: Colossus at the Crossroads* (New York: St. Martin's Press, 1991).

Gregory F. Treverton, *America, Germany and the Future of Europe* (Princeton: Princeton University Press, 1992).

Henry Ashby Turner, Jr., *The Two Germanies Since 1945* (New Haven: Yale University Press, 1987).

Paul J. J. Welfens, ed., *Economic Aspects of German Unification: National and International Perspectives* (Berlin; New York: Springer Verlag, 1992).

Greece

David Close, *The Origins of the Greek Civil War* (London; New York: Longman, 1995).

Kevin Featherstone and Dimitrios K. Kasoudas, eds., *Political Change in Greece: Before and After the Colonels* (London: Croom Helm, 1987).

Robert Frazier, *Anglo-American Relations With Greece: The Coming of the Cold War, 1942–47* (New York: St. Martin's Press, 1991).

Nicholas Gage, *Hellas: A Portrait of Greece* (New York: Villard Books, 1987).

Michael Grant, *A Social History of Greece and Rome* (New York: Scribner: Maxwell Macmillan International, 1992).

Nicholas Stavrou, ed., *Greece Under Socialism: A NATO Ally Adrift* (New Rochelle: A. D. Caratzas, 1988).

Iceland

E. Paul Durrenberger and Gisli Palsson, eds., *Images of Contemporary Iceland: Everyday Lives and Global Contexts* (Iowa City: University of Iowa Press, 1996).

Gudmundur Gunnarsson, *The Economic Growth in Iceland, 1910–1980* (Stockholm: Almqvist & Wiksell International, 1990).

John J. Horton, *Iceland* (Oxford; Santa Barbara: Clio Press, 1983).

Donald Edwin Nuechterlein, *Iceland, Reluctant Ally* (Westport: Greenwood Press, 1975, 1961).

Xan Smiley, "A Survey of the Nordic Countries: Happy Family?" *The Economist* (January 23, 1999).

Richard F. Tomasson, *Iceland: The First New Society* (Minneapolis: University of Minnesota Press, 1980).

Ireland

John Ardagh, *Ireland and the Irish: Portrait of a Changing Society* (London: Penguin Books, 1995).

J. Bowyer Bell, *The Gun in Politics: An Analysis of Irish Political Conflict* (New Brunswick: Transaction Books, 1987).

George D. Boyce, *Making Modern Irish History* (New York: Routledge, 1996).

Niamh Brennan, *Ireland* (London; New York: Routledge in Association with the Institute of Chartered Accountants in England and Wales, 1992).

Sean Byrne, *Third Parties in Northern Ireland: Exacerbation or Amelioration?* (St. Louis: University of Missouri, Center for International Studies, 1994).

Neil Collins, *Irish Politics Today,* 2nd ed. (Manchester; New York: Manchester University Press; distributed exclusively in the United States and Canada by St. Martin's Press, 1992).

John E. Finn, *Constitutions in Crisis: Political Violence and the Rule of Law* (New York: Oxford University Press, 1991).

John Hume, *A New Ireland: Politics, Peace, and Reconciliation* (Niwot, Colorado: Robert Rinehart, 1996).

Independent Study Group, *Ulster After the Ceasefire* (London: Alliance Publishers, Ltd., for the Institute for the European Defence and Strategic Studies, 1994).

Italy

Percy Allum, *Chronicle of a Death Foretold: The First Italian Republic* (Reading: Department of Politics, University of Reading, 1993).

Stanton H. Burnett and Luca Mantovani, *The Italian Guillotine: Operation Clean Hands and the Overthrow of Italy's First Republic* (Washington, D.C.: CSIS, 1998).

Christopher Duggan, *A Concise History of Italy* (New York: Cambridge University Press, 1994).

Francesco Francioni, ed., *Italy and EC Membership Evaluated* (New York: St. Martin's Press, 1992).

Mark Gilbert, *The Italian Revolution: The End of Politics Italian Style?* (Boulder: Westview Press, 1994).

Cheryl Maclachlin, *Bringing Italy Home* (New York, Crown Publishing Group, 1995).

Ray Porter and Mikulas Tiech, eds., *The Renaissance in National Context* (Cambridge; New York: Cambridge University Press, 1992).

Robert D. Putnam, *Making Democracy Work: Civic Traditions in Modern Italy* (Princeton: Princeton University Press, 1993).

Donald Sassoon, *Contemporary Italy: Politics, Economy & Society Since 1945* (London: Longman, 1986).

Frederic Spots and Theodore Wieser, *Italy: A Difficult Democracy* (Cambridge: Cambridge University Press, 1986).

James Walston, *The Mafia and Clientelism: Roads to Rome in Post-War Calabria* (London; New York: Routledge, 1988).

Leonard B. Weinberg and William Lee Eubank, *The Rise and Fall of Italian Terrorism* (Boulder: Westview Press, 1987).

Liechtenstein

Regula A. Meier, *Liechtenstein* (Oxford; Santa Barbara: Clio Press, 1993).

Luxembourg

Harry C. Barteau, *Historical Dictionary of Luxembourg* (Metuchen: Scarecrow Press, 1996).

Ed Needham, *The Countries of Benelux* (London: Gloucester Press, 1994).

Malta

Salvino Busuttil, *The Future of the Mediterranean* (Valetta: University of Malta, 1995).

Desmond Gregory, *Malta, Britain, and the European Powers, 1793–1875* (Madison: Fairleigh Dickerson University Press, 1996).

Adrianus Koster, *Prelates and Politicians in Malta: Changing Power Balances Between Church and State in a Mediterranean Island Fortress, 1800–1976* (Assen: van Gorcum, 1984).

Christopher F. Shores, *Malta: The Hurricane Years: 1940–41* (London: Grub Street, 1987).

Mario Vassallo, *From Lordship to Stewardship: Religion and Social Change in Malta* (The Hague; New York: Mouton, 1979).

Barry York, *Malta, a Nonaligned Democracy in the Mediterranean* (Sydney: A Friend of Malta publication, 1987).

Monaco

Raymond de Vos, *History of the Monies, Medals, and Tokens of Monaco* (Long Island City: Sanford J. Durst, 1978).

Anne Edwards, *The Grimaldis of Monaco* (New York: Morrow, 1992).

The Netherlands

Rudi B. Anderweg and Galen A. Irwin, *Dutch Politics and Government* (New York: St. Martin's Press, 1993).

Herman Bakvis, *Catholic Power in the Netherlands* (Kingston: McGill-Queen's University Press, 1981).

Robert H. Cox, *The Development of the Dutch Welfare State* (Pittsburgh: Pittsburgh University Press, 1993).

H. Entzinger, *Immigrant Ethnic Minorities in the Dutch Labor Market* (Amsterdam: Thesis Publishers, 1994).

Ph. P. Everts, ed., *Controversies at Home: Domestic Factors in the Foreign Policy of the Netherlands* (Dordrecht; Boston: M. Nijhoff, 1985).

Fernando Garrido, *EC and National Regulations on Environment and Agriculture in Denmark, the Netherlands and Spain* (Esbjerg: South Jutland University Press, 1994).

Simon Shama, *The Embarrassment of Riches: An Interpretation of Dutch Culture in the Golden Age* (New York: Knopf, 1987).

Norway

Sherwood S. Cordier, *The Defense of NATO's Northern Front and US Military Policy* (Lanham: University Press of America, 1989).

Erik Damman, *Revolution in the Affluent Society* (London: Heretic Books, 1984), translated by Louis Mackay.

Kurt Feldbakken, *The Honeymoon* (New York: St. Martin's Press, 1987).

Anne Coshen Kiel, ed., *Continuity and Change: Aspects of Contemporary Norway* (Oslo: Scandanavian University Press, 1993).

Arne Selbyg, *Norway Today: An Introduction to Modern Norwegian Society* (Oslo: Norwegian University Press, 1986).

Xan Smiley, "A Survey of the Nordic Countries: Happy Family?" *The Economist* (January 23, 1999).

Kaare Strom and Lars Svansand, eds., *Challenges to Political Parties: The Case of Norway* (Ann Arbor: University of Michigan Press, 1997).

Portugal

Enrique A. Baloyra, ed., *Comparing New Democracies: Transition and Consolidation in Mediterranean Europe and the Southern Cone* (Boulder: Westview Press, 1987).

Thomas C. Bruneau, *Political Parties and Democracy in Portugal* (Boulder: Westview Press, 1997).

Lawrence S. Graham and Douglas Wheeler, *In Search of Modern Portugal* (Madison: University of Wisconsin Press, 1983).

Patrick Lane, "A Survey of Portugal: Half-Way There," *The Economist* (December 2, 2000).

Ulrike Liebert and Maurizio Cotta, eds., *Parliament and Democratic Consolidation in Southern Europe: Greece, Italy, Portugal, Spain, and Turkey* (London; New York: Pinter Publishers, 1990).

Walter C. Opello, *Portugal's Political Development: A Comparative Approach* (Boulder: Westview Press, 1985).

____, *Portugal: From Monarchy to Pluralist Democracy* (Boulder: Westview Press, 1991).

Eric Solsten, ed., *Portugal: A Country Study* (Washington, DC: Library of Congress, 1993).

San Marino

James T. Bent, *A Freak of Freedom or the Republic of San Marino* (Port Washington, NY: Kennikat, 1970).

Spain

Rodrigo Botero, *Reflections on the Modernization of Spain* (San Francisco: ICS Press, 1992).

Raymond Carr and Juan Pablo Fusi, *Spain: Dictatorship to Democracy,* 2nd ed. (Winchester: Allen and Unwin, 1981).

J. H. Elliott, *The Hispanic World: Civilization and Empire* (London: Thames and Hudson, 1991).

James D. Fernandez, *Apology to Apostrophe: The Autobiography and the Rhetoric of Self-Representation in Spain* (Durham: Duke University Press, 1992).

Richard A. Fletcher, *Moorish Spain* (New York: H. Holt, 1992).

Jeffrey R. Franks, *Explaining Unemployment in Spain: Structural Exchange, Cyclical Fluctuations, and Labor Market Rigidities* (Washington, D.C.: International Monetary Fund, 1994).

José A. Gonzalo, *Spain* (London; New York: Routledge, 1992).

Richard Gunther, ed., *Politics, Society, and Democracy. The Case of Spain* (Boulder: Westview Press, 1992).

Stephen Hugh-Jones, "A Survey of Spain: A Country With Many Faces" *The Economist* (November 25, 2000).

Thomas D. Lancaster, *Political Stability and Democratic Change: Energy in Spain's Transition* (University Park: The Pennsylvania State University Press, 1989).

D. S. Morris, *Britain, Spain and Gibraltar, 1945–1990: The Eternal Triangle* (London; New York: Routledge, 1992).

Edward Moxon-Browne, *Political Change in Spain* (London; New York: Routledge, 1989).

Mary Elizabeth Perry and Anne J. Cruz, *Cultural Encounters: The Impact of the Inquisition in Spain and the New World* (Berkeley: University of California Press, 1991).

Andrew J. Richards, "Spain: From Isolation to Integration," in Ronald Tiersky, ed., *Europe Today: National Politics, European Integration, and European Security* (Lanham: Rowman & Littlefield, 1999), pp. 161–196.

Sweden

Ebba Dohlman, *National Welfare and Economic Interdependence: The Case of Sweden's Foreign Trade Policy* (Oxford: Clarendon Press; New York: Oxford University Press, 1989).

Lars Engwall, ed., *Economics in Sweden: An Evaluation of Swedish Research in Economics* (London; New York: Routledge, 1992).

Jonas Frykman and Orvar Lofgren, *Culture Builders: A Historical Anthropology of Middle Class Life* (New Brunswick: Rutgers University Press, 1987), translated by Alan Crozier.

Lars Hultkrantz, *Chernobyl Effects on Domestic and Inbound Tourism in Sweden: A Time Series Analysis* (Umea: University of Umea, 1994).

Michael Maccoby, ed., *Sweden at the Edge: Lessons for American and Swedish Managers* (Philadelphia: University of Pennsylvania Press, 1991).

Michele Micheletti, *Civil Society and State Relations in Sweden* (Brookfield: Ashgate Publishing Co., 1995).

Jonas Pontusson, *The Limits of Social Democracy: Investment Politics in Sweden* (Ithaca: Cornell University Press, 1992).

Bo Rothstein, *The Social Democratic State: The Swedish Model and the Bureaucratic Problem of Social Reform* (Pittsburgh, PA: University of Pittsburgh Press, 1996).

Xan Smiley, "A Survey of the Nordic Countries: Happy Family?" *The Economist* (January 23, 1999).

Tim Tilton, *The Political Theory of Swedish Social Democracy: Through the Welfare State to Socialism* (Oxford: Oxford University Press, 1990).

Switzerland

Janet E. Hilowitz, ed., *Switzerland in Perspective* (New York: Greenwood Press, 1990).

Rolf Keiser and Kurt R. Spillman, eds., *The New Switzerland: Problems and Policies* (Palo Alto: SPROSS, 1996).

Kenneth D. McRae, *Conflict and Compromise in Multilingual Societies* (Waterloo: Wilfrid Laurier University Press, 1983).

Rene Schwok, *Switzerland and the European Common Market* (New York: Praeger, 1991).

Ralph Segelman, *The Swiss Way of Welfare: Lessons for the Western World* (New York: Praeger Publishers, 1986).

Donald Arthur Waters, *Hitler's Secret Ally, Switzerland* (La Mesa: Pertinent Publications, 1992).

The United Kingdom

Anderson Consulting, *The Future of European Health Care* (London: Burston Marsteller, 1993).

Arthur Aughey and Duncan Morrow, eds., *Northern Ireland Politics* (London: Longman, 1996).

Samuel H. Beer, *Britain Against Itself: The Political Contradictions of Collectivism* (New York: W. W. Norton, 1982).

David George Boyce, *The Irish Question and British Politics, 1868–1986* (Basingstoke: Macmillan Education, 1988).

Simon Bulmer, Stephen George, and Andrew Scott, eds., *The United Kingdom and EC Membership Evaluated* (New York: St. Martin's Press, 1992).

Ernest E. Cashmore, *United Kingdom: Class, Race and Gender Since the War* (London; Boston: Unwin Hyman, 1989).

Richard Critchfield, *An American Looks at Britain* (New York: Anchor Books/Doubleday, 1990).

Nelson Antonio Da Costa, *An Impossible Meeting of the Minds: A Rhetorical Analysis of the 1982 Falkland (Malvinas) Conflict Between Argentina and the United Kingdom* (Lawrence: University of Kansas, Communication Studies, 1994).

Peter David, "A Survey of Britain: Undoing Britain?" *The Economist* (November 6, 1999).

Rudi Dornbusch and Richard Layard, eds., *The Performance of the British Economy* (New York: Oxford University Press, 1988).

Patrick Dunleavy et al., eds., *Developments in British Politics 5* (New York: St. Martin's Press, 1997).

Andrew Gamble, *The Free Economy and the Strong State: The Politics of Thatcherism* (London: Macmillian, 1988).

Alfred F. Havighurst, *Britain in Transition: The Twentieth Century* (Chicago: University of Chicago Press, 1985).

Bill Jones and Dennis Kavanagh, *British Politics Today.* 6th ed. (Washington, D.C.: Congressional Quarterly Press, 1998).

Dennis Kavanagh, *British Politics: Continuities and Change,* 4th ed. (Oxford; New York: Oxford University Press, 2000).

____, *Thatcherism and British Politics: The End of Consensus?* 2nd ed. (New York: Oxford University Press, 1990).

Dennis Kavanagh and Anthony Selden, *The Thatcher Effect* (Oxford: Clarendon Press; New York: Oxford University Press, 1989).

Anthony King et al., *New Labour Triumphs: Britain at the Polls* (Chatham: Chatham House, 1997).

Brian Lapping, *End of Empire* (New York: St. Martin's Press, 1985).

William Roger Louis, ed., *Adventures with Britannia: Personalities, Politics, and Culture in Britain* (London: I. B. Tauris, 1995).

Andrew Marr, *Ruling Britannia: The Failure and Future of British Democracy* (London: Michael Joseph, 1995).

Philip Norton, *The British Polity,* 4th ed. (New York: Longman, 2001).

Dawn Oliver, *Government in the United Kingdom: The Search for Accountability, Effectiveness and Citizenship* (Milton Keynes; Philadelphia: Open University Press, 1991).

James S. Olsen et al., *Historical Dictionary of the British Empire* (Westport: Greenwood Publishers, 1996).

Orde, *The Eclipse of Great Britain* (New York: St. Martins, 1996).

Peter Orton, *UK Health Care: The Facts* (Dordrecht; Boston: Kluwer Academic Publishers, 1994).

Giles Radice, *Labour's Path to Power: The New Revisionism* (New York: St. Martin's Press, 1989).

John Rentoul, *Tony Blair* (London: Warner Books, 1996).

Peter Riddell, *Thatcher Decade: How Britain Has Changed During the 1980s* (Oxford: Basil Blackwell, 1989).

____, *The Thatcher Era and Its Legacy* (Oxford; Cambridge: B. Blackwell, 1991).

Alan Sked and Chris Cook, *Post-War Britain: A Political History,* 3rd ed (London: Penguin Books, 1990).

Steve Smith et al., eds., *British Foreign Policy: Tradition, Change, and Transformation* (Winchester: Unwin Hyman, 1988).

Vatican City

Malachi Martin, *Vatican* (New York: Harper & Row, 1986).

Francesco Papafava, ed., *The Vatican* (Firenze: Scala Books; New York: Distributed by Harper & Row, 1984).

Thomas Reese, *Inside the Vatican: The Politics and Organization of the Catholic Church* (Cambridge, MA: Harvard University Press, 1996).

Michael J. Walsh, *Vatican City State* (Santa Barbara: ABC-Clio, 1983).

EUROPEAN UNION

Timoth Bainbridge (with Anthony Teasdale), *The Penguin Companion to European Union* (London: Penguin Books, 1995).

Elizabeth Bomberg, *Green Parties and Politics in the European Union* (London: Routledge, 1998).

Robert Cottrell, "A Survey of Europe: My Continent, Right or Wrong," *The Economist* (October 23, 1999).

Laura Cram et al., eds., *Developments in the European Union* (New York: St. Martin's Press, 1999).

Desmond Dinan, *Ever Closer Union? An Introduction to the European Community,* 2nd ed. (Boulder: Lynne Rienner, 1999).

Stephen George, *Politics and Policy in the European Community,* 3rd ed. (Oxford: Oxford University Press, 1996).

Neill Nugent, *The Government and Politics of the European Union,* 4th ed. (Durham: Duke University Press, 1999).

Derek Urwin, *The Community of Europe: A History of European Integration Since 1945,* 2nd ed. (London: Longman, 1995).

John Van Oudenaren, *Uniting Europe: European Integration and the Post–Cold War World* (Lanham: Rowman & Littlefield, 2000).

Helen Wallace and William Wallace, eds., *Policy-Making in the European Union,* 4th ed. (Oxford: Oxford University Press, 2000).

AN "EMERGING" EUROPE

Attila Agh, *Emerging Democracies in East Central Europe,* (Williston: Edward Elgar, 1999).

Sorin Antohi and Vladimir Tismaneanu, eds., *Between Past and Future: The Revolutions of 1989 and Their Aftermath* (New York: Central European, 1999).

Stan Berglund and Jan Ake Dellebrant, eds., *The New Democracies in Eastern Europe* (Williston: Edward Elgar, 1999).

J. F. Brown, *Eastern Europe and Communist Rule* (Durham: Duke University Press, 1998).

____, *Hopes and Shadows: Eastern Europe After Communism* (Durham: Duke University Press, 1994).

Karen Dawisha and Bruce Parrott, eds., *The Consolidation of Democracy in East-Central Europe* (Cambridge: Cambridge University Press, 1997).

Slovenka Drakulic, *Cafe Europa: Life After Communism* (New York: Penguin, 1997).

Minton F. Goldman, *Revolution and Change in Central and Eastern Europe: Political, Economic and Social Challenges* (Armonk: M. E. Sharpe, 1997).

Joseph Held, ed., *Populism, Racism, and Society in Eastern Europe* (New York: Columbia University Press, 1996).

The Czech Republic

John F. N. Bradley, *Post-Communist Czechoslovakia* (New York: Columbia University Press, 1997).

Carol Skalnick Leff, *The Czech and Slovak Republics: Nation Versus State* (Boulder: Westview Press, 1997).

Hungary

Wendy Hollis, *Democratic Consolidation in Eastern Europe: The Influence of the Communist Legacy in Hungary, the Czech Republic, and Romania* (New York: Columbia University Press, 1999).

Andras Korosenyi, *Government and Politics in Hungary* (New York: Central European, 1999).

Rudolph Tokes, *Hungary's Negotiated Revolution: Economic Reform, Social Change, and Political Succession* (Cambridge: Cambridge University Press, 1996).

Poland

John Fitzmaurice, *Politics and Government in the Visegrad Countries: Poland, Hungary, the Czech Republc, and Slovakia* (New York: St. Martins, 1998).

George Sanford, *Poland: The Conquest of History* (Williston: Harwood, 1999).

SOURCES WITH A GENERAL, COMPARATIVE, OR REGIONAL PERSPECTIVE

Percy Allum, *State and Society in Western Europe* (Cambridge: Polity Press, 1995).

Guido Baglioni and Colin Crouch, eds., *European Industrial Relations: The Challenge of Flexibility* (London; Newbury Park: Sage Publications, 1990).

Richard Batley and Gerry Stoker, eds., *Local Government in Europe: Trends and Development* (New York: St. Martin's Press, 1991).

Hans-Georg Betz, *Radical Right-Wing Populism in Western Europe* (New York: St. Martin's Press, 1994).

Norman Birnbaum, *After Progress: American Social Reform and European Socialism in the Twentieth Century* (New York: Oxford University Press, 2001).

Gillian Bottomley, *From Another Place: Migration and the Politics of Culture* (Cambridge; New York: Cambridge University Press, 1992).

Luciano Cheles, Ronnie Ferguson, and Michalina Vaughan, *The Far Right in Western & Eastern Europe* (London: Longman, 1995).

John Coakley, *Social Origins of Nationalistic Movements: The Contemporary European Experience* (Sage Publications, 1991).

Norman Davies, *Europe: A History* (New York: HarperCollins, 1998).

J. William Derleth, *The Transition in Central and Eastern European Politics* (Upper Saddle River: Prentice Hall, 2000).

Michale Gallagher, Michael Laver, and Peter Maier, *Representative Government in Modern Europe,* 2nd ed. (New York: McGraw-Hill, 1995).

Misha Glenny, *The Rebirth of History: Eastern Europe in the Age of Democracy* (London: Penguin Books, 1990).

Kenneth Hanf and Alf-Inge Jansen, eds., *Governance and Environment in Western Europe: Politics, Policy and Administration* (London: Longman, 1998).

Jan-Erik Lane and Svante Ersson, *Politics and Society in Western Europe,* 4th ed. (London: Sage, 1999).

Voitech Mastny, *The Helsinki Process and the Reintegration of Europe, 1986–1991: Analysis and Documentation* (New York: New York University Press, 1990).

Yves Mény and Andrew Knapp, *Government and Politics in Western Europe,* 3rd ed. (Oxford: Oxford University Press, 1998).

Hans J. Michelmann and Panayotis Soldatos, eds., *Federalism and International Relations: The Role of Subnational Units* (Oxford: Clarendon Press; New York: Oxford University Press, 1990).

Andrew A. Michta, *The Government and Politics of Postcommunist Europe* (Wesport: Praeger, 1994).

William E. Paterson and Alastair H. Thomas, eds., *The Future of Social Democracy: Problems and Prospects of Social Democratic Parties in Western Europe* (New York: Oxford University Press, 1986).

Elizabeth Pond, *The Rebirth of Europe* (Washington, D.C.: Brookings Institution Press, 1999).

Geoffrey Pridham, ed., *Encouraging Democracy: The International Context of Regime Transition in Southern Europe* (New York: St. Martin's Press, 1991).

Paul Anthony Rahe, *Republics Ancient and Modern: Classical Republicanism and the American Revolution* (Chapel Hill: University of North Carolina Press, 1992).

Martin Rhodes et al., eds., *Developments in West European Politics* (New York: St. Martin's Press, 1997).

Michael Roskin, *The Rebirth of East Europe,* 3rd ed. (Upper Saddle River: Prentice Hall, 1997).

Stephen Salter and John Stevenson, eds., *The Working Class and Politics in Europe and America, 1929–1945* (London; New York: Longman, 1990).

Jürg Steiner, *European Democracies,* 4th ed. (New York: Longman, 1998).

Wayne C. Thompson, *Western Europe 1999,* 18th ed. (Harpers Ferry: Stryker-Post Publications, 1999).

Tibor Vasko, ed., *Problems of Economic Transition: Regional Development in Central and Eastern Europe* (Aldershot: Avebury; Brookfield: Ashgate Publishing Company, 1992).

Frank L. Wilson, *European Politics Today,* 3rd ed. (Upper Saddle River: Prentice Hall, 1999).

PERIODICALS AND CURRENT EVENTS

The Christian Science Monitor
One Norway Street
Boston, MA 02115
This newspaper is published 5 days per week, with news coverage, articles, and specific features on world events.

Current History, A World Affairs Journal
Provides focus on geopolitical regions throughout the world.

The Economist
125 St. James's Street
London, United Kingdom
This periodical presents world events from a British perspective.

Europe: Magazine of the European Union
2300 M Street, NW
Washington, D.C. 20037
An illustrated magazine that provides a monthly review and forum regarding European developments, sponsored by the Delegation of the European Commission, Washington, D.C.

European Voice
Rue Montoyerstraat 17–19
1000 Brussels, Belgium
An independent weekly review of the news and commentary regarding the European Union published by the Economist Group.

The Guardian Weekly
75 Farringdon Road
London, EC1M 3HQ
United Kingdom
A weekly review of international and British news with commentary and analysis, which also includes selected articles from *Le Monde* (translated) and *The Washington Post.*

Le Monde (weekly edition, in English)
7 Rue des Italiens
Paris, France
A summary of the previous week's news, with separate sections on various geographical regions.

Multinational Monitor
Ralph Nader's Corporate Accountability Research Group
1346 Connecticut Avenue, NW
Washington, D.C. 20006
This monthly periodical offers editorials and articles on world events and current issues.

The Nation
33 Irving Place
New York, NY 10003
A weekly magazine of commentary on national and international news from a left-of-center perspective.

The New York Times
229 West 43rd Street
New York, NY 10036
A daily newspaper that covers world news through articles and editorials.

The Wall Street Journal
Dow Jones Books
Box 300
Princeton, NJ 08540
Presents broad daily coverage of world news through articles and editorials.

World Press Review
The Stanley Foundation
230 Park Avenue
New York, NY 10169
Each month this publication presents foreign magazine and newspaper stories on political, social, and economic affairs.

Index